Comprehensive Virology 19

Comprehensive Virology

Edited by Heinz Fraenkel-Conrat
University of California at Berkeley

and Robert R. Wagner
University of Virginia

Editorial Board

Comprehensive

Edited by

Heinz Fraenkel-Conrat

Department of Molecular Biology and Virus Laboratory
University of California, Berkeley, California

and

Robert R. Wagner

Department of Microbiology
University of Virginia, Charlottesville, Virginia

Virology

19

Viral Cytopathology

*Cellular Macromolecular Synthesis
and Cytocidal Viruses*

Including a Cumulative Index to the Authors and
Major Topics Covered in Volumes 1–19

PLENUM PRESS • NEW YORK AND LONDON

Library of Congress Cataloging in Publication Data

Main entry under title:

Viral cytopathology.

 (Comprehensive virology; 19)
 Includes bibliographies and index.
 1. Host-virus relationships. 2. Pathology, Cellular. 3. Viruses—Reproduction. I.
Series.
QR357.F72   vol. 19   [QR482] 576′.64 s [574.2′34]   84-16081
ISBN 0-306-41698-0

QR
357
.F72
v.19

© 1984 Plenum Press, New York
A Division of Plenum Publishing Corporation
233 Spring Street, New York, N.Y. 10013

Printed in the United States of America

Foreword

The time seems ripe for a critical compendium of that segment of the biological universe we call viruses. Virology, as a science, having passed only recently through its descriptive phase of naming and numbering, has probably reached that stage at which relatively few new—truly new—viruses will be discovered. Triggered by the intellectual probes and techniques of molecular biology, genetics, biochemical cytology, and high resolution microscopy and spectroscopy, the field has experienced a genuine information explosion.

Few serious attempts have been made to chronicle these events. This comprehensive series, which will comprise some 6000 pages in a total of 19 volumes, represents a commitment by a large group of active investigators to analyze, digest, and expostulate on the great mass of data relating to viruses, much of which is now amorphous and disjointed, and scattered throughout a wide literature. In this way, we hope to place the entire field in perspective, and to develop an invaluable reference and sourcebook for researchers and students at all levels.

This series is designed as a continuum that can be entered anywhere, but which also provides a logical progression of developing facts and integrated concepts.

Volume 1 contains an alphabetical catalogue of almost all viruses of vertebrates, insects, plants, and protists, describing them in general terms. Volumes 2–4 deal primarily, but not exclusively, with the processes of infection and reproduction of the major groups of viruses in their hosts. Volume 2 deals with the simple RNA viruses of bacteria, plants, and animals; the togaviruses (formerly called arborviruses), which share with these only the feature that the virion's RNA is able to act as messenger RNA in the host cell; and the reoviruses of animals and plants, which all share several structurally singular features, the most important being the double-strandedness of their multiple RNA molecules.

Volume 3 addresses itself to the reproduction of all DNA-containing viruses of vertebrates, encompassing the smallest and the largest viruses known. The reproduction of the larger and more complex RNA viruses is the subject matter of Volume 4. These viruses share the property of being enclosed in lipoprotein membranes, as do the togaviruses included in Volume 2. They share as a group, along with the reoviruses, the presence of polymerase enzymes in their virions to satisfy the need for their RNA to become transcribed before it can serve messenger functions.

Volumes 5 and 6 represent the first in a series that focuses primarily on the structure and assembly of virus particles. Volume 5 is devoted to general structural principles involving the relationship and specificity of interaction of viral capsid proteins and their nucleic acids, or host nucleic acids. It deals primarily with helical and the simpler isometric viruses, as well as with the relationship of nucleic acid to protein shell in the T-even phages. Volume 6 is concerned with the structure of the picornaviruses, and with the reconstitution of plant and bacterial RNA viruses.

Volumes 7 and 8 deal with the DNA bacteriophages. Volume 7 concluded the series of volumes on the reproduction of viruses (Volumes 2–4 and Volume 7) and deals particularly with the single- and double-stranded virulent bacteriophages.

Volume 8, the first of the series on regulation and genetics of viruses, covers the biological properties of the lysogenic and defective phages, the phage-satellite system P2–P4, and in-depth discussion of the regulatory principles governing the development of selected lytic phages.

Volume 9 provides a truly comprehensive analysis of the genetics of all animal viruses that have been studied to date. These chapters cover the principles and methodology of mutant selection, complementation analysis, gene mapping with restriction endonucleases, etc. Volume 10 also deals with animal cells, covering transcriptional and translational regulation of viral gene expression, defective virions, and integration of tumor virus genomes into host chromosomes.

Volume 11 covers the considerable advances in the molecular understanding of new aspects of virology which have been revealed in recent years through the study of plant viruses. It covers particularly the mode of replication and translation of the multicomponent viruses and others that carry or utilize subdivided genomes; the use of protoplasts in such studies is authoritatively reviewed, as well as the nature of viroids, the smallest replicatable pathogens. Volume 12 deals with special groups of viruses of protists and invertebrates which show properties that set them apart from the main virus families. These are the lipid-containing phages and the viruses of algae, fungi, and invertebrates.

Volume 13 contains chapters on various topics related to the structure and assembly of viruses, dealing in detail with nucleotide and amino acid sequences, as well as with particle morphology and assembly, and the structure of virus membranes and hybrid viruses. The first complete sequence of a viral RNA is represented as a multicolored foldout.

Volume 14 contains chapters on special and/or newly characterized vertebrate virus groups: bunya-, arena-, corona-, calici-, and orbiviruses, icosahedral cytoplasmic deoxyriboviruses, fish viruses, and hepatitis viruses.

Following Volume 14 is a group of volumes dealing with virus-host interactions. Volume 15 focuses on immunity to viruses: Volume 16 on viral invasion, factors controlling persistence of viruses, responses to viral infection, and certain diseases. Volume 17 contains chapters discussing and evaluating most of the biophysical, biochemical, and serological methods used in virus research. Volume 18 contains chapters on cell receptors of picornaviruses and persistence of lymphocytic choriomeningitis virus, as well as two on the most important neurological diseases known to be caused by viruses.

The current volume is the last in the series and deals extensively with the molecular basis of viral cytopathogenicity. An introductory chapter on historical perspectives of viral cytopathic effects is followed by two chapters on transcriptional and translational strategies of uninfected mammalian cells. The remaining chapters provide in-depth analyses of the mechanisms by which cytocidal viruses shut off cellular macromolecular synthesis leading to cell death.

Our knowledge of certain viruses has advanced greatly since publication of the early volumes of *Comprehensive Virology*, and a second updated edition for each of these was considered. The editors and publishers have decided that instead of such a second edition they would approach the concept of comprehensive coverage of virology in a different manner. A series of books or groups of books, termed *The Viruses*, each dealing with a specific virus family *in extenso*, will be planned and edited by an eminent specialist in the respective field.

Contents

Chapter 2

Transcription by RNA Polymerase II

Ulla Hansen and Phillip A. Sharp

Chapter 3

Regulation of Eukaryotic Translation

Raymond Kaempfer

Chapter 4

Picornavirus Inhibition of Host Cell Protein Synthesis

Ellie Ehrenfeld

Chapter 5

Rhabdovirus Cytopathology: Effects on Cellular Macromolecular Synthesis

Robert R. Wagner, James R. Thomas, and John J. McGowan

Chapter 6

Adenovirus Cytopathology

S. J. Flint

Chapter 7

The Effects of Herpesviruses on Cellular Macromolecular Synthesis

Michael L. Fenwick

Chapter 8

**Poxvirus Cytopathogenicity: Effects on Cellular Macromolecular
Synthesis**

Rostom Bablanian

Chapter 9

**Reovirus Cytopathology: Effects on Cellular Macromolecular
Synthesis and the Cytoskeleton**

Arlene H. Sharpe and Bernard N. Fields

Chapter 10

Inhibition of Host Cell Macromolecular Synthesis following Togavirus Infection

Bunsiti Simizu

CHAPTER 1

Cytopathic Effects of Viruses: A General Survey

Robert R. Wagner

Department of Microbiology
The University of Virginia
Charlottesville, Virginia 22908

1. DEFINING THE PROBLEM

Several years ago, I blythely and naively set out to write a single-authored chapter on the general subject of viral cytopathogenicity with particular emphasis on the molecular events that ostensibly lead to cell death. After collecting several thousand references and spending 2 months merely drafting an outline, I realized the folly of such a herculean undertaking. Hence, this original single chapter has evolved into a full volume of chapters which shall attempt to cover comprehensively the subject of the molecular bases by which the best known pathogenetic animal viruses affect cellular functions. Authors who have generally made the most important and most recent contributions have been selected to write individual chapters on the best-studied families of pathogenic viruses, as well as chapters on the biochemistry of eukaryotic transcription and translation, which appear to be primary targets for invading cytopathic viruses. The purpose of this introductory chapter is to set the stage for the chapters which follow by providing a general survey of the nature of cytopathic effects caused by virulent viruses as well as to provide some historical perspective of a field that represents the origins of virology.

1

The overall purpose of this volume is to analyze in some depth the mechanisms by which certain viruses kill cells and, in so doing, subvert cell functions. An attempt will be made to approach these problems from the standpoint both of the viruses and of the host cells. It has become increasingly clear that there is no single underlying principle by which viruses perturb cell-synthesizing machinery; rather various viral factors and host factors are certainly operative to varying degrees in all virus–cell interactions that lead to cell death. The offending viruses are composed of or synthesize different components, one or more of which can be responsible for one or more effect on cells. It seems likely that certain of the structural components of diverse viruses can produce direct "toxic" effects on cells, including cell surface membrane alterations after virus attachment. Probably more commonly, the diverse replicative functions of various viruses intrude on and disturb finely balanced cell functions. Much of the chapters will be devoted to the transcriptional and replicative strategies of well-studied viruses that appear to be responsible for suppressing cellular macromolecular synthesis; this must include virus genetics as well as phenotypic variations among strains of the same or related virus.

An essential approach to our understanding of viral cytopathogenicity is the host cell and its innumerable variations in susceptibility and resistance to viruses of different types. Little is known about cell genotypes, or for that matter, phenotypes which determine cellular susceptibility to viral infection. An understanding of cell responses must consider the subcellular targets for the action of various viruses. Of great importance in determining those cell targets concerned with viral susceptibility, in no particular rank order, are cell membranes, cytoplasmic organelles (such as lysosomes), ribosomes, and of course, the nucleus. Major consideration has been given in recent years to the macromolecular synthesizing machinery of the cell undergoing viral infections; these include translational machinery that controls cell protein synthesis, transcriptional machinery that controls RNA synthesis, and cell DNA replicative machinery. Viruses also affect the cytoskeletal and lipid biosynthesis and probably other processes as well. Therefore, the study of the pathogenicity of viruses encompasses most, if not all, of molecular cell biology at its most fundamental level. In fact, viruses can serve as incisive probes for understanding cell biology.

Maltzman and Levine (1981) have written an excellent review of "Viruses as Probes for Development and Differentiation" in which they discuss in detail the mechanisms by which certain viruses can be used to explore complicated cell functions. In that review, they discuss the influence of the differentiated state of a cell on viral gene expression and its consequences, virus-mediated reprogramming of cellular gene expression (including arrest of cell development, re-expression of developmental antigens and virus-mediated production of new cell and tissue types), and viral gene organization as a model for eukaryotic gene expression and developmental processes. These fascinating topics, probed largely by temperate, persistent, and on-cogenic viruses, are outside the scope of this chapter and this volume, which is devoted primarily to acute cytopathic effects of virulent viruses that rapidly and drastically alter cellular metabolic functions.

Quite obviously, the general field of viral cytopathology is as complex as the whole science of cell biology. The major problem confronting the investigator, one that has caused considerable concern and possibly erroneous data, is the complexity of the target cell as well as the somewhat lesser complexity of the invading virus. The difficulty in experimental design is to segregate the cell function that is being targeted and the virus component thought to be directed toward the target. This is not a simple task. Quite obviously, for example, cellular replication, transcription, and translation are integrated functions that proceed *pari passu*; many investigations that have focused on translational effects of viruses can be misled by not adequately controlling transcriptional events. Of enormous help of late is the capacity to use enucleated cells and subcellular transcriptional and translational systems. By the same token, most experiments on viral pathogenicity are conducted with whole viruses which must perform several, albeit more limited, functions in the process of infection; particularly difficult are the negative-strand viruses, the replicative strategy of which involves coupled transcription and translation to say nothing of genome reproduction.

Certain investigations of viral cytopathogenicity have profited from the use of chemical inhibitors or by alteration of environmental conditions. Specific inhibitors of cellular macromolecular synthesis, including antibiotics and structural analogues, have been used to isolate various targeted systems; unfortunately, some of these inhibitors also cause side effects which alter the cell target function to be studied. Good use has been made of altered environmental conditions, such as temperature and ionic strength of the supporting medium, to provide insight into specific sites of viral action. Some use has been made of natural cellular products, such as interferons.

It is quite evident that the multiplicity of factors and conditions impose considerable burdens on design of experiments which attempt to pinpoint viral functions and cell targets in the cytopathogenetic processes to be studied. The literature is enormous and some of it must be ignored or short-circuited to make this chapter comprehensible and encompassible. There was no way to undertake in the confines of a single chapter a complete review of the cytopathogenicity of every virus, nor was it feasible to even consider reviewing viral diseases.

2. HISTORICAL BACKGROUND

The dictionary definition of virus (L. slimy liquid, poison) as a cause of disease has long been the basic concept of the nature of these biological entities. In their original descriptions, Ivanovski (1892) and later, independently, Beijerinck (1898) were concerned only with the disease-producing capacity of the original virus, tobacco mosaic virus. Similarly, Twort (1915) and d'Herelle (1917) stumbled upon bacteriophages by their astute observation of lesions in bacteria. The recognition that viruses cause disease in animals, including man, opened up the field of viral pathogenicity. Virologists have pursued this avenue of research in a rather desultory fashion for many decades, the major advance of which has been the burgeoning field of viral immunology. Left somewhat in the lurch has been the critical problem of how viruses cause disease. Perhaps the major stumbling block has been the concept, based on wishful thinking, that all viruses possess a single general disease-causing property and, conversely, that all hosts react to infection in a similar manner. The search for the single underlying principle in viral pathogenicity has probably hampered progress in this field.

The major breakthrough in technology that provides the basis for relating host reactions to viruses was the discovery by Enders *et al.* (1949) that poliovirus is "cytopathogenic" for certain cultured cells. In addition to providing the means to uncover a multitude of heretofore undetectable viruses, this finding stimulated a wealth of research on the cytopathic effects of viruses based largely on gross and microscopic observations of cells in culture. The advent of electron microscopy greatly augmented the investigative methods that helped to define the subcellular sites of viral action. By these means, it was possible for Enders (1954) to define certain of the different "morphological manifestations of viral cytopathogenicity" and to promulgate a classification of "(1) viruses causing only cellular degeneration, (2) viruses causing formation of inclusion bodies and cell degeneration, and (3) viruses causing formation of multinucleated cells or syncytial masses and degeneration with or without inclusion bodies."

No attempt will be made in this chapter to review the enormous early literature on viral cytopathology. Among the many reviews since the pioneering effort of Enders (1954), several provide an excellent background and establish the fundamental principles and terminology of viral cytopathology: these include the review of Walker (1960) on "*in vitro* cell–virus relationships resulting in cell death" which includes a discussion of viral toxin-like materials, the review by Pereira (1961) on "the cytopathic effect of animal viruses" which includes an excellent section on cellular susceptibility and virus classification based on cytopathic effects, and a review by Poste (1970) on "virus-induced polykariocytosis and the mechanism of cell fusion." The most recent analysis and critical evaluation in the literature was made by Bablanian (1975) who provides the first in-depth critique of both structure *and* functional alterations in cultured cells infected with cytocidal viruses." A general discussion of the "specificity of viruses for tissues and hosts" is provided by Bang (1972) and earlier by Bang and Luttrell (1961); tissue tropisms of viruses will not be discussed in this chapter.

2.1. Early Observations on Virus-Induced Cytopathology

It has been realized for some time that viral multiplication can occur with only minimal cytopathic effects (Choppin, 1964). Conversely, considerable cytopathology can result from infection with a virus, the multiplication of which is restricted in that host. Cantell *et al.* (1962) were perhaps the first to describe that L cells could be destroyed by large doses of a UV-irradiated virus, vesicular stomatitis virus, in the absence of a detectable synthesis of viral antigen: this "toxic" activity could not be separated from the viral particle. We shall discuss in a later chapter why this does not prove the absence of viral replicative functions or that preformed virion components are necessarily cytotoxic. However, it is quite clear that early interactions of viruses with cellular membranes can seriously, often temporarily, disturb cellular functions (Kohn, 1979). As will be discussed in greater detail later, virus-induced cell damage can occur in the presence of chemical inhibitors of virus reproduction (Bablanian, 1972). Virus mutants restricted in some functions are also defective in causation of cytopathic effects (Garwes *et al.*, 1975), whereas others retain cytopathogenicity despite restricted replication (Marcus and Sekellick, 1975).

All functions of a cell need not remain intact to retain susceptibility to viral infections. However, loss of certain functions will predispose to resistance to certain viruses and not others. For example, cells enucleated by cytochalasin B will support the replication of vesicular stomatitis virus but not influenza virus, which also fails to inhibit protein synthesis in enucleated cells (Follett *et al.*, 1974).

Cytopathic effects, a term coined by Enders (1954), is defined as a morphological or biochemical effect in cells infected with viruses. In most cases, the morphological change is detected by gross or light microscopic observations of unstained cells. Dying cells fail to exclude certain dyes, such as trypan blue, or fail to incorporate vital stains, such as neutral red. Failure to stain with neutral red forms the basis of the plaque assay method for animal cells growing in monolayers, as designed by Dulbecco (1952). Monolayers of cells killed by viruses tend to round up, ostensibly because of dissolution of cytoskeleton (Dales, 1975; Lenk and Penman, 1979; Lenk *et al.*, 1980) and detach from the glass or plastic surface. Other pathologic evidence of infection with certain viruses, but not others, are inclusion bodies consisting of aggregated viral products in the cytoplasm or nucleus (Rivers, 1928).

One of the problems encountered in studies of viral cytopathic effects is to quantitate the number of cells undergoing changes or dying. Some indication of the number of cells involved can be obtained by estimating those that fail to exclude dyes because of altered membrane impermeability. A more accurate, but somewhat tedious technique was designed by Marcus and Puck (1958). These authors devised a means to study quantitatively the destruction of mammalian cell reproductive capacity by scoring the fraction of monodisperse HeLa cells infected with Newcastle disease virus to survive and form colonies over a period of 8–10 days incubation. They concluded that one virus particle was sufficient to kill a single cell. This method has been extended to other viruses by Marcus (1959) and can be used to monitor cell killing by viruses at the sensitive level of reproduction of the cell population.

2.2. Early Observations on Virus-Induced Alterations in Cell Macromolecular Synthesis

The capability of viruses to preempt the macromolecular synthetic machinery of host cells had probably been suspected from the earliest days of virology. However, this property of viruses has recently taken on the aspects of a separate subdiscipline. Perhaps the earliest observations were made by Ackermann (1958) and Ackermann and his colleagues (1959) who noted that poliovirus altered macromolecular synthesis in HeLa cells; their original observation suggested actual increased rate of overall RNA and protein synthesis in infected cells. More definitive studies soon appeared by Salzman *et al.* (1959) who made the definitive observation that poliovirus-infected HeLa cells exhibit a sharp decline in cellular RNA, DNA, and protein synthesis. These observations were soon confirmed by other laboratories that reported profound inhibition of RNA and protein synthesis in cells infected with picornaviruses, such as encephalomyocarditis virus (Martin *et al.* 1961), mengovirus (Baltimore and Franklin, 1962), and poliovirus (Holland, 1962). The picornaviruses have continued to be the major system for studying virus-induced suppression of cellular macromolecular synthesis, but many other viruses have been found to possess similar properties. Excellent descriptions of the early work in this field are contained in the reviews by Martin and Kerr (1968) and by Roizman and Spear (1969).

Table 1 represents an attempt to provide a summary of all of the vertebrate viruses that appear capable of compromising cellular macromolecular functions. Only a limited number of references to the earliest definitive observations are listed in this table; more details at the molecular level will be provided in subsequent chapters on each virus family.

3. VIRAL CYTOPATHOLOGY: GENERAL PRINCIPLES

Cell death as the end result of viral infection is probably the exception rather than the rule. It cannot be stressed strongly enough that viruses that are virtually indistinguishable genetically and anti-

TABLE 1
Early Definitive Studies on Viruses that Suppress Cellular Macromolecular Synthesis

Virus	Cell type	50% Inhibition (hours)			References
		RNA	DNA	Protein	
A. Picornavirus					
Polio 1	HeLa	3.5		1.5	Zimmerman et al. (1963)
	ERK	2.7		2.7	
Polio 2	HeLa	3.5		<1.5	Bablanian et al. (1965a,b)
EMC	Krebs-II	±2		4.5	Martin et al. (1961)
Mengo	L	0.5		0.5	Baltimore et al. (1963)
ME	L	1.5		2.5	Verwoerd and Hausen, (1963)
Foot-and-mouth	BHK	1.5		0.5	Brown et al. (1966)
B. Togavirus					
Sindbis	Chick			~2	Strauss et al. (1969)
SFV	Chick			~4	Lachmi and Kääriäinen (1977)
VEE	HeLa	<2		1.0	Lust (1966)
WEE	BHK		~1.5		Simizu et al. (1976)
C. Rhabdovirus					
VSV$_{Ind}$	L			~2.5	McAllister and Wagner (1976)
	Krebs-II	~1			Wagner and Huang (1966)
	L	~3			Weck and Wagner (1978)
	BHK	~3			
	MPC-11	~1			
VSV$_{NJ}$	Chick	~2	~4	~8	Yaoi et al. (1970)
Rabies	BHK			+	Madore and England (1975)
D. Paramyxovirus					
NDV	Chick			>6	Hightower and Bratt (1974)
NDV (Texas)	Chick	~7		~7	Wilson (1968)
NDV	L		~6		Hand (1976)
E. Myxovirus					
FPV	Chick			~6	Long and Burke (1970)
F. Bunyavirus					
Bunyamwera	Vero	~6 hr		~6 hr	Lazdins and Holmes (1979)
G. Reovirus					
Reovirus 3	L		~8h		Shaw and Cox (1973)
H. Poxvirus					
Vaccinia	HeLa	~5.5			Salzman et al. (1964)
	L	~5			Becker and Joklik (1964)
	HeLa			0.5–2.0	Moss (1968)
	HeLa		~2		Jungwirth and Launer (1968)
Cowpox	HeLa		~2		
I. Adenovirus					
Adeno 5	KB			9–12[a]	Bello and Ginsberg (1967)
Adeno 2	KB	9–14[b]			Raskas et al. (1970)
	MK		~20	>20	Ledinko (1966)
J. Herpesvirus					
Pseudorabies	RK		~7		Ben Porat and Kaplan (1965)
	RK			~5	Ludwig and Rott (1975)
HSV-1	Hep-2	~3		>11	Roizman et al. (1965)
EAV	L-M		~10		O'Callaghan et al. (1968)
K. CDV[c]					
FV-3	FHM	+	+	1–3	Goorha and Granoff (1974)
	BHK			1–3	Raghow and Granoff (1979)

[a] Specific enzyme activity measured.
[b] Ribosomal RNA.
[c] Cytoplasmic deoxyvirus.

genically can vary enormously in their cytopathic capacity; single mutational events can drastically reduce the cytopathology of a given virus. By the same token, cells differ enormously in their response to viral infection, varying from great susceptibility to complete resistance. In the final analysis, cell susceptibility/resistance to viruses is a genetic determinant. Resistant cell mutants, by the nature of cellular complexity, are much more difficult to come by than are virus mutants of differing pathogenicity. Therefore, investigators comparing susceptibility among cells to a given virus generally depend on cell types of different tissue origin or at different stages of differentiation.

3.1. Cellular Differentiation and Susceptibility to Viral Infection

From the earliest days of virology, there has been general recognition that undifferentiated cells of embryonic or neonatal origin are ordinarily much more susceptible to viral infection than fully differentiated cells from tissues of adult animals. Many of the original studies of cytopathology were performed with primary or secondary cultures of embryonic cells. An excellent example of age-dependent susceptibility to a virus is provided by the limited tissue specificity of a neurotropic strain of influenza virus in adult mice compared with widespread tissue susceptibility of the infant mouse (Wagner, 1955). Moreover, dedifferentiated transplanted tumor cells exhibit greater susceptibility than differentiated cells to the neurotropic influenza virus (Wagner, 1954).

Cellular susceptibility to viral infection obviously occurs at different stages of infection, varying from virus adsorption onto cell surface membrane to final stages of assembly and release. The classical example of cell surface receptors for viral infection is that of the poliovirus which attaches only to primate cell surfaces but the RNA of which can transfect, replicate, and destroy many types of non-primate animal cells (Mountain and Alexander, 1959; DeSomer et al., 1959). Therefore, the first level of cellular susceptibility/resistance is virus adsorption to the cell surface. Equally important, but somewhat more difficult to study, is the second stage of infection: virus penetration into cytoplasm following adsorption. The mechanism by which viruses penetrate into animal cells has been a matter of controversy (Morgan and Rose, 1968; Dales, 1973) for many years; animal viruses do not possess the DNA injection properties of certain bacteriophages

(Hershey and Chase, 1952). The question of mode of penetration of fully encapsidated animal viruses is of some importance to the problem of viral pathogenicity. Following adsorption of a virus, at least two basic mechanisms are possible for virus penetration: (1) the envelope or capsid dissociates at the cell surface and only the viral nucleic acid or nucleoprotein enters the cytoplasm (Morgan and Rose, 1968), or (2) the intact virus enters the cytoplasm by phagocytosis, in which case it is surrounded by a vesicle (phagosome), the membrane of which is inverted (Dales, 1973); following phagocytosis, the vesicle membrane must undergo dissolution in order for the virion nucleic acid or nucleoprotein to function in the intracellular environment and to utilize the machinery provided by the cytoplasm. Detailed discussion of these two possible mechanisms of virus penetration is beyond the scope of this review. Suffice it to say that in all likelihood, both mechanisms are possible depending on the infecting virus and the host cell. There seems little question that the paramyoviruses, the envelope of which often contains a protein ($F_{1,2}$) cell fusion factor, can penetrate cells by dissolution of virion membrane at the cell surface and depositing the nucleocapsin in the cytoplasm (Scheid and Choppin, 1974). However, more recent studies suggest that many, if not most, viruses enter cells by phagosomes, possibly at the region of coated pits which become coated vesicles surrounding the intact virus; this phagocytic process also requires release of the intact virus, possibly by interaction of the coated vesicle with lysosomes (Helenius et al., 1980).

3.2. Methods for Studying Viral Cytopathic Effects

After a prolonged and rather fruitless search for a single or unified principle to explain the cytopathic effect of all viruses, it is now becoming abundantly clear that different viruses, or sometimes the same virus, can cause cytopathic effects by diverse mechanisms. This diversity among viruses was implicit in the old observations that different viruses produce different morphologic changes in tissue culture cells (Pereira, 1961; Barski, 1962). Also, as indicated in the review by Bablanian (1972), "multiplication of a virus in a host is prerequisite for the development of a disease, but virus multiplication does not always lead to disease." Many viruses multiply without causing cytopathic effects (Choppin, 1964) and, in fact, can persist for long periods of time in host tissues or cell culture without stopping cell growth or other discernable functions (Holland *et al.*, 1980; Youngner and Preble, 1980). Many investigators have observed cytopathic effects in the absence of multiplication of virus to the extent of producing infectious progeny (Henle *et al.*, 1955; Appleyard *et al.*, 1962). Viral cytopathogenicity can also be studied by the use of metabolic inhibitors that abort viral replication at different stages of the infectious cycle; viral cytopathology can occur in the presence of antiviral reagents (Ledinko, 1958; Appleyard *et al.*, 1965).

Expression of viral cytopathology is complicated by a variety of factors, some of which were not considered in the early literature. Only recently have investigators been cognizant of these intrinsic factors which involve homogeneity of the virus population. Moreover, extraneous factors in the media and sera will often severely compromise the experimental protocols. Current studies in viral cytopathogenicity and other cellular effects are hopefully done with carefully cloned virus populations to insure homogeneity and, when feasible, cell clones to insure host homogeneity. All virus preparations are now known to contain defective interfering viral particles which interfere with viral replication and, oftentimes, the cytopathic effects (Huang and Baltimore, 1977). It is wise to remember, however, that interference with replication of new progeny does not necessarily inhibit the cytopathic effect of the parental virus. Schnitzlein *et al.* (1983) have recently shown quite conclusively that inhibition of standard VSV replication does not reverse the capacity of VSV to shut off cellular protein synthesis. It is, nevertheless, still incumbent upon the investigator to clone and prepare stocks of the virus by diluted passage to avoid defective-interfering particles. Moreover, many virus and cell populations are contaminated with extraneous viruses or, for that matter, mycoplasma, which can also influence the outcome of a cytopathogenicity experiment. In addition, "nonspecific inhibitors" of viral multiplication and cytopathic effects, often heat stable, are frequently present in the gamma globulin or other components of serum (Pereira, 1961). Also of some concern is a variety of heat *labile* serum factors (Chu, 1951) which have been found to inhibit cytopathic effects of influenza virus (Henry and Youngner, 1957) and poxvirus (Plowright and Ferris, 1958). Lastly, environmental factors such as temperature, pH, and ionic composition of the media are important considerations in determining the outcome of cytopathic effects (see excellent early review of Pereira, 1961).

Many other factors that determine viral cytopathogenicity can be used as probes for studying certain mechanisms by which viruses impair cell functions leading to death. Chemical inhibitors, such as guanidine and 2-(α-hydroxybenzyl) benzimidazole, which both affect replication of picornaviruses, can be used to study the events of viral replication which bear on cytopathology (reviewed by Tamm and Eggers, 1963). Elegant early studies by Lake and Ludwig (1971) have shown that mengovirus-induced cytopathology in L-929 cells is moderated by the host–cell cycle. Cells in late S and G2-M phases resist virus-induced cytolysis and nuclear lesions are found to precede lysosomal damage. All these built-in variables suggest the multitude of factors which enter into attempts to investigate cytopathic effects by isolated, uncontrolled systems. In an attempt to derive a quantitative estimate of "cell killing," Marcus and his colleagues have devised a procedure to score effects by death or nondeath based on the ability of virus-infected monodisperse cells to form colonies when plated in cultures (see review by Marcus, 1976).

The term cytopathic effect has at least two connotations, which can result in considerable confusion and lead to misinterpretations of data. The end result is a sick or dead cell but two distinct phenomena can be involved. First, there is a direct cytopathic effect in which the parental, invading virus *per se* perturbs the cell machinery resulting in a morphological or biochemical derangement, reversible or irreversible. The parental virion need not replicate but either contains structural components or synthesizes such components which subvert cell functions. The term viral cytotoxicity is sometimes applied to this direct parental effect, which merely evades the issue of what causes the pathology. Second, the term cytopathic effect is also used, perhaps its original connotation, to denote that a virion is infectious and gives rise to progeny and late cell death. These are the footprints of the virus by which it can be identified, a powerful diagnostic tool that opened up the field of determinative virology. However, it should be kept in mind that replication is probably not the primary event but actually obscures the outcome because it is not usually possible to tell whether cytopathogenicity was caused by the parental or progeny viruses. Replication of virus may or may not result in a diseased cell, tissue, or organism. To understand the mechanism(s), however, one must dissociate the primary cytopathic effect from secondary cytopathic effects associated with viral replicative events.

3.3. Variability of Cytopathic Effects

With the above caveats in mind, it might be useful to consider a comparative scale of pathogenic properties of the different classes of viruses. Early attempts to classify viruses by their relative capacity to cause cytopathic effects were based, almost entirely, on gross or microscopic morphological alterations. A still useful classification of viruses was that of Lynn and Morgan (1954) who divided viruses into groups: (1) without cytopathogenicity (mumps, influenza, yellow fever, and St. Louis encephalitis), (2) with moderate cytopathogenicity (vaccinia, varicella–zoster, herpes simplex, fowlpox, pseudorabies, and coxsackie), and (3) with marked cytopathogenicity (equine encephalitis and polio). These early reports suffer, of course, from lack of virus classification, including designation of lymphogranuloma and psittacosis as viruses, when they are now known to be intracellular bacteria (*Chlamydia*).

A more useful classification of viruses according to their cytopathic effects is that of Pereira (1961) who used morphologic criteria such as cellular site of multiplication, maturation, and liberation. This classification also has the significant advantage of having better taxonomic criteria for identification of virus families, primarily based on electron microscopy. Also of value in discriminating among viruses are host range and inclusion bodies formed in the nucleus (type A, eosinophilic staining) or the cytoplasm (type B, basophilic). By these and other means, Pereira (1961) was able to designate and separate the general expression of cytopathic effects of poxviruses, herpesviruses, measles–distemper–rinderpest, adenoviruses, tumor viruses, myxoviruses (not yet separate from paramyoviruses), enteroviruses, arboviruses, (mostly togaviruses), and a large not-yet classified group of viruses.

Virtually every vertebrate virus that has ever been isolated has been studied for its capacity to cause death and cytological changes in a variety of cell types. Oftentimes, these viruses have been poorly characterized and their taxonomic designations were yet to be determined when their pathological or physiological properties were being studied. As mentioned above, the studies were frequently performed with uncloned populations of viruses, the homogenicity of which, or even the ancestry of which, was highly suspect. No attempt will be made to review this enormous literature. Instead, the best characterized and best prepared groups of viruses will be analyzed from the standpoint of their capacity to cause morphological change, often leading to death, in well-characterized cell types. I shall concentrate on the virulent viruses that are proven to cause disease, but whose host range is limited or broad. Very little will be said about the replicative strategies of those viruses, a subject which has been covered *in extenso* in previous volumes of this series of *Comprehensive Virology*. The cytopathology of picornaviruses, poxviruses, rhabdoviruses, and adenoviruses has been selected to provide brief examples of widely different viruses that produce acute cytopathic effects that have been well characterized and will be described in detail in subsequent chapters concerned with each virus family.

4. SPECIFIC CYTOPATHIC VIRUSES: AN OVERVIEW

4.1. Picornavirus Cytopathology

The picornaviruses are, without question, the most extensively studied viruses particularly from the standpoint of cytopathology and, as described in later chapters, for their drastic effect on cellular macromolecular synthesis. It is fair to say that the picornaviruses are the prototypes for studying viral pathogenicity as well as being the earliest studied. The Picornaviridae comprise one of the largest of the virus families (see Matthews, 1979); their structure and reproductive strategy have been studied in great detail (see reviews in Levintow, 1974; Rueckert, 1976; Cooper, 1977). Briefly, they are small, icosahedral viruses measuring 23–30 nm in diameter and contain an unsegmented genome of single-stranded RNA that is not polyadenylated nor capped, but probably all contain a covalent-linked protein at the 5' end (Ambros and Baltimore, 1978; Rothberg *et al.*, 1978). The genome has messenger sense, thus being classified as a positive-strand virus. The initial event in infection is translation of the polycistronic genome messenger to form a polypeptide which is then cleaved enzymatically into active proteins, some of which are assembled into progeny virions and two others serve as nonstructural replicases (Rekosh, 1977). The major viruses that have been studied are the human polioviruses (three antigenic types restricted to infection of primate cells which contain receptors for adsorption), the human coxsackie- and echovirus, and the murine encephalomyocarditis (EMC) and mengovirus.

One of the original controlled studies on picornavirus cytopathic effects was performed by Ackermann *et al.* (1954) who, during investigation of the growth characteristics of poliovirus type 3 in HeLa cell cultures, noted a "lack of parallelism in cellular injury and virus replication." These authors reported a latent period (before virus release) of 4–5 hr followed by gradual virus release during the ensuing

5–7 hr. They noted alteration in the staining patterns of the infected cells before the major portion of poliovirus progeny was released. They also reported marked variation among individual cells in culture, particularly marked basophilia in the staining pattern; they also reported that the amino acid analogue, p-fluorophenylalanine, blocked viral multiplication but did not affect the virus-induced cytopathic effects.

Although the majority of investigators have used morphological changes of cells as criteria of cytopathic effects, others have tried to measure other modalities, partially in an attempt to make the observations more quantitative. For example, Bubel (1967) studied protein leakage from mengovirus-infected cells (an established guinea pig spleen line) in serum-free medium; he claimed that leakage of protein could be assayed quantitatively by perchloric acid precipitation. In later studies, Blackman and Bubel (1969) used leakage of specific lysosomal proteins as a presumed quantitative measure of poliovirus-induced injury of cells in suspension cultures; they considered this method more sensitive for measuring early cell membrane damage triggered by release of lysosomal hydrolases which would injure the plasma membrane sufficiently to make it leak other cytoplasmic proteins at a later stage in infection. Another effect of viral infection is increased permeability of the infected-cell membrane to anions and cations (Farnham and Epstein, 1963; Carrasco, 1978); these alterations in cell function will be discussed in a later section. Virus-induced damage to cell membrane has also been studied by preloading cells with ^{51}Cr followed by infection with poliovirus and then measuring release of ^{51}Cr into the medium 5–6 hr postinfection (Koschel, 1971). All these techniques suffer from the obvious drawback that investigators are probably examining a late, possibly terminal, event secondary to the site of initial injury caused by the virus. However, such techniques may have greater use and applicability for studies of cytotoxicity caused by viral components. As evidenced by the following discussion, most investigations of picornavirus cytopathology have utilized morphologic criteria and have attempted to determine these viral effects by use of metabolic inhibitors, variations in cell cycle, and other cell properties as well as by inactivation of the infecting virus.

4.1.1. Morphological Changes in Picornavirus-Infected Cells

Following the lead of Ackermann *et al.* (1954), Dunnebacke (1956*a,b*) studied the effect of all three types of polioviruses on mon-

key kidney, HeLa, human fetal, and human amnion cells; she measured the cytopathic effects by light microscopy of stained cells and
divided the events into four stages: (I) early nuclear pyknosis and
certain cytoplasmic changes, (II) more advanced changes, (III) accumulation of nuclear masses, and (IV) cell disintegration. Dunnebacke (1956a) also reported that virus release does not occur until
after stage III except in human amnion cells from which virus was
released 4–8 hr later and the nuclei developed indistinct nucleoli and
nodules. Significant advances were made by Horne and Nagington
(1959) who devised electron microscopic techniques for studying the
development of poliovirus within infected HeLa cells, particularly the
pioneering technique of negative staining with potassium phospho-
tungstic acid which enabled them to detect complete, incomplete, and
empty virus particles in fragmented cells. Electron microscopic techniques were also applied to the investigation of FL human amnion
cells infected with poliovirus type 1 and coxsackievirus B5 and to
embryonic mouse cells infected with encephalomyocarditis virus
(Hintz *et al.*, 1962); these observations revealed that nuclear changes,
consisting of chromatin margination, nucleolar condensation, and a
dense granular substance in the interchromatin zone, preceded cytoplasmic changes and nuclear shrinkage. Dales and Franklin (1962)
in a more detailed study found similar effects in L cells infected with
mengovirus and encephalomyocarditis virus which also caused the
formation of small vesicles in the centrosphere region commencing
4–6 hr postinfection. These early effects were followed by progressive
degeneration of nucleus and vesiculation of the cytoplasm up to 18–
20 hr; increased numbers of small dense granules, indistinguishable
from ribonucleoprotein, were also observed 18–20 hr postinfection.
Also noted was an early permeability of the mengovirus-infected cell
membrane to the erythrocin dye which is excluded from viable, uninfected cells (Dales and Franklin, 1962). The presence of eosinophilic
and basophilic cytoplasmic inclusions containing virus particles in
crystalline array were found in cells infected with respiratory isolates
of echovirus 11 by Soloviev *et al.* (1967) who also found evidence for
crystalline arrays of empty virus particles in the nucleus; this observation was confirmed by Anzai and Ozaki (1969) who also reported
in FL amnion cells infected with poliovirus type 1, the presence of
intranuclear crystals that stained with fluorescent antibody that appeared to contain incomplete poliovirus particles. Similar ultrastructural and histochemical studies have been reported for human em-

bryonic lung cells (L132) infected with coxsackievirus A13 (Jezequel and Steiner, 1966).

Numerous attempts have been made to relate electron microscopic cytopathic events to the replicative cycle of picornaviruses in infected cells. In one such study of HeLa cells infected with poliovirus 1, Dales *et al.* (1965) found beginning 3 hr after infection the formation of large ribosomal aggregates and dense aggregates of viroplasm which contained progeny particles that later assumed crystalline arrays; as infection progressed, masses of small membrane-enclosed bodies (vesicles, vacuoles, and cisternae) formed, containing some virus particles. Similar findings, also in HeLa cells infected with poliovirus 1, were reported by Mattern and Daniel (1966) who detected nuclear and cytoplasmic changes during the first 4 hr postinfection consisting of nuclear extrusions and cytoplasmic vesicles, leading to arrays of virus particles by 6 hr. Detailed kinetic analyses of cytopathic events in a subline of cloned L cells infected with large and small plaque variants of mengovirus were reported by Amako and Dales (1967*a,b*). These investigators found that the large-plaque mengovirus variant "is more virulent as judged by its capacity to kill animals and cultured cells more rapidly"; they also used the protein-synthesis inhibitor, streptavitacin A, to study mengovirus cytopathology and came to the conclusion that the cytotoxic principle responsible for initiating cell degradation is a late viral protein. Permeability to erythrocin was correlated with later cell death, rupture, and release of progeny virus. An observation of considerable importance for later research was the finding of increased synthesis of phosphatidylcholine in mengovirus-infected L-2 cells coincident with proliferation of membrane cysternae laden with progeny virus particles (Amako and Dales, 1967*b*). Skinner *et al.* (1968) conducted an excellent light and electron microscopic study of single-cycle infection with echovirus 12 of LLC-MK$_2$ cells (a continual line of monkey kidney cells). They found vesicles present 4–5 hr postinfection; these vesicles coalesced and increased in size along with cytopathic effects detected by phase microscopy beginning 5–6 hr postinfection. The presence in the medium of 2-(α-hydroxybenzyl-benzimidazole), an inhibitor of virus replication, at time 0 or 1–2 hr postinfection prevented formation of vesicles and cytopathic effects over the ensuing 7 hr but did not prevent aggregation of polyribosomes.

In a much later study, when biochemical technology was further advanced, Bienz *et al.* (1980) studied more definitively the "kinetics and location of poliovirus macromolecular synthesis in correlation to virus-induced cytopathology.'' These authors found that by 2.5 hr postinfection viral protein synthesis in HEp-2 cells infected with poliovirus 1 was maximal before virus-induced host cell alterations could be detected by electron microscopy. Maximal viral RNA synthesis occurred 30–60 min later (3–3.5 hr postinfection) and seemed to take place exclusively in the newly formed membrane vesicles. Subsequently, the individual viral RNA-synthesizing vesicles coalesce to form large masses of vesicles exhibiting the typical cytopathic effect attributed to all picornaviruses. Bienz *et al.* (1980) concluded that "viral RNA synthesis, but not viral protein synthesis, is *structurally* [italics theirs] tightly connected with the onset of cytopathology." As mentioned elsewhere, this conclusion has certain validity but cannot be accepted at face value.

4.1.2. Effects of Viral Inhibitors

In attempts to elucidate the mechanisms by which picornaviruses induce cytopathic effects, various chemical and biological reagents have been used to block selectively certain viral functions without, hopefully, directly affecting cell functions. Two reagents that have been most widely used are 2-(α-hydroxylbenzyl)-benzimidazole (HBB) and guanidine-HCl; much of the work on these compounds has been reviewed by Bablanian (1975) and more recently by Eggers (1982). HBB and guanidine both inhibit replication, RNA synthesis, and protein synthesis of many, but not all, picornaviruses by mechanisms still unknown. The HBB has little or no effect on cells, whereas guanidine may have some effects depending on the concentration used. One interesting property of HBB and guanidine is the extremely rapid rate at which drug-resistant virus variants arise under selective pressure of the drugs; frequently, HBB-resistant variants will retain susceptibility to guanidine and *vice versa*, suggesting that they can act at different genome sites. In a series of classical papers, Bablanian *et al.* (1965a,b, 1966) studied the mechanism of poliovirus-induced cell damage. They found that guanidine markedly delayed morphological changes of infected cells while blocking virus-induced RNA and protein synthesis as well as virion yield; they hypothesized that viral functions were essential for causing cell morphological changes as determined by quantitative phase microscopy based on scoring vacuolization, retraction, and rounding of cells (Bablanian *et al.* 1965a). Eventually, however, the cells die, presumably owing to an effect on cell RNA and protein synthesis. In a companion kinetic analysis, Bablanian *et al.* (1965*b*) found that HEL (human embryo lung) cells infected with poliovirus 2 at a multiplicity of 100 underwent these morphological changes beginning at 4–4.5 hr postinfection and 97% of cells were affected by 7 hr postinfection. These virus-induced cytopathic effects closely followed virus multiplication. The cytopathic effect of poliovirus was inhibited when guanidine was introduced within 2 hr of infection but not at 3.5 hr; puromycin, on the other hand, inhibited this poliovirus-induced cytopathic effect at both 2 hr and 3.5 hr. The authors postulate that poliovirus coat protein synthesis is required for producing cell morphological changes but they cannot rule out synthesis of other virus-directed proteins. In a subsequent study, Bablanian *et al.* (1966) found that HBB markedly delayed the development of cytopathic effects caused by echovirus 12 and coxsackie-virus B4 infection of monkey kidney cells; once again, the HBB did not prevent ultimate degeneration of infected cells even though viral multiplication was inhibited. Guskey *et al.* (1970) confirmed and extended these studies somewhat by testing the effect of guanidine and HBB on the release of lysosomal enzymes as well as cytopathology of HEp-2 cells infected with poliovirus 1. Poliovirus replication, cytopathology, and lysosomal enzyme release were all inhibited or delayed if the antiviral agents were present up to 2–3 hr postinfection but not thereafter. They also hypothesized, without data, the requirement of a virus-induced protein for producing a cytopathic effect.

Eggers (1982) makes a very good case that picornavirus inhibition of cellular protein synthesis is unrelated to its cytopathic effect. Citing unpublished experiments in 1978 by Rosenwirth and Eggers, evidence is presented that HBB and guanidine have no effect on the inhibition of protein synthesis in green monkey kidney cells caused by infection with echovirus 12; however, HBB and guanidine still inhibit the usual echovirus 12 cytopathic effects or at least delay it from 6–8 hr until 24 hr (Eggers, 1982). HBB and guanidine inhabit RNA and protein synthesis by echovirus 12, which still inhibits cellular protein synthesis. The tentative conclusion from these experiments is that picornavirus protein synthesis is required for early cytopathology but not for inhibition of cellular protein synthesis (Eggers, 1982).

Many other inhibitory substances have been tested for their effect on picornavirus cytopathology. Among these are actinomycin D tested for its effect on poliovirus-induced leakage of protein from HEp-2 cells (Bubel, 1967) or redistribution of lysosomal enzymes as an indicator of cytopathology in HEp-2 cells infected with poliovirus (Guskey and Wolffe, 1974). Not surprisingly, actinomycin D did not affect poliovirus-induced lysosomal enzyme redistribution, which led to the gratuitous conclusion that cytopathology was not a cell-directed function. Other investigators have tested inhibitors of protein synthesis on picornavirus cytopathic effects, including streptovitacin A (Bubel, 1967), puromycin (Bablanian *et al.*, 1965b), and azetidine-2-carboxylic acid, *p*-fluorophenylalanine, and puromycin (Collins and Roberts, 1972). In experiments with L cells infected with mengovirus in the presence of these inhibitors of protein synthesis, it was found that low doses inhibited viral RNA and protein synthesis; cytopathic effects (and stimulation of phosphatidylcholine synthesis) were inhibited by moderate doses and virus-induced inhibition of cellular RNA and protein synthesis were not affected at all except at very high concentrations of the protein-synthesis inhibitors. These findings led the authors to conclude, perhaps justifiably, that mengovirus-induced cytopathology is not a consequence of viral RNA and protein synthesis but the cytopathic effect may be induced by a viral protein precursor (Collins and Roberts, 1972). Previous experiments in which L cells were infected with mengovirus suggested that a virus-directed protein was required to be synthesized between 4–5 hr postinfection in order for the viral cytopathic effect to be expressed (Gauntt and Lockart, 1966).

Interferon, the cellular antiviral agent, has been studied in a limited way as a probe for dissecting the mechanism by which picornaviruses cause cytopathic effects. Gauntt and Lockart (1966) found that large doses of mouse interferon failed to prevent complete destruction of L cells by mengovirus despite reduction in virus yield as measured by plaques and infectious centers. In follow-up experiments, Gauntt and Lockart (1968) reported that interferon inhibited synthesis of mengovirus RNA, hemagglutinin, and infectious virus by 85–95% in the face of viral antigen accumulation in interferon-treated L cells threefold greater than in control L cells; despite the reduction in synthesis of mengovirus components, all cells in an interferon-treated culture underwent cytopathic effects at the same time as did control infected cells. These results were considered compatible with the hypothesis that cell destruction is due to synthesis of a viral protein in the absence of complete viral replication. Subsequent studies by Haase *et al.* (1969) confirmed that interferon at concentrations sufficient to reduce the yield of mengovirus from infected cells did not affect viral cytopathogenicity (or viral inhibition of host–

cell protein synthesis). However, Haase *et al.* (1969) also found that the effect of mouse interferon on mengovirus-induced cytopathic effects, shutoff of host-cell protein synthesis, and production of mature virions in L cells were dependent on the concentration of the interferon used; cytopathic effects and inhibition of host–cell protein synthesis were not affected until concentrations of interferon were increased 100-fold. These experiments, and some to be described later in which interferon was found not to affect cytopathology caused by vesicular stomatitis virus (Yamazaki and Wagner, 1970), indicate that only certain limited viral functions are required to express cytopathic effects.

4.1.3. Influence of the Cell Cycle

A neglected area of research in viral cytopathology is the variability in response of cells at different stages of the cell cycle. The limited experiments that have been reported to date have been directed, quite understandably, to the virulent, well-characterized picornaviruses. In pioneering studies on cellular changes attending mengovirus-induced cytolysis of mouse L cells at various stages of the host–cell cycle, Lake and Ludwig (1971) reported that late in S and G_2 these cells resist virus-induced cytolysis. Moreover, nuclear lesions were found to precede lysosomal damage as measured by acridine orange staining and direct measure of enzyme release. These authors questioned the primary role of the lysosome in picornavirus cytopathology. Bienz *et al.* (1973) compared replication of poliovirus, cytopathology, and lysosomal enzyme response of interphase and mitotic HEp-2 cells selected by colloidal silica density gradient centrifugation. Cells held in mitosis with colcemid were readily infected with poliovirus (type 1) and produced the expected amount of virus but, in sharp contrast to interphase cells, mitotic cells showed no detectable virus-induced cytopathic effect by light microscopy and only slight effects by electron microscopy. Mitotic cells displayed no redistribution of lysosomal enzymes during poliovirus infection as did interphase cells; once again, these results indicate that lysosomes are not the primary target of picornavirus cytopathology. However, in subsequent studies, Bossart and Bienz (1979) also ruled out the central role of the nucleus in cytopathic effects by finding that enucleated HEp-2 cells underwent the same cytopathic affect as nucleated cells even though the enucleated cells did not support complete replication of poliovirus; as a sidelight, they also found that lysosomal enzyme redistribution failed to occur in enucleated cells undergoing cyto-pathology as judged by vesicle formation viewed by electron microscopy of control enterovirus-infected nucleated cells.

All these papers point to principle targets other than the lyso-somes or nucleus for enterovirus cytopathology. Hugentobler and Bienz (1980) carried these experiments somewhat further by studying the influence of poliovirus infection in S phase and mitosis of HEp-2 cells that had been synchronized by double-thymidine block and infected at hourly intervals after release from the second block. These investigators found that the S phase was not prevented by poliovirus infection but its duration increased somewhat when cells were infected 0–4 hr after release from thymidine block; infection 5 hr or later after release did not prolong the S phase. Some DNA synthesis was noted late in S phase but there was no delay in the cytopathic effect in cells infected at any stage of release and, in fact, the cells continued to synthesize DNA despite onset of cytopathology. Mitotic indices did not vary significantly at different stages of infection and viral RNA synthesis and yield of virions were not affected by time of infection after release from thymidine block.

4.2. Poxvirus Cytopathology

In contrast to the Picornaviridae, among the simplest of all viruses, the family Poxviridae are without question the most complex, sharing some properties with lower prokaryotes. The structure and reproductive strategy of the poxviruses have been reviewed in an earlier volume of this series (Moss, 1974). The poxviruses represent a large group of closely related viruses, which even share certain common antigens, the prototype of which is the vaccinia virus. The poxviruses are very large viruses, the genomes of which consist of double-stranded DNA (mol. wt. $\simeq 150 \times 10^6$) encased in a nucleoid core and surrounded by a membranous, complex shell. These viruses replicate and assemble in the cytoplasm and derive their membranes by *de novo* synthesis rather than from preformed cell membranes as is true for other enveloped viruses. The vaccinia virion has been shown to contain a very large number of enzymes, largely concerned with nucleic acid synthesis, including a DNA-dependent RNA polymerase and a DNA replicase (Moss, 1974). Most significantly from the standpoint of this chapter, the poxviruses are generally species specific and virulent for the various animal species which they infect; cytopathic effects caused by these viruses are frequently rapid and decisive.

Two types of cytopathic effects have been described for poxviruses (Joklik, 1966). Infection frequently results in an early cytopathic effect (1.5–2.0 hr postinfection), which is characterized by cell rounding and has been referred to as a "toxic effect" because it occurs with nonreplicating poxviruses. Later cytopathic effects are noted after the onset of poxvirus DNA synthesis, and this is characterized by cell fusion and other effects eventuating in death. A good example of the early, so-called toxic effect, is seen with mutants of rabbit poxvirus designated white pock (μ); these mutants fail to multiply in a continuous (PK-2) pig cell line but nevertheless produce moderate cytopathic changes characterized by early rounding of cells (McClain, 1965). Cytopathic effects caused by vaccinia virus infection of fetal mouse lung cells were also attributed to a "toxic effect" because it occurred early in infection and at high input multiplicity (Brown *et al.*, 1959); this vaccinia "toxic" cytopathic effect was inactivated by heat, formalin, immune serum, and UV irradiation, but it was noted that "toxicity" was more resistant to UV irradiation than was infectivity of the virus. Experiments by Hanafusa (1960), who examined the cytopathic effects on L cells of heat-inactivated and UV-irradiated vaccinia virus, went a long way to solving the nature of this "toxic" effect. He found that the heat-inactivated virus caused only a slow shrinkage of cells and that very high multiplicities were required to kill the cells. In contrast, the cytopathic effect caused by UV-irradiated vaccinia virus closely corresponded to that caused by the unirradiated virus, especially because of early cell rounding; this early killing effect of UV-irradiated virus was heat labile. In follow-up studies Hanafusa (1962) found that not only UV irradiation but also puromycin, fluorodeoxyuridine, metamycin, and actinomycin did not block this early cytopathic effect of vaccinia virus, as did heat; these data were felt to imply that viral protein synthesis is required to cause early cytopathology. The early cytopathology by UV-irradiated viruses was eliminated by heating and the late cytopathic effect of heated virus could be eliminated by UV irradiation of virus. It seemed evident that the early cytopathic effect of vaccinia virus is not due to a toxic component of the input virion as was originally hypothesized.

Similar studies on rabbitpox virus infection of ERK-1 cells by Appleyard *et al.* (1962) also revealed two stages in the cytopathic effect: early rounding at ~ 2 hr followed by late fusion of cells at ~ 8 hr postinfection. Although multiplication of rabbitpox virus was inhibited by UV irradiation of virions or by the presence of azide or isatin-β-thiosemicarbazone in the medium, the early cytopathic effect occurred in the presence of these inhibitors almost as rapidly as it did in their absence. Appleyard and co-workers (1962) hypothesized that the rabbitpox toxic effect was probably due to formation of some of the soluble viral antigens formed early in infected cells.

A detailed analysis of the mechanisms of vaccinia virus-induced cytopathic effects was undertaken by Bablanian (1968, 1970). He found that puromycin at a concentration of 330 μg/ml prevented the early cytopathic effect and multiplication of vaccinia virus in LLC-MK2 cells, but that puromycin at a concentration of 33 μg/ml did not prevent the cytopathic effect although this largely inhibited viral multiplication. When puromycin (330 μg/ml) was removed from infected cell cultures after 4 hr, virus-induced cell damage began almost immediately and cell rounding was complete by 1 hr. Actinomycin D (5.0 μg/ml) protected cells from the early cytopathic effect but 0.5 μg/ml did not, despite inhibition of virus growth. Bablanian (1968) also found that p-fluorophenylalanine, 5-fluorodeoxyuridine, and isatin-β-thiosemicarbazone had no effect on vaccinia virus cytopathology; he considered these findings to rule out a "toxic effect" as explaining the early cytopathic effect of vaccinia virus, and postulated a requirement for early viral protein synthesis. Confirmation of this hypothesis was reported in a later study (Bablanian, 1970) which showed that 1μg/ml of either cycloheximide or streptovitacin A protected 75% of LLC-MK2 cells from vaccinia virus cytopathogenicity. Removal from the medium of cycloheximide (or puromycin) resulted in a rapid onset of the cytopathic effect, but cytopathology was delayed 1 hr after removal of streptovitacin A. In a later study with vaccinia infection of L cells and HeLa cells, Bablanian et al. (1978) found that the cytopathic effect blocked by cycloheximide could be restored only if the cycloheximide is removed 40–120 min after infection. They also reported variability in time of onset of the cytopathic effect in vaccinia-infected LLC-MK2, HeLa, and L cells after removal of cycloheximide; this effect depends on multiplicity of infection, but no multiplicity-dependent variability was noted on removal of puromycin. Late cytopathic effects (cell fusion) occurred in the presence of cycloheximide when cells were infected with vaccinia virus at a multiplicity of 2700 PFU/cell; they concluded that the early cytopathic effect requires viral protein synthesis but the late (cell fusion) cytopathology appears to be mediated by a virion component.

Horak *et al.* (1971) investigated the effect of interferon on vaccinia virus-induced cytopathology in L cells, following up an earlier observation of Joklik and Merigan (1966) that interferon prevented the early cytopathic effect of vaccinia virus but not the drastic late effect on "lysis" of cells. This late cytopathic effect also could be demonstrated by release of ^{51}Cr from interferon-treated cells 4 hr after infection with vaccinia virus; no comparable release of ^{51}Cr could be demonstrated in L cells not pretreated with interferon or in chick embryo fibroblasts. This cytopathic effect, evidenced by ^{51}Cr release from interferon-pretreated cells, was dependent on multiplicity of vaccina virus infection as well as the amount or duration of exposure to interferon. Cowpox virus or inactivated vaccinia virus did not produce late cytopathology or ^{51}Cr release. However, both cytopathology and ^{51}Cr release were prevented in vaccinia virus infected L cells in the presence of DNA and protein inhibitors and actinomycin D (Horak *et al.*, 1971).

These early studies on the cytopathogenicity of poxviruses all point to two separate cytopathic effects. Although not a great deal of evidence was presented, it seems highly likely that the early cytopathic effect of poxviruses is single hit, not multiplicity dependent, and is due to a newly synthesized viral product(s) rather than components of the input virion. Failure to inhibit the early cytopathic effect by low-dose ultraviolet irradiation or by certain metabolic inhibitors, and retention of cytopathic effect by nonreplicating mutants of rabbitpox virus (McClain, 1965), strongly suggest that one or a limited number of early viral functions are responsible for this early cytopathic effect. It is tempting to speculate that late cytopathic effects (cell fusion), induced in the presence or absence of interferon, are due to late viral products, presumably structural proteins, that can be mimicked by high multiplicity of infection with inactivated virus. The chapter in this volume by Bablanian devoted to the effects of poxviruses on cellular macromolecular synthesis helps to explain the nature of poxvirus cytopathology.

4.3. Rhabdovirus Cytopathology

The Rhabdoviridae are enveloped viruses, the genome of which contains single-stranded RNA always tightly complexed with protein to form an infectious nucleocapsid. The reproduction and structure of rhabdoviruses were reviewed in detail in this series (Wagner, 1975). The prototype of Rhabdoviridae, vesicular stomatitis virus, has been extensively studied and represents a subgroup somewhat different from viruses comprising the rabies–virus group. Unlike the positive-strand Picornaviridae, the Rhabdoviridae are negative-strand viruses, the RNA genome of which cannot serve as messenger, but must be transcribed by an endogenous RNA-dependent RNA polymerase (Baltimore *et al.*, 1970; Emerson and Wagner, 1972). Quite obviously, the replicative strategy of the rhabdoviruses is quite different from that of the poxviruses. The best studied of all rhabdoviruses, vesicular stomatitis virus (VSV), has a very wide host range extending to vertebrates and invertebrates (Wagner, 1975) and is almost as virulent as picornaviruses and poxviruses. VSV replicates in the cytoplasm and is released by budding from the cytoplasmic surface membrane (Howatson and Whitmore, 1962) or from internal cytoplasmic membranes in different host cells (Zee *et al.*, 1970).

Two types of viral cytopathology have been described for VSV: (1) a rapid cellular response at high multiplicities (200 PFU/cell) which ostensibly does not require active viral synthetic functions (Baxt and Bablanian, 1967*a*), and (2) a slower response that does require active replication of the virus, usually accompanied by release of progeny virions. An early observation by Cantell *et al.* (1962) of a VSV "cytotoxic effect" was demonstrated by infecting L cells at high multiplicities even with UV-irradiated VSV; early cellular reaction occurred even when the L cells were previously infected with interfering Newcastle disease virus. A detailed description of cytopathic events was made by Wagner *et al.* (1963) who studied clones of large-plaque and small-plaque variants of VSV; the large-plaque variant was more infectious, multiplied more rapidly, and caused a more marked cytopathic effect even though its other biological properties were indistinguishable from those of the small-plaque variant, which more readily established persistent infection.

The sequence of early cytopathic changes caused by VSV infection of fibroblasts or epithelial cell types was studied largely by phase microscopy by David-West and Osunkoya (1971). Even though these investigators found that both cell types underwent similar rounding at the final stage of cytopathology, early changes peculiar to each cell type were noted; the fibroblasts showed a preponderance of membrane pseudopods, whereas, the epithelial cells exhibited more cytoplasmic vacuolization. Both fibroblasts and epithelial cells were arrested in late telophase. In a subsequent study of monolayer cultures of monkey kidney (VERO) cells infected with VSV, Osunkoya and David-West (1972) found a progressive rise in the mitotic index and a marked rise in the proportion of mitotic cells in late telophase; they interpreted this and other findings as evidence that VSV inhibits those cell processes which initiate and perpetuate the G1 phase of the mitotic cycle. Farmilo and Stanners (1972) reported that wild-type VSV (HR strain) blocked DNA synthesis and cell division in serum-stimulated hamster embryo cells, but mutant *ts* viruses capable of synthesizing viral RNA allowed essential normal DNA synthesis and cell division.

The question whether VSV kills cells by two separate mechanisms, an early "cytotoxic effect" at high multiplicity and later cytopathology requiring active VSV synthetic processes, has been somewhat controversial. In favor of the hypothesis that host cells are killed by mere contact with parental metabolically-inactive VSV are the following observations: (1) UV-irradiated VSV kills cells (Cantell *et al.*, 1962; Baxt and Bablanian 1976*b*); (2) cells pretreated with interferon do not undergo morphological changes when exposed to VSV at 10 PFU/cell but do when exposed to 100 PFU/cell (Yamazaki and Wagner, 1970); (3) defective-interfering (DI) VSV particles devoid of synthetic activity will, at very high multiplicity, cause cytopathic effects (Baxt and Bablanian, 1976*a*) as well as inhibiting cellular RNA synthesis (Huang and Wagner, 1965); and (4) inhibition of viral protein synthesis (by cycloheximide or puromycin) or of viral RNA synthesis do not affect the early morphological changes of BHK-21 cells or LLC-MK2 cells but do prevent late cytopathology (Baxt and Bablanian, 1976*a*). Although this evidence that VSV causes early cytotoxicity at high multiplicity infection may be correct, the interpretation that this requires no viral metabolic functions needs careful reexamination in the light of more recent data, as follows: (1) UV irradiation does not inhibit all VSV transcription even at enormous doses because the 47-nucleotide wild-type leader presents such a small target (Collono and Banerjee, 1977) (2) interferon may inhibit secondary transcription but does not inhibit primary VSV transcription (Repik *et al.*, 1974); (3) DI particles are not completely inert but do transcribe a 46-nucleotide leader (Emerson *et al.*, 1977); even so, the major flaw in most experiments with DI particles is frequent contamination with standard B virions, whose effect on cellular functions is not inhibited by DI particles (Schnitzlein *et al.*, 1983) which by themselves have little effect on cellular macromolecular synthesis even at very high multiplicities (McGowan and Wagner, 1981), and (4) there is no evidence that inhibitors of VSV protein and RNA synthesis can affect primary VSV transcription, which is surely required to inhibit cellular nucleic acid synthesis (McGowan and Wagner, 1981) and cellular protein synthesis (McAllister and Wagner, 1976) and may be required as well to kill cells (Marvaldi *et al.*, 1977). Inactivation of VSV transcriptase activity probably explains the loss of the early cytopathic effect produced by heating virions (Baxt and Bablanian, 1976*b*).

The definitive studies on cell killing by VSV have been performed by Marcus and Sekellick (1974, 1975, 1976). Instead of using rather crude microscopic techniques for measuring cytopathic effects, these authors studied lethality of VSV for monodisperse cells quantitated by their plating efficiency assay, as described by Marcus and Puck (1958). These studies led to the important and well-documented conclusion that a single infectious VSV (B) particle can kill a single cell; moreover, the number of cell-killing particles in a purified preparation was five times greater than that of plaque-forming particles, reflecting the usual ratio of morphologic B virions to infectious units. DI particles derived from the 5' end of the genome were completely devoid of cell-killing activity regardless of multiplicity, provided that they were freed of standard B particles by successive gradient centrifugations. Moreover, DI particles did not interfere with the cell-killing activity of standard B virions, which led to the hypothesis that "cell killing by standard VSV is not dependent upon complete viral replication" (Marcus and Sekellick, 1974). These investigators then proceeded to show that intact transcription of VSV is required for cell killing (Marcus and Sekellick, 1975). They found that UV irradiation inactivated viral infectivity much more readily than transcriptase or cell-killing activities; in contrast, heated virions progressively lost infectivity, transcriptase activity, and cell-killing activity at the same rate. A temperature-sensitive RNA$^-$ mutant [*ts*G114(I)] defective in transcription did not kill cells at restrictive temperature, whereas, an RNA$^-$ *ts* mutant with intact transcription activity did kill cells at a nonpermissive temperature, thus, reinforcing the concept that VSV transcription is required for cell killing (Marcus and Sekellick, 1975; Marvaldi *et al.*, 1977).

In confirmation of earlier studies by Yamazaki and Wagner (1970), Marcus and Sekellick (1976) next extended their studies by testing the effect of exogenous interferon or endogenous interferon induced by poly(I):poly(C) on the cell-killing activity of VSV. They concluded that the results of these experiments confirmed their previous data that VSV transcription is essential for expression of cell killing and, in fact, that transcription of the N and/or NS cistrons are required, perhaps by forming double-stranded transcriptive or replicative intermediates. Their data clearly show that moderate doses of interferon do not abort cell killing, whereas, very large doses do. The authors relate these results to the potential effect of interferon on VSV transcription as being akin to that of UV irradiation in a similar dose-dependent manner. However, the prevailing evidence indicates that interferon has little or no effect on VSV primary transcription (Repik *et al.*, 1974). Be that as it may, Marcus and Sekellick (1974, 1975, 1976) have made an important contribution by clearly demonstrating that the principal, if not the only, cytopathic effect of VSV is due to primary viral transcription.

A subsequent chapter in this volume will elaborate on this hypothesis by analyzing in depth the viral products and cell targets responsible for inhibition of cellular macromolecular synthesis by VSV (Wagner *et al.*, 1984).

4.4. Adenovirus Cytopathology

The family Adenoviridae comprises a large, homogeneous group of structurally closely-related viruses, the genetic information of which consists of double-stranded DNA varying in molecular weight from 20×10^6 to 25×10^6. This family of viruses has been extensively reviewed in this series from the standpoint of their reproduction (Philipson and Lindberg, 1974) and genetics (Ginsberg and Young, 1977). By definition, the adenoviruses are nonenveloped, and have icosahedral capsids containing 252 capsomeres arranged as hexons and pentons with fibers. These are the only viruses considered in this chapter on cytopathology which replicate exclusively in the nucleus of their host cells. Adenoviruses are widely distributed, persist for long periods of time in host tissues, and certain types can induce malignant transformation in cells of a variety of species. Certain classes of adenoviruses are more prone than others to be virulent and cause cytopathic effects.

Type 5 adenovirus has been extensively studied, particularly in human HeLa cells, in which it was recognized quite early that they can cause two distinct cytopathic effects. Pereira and Kelly (1957) first described a late effect as a manifestation of infection with low doses of unpurified virus compared with an early cytopathic effect produced by large inocula of the same impure virus preparation, which was called a "toxic" effect; they also noted that this "toxic" property was less readily inactivated by ultraviolet light than was the infectivity of the virus. Rowe *et al.* (1958) observed that HeLa-cell or KB-cell monolayers would detach from glass surfaces within a few hours after exposure to low dilutions of infectious tissue culture fluids in contact with cells infected with several types of adenoviruses; this adenovirus cell-detaching factor was found to be nondialyzable, smaller than the virus particles, sensitive to trypsin and neutralizing antibody, and more resistant to heat and UV irradiation than the virus particle. Pereira (1958) carefully characterized the cytopathic action of type 5 adenovirus as two distinct effects: (1) an early clumping of cells that is reversible and unaccompanied by virus multiplication, and (2) typical nuclear alterations caused by the virus particle itself. The factor responsible for the early event was deemed to be a protein that can be separated from the virus by differential centrifugation as well as being nondialyzable and destroyed by trypsin. These experiments were confirmed by Everett and Ginsberg (1958) who found the toxic factor in lysates of cells infected with adenovirus serotypes other than type 5 and who observed that the early rounding and clumping of cells occurred in 3–4 hr but did not affect the nucleus. Later, a terminology for the adenovirus antigens as hexon, penton, and fiber was accepted by workers in this field (Ginsberg *et al.*, 1966) and the toxic factor in infected cells was identified as the fiber antigen (Levine and Ginsberg, 1967).

The toxicity of adenovirus and other viruses will be discussed in the following section.

4.5. Viral Toxicity

The existence of toxic properties of viruses and subviral toxic components have long been suspected but have not been well characterized or, for that matter, even been identified for many viruses. In fact, it is not easy to arrive at a clear definition of viral toxicity. For purposes of this review article, I shall define viral toxicity as an effect on cells (or on multicellular hosts) that occurs in the absence of detectable viral replication and generally earlier than the time required for viral replication; it is clearly often difficult to rule out certain viral functions that occur before replication can be detected.

It is not easy to trace the historical beginnings of research on viral toxicity. Some of the earliest studies antedate cell culture techniques and were done in my laboratory with influenza virus injected into whole animals. We found that influenza virus injected intravenously causes fever and lymphopenia in rabbits within a few hours (Wagner *et al.*, 1949*a*). Influenza virus does not multiply in rabbits which also become tolerant to this pyrogenic effect long before antibody can be detected in the circulation; this pyrogenic tolerance was virus specific and did not extend to the pyrogenicity of bacterial endotoxins (Wagner *et al.*, 1949*b*). In later studies, a similar toxic effect of influenza virus could be demonstrated in mice which developed convulsions and paralysis, resulting in death within a day or so after intracerebral injection of nonneurotropic influenza virus which does not replicate in mouse brain (Wagner, 1952). Schlesinger (1950) had previously demonstrated that nonneurotropic influenza A and B viruses were able to undergo a "single cycle" of infection in mouse brain resulting in yields of noninfectious hemagglutin and complement-fixing antigen. We also discovered that nonreplicating influenza virus caused hemorrhagic encephalopathy secondary to endothelial lesions in the brains of chicken embryo within hours after intravenous injection (Hook and Wagner, 1958; Hook *et al.*, 1958). Demonstration of these "toxic" effects required inoculation of very large doses of viruses.

Henle *et al.* (1955) were apparently the first to demonstrate a nontransmissible cytopathogenic effect of influenza virus in tissue cultures of nonpermissive HeLa cells; the virus produced a cytopathic effect that could not be passaged in series. They found no increase in infectious virus in the infected HeLa cells, but did note a significant increase in viral hemagglutinin as well as viral and soluble antigens; the virus inoculum retained some cytopathic effect after mild inactivation by heat or ultraviolet light. Henle and her colleagues concluded that HeLa cells are capable of supporting an incomplete reproductive cycle which may be the minimal replicative event for producing a "toxic" cytopathic effect.

Rabbitpox virus also causes a rapid cytopathic effect, akin to cytotoxicity, by 2 hr after injection even after virus multiplication is inhibited by ultraviolet light or by the presence of azide or β-thiosemicarbazone (Appleyard *et al.*, 1962). Another type of cytotoxic effect results from infection of BHK-21 cells with extremely high multiplicities (10^6/cell) of defective-interfering (DI) particles, of vesicular stomatitis virus, ostensibly devoid of infectious B particles

(Doyle and Holland, 1973); in the case of these DI particles of vesicular stomatitis virus, the only replicative event detectable is transcription of a 46-nucleotide leader sequence from the 3' end of the remaining segment of the viral genome (Emerson *et al.*, 1977).

The feature common to the cytotoxic effects brought on by non-replicating influenza virus, poxvirus, and defective-interfering vesicular stomatitis virus is the high multiplicity of infection required. This has led to the assumption that the toxic effect is caused by one or more components of the parental input virion, most likely protein in origin. However, Cordell-Stewart and Taylor (1971, 1973) have provided evidence that the double-stranded viral RNA isolated from cells infected with bovine enterovirus causes a rapid cytopathic effect as determined by trypan-blue uptake or ^{51}Cr release from affected Ehrlich ascites tumor cells or L1210 cells; toxic effects are reduced or do not occur in cells exposed to single-stranded or heat-denatured double-stranded viral RNA and the toxic effect of bovine enteroviral double-stranded RNA is not abolished by inhibitors of protein synthesis such as puromycin or cycloheximide.

However, the better authenticated cytotoxic components of virions are their proteins. As mentioned previously, the fiber antigen of adenoviruses has a marked cytotoxic effect as demonstrated by its capacity to inhibit division of KB cells, and to inhibit multiplication of adenovirus 5 and poliovirus in KB cells, as well as inhibiting RNA, DNA, and protein synthesis in uninfected and adenovirus-infected KB cells (Levine and Ginsberg, 1967). The isolated glycoprotein of vesicular stomatitis virus at exceedingly large concentrations significantly impairs macromolecular synthesis of BHK-21 cells (McSharry and Choppin, 1978) and can be assumed to have a cytotoxic effect.

Baxt and Bablanian (1976*a*) mounted an in-depth study of the morphological alterations of LLC-MK$_2$ and BHK-21 cells infected with VSV in which the major thrust was the use of metabolic inhibitors to block specific viral functions (also see review by Bablanian, 1975). These authors found that $\approx 50\%$ of LLC-MK$_2$ cells infected with VSV at a multiplicity of ≈ 200 PFU/cell exhibited early cytopathic changes (rounding) at 1 hr postinfection. This early effect was not prevented by cycloheximide (300 µg/ml) but the antibiotic does inhibit late morphological changes in these cells. The number of LLC-MK$_2$ cells involved in early cytopathic effects was not increased at higher multiplicity (2900 PFU/cell). BHK-21 cells responded to VSV with early rounding only at the higher multiplicity, the effect of which was also blocked by cycloheximide. Early cell rounding of LLC-MK$_2$ cells was also not prevented by inhibition of RNA synthesis with proflavin or 6-azauridine, or by prior UV irradiation of the virus. Early rounding of cells was also caused by exposure of LLC-MK$_2$ cells to 2000 particles/cell of defective-interfering VSV, assumed to be devoid of transcriptase or protein-synthesizing activity. Prior studies by Yamazaki and Wagner (1970) revealed similar multiplicity-dependent cytopathic effects of VSV in primary rabbit kidney cells exposed to the antiviral action of highly purified rabbit interferon; late cytopathic effects were inhibited by interferon in cells infected with VSV at a multiplicity of 10 PFU/cell but not at 100 PFU/cell.

The above studies generally support the concept that VSV can cause two cytopathic effects, early "cytotoxicity" at high multiplicity and late cytopathic effects at lower multiplicity. However, as will be discussed in later chapters on viral effects on cellular macromolecular synthesis, the concept of two sharply separable effects is too simplistic as indicated by variation in cellular susceptibility reported by Baxt and Bablanian (1976a). Related studies by these authors on VSV effects on cellular macromolecular synthesis (Baxt and Bablanian, 1976b), as well as studies by other investigators will be discussed in a later chapter.

5. SOME SUBCELLULAR TARGETS OF PATHOGENIC VIRUSES

The realization that morphological changes are subjective and difficult to quantitate has led some investigators to study disturbed functions of subcellular structures as a means for probing viral cytopathogenesis. This subject has generated a considerable literature, much of which is beyond the scope of this review. Also much of this work has been superseded by even more precise methodology concerned with cellular macromolecular synthesis, the major objective of this volume. An attempt is made in this section to review some of the research on viral effects on certain cell functions, particularly directed toward cell membranes, lysosomes, and cytoskeleton. Here we have chosen selective viruses and only background literature to exemplify these effects of viruses. It is well to keep in mind that cell membranes, lysosomes, and cytoskeleton may not be primary targets of viral infection, as implied by many investigators, but may simply reflect secondary, and oftentimes delayed, effects of pathogenic viruses. Nevertheless, these effects represent a significant subset of viral cytopathic effects and deserve serious consideration.

5.1. Cellular Membrane Effects of Certain Viruses

It has been known for some time that different viruses have different effects on cell membranes. In particular, the enveloped viruses can cause early cell fusion as they enter the host cell or late cell fusion after replication of the virus. These phenomena apply particularly to the paramyxoviruses and have been dubbed fusion from without and fusion from within (Bratt and Gallaher, 1969). Other enveloped viruses can produce similar but usually less striking effects on cell fusion. Nonenveloped viruses can also cause drastic changes in cell membranes, usually resulting in alteration in the permeability barrier of the cell membrane. Still other viruses, or the same viruses under different conditions, can alter in a more subtle way the active or passive transport mechanisms of cells. Under some conditions, some viruses even stimulate the formation of cell membranes by inducing augmented synthesis of cellular membrane lipids.

5.1.1. Virus-Induced Syncytia Formation

5.1.1a. Paramyxoviruses

The cytopathic effect of paramyxoviruses is often reflected in fusion of adjacent cells into multinucleated syncytia. Paramyxoviruses differ widely in their cytopathogenicity and synctia-forming ability, functions that also vary greatly among host cells. An excellent example is the simian virus 5 (SV5) paramyxovirus which causes little cytopathic effect and, in fact, persists in a latent form in monkey kidney cells but is virulent for baby hamster kidney (BHK-21) cells, which rapidly undergo fusion to form huge syncytia (Holmes and Choppin, 1966). The cell-fusing factor of paramyxoviruses, which is identical to the hemolytic factor, was definitively identified by Scheid and Choppin (1974). When Sendai paramyxovirus was grown in chick embryo cells, it was found to contain the F glycoprotein in a form that induces hemolysis and cell fusion, but not the precursor F_0 which is present in Sendai virus grown in MDBK (Madin-Darby bovine kidney) cells; MDBK-grown Sendai virus with F_0 precursor protein lacks hemolytic and cell fusion activity and cannot infect MDBK cells unless F_0 is converted to F by cleavage with trypsin (Scheid and Choppin, 1974). Nagai *et al.* (1976) extended this observation to the Newcastle disease virus (NDV) paramyxovirus by finding that cleavage of F_o is necessary for cell fusion and hemolysis for virulent strains, whereas, avirulent strains generally have an uncleaved F protein. Of additional interest was the finding that cleavage of the HN glycoprotein parallels enhanced hemagglutinin and neuraminidase activity, but has no effect on assembly of progeny virus into fully formed virions of lower infectivity (Nagai et al., 1976). Sheid and Choppin (1977) next extended these observations by finding that the active fusion-factor (F) glycoprotein in Sendai, NDV, and SV5 viruses consists of two disulfide bonded polypeptide chains $(F_{1,2})$ that acquire hemolytic and cell-fusion activity when cleaved from precursor F_o by endogenous or exogenous proteases.

An interesting observation on the F glycoprotein fusion factor of paramyxoviruses was that electron spin resonance spectroscopy (ESR) detects a significant alteration in the membrane spectrum of human or chicken erythrocytes during hemolysis by Sendai virus (Lyles and Landsberger, 1977). In elegant studies, Maeda et al. (1977) demonstrated that spin-labeled phosphatidylcholine inserted in the membrane of Sendai virus (hemagglutinating virus of Japan, HVJ) was rapidly transferred to interacting erythrocyte membrane as demonstrated by a change in ESR spectrum characteristic for erythrocyte membrane. This ESR transfer effect was noted only if the Sendai virus contains the F protein in the cleaved $F_{1,2}$ form acquired either by growth in permissive cells or by cleavage with trypsin. Influenza virus, on the other hand, did not transfer a membrane-inserted phosphatidylcholine spin label to interacting erythrocytes. These studies provide a dramatic demonstration of transfer of paramyxovirus membrane lipids to the membrane of the erythrocyte with which the virus fuses, an observation that was confirmed in similar studies by Lyles and Landsberger (1979). An electron microscopic confirmation of paramyxovirus membrane fusion with multilayered liposomes was also provided by Haywood (1974). A similar fusion effect of Sendai virus on erythrocyte membrane was demonstrated by fluorescence depolarization as well as by scanning electron microscopy (Fuchs et al., 1978).

5.1.1b.   Myxoviruses and Rhabdoviruses

The membranes of negative-strand viruses lacking an identifiable fusion factor can also fuse with interacting cell membranes but to a much lesser extent than the paramyxoviruses. Influenza A and B viruses, for example, exhibit enhanced infectivity if the hemagglutinin glycoprotein is cleaved by endogenous or exogenous proteases to forms HA_1 and HA_2 (Lazarowitz and Choppin, 1975). The membrane-fusing activity of the cleaved HA form of influenza virus was demonstrated by insertion of the glycoprotein into octylglucoside-dialyzed liposomes which were then allowed to interact with cell membrane; only the cleaved form of the hemagglutinin permitted fusion of liposomes with cell membrane (Huang *et al.*, 1980). Clear evidence of membrane fusion of interacting cells and viruses was demonstrated by deuterium nuclear magnetic resonance (NMR) spectroscopy by Nicolau *et al.* (1979). These investigators found that only the cleaved HA protein of influenza virus and cleaved F protein of Newcastle disease virus altered the dynamics of choline headgroups and fatty acyl chains, suggesting that virus–cell membrane fusion is the first event in initiation of infection by these viruses. Virulent enveloped viruses, such as vesicular stomatitis virus, do not ordinarily elicit fusion of host cell membranes but this can occur under special conditions, such as fusion of mouse neuroblastoma cells infected with VSV mutant *ts*G31(III) at restrictive temperature (Hughes *et al.*, 1979). The VSV glycoprotein can also alter the membrane of the infected cell into which it is inserted, thus making them susceptible to lysis by cytotoxic T lymphocytes, indicating that VSV glycoprotein is needed for histocompatibility H2-restricted lysis of cells by cytotoxic T lymphocytes (Hale *et al.*, 1978). A detailed analysis of these effects of viruses on cellular immunity is presented in a previous volume of *Comprehensive Virology* (Zinkernagel, 1979).

5.1.1c. Herpesviruses and Other Viruses

The cytopathology of the enveloped herpesviruses will be discussed in a later chapter of this volume. There is a very large literature on the membrane effects of herpesviruses, particularly on the cell-fusing activity that leads to polykaryocytosis going back to the original observations of Hoggan and Roizman (1959) and extensively reviewed by Roizman (1962) and in *Comprehensive Virology* by Roizman and Furlong (1974). Among the most interesting aspects of these studies are experiments that implicate certain of the numerous glycoproteins of herpes simplex virus type 1. Manservigi *et al.* (1977) studied mutants defective in synthesis of glycoproteins B2 and C2, as well as a recombinant of these mutants; they concluded from these studies that glycoprotein B2 plays a critical role in promotion of cell fusion but that glycoprotein C2 suppresses the fusion effect leading to polykaryocytosis. At the other end of the spectrum, brain extracts of two poorly categorized but very important neuropathogenic viruses, scrapie and Creutzfeld–Jakob viruses, were found to induce fusion of mouse neuroblastoma and L cells at a rate similar to that of Sendai virus (Kidson *et al.*, 1978). The nature of these cytopathogenic phenomena is yet to be uncovered but should yield to a rapidly developing technology.

5.1.2. Virus-Induced Alterations in Membrane Permeability and Transport

It has been known for some time that infection with various types of viruses impairs the permeability barrier function of the host cell cytoplasmic membrane, allowing ordinarily impermeable large molecules to enter the cell from the surrounding medium and allowing particulate intracellular material to leak out. A good example of this is penetration of supravital dyes, such as trypan blue, to enter cells, usually late after viral infection as a criterion of cell death. In addition, certain viral infections induce earlier and more subtle changes by altering membrane transport of small ions into or out of the infected cell. A few selected examples of altered membrane permeability of the cytoplasmic membrane are cited here as an example of readily measured cytopathic effects of well-known virulent viruses.

5.1.2a. Enveloped Viruses

The paramyxoviruses, largely because of their profound cell-fusing activity, have served as an important model of membrane perturbation by viruses. During Sendai virus-mediated fusion of mouse ascites cells, Pasternak and Micklem (1973) detected loss of intracellular metabolites coincident with inhibition of their accumulation from the medium. This failure to maintain selective permeability did not occur at 0°C and was unaffected by cytochalasin B which inhibits fusion by the virus. Chick embryo fibroblasts infected with Newcastle disease virus were found to release cellular enzymes, such as lactate dehydrogenase, glutamic oxaloacetic transaminase, and lysosomal enzymes (Katzman and Wilson, 1974). These cells also became permeable to external sucrose and dextran which entered NDV-infected cells by 4 hr postinfection, before intracellular enzymes were liberated at 6 hr postinfection. Active infection with continuous insertion of NDV glycoprotein was claimed to be essential to maintain the increased state of membrane permeability (Katzman and Wilson, 1974). Transient inhibition of amino acid uptake, but no inhibition of protein synthesis, was observed in HeLa-S, and L-929 cells infected at high multiplicity with UV-inactivated Sendai virus (Negreanu *et al.*, 1974). Maximum inhibition of amino-acid uptake was also noted at the end of the virus-adsorption period but, within the next few hours, the uptake returned to control levels; this transient change in amino acid uptake was concommitant with enhanced uptake of potassium. Micklem and Pasternak (1977) also reported that infection with Sendai virus increases the exchange rate of membrane-bound Ca^{2+} and cells became sensitive to concanavalin A. These authors concluded that the exchangeability of Ca^{2+} represents an early specific event resulting from attachment of Sendai virus. Previously, Fuchs and Giberman (1973) had noted enhancement of K^+ influx in baby hamster kidney and chick embryo cells during adsorption of Sendai virus. They later found that adsorption to HeLa cells of Sendai virus at very high multiplicities caused immediate changes in the cell-membrane ion barrier, concomitant with morphologic changes (Fuchs *et al.*, 1978). Within minutes of adsorption of Sendai virus, the cells began to leak K^+ and exhibited an intensive influx of ions, concommitant with depolarization of the membrane. These membrane defects began to undergo repair within 20–60 min; scanning electron microscopy showed bridging microvilli and the cells then fused. Another virus, vesicular stomatitis virus, was reported to inhibit uptake of uridine in chick embryo cells (Genty, 1975), but this effect was not observed in experiments with mouse myeloma cells and L cells (Weck and Wagner, 1978).

Carrasco (1978, 1981) reported that nonspecific membrane permeability is increased in cells infected with VSV and Semliki Forest virus (SFV), leading to the hypothesis that "many if not all cytolytic viruses induce membrane leakiness in their host cells during their lifetime" (Carrasco and Smith, 1980). These reports prompted Gray *et al.* (1983) to repeat these studies; they found that Na^+ uptake by BHK-21 cells was unchanged during VSV infection or increased during SFV infection and K^+ content was unchanged by infection with either virus. Moreover, the ability of BHK-21 cells to concentrate 2-deoxy-D-glucose was enhanced for a 2–3 hr period after infection with VSV or SFV. These results by Gray *et al.* (1983) "do not support the current hypothesis that a nonspecific increase in membrane permeability occurs in cells infected with rhabdoviruses or togaviruses."

5.1.2b. Nonenveloped Viruses

Much of the very large literature on membrane permeability and ion transport in cells infected with nonenveloped viruses, particularly picornaviruses, will be dealt with *in extenso* in other chapters of this volume. The purpose of this limited analysis of the subject is to point out its relevance to viral cytopathogenicity. In one study, Egberts *et al.* (1977) noted an increased concentration of intracellular ATP by 3 hr after infection of Ehrlich ascites tumor cells with mengovirus, but due to leakage into the medium through a damaged cell membrane, the intracellular concentration of ATP decreased to 0 by 8 hr postinfection. As ATP leaked into the medium, there was a simultaneous loss of intracellular Mg^{2+}, K^+, and polyamines and an increase in cellular Na^+. In another study, infection of susceptible cells with mengovirus, encephalomyocarditis virus, Semliki Forest virus, or SV40 resulted in greatly increased penetration into cells of the bulky translation-inhibitory antibiotics gougerotin, edeine, and blastocidin S (Contreras and Carrasco, 1979). Fernandez-Puentes and Carrasco (1980) next noted that the ordinarily impermeable A chains of protein toxins (abrin, α-sarcin, mitogillin, and restrictocin) were all able to inhibit in infected cells protein synthesis of encephalomyocarditis virus, adenovirus type 5, and Semliki Forest virus; the A chains of these antibiotics did not inhibit translation in uninfected cells because they cannot penetrate without the B chain. They also provide evidence that virus adsorption to cells immediately renders the cell membrane permeable, allowing uptake of the A chains of the plant toxins, abrin and ricin. In a subsequent study, Carrasco (1981) reported permeability to the aminoglycoside antibiotic hyromycin B early after adsorption of encephalomyocarditis virus. The effect was not reversed by cytoskeleton-disrupting agents (vinblastins, cytochalasin B, or colchicine); nor did concanavalin A or interferon reverse this increase permeability to aminoglycosides caused by viral adsorption.

In studies on the effect of viral infection on ion transport, Nair *et al.* (1979) reported that poliovirus infection of HeLa cells caused increased Na^+ influx late in viral infection, after inhibiton of cellular protein synthesis. Guanidine hydrochloride blocked the poliovirus-induced influx of Na^+ and this was reversed by the guanidine inhibitor, choline. It was felt that expression of one or more late viral functions was essential for intracellular Na^+ accumulation, which could be reversed by cycloheximide. Increased permeability of infected cell membrane appeared to be responsible for this Na^+ accumulation rather than inhibition of cellular Na^+-K^+ ATPase activity. In a follow-up study, Nair (1981) found that Na^+ concentration in poliovirus-infected cells began to increase about 3 hr postinfection while the K^+ concentration decreased progressively, as did total cell monovalent cations and cell density. These observations led to the hypothesis that the morphologic lesions in poliovirus cytopathology might be due to changes in cell tonicity.

In quite a different approach to the effect of viruses on cell membrane, Levanon *et al.* (1977) reported that adsorption of encephalomyocarditis, Semliki Forest, and polyoma viruses to BHK-21 and mouse 3T3 cells resulted in a rapid increase in membrane fluidity as measured by depolarization of the fluorescent dye, diphenylhexatriene. The degree of fluidity increase was virus dose dependent and could be reversed by low temperature or by blocking virus receptors on the host cell surface. These investigators suggest that an increase in membrane fluidity is an early event in viral cytopathogenicity.

A different cascade of effects occurs in CV-1 green monkey kidney cells infected with simian virus 40 (Norkin, 1977). In these cells, infection results in release of mitochondrial malic dehydrogenase by 24 hr, which appears first in the cytoplasm and then in the medium surrounding the cell. The infected cells lose their ability to consume O_2 by 48 hr, as well as electron transport. Phospholipid synthesis is increased by 32 hr. Lactic dehydrogenase release precedes failure to exclude trypan blue, which is noted 6 hr postinfection. Early and late SV40 mutants have much less affect on all these cell functions. This complicated series of events occurring after SV40 infection leads to some of the observations made in sections below but also emphasizes two points: (1) cytopathic effects can be a complicated series of interrelated events, and (2) most of the events are secondary or tertiary and it is not easy to identify the primary event which initiated this cascading cytopathology.

5.1.3. Lipid Synthesis by Virus-Infected Cells

During the course of infection with many viruses, host cells may undergo enhanced synthesis of lipids largely destined for incorporation into cellular membranes but to some extent in the membranes of enveloped viruses. The picornaviruses have been studied more intensively for their effect on cellular lipid synthesis and membrane formation. Phospholipid synthesis was found to increase within 1.5 hr after infection of HeLa cells with Maloney type 1 poliovirus (Cornatzer *et al.*, 1961). In a detailed analysis, Penman (1965) found that poliovirus infection of HeLa cells resulted in an increased rate of uptake of [^{14}C]methylcholine from 2.5 hr postinfection, reaching peak uptake at 5 hr postinfection. Puromycin (100 μg/ml) inhibits this increased choline uptake when added at 2 hr but not at 3 hr postinfection; most of the choline taken up is incorporated into large acid-precipitable membrane structures. Guanidine hydrochloride also inhibited choline uptake, suggesting that specific viral functions were required to augment lipid incorporation into membranes of infected cells. Amako and Dales (1967*b*) next reported that picornavirus-infected host cells undergoing cytopathic degeneration exhibit enhanced synthesis of lipids and develop vacuoles prominent in the centrosphere region. After mengovirus infection, incorporation of choline into phosphatidylcholine was increased and this was found in association with microsomes; these newly developed cisternae appear to provide the focus for picornavirus replication and assembly. Similar results were obtained by Plagemann *et al.* (1970) in studies of choline metabolism and membrane formation in Novikoff hepatoma cells infected with mengovirus. Collins and Roberts (1972) reported that the stimulation of phosphatidylcholine synthesis in L cells infected with mengovirus was inhibited by moderate doses of protein-synthesis inhibitors along with cytopathic effects, but small amounts of the protein-synthesis inhibitors also shut off viral RNA and protein synthesis. Mosser *et al.* (1972) confirmed that [^{14}C]choline incorporation is stimulated in poliovirus-infected cells, resulting in formation of internal smooth membrane structures. These poliovirus-infected cells also incorporate 4–5 times more [^{3}H]glycerol, about 80% of which appears in rough endoplasmic reticulum after a 3-min pulse; the [^{3}H]glycerol-labeled material is then chased into smooth membranes in infected cells but not in uninfected cells. Guanidine hydrochloride had only a slight effect. These findings suggested to Mosser *et al.* (1972) that poliovirus infection enhances lipid synthesis particularly directed toward formation of smooth vesicles to serve as factories for poliovirus replication and assembly.

Many viruses other than picornaviruses also augment lipid synthesis in infected cells. In perhaps the first observation, made as a by-product of biochemical studies on adenovirus multiplication, Green (1959) found a significant stimulation of phospholipid synthesis in cells infected with adenovirus. In a more detailed study, McIntosh *et al.* (1971) examined lipid metabolism in HeLa and KB cells infected with adenovirus by measuring uptake of [^{32}P]orthophosphate and [^{14}C]acetate; they found that cellular lipid synthesis began to increase by 6 hr after adenovirus infection and the peak increase of incorporation of [^{14}C]acetate into triglycerides was noted 12–18 hr postinfection. Perhaps the most interesting observation by McIntosh *et al.* (1971) was the finding that UV-irradiated adenovirus and even purified penton-base capsid subunits also enhanced cellular lipid synthesis. In an early study with still another virus, Gausch and Youngner (1963) noted stimulation of phospholipid synthesis in HeLa cells during the first 6 hr of infection with vaccinia virus. Sendai virus was also found to increase incorporation of [^{32}P]orthophosphate and [^{14}C]glucose into all lipids in abortively infected primary chick embryo cells; a similar effect was noted with more productively infected monkey kidney cells (Blair and Brennan, 1972). These authors also reported increased uptake of [^{14}C]glycerol in Sendai virus-infected cells but only when large amounts were added and uptake of both [^{14}C]glycerol and [^{14}C]glucose was inhibited during short pulses of infected cells.

Somewhat contradictory results were obtained in studies of lipid synthesis in cells infected with two other viruses. Infection of chick embryo cells with Sindbis virus resulted in a marked decline in phospholipid synthesis by 5 hr postinfection (Pfefferkorn and Hunter, 1973). Of considerable interest was the casual finding that pretreatment with interferon did not reverse this inhibition in phospholipid synthesis despite a 98% reduction in Sindbis virus yield. Moreover, specific ^{32}P activity of phospholipids was much lower than cell membrane phospholipids, indicating that Sindbis virus membrane lipids are derived largely from preformed membranes. Still another situation was noted in cells infected with SV40, which was found to lead to increased synthesis of phospholipids reaching a peak at $\simeq 32$ hr postinfection, but this was followed by a severe decline in new membrane synthesis (Norkin, 1977).

The only conclusion that can be drawn from these studies is that most viruses stimulate lipid synthesis in infected cells but new membrane formation depends on the replicative strategy of each individual class of viruses.

5.2. Effects of Viruses on Lysosomes

An extensive literature has been spawned over the years concerning the effects of viral infections on the cell lysosome, the omnipresent membrane-enclosed organelle which is replete with degradative enzymes (DeDuve, 1959). A complete review of this literature cannot be contained within the confines of this chapter. A good, but not very recent, overview of the subject of viruses and lysosomes is presented by one of its protagonists (Allison, 1967), who discusses the role of lysosomes in virus uncoating, cytopathic effects of many viruses (including those of mouse hepatitis, vaccinia, myxoviruses, enteroviruses, herpes virus, and oncogenic viruses), virus hemolysis, formation of polykaryocytes, structural alterations of chromosomes, and carcinogenesis. Allison (1967) warns of "the danger in being entangled in a web of speculation" but it seems, or perhaps seemed, a very attractive all-inclusive hypothesis because of the drastic effects of lysosomal enzymes and the ease with which they can be quantitated as a putative measure of viral cytopathic effects. It is easy from my retrospective vantage point to raise the specter of lysosomal release being a secondary phenomenon rather than the primary effect of viral infection. It seems doubtful that lysosomal enzymes are the cause of viral cytopathogenicity, but it would be unwise to downplay or ignore this phenomenology.

5.2.1. Lysosomal Effects of Picornaviruses

As expected from their high degrees of virulence, the picornaviruses have been studied most extensively for their effect on lysosomes and for release of lysosomal enzymes. During the course of infection of KB cells with poliovirus, the cytoplasmic activities of β-glucuronidase as well as acidic proteases, ribonucleases, deoxyribonucleases, and phosphatases rose by 6 hr to levels two- to fourfold greater than that in uninfected cells, coincident with virus release and cytopathology (Flanagan, 1966). Hydrocortisone had no effect on lysosomal enzymes and infection with herpes simplex virus caused only minimal alterations in KB cell enzymes. The author attributed the release of these lysosomal hydrolases to the effect of poliovirus on lysosomal membrane. To test the hypothesis proposed by Amako and Dales (1967b) that a cytotoxic protein synthesized by mengovirus alters the permeability of lysosomes, Thacore and Wolff (1968) examined cytoplasmic extracts from poliovirus-infected HEp-2 cells and found that they contained twice as much β-glucuronidase from lysosomes than did cytoplasmic extracts of uninfected cells; any interpretation of such an experiment might be considered dangerous. Guskey *et al.* (1970) used guanidine-HCl and HBB to study the relationship between lysosomal enzyme release and cytopathic effects in HEp-2 cells infected with poliovirus type 1. Lysosomal enzyme release, cytopathic effect, and virus release were all inhibited or delayed if the antiviral agents were added up to 2–3 hr postinfection, but not later; an hypothesis implicating a newly synthesized poliovirus protein as the lysosome enzyme-releasing factor was not tested. This laboratory later demonstrated that lysosomal enzymes are released from HEp-2 cells, not suddenly, but gradually over several hours after infection with poliovirus (Heding and Wolff, 1973). Subsequently, they described a lysosomal phosphorylase redistributed in the HEp-2 cell during poliovirus infection, which was considered to play a role in cytopathology (Rice and Wolff, 1975). Bovine and equinine enteroviruses also caused the release of lysosomal acid phosphatase in and from lamb kidney cells at the time that cytopathic effects appear (Ram *et al.*, 1978); no mention is made whether the cytopathic effect could be due to lysosomal release or that lysosomal release could be due to the cytopathic effect.

Bossart and Bienz (1979) showed that poliovirus-infected enucleated HEp-2 cells exhibited the same cytopathic effect as poliovirus-infected nucleated cells. However, enucleated cells did not show the same redistribution of lysosomal enzymes (β-glucuronidase and β-glucosaminidase) into the cytoplasm as do infected nucleate cells; however, it should be noted that enucleated cells do not support the replication of poliovirus as well as do nucleated cells. A temporal analysis of the events occuring in poliovirus-infected cells, as well as their cellular location, was made by Bienz *et al.* (1980). By kinetic analysis, viral protein synthesis was found to reach a maximum at 2.5 hr before cell alterations can even be detected. Poliovirus RNA synthesis reached a peak later (at 3.0–3.5 hr postinfection) when new vacuoles can be seen, although viral RNA synthesis continues as vacuoles coalesce to form the typical poliovirus cytopathic effect. Therefore, these authors consider that viral RNA synthesis and not protein synthesis is more closely related to structural changes; possibly this could also include lysosomal structural changes.

5.2.2. Lysosomal Effects of Viruses Other than Picornaviruses

Allison and Sandelin (1963) examined the activation of the enzymes β-glucorinadase and various acid hydrolases in the lysosomal
supernatant fractions of mouse liver cells infected with mouse hepatitis virus and monkey kidney cells infected with vaccinia virus. In
the infected cells, they easily measured the release of lysosomal enzymes into the supernatant fraction and evidence was presented that
this is not an artifact of homogenization. Allison and Sandelin (1963)
suggested that the release of lysosomal enzymes may explain some
of the biochemical changes found in infected cells and may contribute
to the cytopathic effects of some viruses. Allison and Malucci (1965)
next studied monkey kidney cells and HeLa cells infected with influenza, Newcastle disease, adenovirus, and vaccinia virus by fluorescence and light microscopy and by enzyme analyses. Three stages
were identified in each infectious process: (1) permeability of lysosomal membranes without enzyme release as demonstrated by increased aminoacridine fluorescence and by neutral red staining; recovery could occur at this stage with abortive NDV infection; (2)
lysosomal enzymes (acid phosphatase and 5-bromo-4-chloro-indoxyl
acetate esterase) were released into the cytoplasm, the cells round
up and there was decreased uptake of aminoacridine and neutral red
into lysosomes; and (3) not usually seen in cell culture, lysosomal
enzymes are released from or inactivated in cells. The course of infection of primary rhesus monkey kidney cells with adenovirus 7 and
Sendai virus as well as echovirus 12 was followed by assaying tissue
culture fluid for glutamic oxaloacetic transferase and lactic dehydrogenase; each viral infection exhibited patterns of enzyme activity
distinct for the cytopathic effect of each virus (Gilbert, 1963). Vaccinia virus and vesicular stomatitis virus cause little alteration,
whereas, poliovirus produced significant changes in the distribution
of lysosomal enzymes β-glucuronidase and acid phosphatases in GPS
cells, correlated with cytopathic effects (Wolff and Bubel, 1964). Koschel *et al.* (1974) studied lysosomal enzyme distribution in HeLa cells
by histochemical and biochemical techniques and found that poliovirus-infected cells released lysosomal enzymes into the cytoplasm
starting 3 hr postinfection but vesicular stomatitis showed no such
effect under comparable conditions. Simian virus 40 infection did
cause the release of lysosomal *N*-acetyl-β-glucosaminidase into the
cytoplasm of resistant F-22 transformed cells as it did in highly sus-

ceptible rhesus monkey kidney cells (Norkin and Ouellette, 1976). Rabbit poxvirus did alter the plasma membrane and lysosomal membrane permeability of HeLa cells (Schümperli *et al.*, 1978). Finally, high-multiplicity infection of BSC-1 green monkey kidney cells with SV40 and 3T3 cells infected with herpes virus did result in lysosomal changes accompanied by cytoplasmic vacuolization and extrusion of lysosomal enzymes into the cytoplasm (Allison and Black, 1967).

It seems clear that infections with picornaviruses generally increase permeability of lysosomal membrane that results in extrusion of hydrolases into the cytoplasm, generally late in infection. Infection with other viruses appears to cause variable results. It seems quite likely from these studies that lysosomes are rarely, if ever, primary targets for the infectious process by any viruses. In all likelihood, leaking of lysosomal hydrolases is a late event in the phenomenology of viral infection and has little or no bearing on the critical events that lead to viral cytopathogenicity. Lysosomal damage is more likely to be the end result rather than the causal event in viral cytopathogenicity.

5.3. Effects of Viruses on the Cytoskeleton

Of special interest to cytologists is the effects certain viruses may have on cytoskeletal structures during the course of infection. Of particular significance is the degree to which changes in cytoskeleton components, such as microtubules and microfilaments, determine the cytopathic effects observed after viral infection. Quite understandably, such interest has generated a large body of literature, primarily descriptive, which cannot be covered in this chapter; specific cytoskeletal effects of certain viruses will be dealt with in chapters in this volume devoted to specific viruses. In this introductory chapter, we merely highlight a few aspects of viral effects on cytoskeleton as they apply to cytopathic effects. As mentioned in preceding sections on membrane and lysosomal effects of viruses, it seems likely that cytoskeletal alterations, particularly occurring late in infection, are largely secondary effects rather than providing the primary targets for cytopathogenic viruses.

As expected, electron microscopy has been the principal technique for studying the effects of viruses on subcellular components, including the cytoskeleton. A great deal of the early work was carried out by Dales and his colleagues (see summary by Dales, 1975) who clearly noted by electron microscopy that adenovirus type 5 and reovirus type 3 bind to microtubules during the course of their replication. Association of adenovirus with microtubules and nuclear pores appeared to be an early event in vectorial movement of the inoculum (Dales and Chardonnet, 1973). Luftig and Weihing (1975) tested binding of adenovirus and reovirus to rat brain microtubules spread on electron microscopic grids; they found that 72% of adenovirus but only 32% of reovirus underwent random association with spread microtubules. In a follow-up study, Babiss et al. (1979) detected greater binding *in vitro* of reovirus type 1 (81%) than of reovirus type 3 (56%) to the "edge" of microtubules, regardless of whether they were obtained from chick brain, rabbit brain, or HeLa cells. This greater affinity of reovirus 1 binding to microtubules was attributed to the σl polypeptide (hemagglutinin and neutralization antigen), based primarily on testing recombinant virus particles. These *in vitro* results were also validated by finding eight times as many viral factories associated with microtubules in reovirus type-1 infected cells compared to reovirus type 3-infected cells. In a quantitative study of interaction of adenovirus types 2 and 5 with microtubules inside infected cells, Miles *et al.* (1980) found similar binding 1–6 hr postinfection of adenovirus wild-type and a temperature-sensitive mutant (*ts*1) grown at permissive temperature. In contrast, when cells were infected with *ts*1 grown at restrictive temperature (39°C), most of the particles (74–94%) were found in large vacuoles thought to be lysosomes.

A detailed analysis of the involvement of microtubules in cytopathic effects was made by Ebina *et al.* (1978) who infected cells with poliovirus, Sendai virus, adenovirus, and herpesvirus in order to examine the effect of each virus on the formation of microtubular paracrystals induced by vinblastine sulfate in HeLa-S3 cells. In poliovirus-infected cells, the cytopathic effect (cell rounding) and inhibition of paracrystal formation were both noted at 4 hr postinfection, proceeding in parallel. In the case of Sendai virus infection, no effect on paracrystal formation could be noted despite a syncytial cytopathic effect. In adenovirus- and herpesvirus-infected cells, inhibition of paracrystal formation occurred well before the cytopathic effect and was not blocked by UV irradiation or nucleic acid analogues but was by inhibition of protein synthesis. These findings led Ebina *et al.* (1978) to the hypothesis that early viral proteins are responsible for inhibition of microtubule formation and the cytopathic effect (cell rounding) except that Sendai virus did not cause this type of cytopathology.

Quite different effects on actin-containing structures were found by Rutter and Mannweiler (1977) in BHK-21 cells infected with another paramyxovirus, Newcastle disease virus (NDV) and the rhabdovirus vesicular stomatitis virus (VSV). Studies made by fluorescence labeling of actin with anti-actin showed that in NDV-infected cells, the number of actin filaments increases, some zones which contain virus antigens apparently being in close proximity to actin structures; by contrast, VSV infection results in strong reduction of actin-containing fibers. These widely differing effects on actin-filament cytoskeleton are presumed to be a direct reflection of the different cytopathic effects caused by NDV and VSV.

The question whether the cytoskeleton is involved in reproduction of vesicular stomatitis virus was raised by Genty and Bussereau (1980). These authors found that CER cells infected with VSV showed a morphology similar to that observed after treatment with cytochalasin B, but the cytoskeleton did not appear to be implicated in VSV reproduction. Temperature-sensitive mutants affected in envelope protein maturation had no cytoskeleton effects at restrictive temperature. Vaccinia virus was found to cause a rapid decrease in microfilament bundles of infected 3T3 cells (Meyer *et al.*, 1981). The major disappearance of microfilaments occurred within the first hour after infection, the period of uncoating and early protein synthesis; puromycin blocked this effect of vaccinia virus on microfilaments, presumptive evidence for early viral protein as the microfilament inhibitor. It is also of interest that the small-t protein coded by simian virus 40 causes disruption of actin cables in infected rat cells, whereas, SV40 mutants in which the small-t protein region of the genome is deleted, do not affect actin cables in infected cells (Graessmann *et al.*, 1980). In contrast, reovirus does not disorganize microtubules or microfilaments in monkey kidney CV-1 cells but does cause major disruption of vimentin filaments (Sharpe *et al.*, 1982). Reovirus also causes reorganization of vimentin filaments found in inclusion bodies.

A major advance in technique for studying cytopathic effects of viruses, particularly relating to the cytoskeleton, has come from Penman's laboratory, in which the cytoskeletal framework is prepared by detergent lysis; these suspended cells retain the major features of cell morphology, such as polyribosomes and major structural filaments. Lenk and Penman (1979) found extensive cytoskeletal changes by this technique in poliovirus-infected cells even when guanidine blocks synthesis of most viral products. The skeleton preparation reveals intermediate filaments arranged in a pattern unique to infected cells. This new framework becomes intimately involved in the macromolecular events in poliovirus replication, providing a unique cytopathic pattern in which the cytoskeleton has lost its normal metabolic role. Presumably, a limited number of poliovirus proteins are involved in this transformation to a specific cytopathology. Similar studies on cell architecture during adenovirus infection were performed by Lenk *et al.* (1980). In addition to late nuclear changes, by 6 hr postinfection, these authors noted marked changes in cell architecture as well as the nuclear envelope. Virtually all the early adenovirus proteins are found associated with the nucleus and its appurtenant cytoplasmic matrix. 58K and 40K early proteins were found primarily in association with the cytoskeleton and may be responsible for transformation. The changes in ultrastructure of adenovirus-infected cells observed late in infection involve primarily the nucleus.

6. SUMMARY

The history of virology has its origins in pathology, beginning with the intact host and evolving to the cell. Viral cytopathology became a science with the development of cell culture techniques and has progressed rapidly from the pioneering studies by microscopic description of lesions to biochemistry and molecular cell biology. Viruses vary greatly in their cytopathogenicity and the degree to which they compromise various cell functions. In turn, cells vary greatly in their susceptibility to viruses based to at least some extent on their species and tissue of origins, and their degree of differentiation. Certain viruses and specific structural proteins of the invading virion can cause early cell lethality even in the absence of viral replication, often referred to as viral cytotoxicity. Most cytopathic viruses kill cells only following virus-specific replicative events, including viral transcription and/or translation. Among the subcellular components of infected cells that are targets of cytopathogenic viruses are the membrane, lysosome, and cytoskeleton; these cellular organelles are usually affected late in infection and are probably secondary targets of the primary viral cytopathogenic effect. The original and continuing problem in the study of viral cytopathogenicity is distinguishing the primary from the secondary effects occurring during the course of viral infection of the host cell. Most, if not all, viruses that are acutely cytopathogenic also inhibit cellular RNA, DNA, and/or protein synthesis to varying degrees. These effects of viruses on cellular macromolecular synthesis occur quite early in infection, usually proceeding perturbation of other cell functions and always before cell death. It is tempting to invoke a cause-and-effect relationship between compromised cellular macromolecular synthesis and subsequent death of the virus-infected cell. It is prudent to remember, however, that no evidence is yet available to attribute cell death directly to viral inhibition of cellular RNA, DNA, and/or protein synthesis. Most of the chapters in this volume will be devoted to an in-depth analysis of the mechanisms by which different viruses shut off cellular macromolecular synthesis.

ACKNOWLEDGMENTS

This manuscript was prepared during my tenure of a U.S. Senior Scientist Award from the Alexander von Hunboldt Foundation at the Universities of Giessen and Würzburg. The support and encouragement of the Hunboldt Foundation and Professors Rudolf Rott and Volker ter Meulen are gratefully acknowledged. I am also grateful to Paige Hackney for excellent preparation of this manuscript.

7. REFERENCES

Ackermann, W. W., 1958, Cellular aspects of the cell–virus relationship, *Bacteriol. Rev.* **22**:223.
Ackermann, W. W., Rabson, A., and Kurtz, H., 1954, Growth characteristics of poliovirus in HeLa cell cultures. Lack of parallelism in cellular injury and virus increase, *J. Exp. Med.* **100**:437.
Ackermann, W. W., Loh, P. C., and Payne, F. E., 1959, Studies of the biosynthesis of protein and ribonucleic acid in HeLa cells infected with poliovirus, *Virology* **7**:170.
Allison, A. C., 1967, Lysosomes in virus-infected cells; in virus-directed host response, *Perspect. Virol.* **5**:29.
Allison, A. C., and Black, P. H., 1967, Lysosomal changes in lytic and nonlytic infections with the simian vacuolating virus (SV40), *J. Natl. Cancer Inst.* **39**:775.
Allison, A. C., and Malucci, L., 1965, Histochemical studies of lysosomes and lysosomal enzymes in virus-infected cell cultures, *J. Exp. Med.* **121**:463.
Allison, A. C., and Sandelin, K., 1963, Activation of lysosomal enzymes in virus-infected cells and its possible relationship to cytopathic effects, *J. Exp. Med.* **117**:879.
Amako, K., and Dales, S., 1967a, Cytopathology of mengovirus infection. I. Relationship between cellular disintegration and virulence, *Virology* **32**:184.

Amako, K., and Dales, S., 1967b, Cytopathology of mengovirus infection. II. Proliferation of membrane cisternae, *Virology* **32**:201.

Ambros, V., and Baltimore, D., 1978, Protein is linked to the 5′ end of poliovirus RNA by a phosphodiester linkage to tyrosine, *J. Biol. Chem.* **253**:5263.

Anzai, T., and Ozaki, Y., 1969, Intranuclear cystic formation of poliovirus: Electron microscopic observations, *Exp. Mol. Pathol.* **10**:176.

Appleyard, G., Westwood, J. C. N., and Zwartouw, H. T., 1962, The toxic effect of rabbitpox virus in tissue culture, *Virology* **18**:159.

Appleyard, G., Hume, V. B. M., and Westwood, J. C. N., 1965, The effect of thiosemicarbazone on the growth of rabbit poxvirus in tissue culture, *Ann. N.Y. Acad. Sci.* **130**:92.

Babiss, L. E., Luftig, R. B., Weatherbee, J. A., Weihing, R. R., Ray, U. R., and Fields, B. N., 1979, Reovirus serotypes 1 and 3 differ in their *in vitro* associations with microtubules, *J. Virol.* **30**:863.

Bablanian, R., 1968, The prevention of early vaccinia virus-induced cytopathic effects by inhibition of protein synthesis, *J. Gen. Virol.* **3**:51.

Bablanian, R., 1970, Studies on the mechanisms of vaccinia virus cytopathic effects. Effect of inhibitors of RNA and protein synthesis on early virus-induced cell damage, *J. Gen. Virol.* **6**:221.

Bablanian, R., 1972, Mechanisms of virus cytopathic effects, *Symp. Soc. Gen. Microbiol.* **22**:359.

Bablanian, R., 1975, Structural and functional alterations in cultured cells infected with cytocidal viruses, *Prog. Med. Virol.* **19**:40.

Bablanian, R., Eggers, H. J., and Tamm, I., 1965a, Studies on the mechanism of poliovirus-induced cell damage. I. The relation between poliovirus-induced metabolic and morphological alterations in cultured cells, *Virology* **26**:100.

Bablanian, R., Eggers, H. J., and Tamm, I., 1965b, Studies on the mechanism of poliovirus-induced cell damage. II. The relationship between poliovirus growth and virus-induced morphological changes in cells, *Virology* **26**:114.

Bablanian, R., Eggers, H. J., and Tamm, I., 1966, Inhibition of enterovirus cytopathic effects of 2-(α-hydroxybenzyl)benzimidazole, *J. Bacteriol.* **91**:1289.

Bablanian, R., Baxt, B., and Sonnabend, J. A., 1978, Studies on the mechanism of vaccinia virus cytopathic effects. II. Early cell rounding is associated with virus polypeptide synthesis, *J. Gen. Virol.* **39**:403.

Baltimore, D., and Franklin, R. M., 1962, The effect of mengovirus infection on the activity of the DNA-dependent RNA polmerase of L cells, *Proc. Natl. Acad. Sci. USA* **48**:1383.

Baltimore, D., Franklin, R. M. and Callender, J., 1963, Mengovirus induced inhibition of host ribonucleic acid and protein synthesis, *Biochim. Biophys. Acta* **76**:425.

Baltimore, D., Huang, A. S., and Stampfer, M., 1970, Ribonucleic acid synthesis of vesicular stomatitis virus. II. An RNA polymerase in the virion, *Proc. Natl. Acad. Sci. USA* **66**:572.

Bang, F. B., 1972, Specificity of viruses for tissues and hosts, *Symp. Soc. Gen. Microbiol.* **22**:359.

Bang, F. B., and Luttrell, C. N., 1961, Factors in the pathogenesis of virus diseases, *Adv. Virus Res.* **8**:199.

Barski, G., 1962, The significance of *in vitro* cellular lesions for classification of viruses, *Virology* **18**:152.

Baxt, B., and Bablanian, R., 1976a, Mechanism of vesicular stomatitis virus-induced cytopathic effects. I. Early morphological changes induced by infectious and DI particles, *Virology* **72**:370.

Baxt, B., and Bablanian, R., 1976b, Mechanism of vesicular stomatitis virus-induced cytopathic effects. II. Inhibition of macromolecular synthesis induced by infectious and defective-interfering particles, *Virology* **72**:383.

Becker, Y., and Joklik, W. K., 1964, Messenger RNA in cells infected with vaccinia virus, *Proc. Natl. Acad. Sci. USA* **51**:577.

Beijerinck, M. W., 1898, Ueber ein Contagium vivum fluidum als Ursache der Fleckenkrankheit der Tobaksbläter, Verh. K. Akad. Wetensch, *Amsterdam II* **6**:1.

Bello, L. J., and Ginsberg, H. S., 1967, Inhibition of host protein synthesis in type 5 adenovirus-infected cells, *J. Virol.* **1**:843.

Ben-Porat, T., and Kaplan, A. S., 1965, Mechanism of inhibition of cellular DNA synthesis by pseudorabiesvirus, *Virology* **25**:22.

Bienz, K., Egger, D., and Wolff, D. A., 1973, Virus replication, cytopathology and lysosomal enzyme response of mitotic and interphase HEp-2 cells infected with poliovirus, *J. Virol.* **11**:565.

Bienz, K., Egger, D., Rasser, Y., and Bossart, W., 1980, Kinetics and location of poliovirus macromolecular synthesis in correlation to virus-induced cytopathology, *Virology* **100**:390.

Blackman, K. E., and Bubel, H. C., 1969, Poliovirus-induced cellular injury, *J. Virol.* **4**:203.

Blair, C. D., and Brennan, P. J., 1972, Effect of Sendai virus infection on lipid metabolism in chich embryo fibroblasts, *J. Virol.* **9**:813.

Bossart, W., and Bienz, K., 1979, Virus replication, cytopathology, and lysosomal enzyme response in enucleated HEp-2 cells infected with poliovirus, *Virology* **92**:331.

Bratt, M. A., and Gallaher, W. R., 1969, Preliminary analysis of the requirements for fusion from within and fusion from without by Newcastle disease virus, *Proc. Natl. Acad. Sci. USA* **64**:536.

Brown, A., Mayyasi, S. A., and Officer, J. E., 1959, The "toxic" activity of vaccinia virus in tissue culture, *J. Infect. Dis.* **104**:193.

Brown, F., Martin, S. J., and Underwood, B., 1966, A study of the kinetics of protein and RNA synthesis induced by foot-and-mouth disease virus, *Biochim. Biophys. Acta* **129**:166.

Bubel, H. C., 1967, Protein leakage from mengovirus-infected cells, *Proc. Soc. Exp. Biol. Med* **125**:783.

Cantell, K., Skurska, Z., Paucker, K., and Henle, W., 1962, Quantitative studies on viral interference in suspended L cells. II. Factors afftecting interference by UV-irradiated Newcastle disease virus against vesicular stomatitis virus, *Virology* **17**:312.

Carrasco, L., 1978, Membrane leakiness after viral infection and a new approach to the development of antiviral agents, *Nature* **272**:694.

Carrasco, L., 1981, Modification of permeability induced by animal viruses early in infection, *Virology* **113**:623.

Carrasco, L., and Smith, A. E., 1980, Molecular biology of animal virus infection, *Pharmacol. Therapeut.* **9**:311.

Choppin, P. W., 1964, Multiplication of myxovirus (SV5) with minimal cytopathic effects and without interference, *Virology* **23**:224.

Chu, C. M., 1951, The action of normal mouse serum on influenza virus, *J. Gen. Microbiol.* **5**:739.

Collins, F. D., and Roberts, W. K., 1972, Mechanism of mengovirus-induced cell injury in L cells: Use of inhibitors of protein synthesis to dissociate virus-specific events, *J. Virol.* **10**:969.

Collono, R. J., and Banerjee, A. K., 1977, Mapping and initiation studies on the leader RNA of vesicular stomatitis virus, *Virology* **77**:260.

Contreras, A., and Carrasco, L., 1979, Selective inhibition of protein synthesis in virus-infected mammalian cells, *J. Virol.* **29**:114.

Cooper, P. D., 1977, Genetics of picornaviruses, in *Comprehensive Virology*, Vol. 9 (H. Fraenkel-Conrat and R. R. Wagner, eds.), pp. 133–207, Plenum Press, New York.

Cordell-Stewart, B., and Taylor, M. W., 1971, Effect of double-stranded viral RNA on mammalian cells in culture, *Proc. Natl. Acad. Sci. USA* **68**:1326.

Cordell-Stewart, B., and Taylor, M. W., 1973, Effect of viral double-stranded RNA on mammalian cells in culture: Cytotoxicity under conditions preventing viral replication and protein synthesis, *J. Virol.* **12**:360.

Cornatzer, W. E., Sandstrom, W., and Fischer, R. G., 1961, Effect of poliomyelitis virus type I (Mahoney strain) on the phospholipid metabolism of the HeLa cell, *Biochim. Biophys. Acta* **49**:414.

Dales, S., 1973, Early events in cell-animal virus interactions, *Bacteriol. Rev.* **37**:103.

Dales, S., 1975, Involvement of the microtubule in replication cycle of animal viruses, *Ann. N. Y. Acad. Sci.* **253**:440.

Dales, S., and Chardonnet, Y., 1973, Early events in the interaction of adenovirus with HeLa cells. IV. Association with microtubules and the nuclear pore complex during vectorial movement of the inoculum, *Virology* **56**:465.

Dales, S., and Franklin, R. M., 1962, A comparison of the changes in fine structure of L-cells during single cycle of viral multiplication following their infection with the viruses of mengo and encephalmyocarditis, *J. Cell Biol.* **14**:281.

Dales, S., Eggers, H. J., Tamm, I., and Palade, G., 1965, Electron microscopic study of the formation of poliovirus, *Virology* **26**:379.

David-West, T. S., and Osunkoya, B. O., 1971, Cytopathology of vesicular stomatitis with phase microscopy, *Arch. Ges. Virusforsch.* **35**:126.

DeDuve, C., 1959, Lysosomes, a new group of subcellular particles, in: *Subcellular Particles* (T. Hayashi, ed.), p. 130, Ronald Press, New York.

DeSomer, P., Prinzie, A., and Schonne, E., 1959, Infectivity of poliovirus ribonucleic acid for embryonated eggs and unsusceptible cell lines, *Nature* **184**:652.

d'Herelle, F., 1917, Sur un microbe invisible antagoniste des bacille dysentérique, *Comp. Rend. Acad. Sci.* **165**:373.

Doyle, M., and Holland, J. J., 1973, Prophylaxis and immunization in mice by use of virus-free defective T particles to protect against intracerebral infection by vesicular stomatitis virus, *Proc. Natl. Acad. Sci. USA* **70**:2105.

Dulbecco, R., 1952, Production of plaques in monolayer tissue cultures by single particles of an animal virus, *Proc. Natl. Acad. Sci. USA* **38**:747.

Dunnebacke, T. H., 1956a, Correlation of the stage of cytopathic change with the release of poliomyelitis virus, *Virology* **2**:399.

Dunnebacke, T. H., 1956b, Cytopathic changes associated with poliomyelitis infection in human amnion cells, *Virology* **2**:811.

Ebina, T., Satake, M., and Ishida, N., 1978, Involvement of microtubules in cytopathic effects of animal viruses: Early proteins of adenovirus and herpesvirus inhibit formation of microtubule paracrystals in HeLa cells, *J. Gen. Virol.* **38**:535.

Egberts, E., Hackett, P. B., and Traub, P., 1977, Alteration of the intracellular energetic and ionic conditions by mengovirus of Ehrlich ascites tumor cells and its influence on protein synthesis in the midphase of infection, *J. Virol.* **22**:591.

Eggers, H., 1982, Benzimidazoles. Selective inhibitors of picornavirus replication in cell culture and in the organisms, in: *Handbook of Experimental Pharmacology*, Vol. 61 (P. E. Came and L. A. Caliguiri, eds.), pp. 377–417, Springer-Verlag, Berlin/Heidelberg.

Emerson, S. U., and Wagner, R. R., 1972, Dissociation and reconstitution of the transcriptase and template activities of vesicular stomatitis B and T virions, *J. Virol.* **10**:297.

Emerson, S. U., Dierks, P. M., and Parsons, J. T., 1977, *In vitro* synthesis of a unique RNA species by a T particle of vesicular stomatitis virus, *J. Virol.* **23**:708.

Enders, J. F., 1954, Cytopathology of virus infections, *Annu. Rev. Microbiol.* **8**:473.

Enders, J. F., Weller, T. H., and Robbins, F. C., 1949, Cultivation of the Lansing strain of poliomyelitis virus in cultures of various human embryonic tissues, *Science* **109**:85.

Everett, A. J., and Ginsberg, H. S., 1958, A toxinlike material separable from type 5 adenovirus particles, *Virology* **6**:770.

Farmilo, A. J., and Stanners, C. P., 1972, Mutant of vesicular stomatitis virus which allows DNA synthesis and division in cells synthesizing viral RNA, *J. Virol.* **10**:605.

Farnham, A. E., and Epstein, W., 1963, The influence of EMC virus infection on potassium transport in L-cells, *Virology* **21**:436.

Fernandez-Puentes, C., and Carrasco, L., 1980, Viral infection permeabilizes mammalian cells to protein toxins, *Cell* **20**:769.

Flanagan, J. F., 1966, Hydrolytic enzymes in KB cells infected with poliovirus and herpes simplex virus, *J. Bacteriol.* **91**:789.

Follett, E. A. C., Pringle, C. R., Wunner, W. H. and Skehel, J. J., 1974, Virus replication in enucleate cells: Vesicular stomatitis virus and influenza virus, *J. Virol.* **13**:394.

Fuchs, P., and Giberman, E., 1973, Enhancement of potassium influx in baby hamster kidney cells and chicken erythrocytes during adsorption of parainfluenza (Sendai) virus, *FEBS Lett.* **31**:127.

Fuchs, P., Spiegelstein, M., Hainson, M., Gitelman, J., and Kohn, A., 1978, Early changes in the membrane of HeLa cells adsorbing Sendai virus under conditions of fusion *J. Cell Physiol.* **95**:223.

Garwes, D. J., Wright, P. J., and Cooper, P. D., 1975, Poliovirus temperature-sensitive mutants defective in cytopathic effects are also defective in synthesis of double-stranded RNA, *J. Gen. Virol.* **27**:45.

Gauntt, C. J., and Lockart, R. Z., 1966, Inhibition of mengovirus by interferon, *J. Bacteriol.* **91**:176.

Gauntt, C. J., and Lockart, R. Z., 1968, Destruction of L cells by mengovirus: Use of interferon to study the mechanism, *J. Virol.* **2**:567.

Gausch, C. R., and Youngner, J. S., 1963, Lipids of virus-infected cells. II. Lipid analysis of HeLa cells infected with vaccinia virus, *Proc. Soc. Exp. Biol. Med.* **112**:1082.

Genty, N., 1975, Analysis of uridine incorporation in chicken embryo cells infected with vesicular stomatitis virus and its temperature-sensitive mutants: Uridine transport, *J. Virol.* **15**:8.

Genty, N., and Bussereau, F., 1980, Is cytoskeleton involved in vesicular stomatitis virus reproduction? *J. Virol.* **34**:777.

Gilbert, V. E., 1963, Enzyme release from tissue cultures as an indication of cellular injury by viruses, *Virology* **21**:609.

Ginsberg, H. S., and Young, C. S. H., 1977, Genetics of adenoviruses, in: *Comprehensive Virology*, Vol. 9 (H. Fraenkel-Conrat and R. R. Wagner, eds.), pp. 27–88, Plenum Press, New York.

Ginsberg, H. S., Pereira, H. S., Valentine, R. C., and Wilcox, W. C., 1966, A proposed terminology for the adenovirus antigens and virion morphological subunits, *Virology* **28**:782.

Goorha, R., and Granoff, A., 1974, Macromolecular synthesis in cells infected by frog virus 3. I. Virus-specific protein synthesis and its regulation, *Virology* **60**:237.

Graessmann, A., Graessmann, M., Tjian, R., and Topp, W. C., 1980, Simian virus 40 small-t protein is required for loss of actin cable network, *J. Virol.* **33**:1182.

Gray, M. A., Micklem, K. J., Brown, F., and Pasternak, C. A., 1983, Effect of vesicular stomatitis virus and Semliki Forest virus on uptake of nutrients and intracellular cation concentration, *J. Gen. Virol.* **64**:1449.

Green, M., 1959, Biochemical studies on adenovirus multiplication. I. Stimulation of phosphorus incorporation into deoxyribonucleic acid and ribonucleic acid, *Virology* **9**:343.

Guskey, L. E., and Wolff, D. A., 1974, Effects of actinomycin D on the cytopathology induced by poliovirus in HEp-2 cells, *J. Virol.* **14**:1229.

Guskey, L. E., Smith, P. C., and Wolff, D. A., 1970, Patterns of cytopathology and lysosomal enzyme release in poliovirus-infected HEp-2 cells treated with either 2-(α-hydroxybenzyl)-benzimidazole or guanidine HCl, *J. Gen. Virol.* **6**:151.

Haase, A. T., Baron, S., Levy, H., and Kasel, J. A., 1969, Mengovirus-induced cytopathic effects in L-cells: Protective effect of interferon *J. Virol.* **4**:490.

Hale, A. H., Witte, O. N., Baltimore, D., and Eisen, H. S., 1978, Vesicular stomatitis virus glycoprotein is necessary for H-2 restricted lysis of infected cells by cytotoxic T lymphocytes, *Proc. Natl. Acad. Sci. USA* **75**:970.

Hanafusa, H., 1960, Killing of L cells by heat- and UV-inactivated vaccinia virus, *Bikens J.* **3**:191.

Hanafusa, H., 1962, Factors involved in the initiation of multiplication of vaccinia virus, *Cold Spring Harbor Symp. Quant. Biol.* **27**:209.

Hand, R., 1976, Thymidine metabolism and DNA synthesis in Newcastle disease virus-infected cells, *J. Virol.* **19**:801.

Haywood, A. M., 1974, Fusion of Sendai virus with model membranes, *J. Mol. Biol.* **87**:625.

Heding, L. D., and Wolff, D. A., 1973, The cytochemical examination on poliovirus-induced cell damage, *J. Cell Biol.* **59**:530.

Helenius, A., Kartenbeck, J., Simons, K., and Fries, E., 1980, On the entry of Semliki Forest virus into BHK-21 cells, *J. Cell Biol.* **84**:404.

Henle, G., Girardi, A., and Henle, W., 1955, A non-transmissible cytopathogenic effect in influenza virus in tissue culture accompanied by formation of noninfectious hemagglutinin, *J. Exp. Med.* **101**:25.

Henry, C., and Youngner, J. S., 1957, Influence of normal animal sera on influenza viruses in cultures of trypsin-dispersed monkey kidney cells, *J. Immunol.* **78**:273.

Hershey, A. D., and Chase, M., 1952, Independent functions of viral proteins and nucleic acid in growth of bacteriophage, *J. Gen Physiol.* **36**:39.

Hightower, L. E., and Bratt, M. A., 1974, Protein synthesis in Newcastle disease virus-infected chicken embryo cells, *J. Virol.* **13**:788.

Hintz, R. M., Barski, G., and Bernhard, W., 1962, An electron microscopic study of the development of the encephalomyocarditis (EMC) virus propogated *in vitro*, *Exp. Cell. Res.* **26**:571.

Hoggan, M. D., and Roizman, B., 1959, The isolation and properties of a variant of herpes simplex producing multinucleated giant cells in monolayer cultures in the presence of antibody. *Am. J. Hyg.* **70**:208.

Holland, J. J., 1962, Inhibition of DNA-primed RNA synthesis during poliovirus infection of human cells, *Biochem. Biophys. Res. Commun.* **9**:556.

Holland, J. J., Kennedy, S. I. T., Semler, B. L., Jones, C. L., Roux, L., and Grabau, E. A., 1980, Defective-interfering RNA viruses and the host-cell response. in: *Comprehensive Virology*, Vol. 16 (H. Fraenkel-Conrat and R. R. Wagner, eds.), pp. 137–192, Plenum Press, New York.

Holmes, K. V., and Choppin, P. W., 1966, On the role of the response of the cell membrane in determining virus virulence. Contrasting effects of the parainfluenza virus SV5 in two cell types, *J. Exp. Med.* **124**:501.

Hook, E. W., and Wagner, R. R., 1958, Hemorrhagic encephalopathy in chicken embryos infected with influenza virus. I. Factors influencing the development of hemorrhages, *Bull. Johns Hopkins Hosp.* **103**:125.

Hook, E. W., Luttrell, C. N., and Wagner, R. R., 1958, Hemorrhagic encephalopathy in chicken embryos infected with influenza virus. II. Pathology, *Bull. Johns Hopkins Hosp.* **103**:140.

Horak, I., Jungwirth, C., and Bodo, G., 1971, Poxvirus specific cytopathic effect in interferon-treated L cells, *Virology* **45**:456.

Horne, R. W., and Nagington, J., 1959, Electron microscopic studies of the development and structure of poliomyelitis virus, *J. Mol. Biol.* **1**:333.

Howatson, A. F., and Whitmore, C. F., 1962, The development and structure of vesicular stomatitis virus, *Virology* **16**:466.

Huang, A. S., and Baltimore, D., 1977, Defective interfering animal viruses, in: *Comprehensive Virology*, Vol. 10 (H. Fraenkel-Conrat and R. R. Wagner, eds.), pp. 73-116, Plenum Press, New York.

Huang, A. S., and Wagner, R. R., 1965, Inhibition of cellular RNA synthesis by nonreplicating vesicular stomatitis virus, *Proc. Natl. Acad. Sci. USA* **54**:1579.

Huang, R. T. C., Wahn, K., Klenk, H.-D., and Rott, R., 1980, Fusion between cell membrane and liposomes containing the glycoproteins of influenza virus, *Virology* **104**:294.

Hugentobler, A.-L., and Bienz, K., 1980, Influence of poliovirus infection on S-phase and mitosis of the host cell, *Arch. Virol.* **64**:25.

Hughes, J. V., Dille, B. J., Thimmig, R. L., Johnson, T. C., Rabinowitz, S. C., and DalCanto, M. C., 1979, Neuroblastoma cell fusion by a temperature-sensitive mutant of vesicular stomatitis virus, *J. Virol.* **30**:883.

Ivanovski, D., 1892, Über die Mosaikkrankheit der Tabakspflanze, *Bull. Acad. Inp. Sci. St. Petersburg* **3**:67.

Jezequel, A.-M., and Steiner, J. W., 1966, Some ultrastructural and histochemical aspects of coxsackie virus–cell interactions, *Lab. Invest.* **15**:1055.

Joklik, W. K., 1966, The poxviruses, *Bacteriol. Rev.* **30**:33.

Joklik, W. K., and Merigan, T. C., 1966, Concerning the mechanisms of action of interferon, *Proc. Natl. Acad. Sci. USA* **56**:558.

Jungwirth, C., and Launer, J., 1968, Effect of poxvirus infection on host cell deoxyribonucleic acid synthesis, *J. Virol.* **2**:401.

Katzman, J., and Wilson, D. E., 1974, Newcastle disease virus-induced plasma membrane damage, *J. Gen. Virol.* **24**:101.

Kidson, C., Moreau, M. C., Asher, D. M., Brown, P. W., Coon, H. G., Gajdusek, D. C., and Gibbs, C. J., 1978, Cell fusion induced by scrapie and Creutzfeldt–Jakob virus-infected brain preparations, *Proc. Natl. Acad. Sci. USA* **75**:2969.

Kohn, A., 1979, Early interaction of viruses with cellular membranes, *Adv. Virus Res.* **24**:223.

Koschel, K., 1971, Release of ^{51}chromium from labeled HeLa cells after infection by poliovirus, *Z. Naturforsch.* **B26**:929.

Koschel, K., Aus, H. M., and ter Meulen, V., 1974, Lysosomal enzyme activity in poliovirus-infected and VSV-infected L cells: Biochemical and histochemical analysis with computer-aided techniques, *J. Gen. Virol.* **25**:359.

Lachmi, B.-E., and Kääriäinen, L., 1977, Control of protein synthesis in Semliki-Forest virus infected cells, *J. Virol.* **22**:142.

Lake, R. S., and Ludwig, E. H., 1971, Cellular changes attending mengo-virus-induced cytolysis of mouse L cells, *Biochim. Biophys. Acta* **244**:466.

Lazarowitz, S. C., and Choppin, P. W., 1975, Enhancement of the infectivity of influenza A and B viruses by proteolytic cleavage of the hemagglutinin polypeptide, *Virology* **68**:440.

Lazdins, I., and Holmes, I. H., 1979, Protein synthesis in Bunyamwera virus-infected cells, *J. Gen. Virol.* **44**:123.

Ledinko, N., 1958, Production of non-infectious complement-fixing poliovirus particles in HeLa cells treated with proflavine, *Virology* **6**:512.

Ledinko, N., 1966, Changes in metabolic and enzymatic activities of monkey kidney cells after infection with adenovirus 2, *Virology* **15**:173.

Lenk. R., and Penman, S., 1979, The cytoskeletal framework and poliovirus metabolism, *Cell* **16**:289.

Lenk, R., Storch, T., and Maisel, J. V., Jr., 1980, Cell architecture during adenovirus infection, *Virology* **105**:19.

Levanon, A., Kohn, A., and Inbar, M., 1977, Increase in lipid fluidity of cellular membranes induced by adsorption of RNA and DNA virions, *J. Virol.* **22**:353.

Levine, A. J., and Ginsberg, H. S., 1967, Mechanism by which fiber antigen inhibits multiplication of type 5 adenovirus, *J. Virol.* **1**:747.

Levintow, L., 1974, Reproduction of picornaviruses, in: *Comprehensive Virology,* Vol. 2 (H. Fraenkel-Conrat and R. R. Wagner, eds.), pp. 109–169, Plenum Press, New York.

Long, W. F., and Burke, D. C., 1970, The effect of infection with fowl plaque virus on protein synthesis in chick embryo cells, *J. Gen. Virol.* **6**:1.

Ludwig, H., and Rott, R., 1975, Effect of 2-deoxyglucose on herpes virus-induced inhibition of cellular DNA synthesis, *J. Virol.* **16**:217.

Luftig, R. B., and Weihing, R. R., 1975, Adenovirus binds to rat brain microtubules *in vitro, J. Virol.* **16**:696.

Lust, G., 1966, Alterations of protein synthesis in arbovirus-infected L cells, *J. Bacteriol.* **91**:1612.

Lyles, D. S., and Landsberger, F. R., 1977, Sendai virus-induced hemolysis: Reduction in heterogeneity of erythrocyte lipid bilayer fluidity, *Proc. Natl. Acad. Sci. USA* **74**:1918.

Lyles, D. S., and Landsberger, F. R., 1979, Kinetics of Sendai virus envelope fusion with erythrocyte membrance and virus-induced hemolysis, *Biochemistry* **18**:5088.

Lynn, I. W., and Morgan, H. R., 1954, Cytopathogenicity of animal viruses *in vitro*, *Arch. Pathol.* **57**:301.

Madore, H. P., and England, J. M., 1975, Selective suppression of cellular protein synthesis in baby hamster kidney (BHK-21) cells infected with rabies virus, *J. Virol.* **16**:1351.

Maeda, T., Asano, A., Okada, Y., and Ohnishi, S., 1977, Transmembrane phospholipid motions induced by F glycoprotein in hemagglutinating virus of Japan, *J. Virol.* **21**:232.

Maltzman, W., and Levine, A. J., 1981, Viruses as probes for development and differentiation, *Adv. Virus Res.* **26**:66.

Manservigi, R., Spear, P. G., and Buchan, A., 1977, Cell fusion induced by herpes simplex virus is promoted and suppressed by different viral glycoproteins, *Proc. Natl. Acad. Sci. USA* **74**:3913.

Marcus, P. I., 1959, Symposium on the biology of cells modified by viruses or antigens. IV. Single-cell techniques in tracing virus–host interactions, *Bacteriol. Rev.* **23**:232.

Marcus, P. I., 1976, Cell killing by viruses: Single-cell survival procedure for detecting viral functions required for cell killing, in: *Cancer Biology IV: Differentiation and Carcinogenesis* (C. Borek, C. M. Fenoglio, and D. W. King, eds.), pp. 192–226, Stratton Intercontinental Medical Corp., New York.

Marcus, P. I., and Puck, T. T., 1958, Host–cell interaction of animal viruses. I. Titration of cell-killing by viruses, *Virology* **6**:405.

Marcus, P. I., and Sekellick, M. J., 1974, Cell killing by viruses. I. Comparison of cell-killing, plaque-forming and defective-interfering particles of vesicular stomatitis virus, *Virology* **57**:321.

Marcus, P. I., and Sekellick, M. J., 1975, Cell killing by viruses II. Cell killing by vesicular stomatitis virus: A requirement for virion-derived transcription, *Virology* **63**:176.

Marcus, P. I., and Sekellick, M. J., 1976, Cell killing by viruses. III. The interferon system and inhibition of cell killing by vesicular stomatitis virus, *Virology* **69**:378.

Martin, E. M., and Kerr, I. M., 1968, Virus-induced changes in host cell macromolecula synthesis, *Symp. Soc. Gen. Microbiol.* **18**:15.

Martin, E. M., Malec, J., Sved, S., and Work, T. S., 1961, Studies on protein and nucleic acid metabolism in virus-infected mammalian cells. I. Encephalomyocarditis virus in Krebs-II mouse ascites tumour cells, *Biochem. J.* **80**:585.

Marvaldi, J. L., Lucas-Lenard, J., Sekellick, M. J., and Marcus, P. I., 1977, Cell killing by viruses. IV. Cell killing and protein synthesis inhibition by vesicular stomatitis virus require the same gene functions, *Virology* **79**:267.

Mattern, C. F. T., and Daniel, W. A., 1966, Replication of poliovirus in HeLa cells: Electron microscopic observations, *Virology* **26**:646.

Matthews, R. E. F., 1979, Classification and nomenclature of viruses. Third report of the International Committee on Taxonomy of Viruses, *Intervirology* **12**:129.

McAllister, P. E., and Wagner, R. R., 1976, Differential inhibition of host protein synthesis in L cells infected with RNA⁻ temperature-sensitive mutants of vesicular stomatitis virus, *J. Virol.* **18:**550.

McClain, M., 1965, The host range and plaque morphology of rabbitpox virus (RPu⁺) and its u mutants on chick fibroblast, PK-2a, and L-929 cells, *Austr. J. Exp. Biol. Med. Sci.* **43:**31.

McGowan, J. J., and Wagner, R. R., 1981, Inhibition of cellular DNA synthesis by vesicular stomatitis virus, *J. Virol.* **38:**356.

McIntosh, K., Payne, S., and Russell, W. C., 1971, Studies on lipid metabolism in cells infected with adenovirus, *J. Gen. Virol.* **10:**251.

McSharry, J. J., and Choppin, P. W., 1978, Biological properties of the VSV glycoprotein. I. Effects of the isolated glycoprotein on host macromolecular synthesis, *Virology* **84:**172.

Meyer, R. K., Burger, M. M., Tschannen, R., and Schäfer, R., 1981, Actin filament bundles in vaccinia virus infected fibroblasts, *Arch. Virol.* **67:**11.

Micklem, K. J., and Pasternak, G. A., 1977, Surface components involved in virally mediated membrane changes, *Biochem. J.* **162:**405.

Miles, B. D., Luftig, R. B., Weatherbee, S. A., Weihing, R. R., and Weber, J., 1980, Quantitation of the interaction between adenovirus types 2 and 4 and microtubules inside infected cells, *Virology* **105:**265.

Morgan, C., and Rose, H. M., 1968, Structure and development of viruses as observed in the electron microscope. VIII. Entry of influenza virus, *J. Virol.* **2:**925.

Moss, B., 1968, Inhibition of HeLa cell protein synthesis by the vaccinia virion, *J. Virol.* **2:**1028.

Moss, B., 1974, Reproduction of poxviruses, in: *Comprehensive Virology*, Vol. 3 (H. Fraenkel-Conrat and R. R. Wagner, eds.), pp. 405–474, Plenum Press, New York.

Mosser, A. G., Caliguiri, L. A., and Tamm, I., 1972, Incorporation of lipid precursors into cytoplasmic membranes of poliovirus-infected HeLa cells, *Virology* **47:**39.

Mountain, I. M., and Alexander, H. E., 1959, Infectivity of ribonucleic acid (RNA) from type I poliovirus in embryonated egg, *Proc. Soc. Exp. Biol. Med.* **81:**513.

Nagai, T., Klenk, H.-D., and Rott, R., 1976, Proteolytic cleavage of the viral glycoprotein and its significance for the virulence of Newcastle disease virus, *Virology* **72:**494.

Nair, C. N., 1981, Monovalent cation metabolism and cytopathic effects of poliovirus-infected HeLa cells, *J. Virol.* **37:**268.

Nair, C. N., Stowers, J. W., and Singfield, B., 1979, Guanidine-sensitive Na⁺ accumulation by poliovirus-infected HeLa cells, *J. Virol.* **31:**184.

Negreanu, Y., Reinhertz, Z., and Kohn, A., 1974, Effects of adsorption of UV-inactivated parainfluenza (Sendai) virus on the incorporation of amino acids in animal host cells, *J. Gen. Virol.* **22:**265.

Nicolau, C., Klenk, H.-D., Reiman, A., Hilderbrand, K., and Bauer, H., 1979, Molecular events during the interaction of envelopes of myxo- and RNA-tumor viruses with cell membranes. A 270 MHz ¹H nuclear magnetic resonance study, *Biochim. Biophys. Acta* **511:**83.

Norkin, L. C., 1977, Cell killing by SV40: Impairment of membrane formation and function, *J. Virol.* **21:**872.

Norkin, L. C., Ouellette, J., 1976, Cell killing by simian virus 40: Variation in the pattern of lysosomal enzyme release, cellular enzyme release, and cell death during productive infection of normal and simian virus 40-transformed simian cell lines, *J. Virol.* **18**:48.

O'Callaghan, D. J., Cheevers, D. J. W., Gentry, G. A., and Randall, C. C., 1968, Kinetics of cellular and viral DNA synthesis in equine abortus (herpes) virus infection of L-M cells, *Virology* **36**:104.

Osunkoya, B. O., and David-West, T. S., 1972, Telephase arrest of cultured cells by vesicular stomatitis vius, *Arch. Ges. Virusforsch.* **38**:228.

Pasternak, C. A., and Micklem, K. J., 1973, Permeability changes during cell fusion, *J. Membr.Biol.* **14**:293.

Penman, S., 1965, Stimulation of the incorporation of choline in poliovirus-infected cells, *Virology* **25**:148.

Pereira, H. G., 1958, A protein factor responsible for the early cytopathic effect of adenovirus, *Virology* **6**:601.

Pereira, H. G., 1961, The cytopathic effect of animal viruses, *Adv. Virus Res.* **8**:245.

Pereira, H. G., and Kelly, B., 1957, Dose-response curves of toxic and infective actions of adenovirus in HeLa cells, *J. Gen. Microbiol.* **17**:517.

Pfefferkorn, E. R., and Hunter, H. S., 1973, The source of the ribonucleic acid and phospholipid of Sindbis virus, *Virology* **20**:446.

Philipson, L., and Lindberg, C., 1974, Reproduction of adenoviruses, in: *Comprehensive Virology*, Vol. 3 (H. Fraenkel-Conrat and R. R. Wagner, eds.), pp.143–227, Plenum Press, New York.

Plagemann, P. G. W. Cleaveland, P. H., and Shea, M. A., 1970, Effect of mengovirus replication on choline metabolism and membrane formation in Novikoff hepatoma cells, *J. Virol.* **6**:800.

Plowright, W., and Ferris, R. D., 1958, The growth and cytopathogenicity of sheeppox virus in tissue cultures, *Brit. J. Exp. Pathol.* **39**:424.

Poste, G., 1970, Virus-induced polykaryocytosis and the mechanism of cell fusion, *Adv. Virus Res.* **16**:303.

Raghow, R., and Granoff, A., 1979, Macromolecular synthesis in cells infected by frog virus 3. 10. Inhibition of cellular protein synthesis by heat-inactivated virus, *Virology* **98**:319.

Ram, G. C., Jain, N. C., and Sharma, V. K., 1978, Viral induced cytopathic effect and release of lysosomal enzyme, *Indian J. Exp. Biol.*, **16**:1302.

Raskas, H. J., Thomas, D. C., and Green, M., 1970, Biochemical studies on adenovirus multiplication. XVII. Ribosomal synthesis in uninfected and infected KB cells, *Virology* **40**:893.

Rekosh, D. M. K., 1977, The molecular biology of picornaviruses, in: *The Molecular Biology of Animal Viruses*, (D. P. Nayak, ed.), pp. 63–110, Marcel Dekker, New York.

Repik, R., Flamand, A., and Bishop, D. H. L., 1974, Effect of interferon upon the primary and secondary transcription of VSV and influenza viruses, *J. Virol.* **14**:1169.

Rice, J. M., and Wolff, D. A., 1975, Phospholipase in the lysosomes of HEp-2 cells and its release during poliovirus infection, *Biochim. Biophys. Acta* **381**:17.

Rivers, T. M., 1928, Some general aspects of pathological conditions caused by filterable viruses, *Am. J. Pathol.* **4**:91.

Roizman, B., 1962, Polykaryocytosis induced by virus. *Proc. Natl. Acad. Sci. USA* **48**:228.

Roizman, B., and Furlong, D., 1974, The replication of herpes viruses, in: *Comprehensive Virology*, Vol. 3 (H. Fraenkel-Conrat and R. R. Wagner, eds.), pp. 229–403, Plenum Press, New York.

Roizman, B., and Spear, P. G., 1969, Macromolecular biosynthesis in animal cells infected with cytolytic viruses, *Curr. Top. Dev. Biol.* **4**:79.

Roizman, B., Borman, G. S., and Rousta, M. K., 1965, Macromolecular synthesis in cells infected with herpes simplex virus, *Nature* **206**:1374.

Rothberg, P. G., Harris, T. J. R., Nomato, A., and Wimmer, E., 1978, O^4-(5'-uridyl)tyrosine is the bond between the genome linked protein and the RNA of poliovirus, *Proc. Natl. Acad. Sci. USA* **75**:4868.

Rowe, W. P., Hartley, J. W., Roizman, B., and Levy, H. B., 1958, Characterization of a factor formed in the course of adenovirus infection of tissue cultures causing detachment of cells from glass, *J. Exp. Med.* **108**:713.

Rueckert, R. R., 1976, On the structure and morphogenesis of picornaviruses, in: *Comprehensive Virology*, Vol. 6 (H. Fraenkel-Conrat and R. R. Wagner, eds.), pp. 131–213, Plenum Press, New York.

Rutter, G., and Mannweiler, R., 1977, Alterations of actin containing structures in BHK-21 cells infected with Newcastle disease virus and vesicular stomatitis virus, *J. Gen. Virol.* **37**:233.

Salzman, N. P., Lockart, R. Z., and Sebring, E. D., 1959, Alterations in HeLa cell metabolism resulting from poliovirus infection, *Virology* **9**:244.

Salzman, N. P., Shatkin, A. J., and Sebring, E. D., 1964, The synthesis of DNA-like RNA in the cytoplasm of HeLa cells infected with vaccinia virus, *J. Mol. Biol.* **8**:405.

Scheid, A., and Choppin, P. W., 1974, Identification of biological activities of paramyxovirus glycoproteins. Activation of cell fusion, hemolysis and infectivity by proteolytic cleavage of an inactive precursor protein of Sendai virus, *Virology* **57**:475.

Scheid, A., and Choppin, P. W., 1977, Two disulfide-linked polypeptide chains constitute the active F protein of paramyxoviruses, *Virology* **80**:54.

Schlesinger, R. W., 1950, Incomplete growth cycle of influenza virus in mouse brain, *Proc. Soc. Exp. Biol. Med.* **74**:541.

Schnitzlein, W. M., O'Banion, M. K., Poirot, M. K., and Reichmann, M. E., 1983, Effect of intracellular vesicular stomatitis virus mRNA concentration on the inhibition of host cell protein synthesis, *J. Virol.* **45**:206.

Schümperli, D., Peterhans, E., and Wyler, R., 1978, Permeability changes of plasma membrane and lysosomal membranes in HeLa cells infected with rabbit pox virus, *Arch. Virol.*, **58**:203.

Sharpe, A. H., Chien, L. B., and Fields, B. N., 1982, The interaction of mammalian reoviruses with the cytoskeleton of monkey kidney CV-1 cells, *Virology* **120**:399.

Shaw, J. E., and Cox, D. C., 1973, Early inhibition of cellular DNA synthesis by high multiplicities of infectious and UV-inactivated reovirus, *J. Virol.* **12**:704.

Simizu, B., Wagatsuma, B., Oya, M., Hanaoka, F., and Yamada, M., 1976, Inhibition of cellular DNA synthesis in hamster kidney cells infected with western equine encephalitis virus, *Arch. Virol.* **51**:251.

Skinner, M. S., Halperen, S., and Harkin, J. C., 1968, Cytoplasmic membrane-bound vesicles in echovirus 12-infected cells, *Virology* **36**:241.

Soloviev, V. D., Gutman, N. R., Amichenkova, A. M., Goltsen, G. G., and By-
    kovsky, A. F., 1967, Fine structure of cells infected with respiratory strains of
    echovirus 11, *Exp. Mol. Pathol.* **6**:382.
Strauss, J. H., Jr., Burge, B. W., and Darnell, J. E., 1969, Effect of Sindbis virus
    on host protein synthesis, *Virology* **37**:367.
Tamm, I., and Eggers, H. J., 1963, Specific inhibition of replication of animal viruses,
    Science **142**:24.
Thacore, H., and Wolff, D. A., 1968, Activation of lysosomes by poliovirus-infected
    cell extracts, *Nature* **218**:1063.
Twort, F. W., 1915, An investigation of the nature of ultramicroscopic viruses, *Lan-
    cet* **189**(2):1241.
Verwoerd, D. W., and Hausen, P., 1963, Studies on the multiplication of a member
    of the Columbia SK group (ME virus) in L cells, IV. Role of "early proteins" in
    virus-induced metabolic changes, *Virology* **21**:628.
Wagner, R. R., 1952, Acquired resistance in mice to the neurotoxic action of influ-
    enza viruses, *Br. J. Exp. Pathol.* **33**:157.
Wagner, R. R., 1954, Influenza virus infection of transplated tissues. I. Multiplication
    of a "neurotopic" strain and its effect on solid neoplasms, *Cancer Res.* **14**:377.
Wagner, R. R., 1955, A pantropic strain of influenza virus. Generalized infection
    and viremia in the infant mouse, *Virology* **1**:497.
Wagner, R. R., 1975, Reproduction of rhabdovirus, in: *Comprehensive Virology*,
    Vol. 4 (H. Fraenkel-Conrat and R. R. Wagner, eds.), pp. 1–94, Plenum Press,
    New York.
Wagner, R. R., and Huang, A. S., 1966, Inhibition of RNA and interferon synthesis
    in Krebs-2 cells infected with vesicular stomatitis virus, *Virology* **28**:1.
Wagner, R. R., Bennett, I. L., Jr., and LeQuire, V. S., 1949a, The production of
    fever by influenza viruses. I. Factors influencing the febrile response to single
    injections of virus *J. Exp. Med.* **90**:321.
Wagner, R. R., Bennett, I. L., Jr., and LeQuire, V. S., 1949b, The production of
    fever by influenza viruses. II. Tolerance in rabbits to the pyrogenic effect of in-
    fluenza viruses, *J. Exp. Med.* **90**:335.
Wagner, R. R., Levy, A. H., Snyder, R. M., Ratcliff, G. A., and Hyatt, D. F., 1963,
    Biological properties of two plaque variants of vesicular stomatitis virus (Indiana
    serotype), *J. Immunol.* **91**:112.
Wagner, R. R., Thomas, J. R., and McGowan, J. J., 1984, Rhabdovirus inhibition
    of cellular macromolecular synthesis, in: *Comprehensive Virology*, Vol. 19 (H.
    Fraenkel-Conrat and R. R. Wagner, eds.), pp. 223–295, Plenum Press, New York.
Walker, D. L., 1960, *In vitro* cell–virus relationships resulting in cell death, *Annu.
    Rev. Microbiol.* **14**:177.
Weck, P. K., and Wagner, R. R., 1978, Inhibition of RNA synthesis in mouse mye-
    loma cells infected with vesicular stomatitis virus, *J. Virol.* **25**:770.
Wilson, D. E., 1968, Inhibition of host-cell protein and ribonucleic acid synthesis
    by Newcastle disease virus, *J. Virol.* **2**:1.
Wolff, D. A., and Bubel, H. C., 1964, The disposition of lysosomal enzymes as
    related to specific viral cytopathic effects, *Virology* **24**:502.
Yamazaki, S., and Wagner, R. R., 1970, Action of interferon: Kinetics and differ-
    ential effects on viral functions, *J. Virol.* **6**:421.
Yaoi, Y., Mitsui, H., and Amano, M., 1970, Effect of UV-irradiated vesicular sto-
    matitis virus on nucleic acid synthesis in chick embryo cells, *J. Gen. Virol.* **8**:165.

Youngner, J. S., and Preble, O. T., 1980, Viral persistence: Evolution of viral populations, in: *Comprehensive Virology*, Vol. 16 (H. Fraenkel-Conrat and R. R. Wagner, eds.), pp. 73–135, Plenum Press, New York.

Zee, Y., Hackett, A. J., and Talens, L., 1970, Vesicular stomatitis virus maturation sites in six different host cells, *J. Gen. Virol.* **7**:95.

Zimmerman, E. F., Heeter, M., and Darnell, J. E., 1963, RNA synthesis in poliovirus-infected cells, *Virology* **19**:400.

Zinkernagel, R. M., 1979, Cellular immune responses to viruses and the biological role of polymorphic major transplantation antigens, in: *Comprehensive Virology*, Vol. 15 (H. Fraenkel-Conrat and R. R. Wagner, eds.), pp. 171–204, Plenum Press, New York.

CHAPTER 2

Transcription by RNA Polymerase II

Ulla Hansen* and Phillip A. Sharp

Center for Cancer Research
and
Department of Biology
Massachusetts Institute of Technology
Cambridge, Massachusetts 02139

1. INTRODUCTION

Gene expression in eukaryotic cells is often regulated at the level of transcription of the gene (see Darnell, 1982). In order to understand viral growth, the response of cells to external stimuli, and the processes of differentiation and development, it is of importance to understand what controls the initiation of transcription at a gene. The enzyme of major interest in this regard is the eukaryotic DNA-dependent RNA polymerase (pol II), which is responsible for all cellular messenger RNA synthesis as well as messenger RNA synthesis from many DNA viruses and proviral forms of RNA viruses.

The aim of this review is to summarize recent results concerning the DNA sequences that constitute the promoter or transcriptional

* Present address: The Dana Farber Cancer Institute, Boston, Massachusetts 02115.

control region* of polymerase II genes and the proteins which are required to transcribe and regulate transcription from these promoters. Where possible, these data will be integrated to indicate how the proteins might interact with various DNA sequences to perform the specific functions required for the initiation of transcription.

Whether a gene is positioned in active or inactive chromatin certainly also plays a large role in whether or not transcription can occur. The various elements which define active chromatin (sensitivity and hypersensitivity to nucleases, presence of HMG proteins, modifications of both histone and nonhistone chromosomal proteins, DNA methylation) have recently been reviewed (Weisbrod, 1982; Razin and Riggs, 1980). With a few exceptions, there is no clear idea how the DNA sequences and proteins which are known to modulate transcription relate to the formation of active chromatin. Such a discussion must therefore be deferred to a later date.

2. POLYMERASE II CONTROL REGIONS: *IN VIVO*

By using current recombinant DNA technologies, the DNA sequences required for accurate and efficient transcription *in vivo* have been determined for a variety of viral and cellular genes. Deletion mutations, multiple point mutations (either "linker scanning" mutants, with clustered mutations within a ten base pair region, or mutants generated by bisulfite mutagenesis, with scattered mutations over a 20 to 50 base pair region), single point mutations, or inversions of portions of control regions have been generated. These mutant promoters, linked either to their original gene or to a marker gene, have been reintroduced into eukaryotic cells via DNA transfection, DNA microinjection, or infection by reconstructed virus. The activity of the control region has then been monitored either in a "transient" assay, within 72 hr after introduction of the DNA into cells, or in a long-term assay, following stable transformation of the cells. Stable transformation generally involves integration of the DNA into cellular sequences; however, with the bovine papilloma virus as a vector, the DNA remains episomal.

---

* In the recent review by Shenk (1981), it was suggested that the term promoter should be avoided, due to the present lack of understanding of the mechanism by which the transcription machinery binds to the DNA and arrives at the initiation site of transcription. Recognizing these ambiguities, the term promoter will be used merely to refer to sequences in the vicinity of a gene that influence initiation of transcription by pol II.

Despite the variety of techniques and cell types utilized, a consistent picture has emerged for the structure of polymerase II promoters *in vivo*. The only activities which seem to vary greatly, depending on the cell species and type used, are enhancer activities (see below). In this review, we have excluded yeast promoters. Although gross similarities exist between yeast promoters and those of higher eukaryotes (particularly with respect to enhancer elements), the distances between and structures of the various functional elements may be quite different. For example, potential TATA sequences in yeast promoters are 35–180 base pairs upstream of cap sites, compared to the 30 base pairs seen in higher eukaryotes. In fact, even promoters of different species of yeast may prove to be considerably different from one another (Russell, 1983; Sentenac and Hall, 1982).

The composite promoter, discussed in detail below, can be divided into three basic domains: (1) the initiation region, extending from approximately -50 to $+10$,* including the TATA sequences and the cap sites, (2) the immediate upstream sequences, extending from approximately -50 to -110, and (3) the enhancer (activator, potentiator) sequences, located in widely varying positions, depending on the gene.

2.1. Initiation Regions

RNA polymerase II generally initiates transcription at a single nucleotide or in a small cluster of nucleotides within a stretch of about ten base pairs (Baker and Ziff, 1981; Manley, 1983). The first nucleotide is generally a purine, however, initiation on some genes can occur with either a uridine or a cytidine (see Manley, 1983). The site(s) at which the polymerase initiates transcription can apparently be determined by the following mechanisms: by measuring a distance of about 25 base pairs from a TATA sequence, and/or by recognizing the DNA sequence immediately surrounding the initiation site as being optimal for initiation.

2.1.1. Positioning of Initiation Sites

Many polymerase II promoters contain TATA sequences between nucleotides -25 and -35 (Breathnach and Chambon, 1981).

---

* The initiation site for transcription is designated $+1$, with the transcribed sequences being positive numbers, and the upstream sequences being negative numbers.

That the TATA sequences can position the 5′ termini of the RNA was first demonstrated in a study on the early SV40 promoter. Upon deletion of sequences between the TATA sequence and the normal initiation sites, transcription proceeded from novel initiation sites, still approximately 25 base pairs downstream from the TATA sequences (Ghosh *et al.*, 1981; Benoist and Chambon, 1981). Similar results have been obtained from studies on the adenovirus EIA (Hearing and Shenk, 1983*b*) and on the rabbit β-globin promoters (Grosveld *et al.*, 1982*a*). Furthermore, the deletion of TATA sequences from the SV40 early, the adenovirus EIA, or the sea urchin H2A promoter resulted in the creation of heterogeneous initiation sites over a region of 30 or more base pairs (Benoist and Chambon, 1981; Hearing and Shenk, 1983*b*; Grosschedl and Birnstiel, 1980*a*). Sometimes removal of a TATA sequence resulted in prominent initiation at one or two sites besides the less efficient initiation at heterogeneous sites (herpes simplex virus thymidine kinase promoter; McKnight and Kingsbury, 1982; SV40 early promoter; Fromm and Berg, 1982). This argues for the ability of the transcriptional machinery to select specific initiation sites in the absence of an obvious TATA sequence. Indeed, deletion of TATA sequences from the promoters of the adenovirus major late genes or, by some investigators, of the rabbit β-globin gene did not affect the position of the initiation site for transcription (Hen *et al.*, 1982; Dierks *et al.*, 1983), although the efficiency of transcription decreased significantly (see below). In addition, several promoters which do not contain sequences homologous to the TATA sequence, such as those of the adenovirus EIIA and IV A2 genes, still initiate transcription from a single or small cluster of nucleotides (Baker and Ziff, 1981). We would suggest that, on many promoters which contain TATA sequences, those sequences and the local DNA sequence at the initiation site constitute redundant signals to direct the transcriptional machinery to initiate at a given site. The redundancy of signals may increase the efficiency of initiation at that site.

2.1.2. Efficiency of Initiation

The notion that either TATA sequences or local DNA sequence can direct initiation of transcription is supported when one examines how both these sites affect the efficiency of initiation. Three promoters have been carefully examined for the effect of small internal deletions covering either the TATA or initiation sequences on the level of RNA transcription *in vivo*. For the sea urchin H2A gene, deletion of either region reduced RNA levels in microinjected Xenopus oocytes to about 23% that of wild-type levels (Grosschedl and Birnstiel, 1982); for the adenovirus EIA gene, deletions of either region reduced levels in cells infected with a reconstructed virus to about 65% that of wild-type levels (Hearing and Shenk, 1983*b*); and for the rabbit β-globin gene, deletions of the TATA sequence reduced levels in cells transfected with the DNA to 1–17% that of wild-type levels, whereas deletions of the normal initiation site altered RNA levels from 50–300% that of wild-type levels, depending on the mutation (Grosveld *et al.*, 1982*a*; Dierks *et al.*, 1983). The reduction in transcriptional activity with alterations in the TATA or initiation site sequences suggests that this step in initiation can become rate limiting *in vivo*.

2.2. Immediate Upstream Sequences

For those genes whose promoters have been most exhaustively dissected, transcriptional requirements for sequences between -110 and -50 are exceedingly common (see Table 1). These sequences mediate the efficiency of transcription *in vivo*, sometimes through inducible signals (the heat shock hsp70 and mouse metallothionein I promoters). Only two promoters to date can dispense with these immediate upstream regions and maintain efficient transcription. Transcription from the adenovirus EIA promoter actually increases 1.4-fold when sequences between -145 and -44 are deleted (Hearing and Shenk, 1983*a*), and transcription from the sea urchin H2A promoter increases 1.9-fold upon removal of sequences between -111 and -55 (Grosshedl and Birnstiel, 1982). It might be worthwhile to note, however, that whereas both the adenovirus EIA and sea urchin H2A genes contain putative enhancer sequences (see below), only one of the nine genes listed in Table 1, the SV40 early gene, includes such a sequence. Thus, it is entirely possible that efficient transcription *in vivo* requires either an immediate upstream sequence or an enhancer sequence as part of the promoter.

As each of these identified immediate upstream sequences seems to be unique (although potential homologies can be identified), each will be independently discussed. Finally, we will only deal with those sequences which have been demonstrated by mutational analysis to have effects on transcription.

TABLE 1
Immediate Upstream Sequences

A. GC-Regions

Promoter	C-orientation	G-orientation
Rabbit β-globin[a]		
−120 to −110	GTCATCACCCA	
−111 to −98	CAGA-CCTCACCCTG	
−97 to −83	CAGAGCCACACCCTG	
Herpes simplex virus thymidine kinase		
−105 to −81	CCCCGCCCAGCGTCTTGTCATTGGC	
−61 to −48		CAGTCGGGGCGGCG
Simian virus 40 early		
−113 to −93	TCCCGCCCCTAACTCCGCCCA	
−92 to −72	TCCCGCCCTAACTCCGCCCA	
−71 to −50	GTTCCGCCC–ATTCTCCGCCCCA	
Simian virus 40 late		
−285 to −264		TGGGGCGGAGAAT–GGGCGGAAC
−263 to −243		TGGCCGGAGTTAGGGGCGGGA
−242 to −222		TGGGCGGAGTTAGGGGCGGGA
Human α₁-globin[b]		
−81 to −70	TCCGGCCAGCC	
−66 to −55	GAGCGCGCCCGGCC	

B. CCAAT and other regions

Promoter		
Rabbit β-globin[a]		
−82 to −66	GTGTTGGCCAATCTACA	
Adenovirus 2 major late[b]		
−68 to −54	GTGTAGGCCACGTGA	
Human α₁-globin[b]		
−78 to −62	GCGCCAGCCAATGAGCG	
Heat shock hsp 70[a]		
−68 to −34	GCGGCGCCTCGAATGTTCGGAAAAGAGCGCCGGAG	
Mouse metallothionein I		
−69 to −34	CTGCTGGGTGCAAACCCTTTGCGCCCGGACTCTCCC	

[a] The italicized sequences have been demonstrated to be of particular importance (see text).
[b] These particular sequences have not yet been proven to be critical for biological activity (see text).

2.2.1. Rabbit β-Globin Promoter

An extensive analysis of the effects of both deletion and point mutations on the promoter of the rabbit β-globin gene has been conducted by several laboratories (Dierks *et al.*, 1983; Grosveld *et al.*, 1982*a,b*). Two distinct regions within the immediate upstream sequences have been identified as capable of affecting the efficiency of transcription. One, extending from −77 to −71 consists of the sequence *GGCCA*AT.* This "CCAAT" sequence was identified several years ago as a region of homology common to many pol II promoters (Benoist *et al.*, 1980). Although point mutations in this sequence can decrease the efficiency of β-globin transcription four- to nine-fold, regions of homology have not borne out to be of significance in other promoters (sea urchin histone H2A, herpes simplex virus TK, see below).

The second immediate upstream region affecting the efficiency of β-globin transcription lies between nucleotides −111 and −83. This −100 region contains an imperfect repeat of a 14 or 15 bp sequence and, including the 5 bp adjacent and upstream, three repeats of the sequence ^T_ACACCC. Analysis of internal deletions in the region suggests that it is comprised of redundant signals, which probably include the 6-bp sequence denoted above. A single-point mutation of the italicized C of the repeat nearest the initiation site reduced the efficiency of transcription at least five-fold. In addition, a single point mutation within the same ACACCC sequence from the promoter of the related human β-globin gene may be responsible for a β-thalassemia phenotype. This latter mutation results in a ten-fold lower level of RNA when the gene is transfected into HeLa cells (Treisman *et al.*, 1983).

2.2.2. Herpes Simplex Virus Thymidine Kinase Promoter

Elegant studies by McKnight and Kingsbury (1982) using "linker scanning" mutants (mutants with clustered point alterations in a region covering 10 bp), have demonstrated the requirement of two distinct immediate upstream sequences for efficient transcription from the herpes simplex virus thymidine kinase promoter. The regions from −105 to −80 and from −61 to −47 may actually comprise part of a single element, since double linker scanning mutations in both re-

---

* The italicized bases are those shown to be required for full activity.

gions retain about the same transcriptional activity as either single mutation (McKnight, 1982). Mutations can be made in sequences between these two regions without significant effects on activity, however, despite the occurrence of a strongly homologous CAAT sequence from position -76 to -81.

Both the -100 and -50 regions consist in part of stretches of GC base pairs. In fact, the sequences are largely inverted repeats of one another. The GC-rich portions show some homology to the -100 region of β-globin control region (see Table 1). However, sequences beyond the GC-rich stretches are definitely required for activity.

2.2.3. SV40 Early Promoter

The SV40 early genes have long been known to require an enhancer sequence for efficient expression of transcription (see below). More recently, however, sequences between -113 and -50 have also been shown to be essential for transcription (Fromm and Berg, 1982; Hartzell et al., 1983; Everett et al., 1983; Byrne, et al., 1983). Within this region lie two adjacent direct 21 bp repeats and a third, imperfect 22 bp repeat, also in the same orientation. Each of these repeats, in turn, is comprised in part of two direct repeats of the sequence PyPyCCGCCC. By analysis of deletion mutations, the amount of early transcription decreases with the number of 8 bp GC-rich repeats remaining. Multiple-point mutations seem also to indicate the importance of the repeat structure. Once again, partial sequence homology between the GC-rich repeat region and the -100 region of the β-globin control region can be drawn (see Table 1).

2.2.4. SV40 Late Promoter

Initiation of transcription of the SV40 late genes is somewhat more complex than for many other genes. Initiation sites are scattered over a 200 bp stretch of DNA, without any obvious TATA sequences upstream of any of these sites. The major late initiation site which is the one furthest downstream may, however, have a TATA-like element positioned at about -30, as determined by mutational analysis (Brady et al., 1982). Another unusual feature of this control region is that deletion mutations including or extending to within 10 bp of prominent initiation sites often shift initiation to sites further upstream (for example, Ghosh *et al.*, 1982). Thus, the SV40 late promoter seems to consist of multiple initiation regions, as described above. These initiation regions must somehow be scanned by the polymerase before it initiates an RNA chain.

The efficiency of transcription from the SV40 late promoter is mainly determined by upstream DNA sequences positioned at -285 to -222 from the major late initiation site (Fromm and Berg, 1982). (These sequences are closer to, and sometimes overlapping with, the other minor late initiation sites). The late upstream region is actually identical to the immediate upstream region of the SV40 early promoter, from which transcription is initiated in the opposite direction. Thus, the immediate upstream sequences of the late promoter is comprised of six direct GGGCGGPuPu repeats embedded within three 21 or 22 bp repeats.

2.2.5. Other Uninduced Promoters

Deletion analyses of two other promoters have also indicated the importance of immediate upstream sequences for efficient transcription. For the adenovirus 2 major late promoter, deletion of upstream sequences to position -97 retained a wild-type level of RNA production after DNA transfection into HeLa cells, however, deletion of sequences to -62 resulted in a three-fold decrease in RNA production (Hen *et al.*, 1982). For the human α1-globin promoter, deletion of upstream sequences to -87 retained wild-type level of RNA production, however, deletion to -55 resulted in a 10- to 20-fold decrease in RNA production (Mellon *et al.*, 1981). Both the adenovirus major late and the human α1-globin promoters contain sequences at least partially within these specified regions which are homologous to the rabbit β-globin CCAAT region (see Table 1). However, the α-globin immediate upstream region also contains sequences homologous to the GC-rich sequences of the β-globin, Herpes simplex virus thymidine kinase, and SV40 promoters. Thus, a determination of which DNA sequences are crucial within these immediate upstream regions must await further mutational analysis.

2.2.6. Inducible Control Regions

A large variety of eukaryotic genes have been demonstrated to be inducible by treatment of cells with various stimuli: hormones, metals, heat, etc. For many of these systems, inducibility is controlled at the level of initiation of transcription of the gene, and investigators have therefore begun to determine which DNA sequences in the promoters of these genes respond to the inducing agent. A recent review discusses most of the findings in this field (Kessel and Khoury, 1983). We will mention here four genes where such experiments have pinpointed small regions of DNA in the immediate upstream region which are responsible for the inducible phenotype: the *Drosophila* heat shock gene hsp70, the mouse metallothionein I gene, the human interferon-β gene, and the human interferon-α_1 gene.

The *Drosophila* heat shock gene hsp70, at high copy number, can be induced both by heat and arsenite after either transfection into COS cells (monkey cell line expressing the SV40 T antigen) or microinjection into *Xenopus* oocytes.* Deletion analysis of this promoter indicated that sequences between -10 to -66 were required for inducibility, with at least part of the required sequences residing between -66 and -53 (Pelham, 1982; Bienz and Pelham, 1982; Mirault *et al.*, 1982). Upon comparing this region upstream of the hsp70 gene with upstream sequences of six other *Drosophila* heat shock genes, Pelham and Bienz (1982) derived a consensus sequence for what might constitute the heat shock signal: CT-GAA--TTC-AG. They then demonstrated that a synthetic consensus sequence indeed conferred heat inducibility on the Herpes simplex virus TK gene (which is not normally heat inducible) when the consensus sequence replaced the normal TK immediate upstream region.

The mouse metallothionein I gene, in its normal location in mouse cells, is inducible both by heavy metals and glucocorticoids. Only the heavy metal inducibility is retained when the gene is removed from its normal environment. Deletion analysis indicated that DNA sequences required for this inducibility reside within the first 90 bp of the control region (Brinster *et al.*, 1982). Further experiments have located the inducible signal between -70 and -34 (Carter *et al.*, 1983). Linker scanning mutations in this region abolish inducibility without altering the normal basal level of transcription.

The human interferon-β gene is induced both by double-stranded RNA, e.g., poly(rI)·poly(rC), and by Newcastle's Disease Virus (NDV). One report suggests that coding sequences alone are sufficient for induction (see Kessel and Khoury, 1983). It is clear, however,

---

* That *Drosophila* sequences elicit an accurate response in such widely divergent species indicates an amazing conservation through evolution.

that DNA upstream of the coding sequences of the gene, when linked to thymidine kinase coding sequences, allows induction of thymidine kinase by NDV (Ohno and Taniguchi, 1983). The most careful deletion analysis of the interferon-β control region, in which RNA with the appropriate 5'-terminus was monitored, suggests that sequences between -77 and -73 contain at least part of the DNA element which responds to poly(rI)·poly(rC) (Zinn et al., 1983). However, mutants retaining only sequences 3' of -40 in the interferon control region are still somewhat inducible (Maroteaux et al., 1983; Zinn et al., 1983). This may be due in part to messenger RNA stability, or it may indicate that more than one DNA sequence in and/or around the interferon-β gene can respond to the inducing agent to stimulate transcription of the gene.

The human interferon-α_1 is induced by the same stimuli as those which induce the interferon-β gene. The DNA sequences required for induction by NDV have been localized by 5'-deletion analysis to between -117 and -74. However, in this case, no transcription at all was detectable from the mutant deleting all upstream sequences beyond -74 (Ragg and Weissmann, 1983).

2.2.7. Position Effects

The above discussion should convince the reader of the importance and variety of immediate upstream sequences. However, the delineation of these sequences indicates little concerning their mechanism of action. A few experiments have been done to broach that topic, mainly concerning the positioning and directionality of some of the immediate upstream sequences.

In general, some flexibility in distance is allowed between immediate upstream sequences and the initiation region. The most careful study in that regard has been done on the herpes simplex virus TK promoter (McKnight, 1982). The insertion of 9–30 bp between the initiation and immediate upstream regions had little effect on the efficiency of transcription. However, insertion of 50 or more bp abolished the stimulatory effect of the upstream sequences. The molecular mechanism by which flexibility in distance of 30 bp, but not of 50 bp, would be tolerated is rather unclear. It is of course always difficult to demonstrate that the particular inserted sequences have a totally neutral effect on transcription. In a second instance, the ability of the synthetic consensus heat shock sequence to induce transcription in response to heat was not altered by differences in spacing of 6 bp (between 13–19 bp between the consensus sequence and the TATA sequence; Pelham and Bienz, 1982). Finally, a quick glance at Table 1 will indicate that no strict distance to the initiation site is required for the potentially homologous elements from different promoters. And in the case of the SV40 late promoter, a single upstream element seems to be required for initiation sites varying over 200 bp.

In promoters containing two different upstream sequences, some flexibility between these sequences is also tolerated. In the case of the rabbit β-globin promoter, insertions of 8–50 bp between the CCAAT and − 100 regions result only in moderate two- to five-fold decreases in transcription (Grosveld *et al.*, 1982*b*). In the herpes simplex virus TK upstream region, 10 bp can be inserted or removed from between the two related upstream sequences without effect. Insertions of 36 or more bp in the intervening region abolished stimulation by the upstream sequences, however (McKnight, 1982).

Finally, with at least one immediate upstream sequence, that of the SV40 promoters, the same region of DNA can act to stimulate transcription in both directions (Fromm and Berg, 1982). In fact, the SV40 immediate upstream region, when inverted, retains the ability to support SV40 early transcription at wild-type levels (Everett *et al.*, 1983). Thus, the complementary C-rich and G-rich elements listed in Table 1 may actually represent the same class of upstream elements.

This flexibility in distances between DNA sequence elements required for transcription has also been observed in eukaryotic RNA polymerase III control regions (which lie within the structural genes). However, it is unlike the bacterial promoters, where sequences are fixed with relationship to one another to allow direct and well-defined interactions between the proteins (generally RNA polymerase and a positive activator such as the lambda cII, cI proteins) bound at these sites. Of course, within the actual *E. coli* RNA polymerase binding site, which is composed of two consensus sequences around positions − 35 and − 10, a very strict optimal distance is maintained between these two regions to allow for direct contact of the polymerase simultaneously with both regions. Certainly, double-stranded B-form DNA does not contain sufficient flexibility to allow widely varying spacings between protein binding sites where the proteins must interact with one another.

Thus, we would suggest that the immediate upstream sequences either provide a mechanism for the polymerase and other initiation factors to initially bind the DNA, after which they must scan along the DNA for an appropriate initiation region (the entry model), or provide a mechanism for altering the DNA structure in the region which then allows promotion of transcription from the region, either by direct protein–protein contacts, or by recognition of the altered DNA structure by the initiation machinery.

2.3. Enhancer Regions

In recent years, a new class of control elements has been described with quite unusual properties (for review, see Khoury and Gruss, 1983). These elements act in *cis* to stimulate transcription of pol II genes and can act on heterologous control regions. But they function in an orientation- and position-independent manner. That is, they stimulate transcription when positioned either upstream or downstream from an initiation region and up to several thousand bp away. Although enhancers can stimulate transcription over very large distances, the greatest effect is generally observed on the most proximal initiation regions (Wasylyk *et al.*, 1983). Such elements have been identified and characterized in many viruses: simian virus 40 (Moreau *et al.*, 1981; Banerji *et al.*, 1981; Gruss *et al.*, 1981; Fromm and Berg, 1982; Fromm and Berg, 1983*a,b*; Wasylyk *et al.*, 1983), polyoma virus (de Villiers and Schaffner, 1981; Tyndall *et al.*, 1981; Ruley and Fried, 1983), BK virus (G. Khoury, personal communication), bovine papilloma virus (Lusky *et al.*, 1983), Rous sarcoma virus (Luciw *et al.*, 1983), Moloney murine sarcoma virus (Levinson *et al.*, 1982; Blair *et al.*, 1981), Harvey murine sarcoma virus (Chang *et al.*, 1980; Kriegler and Botchan, 1983), other murine retroviruses (Jolly *et al.*, 1983), adenovirus (the EIA control region, Hearing and Shenk, 1983*a*), and, potentially, mouse mammary tumor virus (Chandler *et al.*, 1983; see below). In all these cases but one (that of bovine papillomavirus), the enhancer elements are normally situated upstream of the rest of the promoter.

The first potential cellular enhancer was identified in the promoter of the sea urchin histone H2A gene (Grosschedl and Birnstiel, 1980b). Subsequently, cellular sequences which can function as enhancer regions have been isolated, either by their ability to hybridize to known viral enhancers or by a direct activity assay (Conrad and Botchan, 1983; Fried *et al.*, 1983). Finally, definitive proof was obtained by several laboratories of an enhancer of a cellular gene, the immunoglobulin heavy chain gene (Gillies *et al.*, 1983; Banerji *et al.*,

1983; Neuberger, 1983; Mercola *et al.*, 1983). This enhancer is normally positioned over a thousand bp downstream of the initiation region, within the major intron. Enhancers within introns of cellular genes may also be present in the immunoglobulin K light chain gene (Queen and Baltimore, 1983).

2.3.1. Sequences

The precise DNA sequences within these enhancer regions which are crucial to activate transcription are beginning to be defined by mutational analysis. An enhancer in the adenovirus 5 EIA control region has been identified by Hearing and Shenk (1983*a*) to be the sequence AGGAAGTGACA. Homologies to this sequence can be found within other viral enhancer sequences. In addition, Weiher *et al.* (1983) have determined the importance of the sequence GTGTGGAAAG within the SV40 72 bp repeats. They also discovered homologous sequences within several of the other viral enhancer regions and arrived at a consensus sequence of $(G)TGG^{AAA}_{TTT}(G)$. Thus, at least two different consensus sequences are capable of eliciting the enhancer phenotype.

Although the SV40 enhancer region requires the identified core sequence for function, these sequences are probably not sufficient. Nordheim and Rich (1983) have identified sequences within the SV40 enhancer that are capable of forming Z-DNA. They have also obtained point mutations within these potential Z-DNA-forming segments and have demonstrated that such mutants are significantly decreased in their enhancer activity (A. Nordheim, personal communication). Whether formation of Z-DNA is actually involved in stimulating transcription remains to be proven. A Z- to B-DNA transition, however, is an intriguing method of providing negative superhelicity in the DNA.

2.3.2. Regulation of Enhancer Activity

That enhancer regions may be more or less effective in different cell types was originally suggested by studies on polyoma virus. Wild-type virus normally multiplies well in mouse cells of many types but will not propagate in undifferentiated embryonal carcinoma cell lines due to a block in the initiation of viral transcription (Dandolo *et al.*,

1983). Mutants of the virus can be isolated which allow growth in these cell lines; the mutations all map to the enhancer region of the polyoma DNA (Katinka *et al.*, 1980; 1981; Fujimara *et al.*, 1981; Sekikawa and Levine, 1981; Tanaka *et al.*, 1982). The polyoma enhancer region is quite complex, containing two potential homologies with the adenovirus EIA enhancer sequence and two potential homologies with the SV40 enhancer sequence. The mutations of the viruses with increased host range lie within different portions of this region, depending on the cell lines on which they were selected. No clear picture emerges as to certain consensus sequences being either important or detrimental in certain cell lines, but the mutations and enhancer region are definitely related. These results suggest that certain enhancers might be activated during differentiation of cells.

Evidence has been also accumulating for viral enhancer regions being more or less effective depending on the species of the cell line tested (de Villiers *et al.*, 1982; Laimins *et al.*, 1982; Kriegler and Botchan, 1983; Spandidos and Wilkie, 1983; Lusky *et al.*, 1983; Berg *et al.*, 1983). In general, viral enhancer regions are most active in cell lines derived from their normal host. They have presumably evolved to take advantage of the response of their normal host.

An effect of tissue specificity on enhancer activity has been clearly demonstrated for the enhancer of the cellular heavy immunoglobulin gene. This enhancer is positioned next to the immunoglobulin heavy chain constant region exon such that it is retained in all the normal DNA rearrangements of the gene. In this position, it was surmised that the enhancer would stimulate transcription from the variable gene promoter only subsequent to rearrangement of the variable region exon near the constant region exon. However, an even tighter control of this stimulation by the enhancer was evidenced when its activity was compared in cells of different origin. Stimulation by the enhancer was observed in both mouse and human lymphoid cells, but not in cells of nonlymphoid origin (Gillies *et al.*, 1983; Banerji *et al.*, 1983; Neuberger, 1983). Such regulation has broad biological significance as a general method of specifying gene expression in different tissues.

Finally, the hormonal stimulation of transcription of the mouse mammary tumor virus (MMTV) might result from the regulation of enhancer activity. Many laboratories have carried out mutation analysis to determine which DNA sequences in the control region of MMTV are responsible for induction of viral transcription by glucocorticoids. A composite from the various laboratories indicates that sequences between − 190 and − 109 are necessary and sufficient for hormone stimulation from the MMTV control region (Buetti and Diggelmann, 1983; Hynes *et al.*, 1983; Chandler *et al.*, 1983; Majors and Varmus, 1983; F. Lee, personal communication). This region of the DNA also binds the glucocorticoid receptor, along with other regions of the MMTV genome (see Kessel and Khoury, 1983; Scheidereit *et al.*, 1983). The recent, striking observation made by Chandler *et al.* (1983) was that these same MMTV DNA sequences will confer inducibility to dexamethasone on a heterologous promoter (that of the Herpes simplex virus TK gene), when positioned at two places upstream of the gene, and in either orientation at one of these positions. These characteristics are similar to those of an enhancer region. Thus, MMTV might contain a hormone-inducible MMTV enhancer activity.

3. POLYMERASE II PROMOTERS: *IN VITRO*

Cell free extracts for investigating the specific initiation of transcription by pol II have been available for several years (Manley *et al.*, 1980; Weil *et al.*, 1979). In that time, many promoters have been dissected *in vitro* to determine which DNA sequences are required for such transcription. Whereas, the *in vitro* and *in vivo* results diverge in many instances, increasing examples of similarities are becoming apparent.

3.1. Initiation Region

In many cases, the initiation region, spanning both the TATA sequence and the initiation site, has been shown to be of overriding importance for efficient transcription *in vitro*. In some cases, these seem to be the only signals to which the transcriptional machinery responds *in vitro*: rabbit β-globin (Grosveld *et al.*, 1981), human α-globin (Talkington and Leder, 1982), adenovirus EIA (Hearing and Shenk, 1983*b*), polyoma early (Jat *et al.*, 1982), avian sarcoma virus (Mitsialis *et al.*, 1983), ovalbumin (Tsai *et al.*, 1981*b*), conalbumin (Corden *et al.*, 1980), and silk fibroin (Tsujimoto *et al.*, 1981). Thus, the immediate upstream sequences of the β- and α-globin promoters and the enhancer sequences of the adenovirus EIA, polyoma early, and avian sarcoma virus promoters (see above) are nonfunctional in the existing extracts. This may indicate that the rate-limiting step for

transcription from these promoters is different *in vitro* from what it
is *in vivo*.

For three other promoters, the initiation region drastically affects
the efficiency of transcription, but is not the sole effector: adenovirus
major late (Hen *et al.*, 1982; Hu and Manley, 1981; Manley, 1983),
adenovirus EIII (Lee *et al.*, 1982), and sea urchin H2A (Grosschedl
and Birnstiel, 1982). Finally, for the SV40 early genes, the initiation
region is required mainly for accurate placement of 5'-termini *in vitro*,
and affects the efficiency of transcription little, if at all (Mathis and
Chambom, 1981; Myers *et al.*, 1981; Hansen and Sharp, 1983). This
promoter is also unique *in vivo* in that the initiation region does not
seem to affect the efficiency of transcription (Benoist and Chambon,
1981; Fromm and Berg, 1982; 1983*a*).

3.2.   Immediate Upstream Region

In four cases, immediate upstream sequences do have effects on
the efficiency of transcription *in vitro*. Sequential upstream deletions
of the adenovirus major late promoter indicated that sequences between -97 and -62 exerted a threefold effect on the efficiency of
transcription *in vitro* (Hen *et al.*, 1982). A similar analysis of the
adenovirus EIII control region indicated that sequences upstream of
-70 also exerted a threefold effect on transcription *in vitro* (Lee *et
al.*, 1982). The dependence on immediate upstream sequences *in vitro*
is much more striking, however, for the SV40 early and late promoters. SV40 early transcription from the major initiation site is undetectable *in vitro* in the absence of sequences between -50 and
-113 (Myers *et al.*, 1981; Lebowitz and Ghosh, 1982; Hansen and
Sharp, 1983). The amount of transcription correlates with the number
of C-rich repeats remaining in this region (see Table 1; Hansen and
Sharp, 1983). Transcription from the SV40 late promoter *in vitro* is
also dependent on this region of DNA, although transcription proceeds in the opposite direction (see above, Table 1; Hansen and
Sharp, 1983). A dependence of greater than 100-fold is seen for transcription from the most proximal initiation sites and a dependence of
three- to five-fold is seen for the most distal initiation sites (approximately 200 bp away). In all four of these cases, the quantitative effects of immediate upstream sequences *in vitro* are similar to those
observed *in vivo*.

3.3. Enhancer Region

The only effect of a potential enhancer on transcription *in vitro* has been obtained for the sea urchin histone H2A promoter. Deletion of sequences between -111 and -452 reduced transcription *in vitro* four- to five-fold (Grosschedl and Birnstiel, 1982). This deletion *in vivo* reduced transcription an equivalent 15- to 20-fold. However, inversion of these sequences produced a 1.6-fold decrease *in vitro*, in marked contrast to the five-fold increase *in vivo*. Unfortunately, in no other promoter has even the deletion of enhancer regions had a marked effect on the efficiency of transcription *in vitro*.

4. RNA POLYMERASE II AND FACTORS

7.1. RNA Polymerase II

RNA polymerase II is typically distinguished from RNA polymerases I and III by its sensitivity to low concentrations of the toxin α-amanitin (Roeder, 1976). This enzyme is a complex protein composed of ten polypeptides (Table 2). The subunit structure is remarkably conserved in divergent species, suggesting essential roles for each polypeptide. In fact, the subunit structure of all three eukaryotic RNA polymerases is similar, and three polypeptides are thought to be held in common. This suggests that the three types of polymerases evolved from a single enzyme and share common constraints in structure (Huet *et al.*, 1982).

Little is known about the structure of genes encoding pol II polypeptides. DNA sequences presumably encoding a structural gene for a pol II subunit of *Drosophila* have recently been isolated (Searles *et al.*, 1982). *Drosophila* mutants with an altered sensitivity to α-amanitin have mutations in this gene (Searles *et al.*, 1982). Similar mutants have been isolated in mammalian cells (Chan *et al.*, 1972), and the properties of these mutants have been transferred via DNA transfection (Ingles and Shales, 1982). The genes specifying most of the subunits of yeast pol II will probably soon be isolated by techniques involving specific antibodies (Young and Davis, 1983).

α-Amanitin resistant variants of Chinese hamster ovary (CHO) cells (Ama^r) yield pol II which is less sensitive to inhibition by the toxin (Lobban *et al.*, 1976). The Ama^r phenotype both in *Drosophila* and mammalian cells may be due to mutations in the gene encoding

TABLE 2

Polypeptide Structure of RNA Polymerase II from Various Eukaryotic Organisms[a]

Yeast	Drosophila	Mouse	Calf
220	215	240 (II_0)	240
		205 (II_A)	214
185	175	170 (II_B)	180
150	140	140	140
44.5		41	34
32	32		
[27]	25	[29]	[25]
[23]	20	27	20.5
	18	22	18
	17.5	[19]	17.5
16	15	16	[16.5]
[14.5]	<15		
12.6	<15		
10	<15		

[a] The molecular weights of RNA polymerase II polypeptides ($\times 10^{-3}$) are reported for yeast (Dezelee et al., 1976), Drosophila (Kramer and Bautz, 1981), mouse myeloma plasmacytoma (Roeder, 1976), and calf thymus (Hodo and Blatti, 1977; Dahmus, 1981a,b). The boxed polypeptides are thought to be common of polymerases I, II, and III (Huet et al., 1982). The three largest subunits of the mouse polymerase are labeled II_0, II_A, and II_B (Schwartz and Roeder, 1975). A multisubunit polymerase only contains one of these polypeptides; II_A and II_B are thought to be related by proteolytic cleavage.

the 140,000 dalton subunit of pol II, as only this polypeptide can be crosslinked to the drug (Brodner and Weiland, 1976). In these systems, mutation to an Ama[r] phenotype often produces a ts pol II (Ingles, 1978). Growth of Ama[s] mammalian cells in the presence of α-amanitin results in increased rates of degradation of pol II proteins. When hybrid CHO cells which are pseudohexaploid (Ama[s]/Ama[s]/Ama[r]) are grown in the presence of the toxin, the levels of the sensitive and resistant polymerases decrease and increase, respectively (Guialis et al., 1977; 1979). The total level of pol II remains constant. This suggests that the level of polymerase II in mammalian cells is autoregulated, perhaps at the stage of transcription.

Chemical modification of enzymes is frequently correlated with changes in levels of activity. Mammalian RNA polymerase II subunits are phosphorylated in vivo (Bell et al., 1977; Dahmus, 1981a). Labeling of HeLa cells with $^{32}P_i$ results in phosphate incorporation into pol II polypeptides of 240,000, 214,000, and 20,500 daltons (Dahmus, 1981a). Purified pol II is a substrate for both casein kinase I and II and for the cyclic AMP independent nuclear protein kinase NII. The major sites of phosphorylation by casein kinase II are on the 240,000 dalton polypeptide where it is estimated that some 20 phosphates might be added (Dahmus, 1981a,b). Both the casein kinase I and II activities modified the 214,00 and 20,500 polypeptides of pol II to a significant level (Dahmus, 1981a,b). The nuclear protein kinase NII modified the 214,000, 140,000, and 21,000 (same as 20,500) dalton subunits of pol II (Stetler and Rose, 1982). Interestingly, in the latter study, phosphorylated pol II had a four- to eight-fold higher level of activity on duplex DNA. However, phosphorylation of pol II from various sources by casein kinases gave varying degrees of stimulation of transcription. It is possible that phosphorylation of pol II could be a signal for modulation of activity *in vivo*; however, few studies of this possibility have been attempted. With the recent development of monoclonal and polyclonal antibodies to mammalian pol II, it should be possible to test whether there is a correlation of modification of pol II and physiological changes in transcription (Dahmus and Kedinger, 1983; Carroll and Stollar, 1982).

Purified pol II transcribes duplex DNA poorly and nonspecifically. Single-stranded DNA is the preferred substrate and the presence of nicks and gaps in duplex DNA stimulates activity. Once initiated, the rate of chain elongation *in vitro* by pol II on duplex DNA is relatively slow, seven nucleotides/second, as compared to a rate *in vivo* of 50–100 nucleotides/second (Kadesch and Chamberlin, 1982). The elongating pol II-DNA-RNA complex is remarkably stable, withstanding low concentrations of detergent and high salts. This stability is exploited in experiments designed to measure the density of active polymerases on a given segment via elongation *in vitro* in the presence of detergent (Ferdinand *et al.*, 1977).

4.2. Termination of Transcription

Termination of transcription by pol II has not been well studied. Until recently, specific sites for termination *in vivo* by pol II had not been identified. Apparently, pol II transcribes some 1000 nucleotides (N) beyond the polyadenylation site of the β-globin gene of mouse before terminating within a specific region (Hofer *et al.*, 1982). The nature of the responsible sequences at this site is under investigation. A site for termination of transcription in the major late transcription unit of adenovirus has also been suggested; again, no particular sequence motif has been identified (Fraser *et al.*, 1979a). Pol II does seem to prematurely terminate transcription *in vivo* shortly down-stream of some initiation sites (Fraser *et al.*, 1979*b*). Such termination has not been related to regulation or any physiological function. It is interesting to note that the drug DRB, a nucleoside analogue (5,6-dichloro-1-β-ribofuranosylbenzimidazole), enhances this premature termination and thus blocks synthesis of mRNAs in mammalian cells (Fraser *et al.*, 1979*b*). DRB has recently been shown to inhibit tran-scription *in vitro* in systems that exhibit accurate initiation (Zando-meni *et al.*, 1982, 1983). The drug probably acts without being charged or modified and accentuates a poorly understood event in specific transcription. Mammalian cell variants resistant to DRB inhibition have been isolated but the mutated gene product has not been iden-tified (Gupta and Siminovitch, 1980; Mittleman *et al.*, 1983).

4.3. Factors Necessary for Accurate Transcription

As described above, studies of deletion and point mutations sug-gest that an extensive set of sequences is important in promoter rec-ognition. Each component in these sequences that influences the ac-tivity of a promoter *in vivo* must be recognized by one or more transcription factors. In addition to sequence recognition factors, other components might be required to interact with pol II to permit initiation. Obviously, modulation of activity of these factors could either enhance or suppress the rate of transcription. The specificity of such regulatory signals would depend on whether a factor was specific for only a subset of promoters.

Although RNA polymerase II will not accurately initiate tran-scription on duplex DNA, when purified pol II is supplemented with an S100 extract, accurate initiation can be demonstrated on a variety of duplex DNA templates (Weil *et al.*, 1979). The endogenous pol II in extracts of HeLa cells will also accurately initiate on many exo-genously added templates (Manley *et al.*, 1980). Both results suggest that cellular factors must interact with pol II and promoter sequences for initiation *in vitro*. Assays can be developed for such factors using purified pol II and DNA templates; typically, the amount of accurate initiation is measured by the radioactivity in a specific length "runoff" RNA product. The nature of the factors purified with this assay might depend on the particular template and the type of initiation reaction responsible for the *in vitro* synthesis. In general, factors involved in elongation will probably not be identified by those assays as purified pol II alone will elongate at reasonable rates over long lengths of duplex DNA (Kadesch and Chamberlin, 1982).

A comparison of the *in vitro* transcription of mutants of several different promoters suggests that the rate of initiation is primarily dependent upon sequences between approximately -50 and $+10$ (see above). In fact, the reaction *in vitro* on most promoters is critically responsive to sequences in the TATA consensus region. For example, converting the TATAAA sequence in the conalbumin promoter to TACAAA reduced the transcription activity *in vitro* ten-fold (Corden *et al.*, 1980). In contrast, promoter activity *in vivo* is affected by mutations in the initiation region, immediate upstream region, and enhancer region. The simplest interpretation of these results is that the *in vitro* reaction generally reflects only one aspect of a more complicated process *in vivo*.

Purification of factors using a template such as the major late promoter of adenovirus (MLP) should identify components that promote initiation through recognition of sequences near TATA. These factors would probably also be required for initiation at a majority, if not all, promoter sites. Under normal physiological conditions, however, their activities may not be rate limiting for initiation. Fractionations of extracts of HeLa cells by chromatography on ion exchange columns have resolved several fractions that must be added in addition to pol II and the MLP template to reconstitute accurate initiation (Matsui *et al.*, 1980; Samuels *et al.*, 1982). Typically, these fractions elute from a phosphocellulose column in the flow-through at low salt, in the 0.35–0.6 M KCl step and in the 0.6–1.0 M KCl step. The 0.35–0.6 M KCl fraction can be further resolved into two fractions by chromatography on a heparin column (Samuels *et al.*, 1982). A reconstituted reaction containing three of these four fractions plus pol II and template forms an assay which can be used to titrate the activity in the fourth fraction. (In general, such assays resond linearly to the added fraction). Assays of this type have been used to follow the various activities during sedimentation in sucrose gradients and during further chromatography. At present, the activities in these four fractions have not been resolved as purified proteins; thus, it is possible that the number of essential components could change upon further purification.

Transcription factors have also been fractionated from a crude extract of chick oviduct cells (Tsai *et al.*, 1981*a*). Fractions from these cells reconstituted accurate initiation in a reaction containing the homologous ovalbumin gene as template. Interestingly, fractions prepared from HeLa cells were interchangeable with those from chick cells.

Preincubation of template with fractions containing some of the transcription factors allows the formation of a complex which is stable when challenged either with excess nonspecific competitor DNA or with promoter-containing DNA (Davison *et al.*, 1983; Fire *et al.*, 1983). Such a complex can be utilized for accurate transcription when purified pol II and the other factors are added. It is likely that one or more components in this preincubation reaction recognize TATA sequences, since complex formation is decreased by mutations in these sequences (Davison *et al.*, 1983). At the moment, there is no other specific assay for the function of the other two fractions that must be added before pol II will accurately initiate transcription at the MLP. Interestingly, preincubation of pol II and factors in the absence of nucleoside triphosphates results in an activated complex that will rapidly utilize specific dinucleotides for priming of transcription (Samuels *et al.*, 1983). The priming dinucleotides must be complementary to sequences within $+4$ to -4 of the initiation site. The transcript produced by initiation with a dinucleotide is not modified by capping, suggesting that neither a triphosphate terminus nor a cap structure is necessary for elongation by RNA polymerase II.

The process of initiation by pol II may be considerably more complex than that of the well-characterized bacterial RNA polymerase. For example, the efficiency of transcription in a whole cell extract system is more sensitive to low ATP concentrations than to limiting concentrations of the other three nucleoside triphosphates (Bunick *et al.*, 1982). However, nascent chains will elongate at low ATP concentrations suggesting that this nucleoside triphosphate may have an additional role of initiation. Specific initiation of transcription does not occur when β-γ methylene adenosine triphosphate is added in lieu of ATP even though this triphosphate is utilized for elongation of nascent chains (Bunick *et al.*, 1982). Interestingly, addition of dATP will stimulate transcription at limiting ATP concentration and may substitute as a cofactor for initiation. Thus, hydrolysis of ATP may play an essential role in the accurate initiation by pol II in addition to its utilization as a substrate for elongation.

4.4. Promoter-Specific Factors

The transcription factors described above have been shown to direct *in vitro* initiation on many different promoters. Firm evidence

for a promoter-specific factor acting *in vitro* has only been obtained
for the early and late SV40 promoters (Dynan and Tjian, 1983; Hansen
and Sharp, 1983). Fractionation of a whole cell extract using transcription of the SV40 early promoter as an assay yielded a factor
which was essential for transcription of this template but did not stimulate transcription of other promoters (Dynan and Tjian, 1983). Previous *in vivo* studies of mutants of SV40 DNA revealed that a region
between -113 to -50 was essential for early transcription (see
above). This region contains six repeats of the sequence PyPy-
CCGCCC, which form part of three tandem 21 or 22 bp repeats. Partial
deletion of these repeats results in lower levels of early RNA *in vivo*
(Fromm and Berg, 1982). Transcription *in vitro* of a series of SV40
deletion mutants also shows a striking dependence on the GC repeat
region (Hansen and Sharp, 1983; Myers *et al.*, 1981; Lebowitz and
Ghosh, 1982). It is highly probable that the SV40-specific transcription factor mediates its effect on early and late SV40 transcription
through recognition of these repeats. The striking dependence on this
factor of *in vitro* transcription from the SV40 early promoter may be
due to the peculiar structure of its initiation region. Segments containing only the TATA sequences of the SV40 early promoter do not
support *in vitro* transcription, in contrast to comparable segments
from other promoters (Hansen and Sharp, 1983; Mathis and Chambon, 1981). This unique feature of the SV40 early TATA sequences
may partially explain the promoter's absolute dependence on upstream sequences. Thus, other promoters may depend on the same
factor *in vivo* but this dependence may not be detectable *in vitro* due
to the efficiency of TATA-mediated initiation.

Comparison of the relative efficiency of *in vitro* transcription of
two promoters with extracts from the gland of silk moth and HeLa
cells has shown preferential transcription in silk moth extracts of a
homologous silk moth fibroin gene (Tsuda and Suzuki, 1981). As yet,
a specific factor has not been identified using this assay. It might be
anticipated that further work with this and other assays will lead to
identification of a number of factors each specific for a subset of
promoters.

Enhancer elements that stimulate transcription *in vivo* at adjacent
promoters are thought to be important in regulation of genes during
differentiation. The mechanism by which such an element can enhance transcription at promoters positioned over 1000 base pairs away
is not known. Some have suggested that enhancer elements may be
sites of entry for pol II, it being subsequently transferred to the ini-

tiation site (Wasylyk *et al.*, 1983). Alternatively, they may be topo-isomerase-binding sites or protein-binding sites leading to chromatin alterations. As yet, the activity of an enhancer element has not been observed in reactions *in vitro*. The only purified factor which potentially stimulates transcription *in vivo* by recognition of an enhancer element is the glucocorticoid receptor (see above; Chandler *et al.*, 1983). Similar factors must recognize other cell type specific enhancer elements but these await identification (Gillies *et al.*, 1983; Banerji *et al.*, 1983).

5.   CLOSING COMMENTS

Alterations in RNA polymerase II transcription are common during replication of most viruses. In some cases, the cellular RNA polymerase is a competitor for the free pool of nucleoside triphosphates and, thus, inhibition of its activity benefits viral replication. In other cases, viral DNA is transcribed by the cellular polymerase II, thus, redistribution of its activity is necessary. Factors necessary for transcription by polymerase II in cells are currently being identified and subsequently purified. The most critical aspect of this work is to identify factors that recognize the sequence elements known to affect transcription *in vivo*. Alterations in activities of either the factors or the polymerase will probably be a common event in virus-infected cells. In fact, understanding how various viruses modulate transcription by RNA polymerase II will probably be the primary means of establishing the *in vivo* function of the factors isolated by *in vitro* assays. Research in this area has only recently begun and its future appears to be rich with new discoveries and insights.

ACKNOWLEDGMENTS

We are grateful to Mark Samuels and Robert E. Kingston for their critical reading of this paper. U. H. acknowledges support from the Jane Coffin Childs Memorial Fund. This work was supported by Grants PCM-7823230 (currently PCM-8200309) from the National Science Foundation, No. PO1-CA26717 from the National Institutes of Health to P.A.S., and partially from the Center for Cancer Biology at MIT (Core) Grant No. NIH-PO1-CA14051.

6. REFERENCES

Baker, C. C., and Ziff, E. B., 1981, Promoters and heterogeneous 5' termini of messenger RNAs of adenovirus serotype 2, *J. Mol. Biol.* **149**:189–221.

Banerji, J., Rusconi, S., and Schaffner, W., 1981, Expression of a β-globin gene is enhanced by remote SV40 DNA sequences, *Cell* **27**:299–308.

Banerji, J., Olson, L., and Schaffner, W., 1983, A lymphocyte-specific cellular enhancer is located downstream of the joining region in immunoglobulin heavy chain genes, *Cell* **33**:729–740.

Bell, G. I., Valenzuela, P., and Rutter, W. J., 1977, Phosphorylation of yeast DNA-dependent RNA polymerase *in vivo* and *in vitro, J. Biol. Chem.* **252**:3082–3091.

Benoist, C., and Chambon, P., 1981, *In vivo* sequence requirements of the SV40 early promoter region, *Nature* **290**:304–310.

Benoist, C., O'Hare, K., Breathnach, R., and Chambon, P., 1980, The ovalbumin gene-sequence of putative control regions, *Nucl. Acids Res.* **8**:127–142.

Berg, P. E., Yu, J.-K., Popovic, Z., Schumperli, D., Johansen, H., Rosenberg, M., and Anderson, W. F., 1983, Differential activation of the mouse β-globin promoter by enhancers, *Mol. Cell. Biol.* **3**:1246–1254.

Bienz, M., and Pelham, H. R. B., 1982, Expression of a *Drosphilia* heat-shock protein in *Xenopus* oocytes: Conserved and divergent regulatory signals, *EMBO J.* **1**:1583–1588.

Blair, D. G., Oskarsson, M., Wood, T. G., McClements, W. L., Fischinger, P. J. and Van de Woude, G. G., 1981, Activation of the transforming potential of a normal cell sequence: A molecular model for oncogenesis, *Science* **212**:941–943.

Brady, J., Radonovich, M., Vodkin, M., Natarajan, V., Thoren, M., Das, G., Janik, J., and Salzman, N. P., 1982, Site-specific base substitution and deletion mutations that enhance or suppress transcription of the SV40 major late RNA, *Cell* **31**:625–633.

Breathnach, R., and Chambon, P., 1981, Organization and expression of eucaryotic split genes coding for proteins, *Annu. Rev. Biochem.* **50**:349–383.

Brinster, R. L., Chen, H. Y., Warren, R., Sarthy, A., and Palmiter, R., 1982, Regulation of metallothionein-thymidine kinase fusion plasmids injected into mouse eggs, *Nature* **296**:39–42.

Brodner, O. G., and Weiland, T., 1976, Identification of the amatoxin binding subunit of RNA polymerase B by affinity labeling experiments, *Biochemistry* **15**:3480–3484.

Buetti, E., and Diggelmann, H., 1983, Glucocorticoid regulation of mouse mammary tumor virus: Identification of a short essential DNA region, *EMBO J.* **2**:1423–1429.

Bunick, D., Zandomeni, R., Ackerman, S., and Weinmann, R., 1982, Mechanism of RNA polymerase II-specific initiation of transcription *in vitro*: ATP requirement and uncapped runoff transcript, *Cell* **29**:877–886.

Byrne, B. J., Davis, M. S., Yamaguchi, J., Bergsma, D. K., and Subramanian, K. N., 1983, Definition of the SV40 early promoter region and demonstration of a host range bias in the enhancement effect of the SV40 72 bp repeat, *Proc. Natl. Acad. Sci. USA* **80**:721–725.

Carroll, S. B., and Stollar, D. B., 1982, Inhibitory monoclonal antibody to calf thymus RNA polymerase II blocks formation of enzyme-DNA complexes, *Proc. Natl. Acad. Sci. USA* **79**:7233–7237.

Carter, A. D., Walling, M. J., and Hamer, D. H., 1983, Distinct promoter and control sequences of an inducible metallothionein gene, in: *Enhancer and Eukaryotic Gene Expression* (Y. Gluzman and T. Shenk, eds.), pp. 170–174, Cold Spring Harbor Laboratory, Cold Spring Harbor, New York.

Chan, V. L., Whitmore, G. F., and Siminovitch, L., 1972, Mammalian cells with altered forms of RNA polymerase II, *Proc. Natl. Acad. Sci. USA* **69:**3119–3123.

Chandler, V. L., Maler, B. A., and Yamamoto, K. R., 1983, DNA sequences bound specifically by glucocorticoid receptor *in vitro* render a heterologous promoter hormone responsive *in vivo*, *Cell* **33:**489–499.

Chang, E. H., Ellis, R. W., Scolnick, E. M., and Lowy, D. R., 1980, Transformation by cloned Harvey murine sarcoma virus DNA. Efficiency increased by long terminal repeat DNA, *Science* **210:**1249–1251.

Conrad, S. E., and Botchan, M. R., 1983, Isolation and characterization of human DNA fragments with nucleotide sequence homologies with the simian virus 40 regulatory region, *Mol. Cell. Biol.* **2:**949–965.

Corden, J., Wasylyk, B., Buchwalder, A., Sassone-Corsi, P., Kedinger, C., and Chambon, P., 1980, Promoter sequences of eukaryotic protein coding genes, *Science* **209:**1406–1414.

Dahmus, M. E., 1981a, Phosphorylation of eukaryotic DNA-dependent RNA polymerase, *J. Biol. Chem.* **256:**3332–3339.

Dahmus, M. E., 1981b, Calf thymus RNA polymerases I and II do not contain subunits structurally related to casein kinases I and II, *J. Biol. Chem.* **256:**11239–11243.

Dahmus, M. E., and Kedinger, C., 1983, Transcription of adenovirus-2 major late promoter inhibited by monoclonal antibody directed against RNA polymerase II_o and II_A, *J. Biol. Chem.* **258:**2303–2307.

Dandolo, L., Blangy, D., and Kamen, R., 1983, Regulation of polyoma virus transcription in murine embryonal carcinoma cells, *J. Virol.* **47:**55–64.

Darnell, J. E., 1982, Variety in the level of gene control in eukaryotic cells, *Nature* **297:**365–371.

Davison, B. L., Egly, J. M., Mulvihill, E. R., and Chambon, P., 1983, Formation of stable preinitiation complexes between eukaryotic class B transcription factors and promoter sequences, *Nature* **301:**680–686.

Dezelee, S., Wyers, F., Sentenac, A., and Fromageot, P., 1976, Two forms of RNA polymerase B in yeast: Proteolytic conversion *in vitro* of enzyme B_I into B_{II}, *Eur. J. Biochem.* **65:**543–552.

Dierks, P., van Ooyen, A., Cochran, M. D., Dobkin, C., Reiser, J., and Weissman, C., 1983, Three regions upstream from the cap site are required for efficient and accurate transcription of the rabbit β-globin gene in mouse 3T6 cells, *Cell* **32:**695–706.

Dynan, W. S., and Tjian, R., 1983, Isolation of transcription factors that discriminate between different promoters recognized by RNA polymerase II, *Cell* **32:**669–680.

Everett, R. D., Baty, D., and Chambon, P., 1983, The repeated GC-rich motifs upstream from the TATA box are important elements of the SV40 early promoter, *Nucl. Acids Res.* **11:**2447–2464.

Ferdinand, F. J., Brown, M., and Khoury, G., 1977, Synthesis and characterization of late lytic simian virus 40 RNA from transcriptional complexes, *Virology* **78:**150–161.

Fire, A., Samuels, M., and Sharp, P. A., 1984, Interactions between RNA polymerase II, factors, and template leading to accurate transcription, *J. Biol. Chem.* **259:**2509–2516.

Fraser, N. W., Nevins, J. R., Ziff, E., and Darnell, J. E., 1979*a*, The major late adenovirus type-2 transcription unit: Termination is downstream from the last poly(A) site, *J. Mol. Biol.* **129:**643–656.

Fraser, N. W., Sehgal, P. B., and Darnell, J. E., 1979*b*, Multiple discrete sites for premature RNA chain termination in adenovirus-2 infection: Enhancement by 5,6-dichloro-1-β-ribofuranosylbenzimidazole, *Proc. Natl. Acad. Sci. USA* **76:**2571–2575.

Fried, M., Griffiths, M., Davies, B., Bjursell, G., La Mantia, G., and Lania, L., 1983, Isolation of cellular DNA sequences that allow expression of adjacent genes, *Proc. Natl. Acad. Sci. USA* **80:**2117–2121.

Fromm, M., and Berg, P., 1982, Deletion mapping of DNA regions required for SV40 early region promoter function *in vivo, J. Mol. Appl. Gen.* **1:**457–481.

Fromm, M., and Berg, P., 1983*a*, Transcription *in vivo* from SV40 early promoter deletion mutants without repression by large T antigen, *J. Mol. Appl. Gen.* **2:**127–135.

Fromm, M., and Berg, P., 1983*b*, Simian virus 40 early- and late-region promoter functions are enhanced by the 72-base-pair repeat inserted at distant locations and inverted orientations, *Mol. Cell. Biol.* **3:**991–999.

Fujimura, F. K., Deininger, P. L., Friedman, T., and Linney, E., 1981, Mutation near the polyoma DNA replication origin permits productive infection of F9 embryonal carcinoma cells, *Cell* **23:**809–814.

Ghosh, P. K., Lebowitz, P., Frisque, R. J., and Gluzman, Y., 1981, Identification of a promoter component involved in positioning the 5′ termini of SV40 early RNAs, *Proc. Natl. Acad. Sci. USA* **78:**100–104.

Ghosh, P. K., Piatak, M., Mertz, J. E., Weissman, S. M., and Lebowitz, P., 1982, Altered utilization of splice sites and 5′ termini in late RNAs produced by leader region mutants of SV40, *J. Virol.* **44:**610–624.

Gillies, S. D., Morrison, S. L., Oi, V. T., and Tonegawa, S., 1983, A tissue-specific transcription enhancer element is located in the major intron of a rearranged immunoglobulin heavy chain gene, *Cell* **33:**717–728.

Grosschedl, R., and Birnstiel, M. L., 1980*a*, Identification of regulatory sequences in the prelude sequences of an H2A histone gene by the study of specific deletion mutants *in vivo, Proc. Natl. Acad. Sci. USA* **77:**1432–1436.

Grosschedl, R., and Birnstiel, M. L., 1980*b*, Spacer DNA sequences upstream of the TATAAATA sequence are essential for promotion of H2A histone gene transcription *in vivo, Proc. Natl. Acad. Sci. USA* **77:**7102–7106.

Grosschedl, R., and Birnstiel, M. L., 1982, Delimitation of far upstream sequences required for maximal *in vitro* transcription of an H2A histone gene, *Proc. Natl. Acad. Sci. USA* **79:**297–301.

Grosveld, G. C., Shewmaker, C. K., Jat, P., and Flavell, R. A., 1981, Localization of DNA sequences necessary for transcription of the rabbit β-globin gene *in vitro, Cell* **25:**215–226.

Grosveld, G. C., de Boer, E., Shewmaker, C. K., and Flavell, R. A., 1982*a*, DNA sequences necessary for transcription of the rabbit β-globin gene *in vivo, Nature* **295:**120–126.

Grosveld, G. C., Rosenthal, A., and Flavell, R. A., 1982b, Sequence requirements for the transcription of the rabbit β-globin gene *in vivo*: The −80 region, *Nucl. Acids. Res.* **10**:4951–4971.

Gruss, P., Dhar, R., and Khoury, G., 1981, Simian virus 40 tandem repeated sequences as an element of the early promoter, *Proc. Natl. Acad. Sci. USA* **78**:943–947.

Guialis, A., Beatty, B. G., Ingles, C. J., and Crerar, M. M., 1977, Regulation of RNA polymerase II activity in α-amanitin-resistant CHO hybrid cells, *Cell* **10**:53–60.

Guialis, A., Morrison, K. E., and Ingles, C. J., 1979, Regulated synthesis of RNA polymerase II polypeptides in Chinese hamster ovary cell lines, *J. Biol. Chem.* **254**:4171–4176.

Gupta, R. S., and Siminovitch, L., 1980, DRB resistance in Chinese hamster and human cells: Genetic and biochemical characteristics of the selection system, *Somat. Cell Genet.* **6**:151–170.

Hansen, U., and Sharp, P. A., 1983, Sequences controlling *in vitro* transcription of SV40 promoters, *EMBO J.* **2**:2293–2303.

Hartzell, S. W., Yamaguchi, J., and Subramanian, K. N., 1983, SV40 deletion mutants lacking the 21 bp repeated sequences are viable, but have noncomplementable deficiencies, *Nucl. Acids Res.* **11**:1601–1616.

Hearing, P., and Shenk, T., 1983a, The adenovirus type 5 EIA transcriptional control region contains a duplicated enhancer element, *Cell* **33**:695–703.

Hearing, P., and Shenk, T., 1983b, Functional analysis of the nucleotide sequence surrounding the cap site for adenovirus type 5 region EIA messenger RNAs, *J. Mol. Biol.* **167**:809–822.

Hen, R., Sassone-Corsi, P., Corden, J., Gaub, M. P., and Chambon, P., 1982, Sequences upstream from the TATA box are required *in vivo* and *in vitro* for efficient transcription from the Ad2 major late promoter, *Proc. Natl. Acad. Sci. USA* **79**:7132–7136.

Hodo, H. G., and Blatti, S. P., 1977, Purification using polyethyleneimine precipitation and low molecular weight subunit analyses of calf thymus and wheat germ DNA-dependent RNA polymerase II, *Biochemistry* **16**:2334–2343.

Hofer, E., Hofer-Warbinek, R., and Darnell, J. E., 1982, Globin RNA transcription: A possible termination site and demonstration of transcriptional control correlated with altered chromatin structure, *Cell* **29**:887–893.

Hu, S.-L., and Manley, J. L., 1981, DNA sequence required for initiation of transcription *in vitro* from the major late promoter of adenovirus 2, *Proc. Natl. Acad. Sci. USA* **78**:820–824.

Huet, J., Sentenac, A., and Fromageot, P., 1982, Spot-immunodeletion of conserved determinants in eukaryotic RNA polymerases, *J. Biol. Chem.* **257**:2613–2618.

Hynes, N., van Ooyen, A. J. J., Kennedy, N., Herrlich, P., Ponta, H., and Groner, B., 1983, Subfragments of the large terminal repeat cause glucocorticoid-responsive expression of mouse mammary tumor virus and of an adjacent gene, *Proc. Natl. Acad. Sci. USA* **80**:3637–3641.

Ingles, C. J., 1978, Temperature sensitive RNA polymerase II mutations in Chinese hamster ovary cells, *Proc. Natl. Acad. Sci. USA* **75**:405–409.

Ingles, C. J., and Shales, M., 1982, DNA-mediated transfer of an RNA polymerase II gene: Reversion of the temperature-sensitive hamster cell cycle mutant TSAF8 by mammalian DNA, *Mol. Cell. Biol.* **2**:666–673.

Jat, P., Novak, U., Cowie, A., Tyndall, C., and Kamen, R., 1982, DNA sequences required for specific and efficient initiation of transcription at the polyoma virus early promoter, *Mol. Cell. Biol.* **2**:737–751.

Jolly, D. J., Esty, A. C., Subramani, S., Friedmann, T., and Verma, I. M., 1983, Elements in the long terminal repeat of murine retroviruses enhance stable transformation by thymidine kinase gene, *Nucl. Acids Res.* **11**:1855–1872.

Kadesch, T. R., and Chamberlin, M. J., 1982, Studies of *in vitro* transcription by calf thymus RNA polymerase II using a novel duplex DNA template, *J. Biol. Chem.* **257**:5286–5295.

Katinka, M., Yaniv, M., Vasseur, M., and Blangy, D., 1980, Expression of polyoma early functions in mouse embryonal carcinoma cells depends on sequence rearrangements in the beginning of the late region, *Cell* **20**:393–399.

Katinka, M., Vasseur, M., Montreau, N., Yaniv, M., and Blangy, D., 1981, Polyoma DNA sequences involved in control of viral gene expression in murine embryonal carcinoma cells, *Nature* **290**:720–722.

Kessel, M., and Khoury, G., 1983, Induction of cloned genes after transfer into eukaryotic cells, in: *Expression of Cloned Genes in Prokaryotic and Eukaryotic Cells*, Vol. 3, (T. S. Papas, M. Rosenberg, and J. G. Chirikjian, eds.), pp. 233–260, Elsevier, New York.

Khoury, G., and Gruss, P., 1983, Enhancer elements, *Cell* **33**:313–314.

Kramer, A., and Bautz, E. K. F., 1981, Immunological relatedness of subunits of RNA polymerase II from insects and mammals, *Eur. J. Biochem.* **117**:449–455.

Kriegler, M., and Botchan, M., 1983, Enhanced transformation by a simian virus 40 recombinant virus containing a Harvey murine sarcoma virus long terminal repeat, *Mol. Cell. Biol.* **3**:325–339.

Laimins, L. A., Khoury, G., Gorman, C., Howard, G., and Gruss, P., 1982, Host-specific activation of transcription by tandem repeats from simian virus 40 and Moloney murine sarcoma virus, *Proc. Natl. Acad. Sci. USA* **79**:6453–6457.

Lebowitz, P., and Ghosh, P. K., 1982, Initiation and regulation of SV40 early transcription *in vitro*, *J. Virol.* **41**:449–461.

Lee, D. C., Roeder, R. G., and Wold, W. S. M., 1982, DNA sequences affecting specific initiation of transcription *in vitro* from the EIII promoter of Ad2, *Proc. Natl. Acad. Sci. USA* **79**:41–45.

Levinson, B., Khoury, G., Van de Woude, G., and Gruss, P., 1982, Activation of SV40 genome by 72-base pair tandem repeats of Moloney sarcoma virus, *Nature* **295**:568–572.

Lobban, P. E., Siminovitch, L., and Ingles, C. J., 1976, The RNA polymerase II of an α-amanitin resistant Chinese hamster ovary cell line, *Cell* **8**:65–70.

Luciw, P. A., Bishop, J. M., Varmus, H. E., and Capecchi, M. R., 1983, Location and function of retroviral and SV40 sequences that enhance biochemical transformation after microinjection of DNA, *Cell* **33**:705–716.

Lusky, M., Berg, L., Weiher, H., and Botchan, M., 1983, Bovine papilloma virus contains an activator of gene expression at the distal end of the early transcription unit, *Mol. Cell. Biol.* **3**:1108–1122.

Majors, J., and Varmus, H. E., 1983, A small region of the mouse mammary tumor virus long terminal repeat confers glucocorticoid hormone regulation on a linked heterologous gene, *Proc. Natl. Acad. Sci. USA* **80**:5866–5870.

Manley, J. L., 1983, Analysis of the expression of genes encoding animal mRNA by *in vitro* techniques, *Prog. Nucl. Acid Res. Mol. Biol.* **30**, in press.

Manley, J. L., Fire, A., Cano, A., Sharp, P. A., and Gefter, M. L., 1980, DNA-dependent transcription of adenovirus genes in a soluble whole-cell extract, *Proc. Natl. Acad. Sci. USA* **77**:3855–3859.

Maroteaux, L., Kahana, C., Mory, Y., Groner, Y., and Revel, M., 1983, Sequences involved in the regulated expression of the human interferon-β_1 gene in recombinant SV40 DNA vectors replicating in monkey cells, *EMBO J.* **2**:325–332.

Mathis, D. J., and Chambon, P., 1981, The SV40 early region TATA box is required for accurate *in vitro* initiation of transcription, *Nature* **290**:310–315.

Matsui, T., Segall, J., Weil, P. A., and Roeder, R. G., 1980, Multiple factors required for accurate initiation of transcription by purified RNA polymerase II, *J. Biol. Chem.* **225**:11992–11996.

McKnight, S. L., 1982, Functional relationships between transcriptional control signals of the thymidine kinase gene of herpes simplex virus, *Cell* **31**:355–365.

McKnight, S. L., and Kingsbury, R., 1982, Transcriptional control signals of a eukaryotic protein-coding gene, *Science* **217**:316–324.

Mellon, P., Parker, V., Gluzman, Y., and Maniatis, T., 1981, Identification of DNA sequences required for transcription of the human α1-globin gene in a new SV40 host-vector system, *Cell* **27**:279–288.

Mercola, M., Wang, X.-F., Olsen, J., and Calame, K., 1983, Transcriptional enhancer elements in the mouse immunoglobulin heavy chain locus, *Science* **221**:663–665.

Mirault, M.-E., Southgate, R., and Delwart, E., 1982, Regulation of heat-shock genes: A DNA sequence upstream of *Drosophilia* hsp70 genes is essential for their induction in monkey cells, *EMBO J.* **1**:1279–1285.

Mitsialis, S. A., Manley, J. L., and Guntaka, R. V., 1983, Localization of active promoters for eukaryotic RNA polymerase II in the long terminal repeat of avian sarcoma virus DNA, *Mol. Cell. Biol.* **3**:811–818.

Mittleman, B., Zandomeni, R., and Weinmann, R., 1983, Mechanism of action of 5,6-Dichloro-1-β-D-ribofuranosylbenzimidazole II: A resistant human cell mutant with an altered transcriptional machinery, *J. Mol. Biol.* **165**:461–473.

Moreau, P., Hen, R., Wasylyk, B., Everett, R., Gaub, M. P., and Chambon, P., 1981, The SV40 72 base pair repeat has a striking effect on gene expression both in SV40 and other chimeric recombinants, *Nucl. Acids Res.* **9**:6047–6068.

Myers, R. M., Rio, D. C., Robbins, A. K., and Tjian, R., 1981, SV40 gene expression is modulated by the cooperative binding of T antigen to DNA, *Cell* **25**:373–384.

Neuberger, M. S., 1983, Expression and regulation of immunoglobulin heavy chain gene transfected into lymphoid cells, *EMBO J.* **2**:1373–1378.

Nordheim, A., and Rich, A., 1983, Negatively supercoiled simian virus 40 DNA contains Z-DNA segments within transcriptional enhancer sequences, *Nature* **303**:674–679.

Ohno, S., and Taniguchi, T., 1983, The 5′-flanking sequence of human interferon-β_1 gene is responsible for viral induction of transcription, *Nucl. Acids Res.* **11**:5403–5412.

Pelham, H. R. B., 1982, A regulatory upstream promoter element in the *Drosophila* Hsp70 heat-shock gene, *Cell* **30**:517–528.

Pelham, H. R. B., and Bienz, M., 1982, A synthetic heat shock promoter element confers heat-inducibility on the herpes simplex virus thymidine kinase gene, *EMBO J.* **1**:1473–1477.

Queen, C., and Baltimore, D., 1983, Immunoglobulin gene transcription is activated by downstream sequence elements, *Cell* **33**:741–748.

Ragg, H., and Weissman, C., 1983, Not more than 117 base pairs of 5' flanking sequence are required for inducible expression of a human IFN-α gene, *Nature* **303**:439–442.

Razin, A., and Riggs, A. D., 1980, DNA methylation and gene function, *Science* **210**:604–610.

Roeder, R. G., 1976, Eukaryotic nuclear RNA polymerases, in: *RNA Polymerase* (R. Losick and M. Chamberlin, eds.), pp. 285–329, Cold Spring Harbor Laboratory, Cold Spring Harbor, New York.

Ruley, H. E., and Fried, M., 1983, Sequence repeats in a polyoma virus DNA region important for gene expression, *J. Virol.* **47**:233–237.

Russell, P. R., 1983, Evolutionary divergence of the mRNA transcription initiation mechanism in yeast, *Nature* **301**:167–169.

Samuels, M., Fire, A., and Sharp, P. A., 1982, Separation and characterization of factors mediating accurate transcription by RNA polymerase II, *J. Biol. Chem.* **257**:14419–14427.

Samuels, M., Fire, A., and Sharp, P. A., 1984, Dinucleotide priming of transcription mediated by RNA polymerase II, *J. Biol. Chem.* **259**:2517–2525.

Scheidereit, C., Geisse, S., Westphal, H. M., and Beato, M., 1983, The glucocorticoid receptor binds to defined nucleotide sequences near the promoter of mouse mammary tumour virus, *Nature* **304**:749–752.

Schwartz, L. B., and Roeder, R. G., 1975, Purification and subunit structure of deoxyribonucleic acid-dependent ribonucleic acid polymerase II from the mouse plasma cytoma, MOPC315, *J. Biol. Chem.* **250**:3221–3228.

Searles, L. L., Jokerst, R. S., Bingham, P. M., Voelker, R. A., and Greenleaf, A. L., 1982, Molecular cloning of sequences from a Drosophila RNA polymerase II locus by P element transposon tagging, *Cell* **31**:585–592.

Sekikawa, K., and Levine, A. J., 1981, Isolation and characterization of polyoma host range mutants that replicate in nullipotential embryonal carcinoma cells, *Proc. Natl. Acad. Sci. USA* **78**:1100–1104.

Sentenac, A., and Hall, B., 1982, Yeast nuclear RNA polymerases and their role in transcription, in: *The Molecular Biology of the Yeast Saccharomyces: Metabolism and Gene Expression* (J. N. Strathern, E. W. Jones, and J. R. Broach, eds), pp. 561–606, Cold Spring Harbor Laboratory, Cold Spring Harbor, New York.

Shenk, T., 1981, Transcriptional control regions: Nucleotide sequence requirement for initiation by RNA polymerase II and III, *Curr. Top. Microbiol. Immunol.* **93**:25–46.

Spandidos, D. A., and Wilkie, N. M., 1983, Host specificities of papillomavirus, Moloney murine sarcoma virus, and SV40 enhancer sequences, *EMBO J.* **2**:1193–1199.

Stetler, D. A., and Rose, K. M., 1982, Phosphorylation of deoxyribonucleic acid dependent RNA polymerase II by nuclear protein kinase NII: Mechanism of enhanced ribonucleic acid synthesis, *Biochemistry* **21**:3721–3728.

Talkington, C. A., and Leder, P., 1982, Rescuing the *in vitro* function of a globin pseudogene promoter, *Nature* **298**:192–195.

Tanaka, K., Chowdhury, K., Chang, K. S. S., Israel, M., and Ito, Y., 1982, Isolation and characterization of polyoma virus mutants which grow in murine embryonal carcinoma and trophoblast cells, *EMBO J.* **1**:1521–1527.

Treisman, R., Orkin, S. H., and Maniatis, T., 1983, Specific transcription and RNA splicing defects in five cloned β-thalassaemia genes, *Nature* **302**:591–596.

Tsai, S. Y., Tsai, M., Kops, L. E., Minghetti, P. P., and O'Malley, B. W., 1981*a*, Transcription factors from oviduct and HeLa cells are similar, *J. Biol. Chem.* **256**:13055–13059.

Tsai, S. Y., Tsai, M.-J., and O'Malley, B. W., 1981*b*, Specific 5'-flanking sequences are required for faithful initiation of *in vitro* transcription of the ovalbumin gene, *Proc. Natl. Acad. Sci. USA* **78**:879–883.

Tsuda, M., and Suzuki, Y., 1981, Faithful transcription initiation of fibroin gene in a homologous cell-free system reveals an enhancing effect of 5'-flanking sequence far upstream, *Cell* **27**:175–182.

Tsujimoto, Y., Hirose, S., Tsuda, M., and Suzuki, Y., 1981, Promoter sequence of fibroin gene assigned by *in vitro* transcription system, *Proc. Natl. Acad. Sci. USA* **78**:4838–4842.

Tyndall, C., La Mantia, G., Thacker, C. M., Favalero, J., and Kamen, R., 1981, A region of the polyoma virus genome between the replication origin and late protein coding sequences is required in *cis* for both early gene expression and viral DNA replication, *Nucl. Acids Res.* **9**:6231–6250.

de Villiers, J., and Schaffner, W., 1981, A small segment of polyoma virus DNA enhances the expression of a cloned β-globin gene over a distance of 1400 base pairs, *Nucl. Acids Res.* **9**:6251–6264.

de Villiers, J., Olson, L., Tyndall, C., and Schaffner, W., 1982, Transcriptional enhancers from SV40 and polyoma virus show a cell type preference, *Nucl. Acids Res.* **10**:7965–7976.

Wasylyk, B., Wasylyk, C., Augereau, P., and Chambon, P., 1983, The SV40 72 bp repeat preferentially potentiates transcription starting from proximal natural or substitute promoter elements, *Cell* **32**:503–514.

Weiher, H., König, M., and Gruss, P., 1983, Multiple point mutations affecting the SV40 enhancer, *Science* **219**:626–631.

Weil, P. A., Luse, D. S., Segall, J., and Roeder, R. G., 1979, Selective and accurate initiation of transcription at the Ad2 major late promoter in a soluble system dependent on purified RNA polymerase II and DNA, *Cell* **18**:469–484.

Weisbrod, S., 1982, Active chromatin, *Nature* **297**:289–295.

Young, R. A., and Davis, R. W., 1983, Efficient isolation of genes using antibody probes, *Proc. Natl. Acad. Sci. USA* **80**:1194–1198.

Zandomeni, R., Mittleman, B., Bunick, D., Ackerman, S., and Weinmann, R., 1982, Mechanism of action of dichloro-β-D-ribofuranosylbenzimidazole: Effect on *in vitro* transcription, *Proc. Natl. Acad. Sci. USA* **79**:3167–3170.

Zandomeni, R., Bunick, D., Ackerman, S., Mittleman, B., and Weinmann, R., 1983, Mechanism of action of DRB III: Effect on specific *in vitro* initiation of transcription, *J. Mol. Biol.* **167**:561–574.

Zinn, K., DiMaio, D., and Maniatis, T., 1983, Identification of two distinct regulatory regions adjacent to the human β-interferon gene, *Cell* **34**:865–879.

CHAPTER 3

Regulation of Eukaryotic Translation

Raymond Kaempfer

Department of Molecular Virology
The Hebrew University-Hadassah Medical School
91010 Jerusalem, Israel

1. INTRODUCTION

There is no doubt about the importance of transcriptional control for eukaryotic gene expression. Modern approaches of reversed genetics, involving analysis of the expression of eukaryotic gene sequences contained in plasmid vectors upon their introduction into eukaryotic cells, have provided a powerful and convenient tool to dissect this manner of control. Because the experimental design of studies of translational control is less straightforward, today, considerably less is understood about its mechanisms. Yet, there is increasing evidence that translational control mechanisms strongly influence the final level of expression of specific genes. Examples of this type of control are encountered in normal growth, in cell differentiation, and in virus infection. Moreover, in a variety of physiological stress conditions, it is the translation process that is most immediately affected. Much valuable information on the subject is summarized in Perez-Bercoff (1982) and in reviews by Austin and Kay (1982) and Maitra *et al.* (1982). Earlier reviews by Lodish (1976), Revel and Groner (1978), and Ochoa and de Haro (1979) remain useful.

Recently, the nature of certain key proteins involved in translational control and the regulation of their activity has been clarified considerably. Moreover, specific interactions between mRNA and proteins of the translational machinery have emerged that may provide a molecular basis for what is perhaps the most important question in translational control, that is, why the translation yield of individual mRNA species can differ as much as 100-fold.

2. THE RATE-LIMITING STEP IN TRANSLATION

It is generally assumed that elongation and termination of all polypeptide chains occur at the same rate in intact cells, and this point has been documented directly for the α- and β-globins and for non-globin chains synthesized in reticulocytes (Lodish and Jacobsen, 1972; Hunt, 1974; Palmiter, 1972, 1974). The translation of viral mRNA, particularly the formation of long polyproteins, for example, on picornavirus RNA templates (Summers and Maizel, 1968; Jacobson and Baltimore, 1968), can in principle be limited by the availability of certain infrequent tRNA species in the cell. Thus, in reticulocyte lysates, which contain a tRNA spectrum optimal for globin mRNA translation, the synthesis of mengovirus proteins is rate limited at elongation, but when mouse liver tRNA is provided, initiation becomes the rate-limiting step (Rosen *et al.*, 1981*a*). In whole cells, initiation is generally considered to be the rate-limiting step in translation. This follows from the observation that slowing down of the elongation rate with low concentrations of inhibitors of polypeptide bond formation, such as cycloheximide, causes a shift towards greater packing density of ribosomes on mRNA. Lodish (1971) used this approach to show that, in reticulocyte lysates synthesizing protein at virtually the same rate as the intact cell (Jackson, 1982), synthesis of α- and β-globin is indeed rate limited at initiation.

It makes good sense that initiation is the rate-limiting step in mRNA translation. Once synthesis of a protein chain has been initiated, the remainder of its synthesis follows more or less automatically. The crucial step for control, then, is that involving the attachment of an mRNA molecule to a 40 S ribosomal subunit. Indeed, most cases of translational control concern this step. In general, once mRNA has entered a 40 S initiation complex, formation of the complete encoded polypeptide chain is virtually assured. Hence, this review will concentrate on the properties of mRNA and the translation components involved in the events leading up to this complex.

3. FUNCTIONAL DOMAINS IN MESSENGER RNA

3.1. AUG Initiation Codon

Although AUG is the initiation codon normally used in bacteria, studies of binding of fMet-tRNA$_f$ to ribosomes (Clarck and Marcker, 1966) or translation of synthetic polynucleotides (Thach *et al.*, 1966) have revealed that GUG and, to a lesser extent, UUG are also functional initiation codons. A single case of initiation at an AUU codon is also known for the *Escherichia coli* initiation factor IF-3 protein (Sacerdot *et al.*, 1982). By contrast, a compilation of over 200 ribosome binding site sequences in eukaryotic mRNA (Kozak, 1981*a*, 1983) has yielded only AUG as initiation codon. It appears, therefore, that initiation of eukaryotic translation occurs exclusively at AUG codons.

Usually, but not always, the functional AUG initiation codon is the first AUG codon encountered when reading the mRNA from its 5' end. This was first pointed out by Kozak (1978, 1981*a*) who could find, among 200 mRNA sequences, only 18 that contained AUG triplets upstream from the functional initiation site (Kozak, 1983). This led to the proposal that 40 S ribosomal subunits may scan the mRNA from its 5' end and simply initiate at the first AUG codon they encounter (Kozak, 1978). In poliovirus RNA, however, no fewer than seven AUG codons precede the 5'-proximal initiation site for translation (Kitamura *et al.*, 1981). To account for the exceptions, Kozak (1981*b*) has proposed that sequences flanking the functional AUG codon may be important; a purine, usually A, frequently occurs three residues before the AUG codon, while a purine, usually G, often follows the AUG sequence. Either of these features is lacking in some, but not all, of the upstream AUG codons in the exceptional mRNA species.

As a rule, eukaryotic mRNA contains only a single functional initiation site for translation (Jacobson and Baltimore, 1968). This appears to be true for cellular mRNA and for many viral mRNA species. In the late 16 S mRNA of SV40, however, a 62-amino acid polypeptide, the agnoprotein, is encoded upstream from the capsid protein VP1 (Jay *et al.*, 1981). Since this mRNA template expresses both proteins, late 16 S mRNA of SV40 is a true polycistronic mRNA containing two independent initiation sites. So far, it is the only unequivocal exception to the observation that eukaryotic mRNA is monocistronic, although evidence in support of the existence of two initiation sites for translation has also been provided for poliovirus RNA and mengovirus RNA (see Section 5.1.3). Moreover, in the genomic RNA of Semliki Forest virus (Wengler *et al.*, 1979; Lehtovaara *et al.*, 1982) and in porcine gastrin mRNA (Yoo *et al.*, 1982), AUG codons occur immediately following the cap structure, and are followed by in-phase termination codons that precede the AUG of the major protein encoded, leaving open the possibility that here, too, ribosomes may initiate at successive functional sites.

A number of viral mRNA species exhibit a polycistronic nature, that is, one nucleotide sequence needed to encode a complete viral polypeptide chain is followed by one or more such sequences. The genomic RNA of Semliki Forest virus (Glanville *et al.*, 1976), Sindbis virus (Collins *et al.*, 1982; Cancedda *et al.*, 1975), Rous sarcoma virus (Weiss *et al.*, 1977), tobacco mosaic virus (Hunter *et al.*, 1976), as well as brome mosaic virus (BMV) RNA-3 (Shih and Kaesberg, 1973), polyoma virus late 19 S mRNA (Smith *et al.*, 1976) and a number of adenovirus late mRNAs (Anderson *et al.*, 1974) possess such a structure. Yet, in all these cases, the polypeptide chain expressed is that starting at the 5′-proximal AUG initiation codon, while the second one remains silent. This second initiation codon becomes active only in shorter derivatives of these RNA species, in which it is now in a 5′-proximal location. Although this finding tended to support the rule of initiation at the first AUG codon (Kozak, 1978), it remained equally possible that the second AUG initiation codon is shielded by secondary and tertiary structure of the intact polycistronic RNA molecule, but becomes accessible to ribosomes in the novel RNA conformation of shorter derivatives. The observation that extensive denaturation of reovirus mRNA (Kozak, 1980*a*) or BMV RNA-3 (Zagorska *et al.*, 1982) does not lead to the activation of additional initiation sites revealed, however, that the complete removal of secondary structure in mRNA is not sufficient to create new initiation sites. Nevertheless, it remains entirely possible that, in addition to location, a certain degree of RNA secondary structure is an important parameter in determining if an AUG codon will function in initiation.

This concept is supported by the findings that two functional, overlapping reading frames occur in influenza B virus RNA 6 (Shaw *et al.*, 1983) and in an mRNA of human adenovirus (Bos *et al.*, 1981). In both cases, two different polypeptide chains are encoded, starting at different AUG triplets.

Finally, the issue of *where* initiation occurs in mRNA should be clearly separated from the question, *how often* initiation takes place.

Clearly, this latter point is far more complex, for two mRNA species can vary widely in translation initiation efficiency, even if initiation occurs at the first AUG codon in both.

3.2. The 5′-Terminal Cap Structure

The 5′-terminal cap structure consists of a 7-methyl-guanosine moiety linked via a 5′-5′ triphosphate bridge to the first coded nucleotide in mRNA (Shatkin, 1976, 1982). All cellular mRNAs that are active in protein synthesis carry a cap structure of the general sequence m^7 Gppp$N$pN', where N is either A or G and N' is often 2-O-methylated (Breathnach and Chambon, 1981). The presence of the cap is of dual functional significance. It enhances both the stability of mRNA (Furuichi et al., 1977) and the binding of mRNA to 40 S ribosomal subunits and, hence, initiation of translation (Shatkin, 1976, 1982; Banerjee, 1980). Removal of the m^7G moiety (Lodish and Rose, 1977; Wodnar-Filipowicz et al., 1978) or introduction of cap analogues containing the minimal structure m^7Gp . . . (Hickey et al., 1976; Weber et al., 1978) cause a decrease in translation activity, the extent of which is characteristic for the mRNA under study. This shows that individual mRNA species differ in their dependence on the cap for initiation of translation. Indeed, the existence of viral RNA species that are not capped yet are often translated even more effectively than (capped) cellular mRNA, shows clearly that the function of the cap structure is not absolutely required for translation. Usually, the cap acts as an enhancer of mRNA binding to ribosomes, in addition to other features in the RNA molecule. Apparently, these other features can surpass the contribution of the cap, for the uncapped RNAs of satellite tobacco necrosis virus (STNV; Wimmer et al., 1968) and picornaviruses (see Fellner, 1979) have been shown to out-compete capped cellular and viral mRNAs during translation (Herson et al., 1979; Lawrence and Thach, 1974; Rosen et al., 1982) and are among the most efficiently initiating mRNA species known (Kaempfer et al., 1984).

3.3. 5′-Leader Sequence

The 5′ leader in eukaryotic mRNA is defined as the sequence between the cap and the AUG initiation codon. The most striking properties of the leader are its lack of uniform features in terms of either length or sequence. Thus, leaders range from three nucleotides for an immunoglobulin mRNA (Kelley *et al.*, 1982) to 742 nucleotides for poliovirus RNA (Kitamura *et al.*, 1981), with most cellular mRNAs carrying a leader of about 50 nucleotides.

The length of the leader sequence does not seem to affect the efficiency of translation of mRNA in any simple manner. Thus, efficiently translated mRNA species may possess a short leader, such as ten nucleotides for vesicular stomatitis virus (VSV) NS or L protein mRNAs (Rose, 1980), 29 for STNV RNA (Ysebaert *et al.*, 1980), or a long leader comprising hundreds of nucleotides, as for foot-and-mouth disease virus RNA (FMDV; Sangar *et al.*, 1980) or mengovirus RNA (Perez-Bercoff and Kaempfer, 1982). On the other hand, the RNA of another picornavirus with along leader, poliovirus, is far less efficient as a template for translation (Daniels-McQueen *et al.*, 1983) than are the RNAs from the cardioviruses: FMDV, encephalomyocarditis (EMC) virus, and mengovirus. Although it could be argued that a long leader sequence may give rise to secondary structure forms that could impede initiation (Kozak, 1983), there is no evidence to support this contention and, indeed, EMC and mengovirus RNA are among the most efficiently translated mRNAs described to date. Rabbit α-globin mRNA is a less efficient template for initiation than is β-globin mRNA (Lodish, 1971); their leader sequences have a length of 36 and 53 nucleotides, respectively (Heindell *et al.*, 1978; Efstratiadis *et al.*, 1977).

The sequences of the leader in eukaryotic mRNA are highly divergent and notably free of identifiable features such as the Shine–Dalgarno sequence in prokaryotes (Shine and Dalgarno, 1974). Indeed, the CCUCCU sequence located close to the 3′ end of bacterial small subunit ribosomal RNA (rRNA) is deleted in the eukaryotic rRNA equivalent, yet sequences just preceding that portion of the rRNA molecule show definite conservation between various prokaryotes and eukaryotes (Kozak, 1983). In this context, it is worth pointing out that eukaryotic mRNA species as a rule cannot be translated in prokaryotic cell-free systems, where their initiation sites are not recognized. An interesting exception is provided by satellite tobacco necrosis virus (STNV) RNA, which acts as a highly efficient template in both prokaryotic and eukaryotic cell-free systems (Klein *et al.*, 1972; Hickey *et al.*, 1976; Herson *et al.*, 1979). The 5′ end of STNV RNA, shown in Fig. 1, is seen to contain in positions 10–13 an AGGA sequence that, in principle, may base-pair with the 3′ terminus of bacterial 16 S rRNA, even though it is not located within a distance of ten nucleotides preceding the AUG codon, as is usually seen in prolaryotic mRNA. Only two other eukaryotic mRNA species are known to be translated correctly in extracts of *E. coli*. One of these is tobacco mosaic virus RNA, whose coat protein cistron is recognized (Glover and Wilson, 1982). Here, too, a potential Shine–Dalgarno sequence occurs in the region just preceding the AUG initiation codon. The other is alfalfa mosaic virus (AMV) RNA which lacks a potential Shine–Dalgarno sequence and is recognized by *E. coli* ribosomes at two sites (Castel *et al.*, 1979).

Although a eukaryotic equivalent of the Shine—Dalgarno interaction has been proposed (Hagenbüchle *et al.*, 1978), the required complementary sequence is not conserved in eukaryotic mRNA at a statistically significant level (De Wachter, 1979). On the other hand, Sargan *et al.* (1982) noticed that some form of the consensus sequence AUCCACC immediately precedes the AUG initiation codon in many cellular mRNA sequences. That sequence could, in principle, base-pair with an exactly complementary sequence formed from the nucleotides that flank a highly conserved hairpin loop located at positions 10–31 from the 3' terminus of 18 S rRNA. Many viral mRNAs, however, do not fit this consensus sequence, indicating that the postulated interaction, if it exists, can be compensated by other mRNA features. Evidence is needed to demonstrate that the mRNA–rRNA interaction proposed by Sargan *et al.* actually occurs.

Another common sequence of the leader of eukaryotic mRNA is CUUPyUG (Baralle and Brownlee, 1978), found in β-like globin mRNA at seven nucleotides downstream from the cap (Efstratiadis *et al.*, 1980). Conserved sequences close to the cap may, however, form part of the promoter for transcription (Talkington and Leder, 1982).

3.4. 3'-Terminal Poly (A)

Cellular mRNA molecules carry a poly (A) tail of between 30–200 residues in length (for review, see Brawerman, 1981). In intact mRNA, the poly (A) sequence is masked by proteins (Bergmann and Brawerman, 1977). There is no absolute requirement for poly (A), for mRNA species exist that lack it, such as sea urchin histone mRNA (Grunstein and Schedl, 1976). Moreover, globin mRNA from which the poly (A) is removed is as active a template for *in vitro* translation as is native globin mRNA (Soreq *et al.*, 1974; Sippel *et al.*, 1974).

Upon microinjection into *Xenopus laevis* oocytes, native globin mRNA molecules remain stable and direct the synthesis of globin chains; by contrast, little synthesis occurs when poly (A)-free globin mRNA is injected, and this RNA is rapidly degraded (Nudel *et al.*, 1976). mRNA molecules carrying 30 A residues at the 3′ end are as stable as native molecules, while those carrying only 16 A residues are as unstable as poly (A)-free mRNA. This finding suggests that a critical length of poly (A) is needed to protect the mRNA against degradation, possibly by allowing interaction of the 3′ end with specific protein molecules. Because poly (A) becomes progressively shorter during the lifetime of an mRNA molecule (Brawerman, 1981), it is attractive to assume that the poly (A) tail promotes the stability of an mRNA molecule, thereby allowing it to complete more rounds of translation. Beyond this effect on half-life, there are no convincing indications as yet that poly (A) is directly involved in any aspect of translational control.

3.5. 3′-Untranslated Sequence

Usually, the 3′-untranslated region in eukaryotic mRNA is from 50–150 nucleotides in length, but in STNV RNA it contains 622 nucleotides (Ysebaert *et al.*, 1980) and in β-neo-endorphin mRNA over 1000 nucleotides (Kakidani *et al.*, 1982). Not only are the sequences in the 3′-untranslated region highly variable, but the mRNA of a given gene may display size heterogeneity in the 3′ end. Thus, mRNA encoding dihydrofolate reductase in mouse cells is heterodisperse, with 3′-untranslated regions ranging from 80 to 900 nucleotides, yet these mRNA forms are all active templates for translation *in vitro* (Setzer *et al.*, 1980). Removal of most of the 3′-untranslated region from globin mRNA (Kronenberg *et al.*, 1979; Kaempfer *et al.*, 1979a) or from β-interferon mRNA (Soreq *et al.*, 1981) causes no loss in the ability of these mRNAs to direct the synthesis of authentic proteins. Nevertheless, the 3′-noncoding sequences must have a role, for even though they are less conserved than coding sequences (see Kozak, 1983), they evolve more slowly than nonfunctional sequences in DNA (Miyata *et al.*, 1980; Martin *et al.*, 1981). One obvious role for the 3′-untranslated region is to serve as acceptor for poly (A) addition, and hence to be indirectly involved in control of mRNA stability. Beyond that, no convincing role for this mRNA segment in translational control has yet been revealed.

4. INITIATION OF TRANSLATION

4.1. Dissociation of Ribosomes into Subunits

Three forms of ribosomal particles are generally observed in extracts of both prokaryotic and eukaryotic cells: polysomes, single ribosomes, and ribosomal subunits. Heavy isotope transfer experiments have shown that in intact bacteria (Kaempfer *et al.*, 1968) as well as in yeast (Kaempfer, 1969), ribosomes undergo continual subunit exchange. This exchange is tightly coupled to protein synthesis: a ribosome dissociates into its two subunits after every round of translation on mRNA (Kaempfer, 1968; for review, see Kaempfer, 1974*a*). These results imply that subunits, not ribosomes, initiate translation (Guthrie and Nomura, 1968). In mammalian cell-free systems, exchange of subunits in ribosomes during protein synthesis was shown by Falvey and Staehelin (1970) and Howard *et al.* (1970).

Because the two ribosomal subunits have a very high tendency to associate, forming single ribosomes that do not participate in protein synthesis (Kaempfer, 1970), it is necessary to stabilize the pair of ribosomal subunits generated at termination of translation, before another round of initiation can take place. In bacteria, this task is fulfilled by IF-3, a small (21,000 M_r) protein that binds with high affinity to the 30 S ribosomal subunit (Sabol and Ochoa, 1971) and prevents its association with 50 S subunits (Kaempfer, 1971, 1972). The situation in eukaryotic translation is less clear at present. A complex initiation factor, eIF-3, containing numerous (about nine) components, sedimenting at about 16 S, and possessing a mass of about 500,000 daltons (Schreier *et al.*, 1977; Benne and Hershey, 1976; Hershey, 1982*a*) binds stoichiometrically to the 40 S subunit, shifting the equilibrium between single ribosomes and subunits in favor of the latter. In the presence of 60 S subunits, eIF-3 permits the formation of initiation complexes that is otherwise inhibited by single ribosome formation (Trachsel *et al.*, 1977; Benne and Hershey, 1976; Trachsel and Staehelin, 1979). Two forms of eIF-3, one only 51,000 M_r in molecular weight, have been reported by Jones *et al.* (1980), but the significance of their findings is not yet clear. One problem complicating the interpretation of results with eIF-3 is the observation that, even though eIF-3 migrates as a single, homogeneous component in gels, density gradients, or ion exchange columns (Benne and Hershey, 1976), many of its component polypeptides are related in primary structure, making it possible that, either in the cell or during isolation, multiple forms are generated by proteolytic digestion (Meyer *et al.*, 1981). The actual structure of eIF-3 may, therefore, turn out to be simpler than first thought.

Another initiation factor, eIF-4C (17,000 M_r) also binds to 40 S subunits and somewhat reduces their association with 60 S subunits (Thomas *et al.*, 1980*a*).

eIF-3 of comparable activity has been isolated from mammalian sources, such as reticulocytes, liver, or ascites cells (Trachsel *et al.*, 1979). Curiously, wheat germ eIF-3 does not stabilize 40 S subunits nor prevent their entry into single ribosomes (Checkley *et al.*, 1981). Instead, wheat germ contains a 23,000 M_r protein, called eIF-6, that binds to 60 S subunits and prevents their association with 40 S subunits (Russell and Spremulli, 1979, 1980). A similar activity has been isolated from liver by Valenzuela *et al.* (1982). Since the 60 S subunit must join the 40 S subunit, once the latter has bound an mRNA molecule, in order to form a functional ribosome, any factor that acts on the 60 S subunit must be released at this point, but not earlier. In the case of eIF-3, this factor remains on the 40 S subunit until after mRNA has been bound, being released just before or concomitant with the junction of a 60 S subunit (Benne and Hershey, 1978; Hershey, 1982*a*). While joining of mRNA to the 40 S subunit could make eIF-3 susceptible to displacement by a 60 S subunit, in the case of eIF-6, one would have to postulate that a 40 S subunit carrying mRNA, but not one lacking mRNA, can displace eIF-6 from the 60 S subunit.

The dissociation step in the ribosome cycle and the resulting need to stabilize the generated pair of ribosomal subunits against entry into a side-track pool of inactive single ribosomes have important consequences for translational control. Because the number of IF-3 or eIF-3 molecules is an order of magnitude smaller than the number of ribosomes in a cell, the size of the pool of stable ribosomal subunits that is capable of initiation must be relatively small and, indeed, is remarkably constant (see Kaempfer, 1974*a*). This property of the ribosome cycle provides the cell with a means of regulating the number of ribosomes active in protein synthesis in response to physiological changes. During maximal rates of protein synthesis, the small subunits generated from polysomes will combine immediately with (e)IF-3 and enter new initiation complexes, while (e)IF-3 is subsequently recycled. Though present in limiting quantity, the cycling amount of (e)IF-3 suffices to maintain the flow of subunits cycling between polysomes. However, as soon as the number of subunits leaving polysomes begins to exceed the number entering polysomes at the initiation site, as during slowing protein synthesis, an imbalance is created; while more small subunits accumulate, (e)IF-3 is recycled less frequently. Free subunit pairs not intercepted by (e)IF-3 will associate with high affinity to form inactive single ribosomes, and this process will continue until the number of small subunits leaving polysomes returns to the number of available (e)IF-3 molecules. The net effect is a reduction in the number of ribosomes translating mRNA, and an increase in the pool of single ribosomes. Conversely, an increase in the rate of initiation will lead to a temporary excess of recycling (e)IF-3 molecules over small subunits leaving polysomes; free (e)IF-3 can then effect a net shift in the equilibrium between single ribosomes and subunits in favor of the latter, until a new steady state is reached (Kaempfer, 1974a).

4.2. Recognition of Met-tRNA$_f$ by eIF-2

Binding of mRNA to the 40 S ribosomal subunit cannot take place unless the initiator tRNA, Met-tRNA$_f$, is first bound (Schreier and Staehelin, 1973; Darnbrough $et\ al.$, 1973). This means that the recognition and binding of Met-tRNA$_f$ are an integral part of the mRNA-binding process.

Met-tRNA$_f$ is recognized with absolute specificity by initiation factor eIF-2; Met-tRNA$_m$, the species utilized in chain elongation, is not bound (Chen $et\ al.$, 1972). This binding requires GTP and leads to the formation of a ternary complex, Met-tRNA$_f$/eIF-2/GTP (Levin $et\ al.$, 1973; Chen $et\ al.$, 1972). Complex formation is conveniently assayed by the GTP-dependent retention of labeled Met-tRNA$_f$ on nitrocellulose membranes in the presence of eIF-2. The GTP is not hydrolyzed, for nonhydrolyzable GTP analogues can be substituted. GDP binds about ten-fold more tightly to eIF-2 than does GTP, and it is a powerful competitive inhibitor of ternary complex formation (Walton and Gill, 1976). Indeed, purified preparations of eIF-2 may contain up to 0.5 mole GDP per mole of eIF-2 (Hershey, 1982a; Siekierka $et\ al.$, 1983).

eIF-2 is thought to be composed of one copy each (Lloyd $et\ al.$, 1980) of three nonidentical subunits: α (32,000–38,000 M_r), β (35,000–52,000 M_r), and γ (52,000–55,000 M_r). The α-subunit is the substrate for various specific eIF-2 kinases that play a pivotal role in translational control (see Section 7), and it has been reported that this subunit binds GTP (Barrieux and Rosenfeld, 1977).

The ternary complex can be stabilized by another protein (25,000 M_r) termed eIF-2A (Dasgupta et al., 1978; Gupta, 1982). Its importance for translation was elegantly demonstrated by the fact that antibodies against homogeneous preparations of either eIF-2 or eIF-2A strongly inhibit translation in reticulocyte lysates and that the inhibition can be reversed only by addition of the corresponding factor (Ghosh-Dastidar et al., 1980).

As will be reviewed in detail in Section 5.1, eIF-2, in addition to binding Met-tRNA$_f$, can bind to mRNA (Kaempfer, 1974b; Kaempfer et al., 1978a,b; Barrieux and Rosenfeld, 1978). The GTP-dependent binding of Met-tRNA$_f$ to eIF-2 is inhibited competitively by mRNA (Kaempfer et al., 1978b; Barrieux and Rosenfeld, 1978; Rosen et al., 1981a; Chaudhuri et al., 1981). In the presence of eIF-2A, however, this inhibition of ternary complex formation by mRNA is less pronounced (Roy et al., 1981). Possibly, therefore, eIF-2A may act to prevent the interaction between mRNA and eIF-2 until after ternary complex formation has occurred.

Because eIF-2 prefers GDP to GTP (Walton and Gill, 1976), and because GTP is cleaved during the initiation process (Trachsel et al., 1977; Benne and Hershey, 1978), one would expect eIF-2 to be released in a complex with GDP after every cycle of initiation. As in the case of the Tu-Ts cycle in elongation of prokaryotic polypeptide synthesis (Kaziro, 1978), one should hence expect to find a protein that catalyzes exchange of GDP for GTP in eIF-2, in order to permit catalytic utilization of the cell's eIF-2 molecules. Such an exchange factor has now been identified and is variously called Co-eIF-2B (Majumdar et al., 1977), ESP (de Haro and Ochoa, 1979), anti-HRI (Amesz et al., 1979), and eIF-2B (Konienczny and Safer, 1983). The latter have shown that eIF-2B complexes with eIF-2 and lowers the affinity of this initiation factor for GDP, permitting its exchange for GTP. eIF-2B apparently leaves eIF-2 after a ternary complex with Met-tRNA$_f$ has been formed. eIF-2B is a complex containing five proteins ($M_r = 26, 39, 58, 67,$ and 82×10^3; Benne et al., 1980; Konienczny and Safer, 1983).

More recently, it was found that phosphorylation of the α-subunit of eIF-2 prevents the GDP/GTP exchange reaction and hence the recycling of eIF-2 in translation (Clemens et al., 1982; Siekierka et al., 1982; see Section 7). That suggests that the α-subunit of eIF-2 may interact with eIF-2B, and this, in turn, fits the finding that the α-subunit binds guanine nucleotides (Barrieux and Rosenfeld, 1977).

Met-tRNA$_f$, on the other hand, is bound by the β-subunit of eIF-2 (Barrieux and Rosenfeld, 1977; Nygard *et al.*, 1980).

4.3.   Binding of Met-tRNA$_f$/eIF-2/GTP to 40 S Subunits

The ternary complex is an obligatory intermediate for the binding of Met-tRNA$_f$ to the 40 S ribosomal subunit. The ternary complex binds readily to the 40 S subunit in the absence of mRNA, and this binding is stable in sucrose gradients (Benne *et al.*, 1976; Trachsel *et al.*, 1977; Peterson *et al.*, 1979). The binding is stimulated in the presence of eIF-3 and eIF-4C, probably because these factors stabilize the resulting complex. Using radioactive factors, the presence in the 40 S/Met-tRNA$_f$/GTP complex of all of the subunit polypeptides of eIF-2, eIF-3, and eIF-4C, but no other initiation factors, could be demonstrated (Benne and Hershey, 1978; Peterson *et al.*, 1979).

4.4.   The mRNA-Binding Step

As already mentioned, the key event for translational control in terms of differential gene expression can be designated as the step where mRNA binds to the 40 S ribosomal subunit. Since mRNA fails to bind in the presence of all components for initiation except Met-tRNA$_f$, binding of Met-tRNA$_f$ to 40 S subunits is a necessary prerequisite for the subsequent binding of mRNA (Darnbrough *et al.*, 1973; Schreier and Staehelin, 1973; Trachsel *et al.*, 1977; Benne and Hershey, 1978). Conversely, the participation of mRNA is not needed for binding of Met-tRNA$_f$ to 40 S subunits.

Because of its great importance for translational control, the role of initiation factors in binding of mRNA will be discussed separately, in Section 5. The net result of joining of mRNA is formation of a 40 S initiation complex that contains mRNA, Met-tRNA$_f$, GTP, and stoichiometric amounts of eIF-2, eIF-3, and eIF-4C (Benne and Hershey, 1978; Thomas *et al.*, 1980*a*). It is thought that in the 40 S initiation complex, mRNA and Met-tRNA$_f$ are already base-paired at the AUG codon.

In order to complete the initiation process, a 60 S subunit must join to form a complete ribosome, and the initiation factors eIF-2, eIF-3, and eIF-4C must be released. This is indeed the case, and as said above, eIF-2 is released in a complex with GDP, implying that

GTP hydrolysis must have occurred. Indeed, nonhydrolyzable analogues of GTP block joining of the 60 S subunit (Trachsel *et al.*, 1977; Benne and Hershey, 1978). These reactions require the presence of eIF-5, the 60 S joining factor (160,000 M_r) (Trachsel *et al.*, 1977; Benne and Hershey, 1978). The resulting 80 S initiation complex is apparently free of initiation factors, and its Met-tRNA$_f$ is reactive to puromycin, meaning that it is capable of entering the elongation phase of polypeptide synthesis.

5. ROLE OF INITIATION FACTORS IN BINDING OF mRNA

5.1. Initiation Factor 2

5.1.1. The mRNA-Binding Function of eIF-2

As we have seen, eIF-2 is absolutely required for the formation of the ternary complex, Met-tRNA$_f$/eIF-2/GTP, that subsequently binds to the 40 S ribosomal subunit. Only when Met-tRNA$_f$ is bound to this subunit can binding of mRNA take place. Thus, the unique property of providing Met-tRNA$_f$ already imparts on eIF-2 a crucial role in the binding of mRNA. It is important to remember that, while additional initiation factors participate in the stable binding of mRNA (see Section 5.2), none can act in the absence of eIF-2.

In addition to binding Met-tRNA$_f$, eIF-2 itself can bind to mRNA (Kaempfer, 1974*b*; Barrieux and Rosenfeld, 1977, 1978; Kaempfer *et al.*, 1978*a*; 1979*a*; Chaudhuri *et al.*, 1981). This binding is specific in that all mRNA species tested possess an effective binding site for eIF-2, including mRNA species lacking the 5'-terminal cap or 3'-terminal poly (A) moieties (Kaempfer *et al.*, 1978*a*), while RNA species not serving as mRNA, such as tRNA (Barrieux and Rosenfeld, 1977, 1978; Kaempfer *et al.*, 1978*a*, 1979*a*; Rosen and Kaempfer, 1979), rRNA (Barrieux and Rosenfeld, 1977; Kaempfer *et al.*, 1981), or negative-strand viral RNA (Kaempfer *et al.*, 1978*a*) bind far more weakly.

The fact that eIF-2 binds to RNA molecules in general, even though it prefers mRNA, initially raised some question as to the specificity of the interaction with mRNA (Hershey, 1982*a*). However, the *lac* repressor, a protein that binds with very high specificity to its operator sequence in *E. coli* DNA, will also bind to any other DNA, albeit with lower affinity (Lin and Riggs, 1972, 1975*a*; Zingsheim *et al.*, 1977), and the latter property is relevant for the *in vivo* regulation of the *lac* operon (Lin and Riggs, 1975*b*; von Hippel *et al.*, 1974). Indeed, a common property of proteins that recognize specific sites in nucleic acids is their tendency to bind with low affinity to non-specific sequences. In the case of eIF-2, specificity of its interaction with mRNA is now supported by structural and functional evidence. As will be reviewed in more detail below, the protein recognizes specifically the ribosome-binding site sequence in the mRNAs that were examined (Kaempfer *et al.*, 1981; Perez-Bercoff and Kaempfer, 1982). Different mRNA species bind to eIF-2 with a characteristically different affinity that correlates closely with their ability to compete in translation with other types of mRNA (Di Segni *et al.*, 1979; Rosen *et al.*, 1982). Agents that affect the ability of eIF-2 to function in translation, such as salts (Di Segni *et al.*, 1979) or double-stranded RNA (Rosen *et al.*, 1981*a*), differentially affect the translation of individual mRNA species and exert a commensurate effect on the interaction of eIF-2 with corresponding mRNA.

The mRNA-binding property is a feature of eIF-2 itself. Thus, eIF-2 is preferentially retained on mRNA-cellulose columns and upon elution is active in translation and in GTP-dependent binding of Met-tRNA$_f$ (Kaempfer *et al.*, 1978*a*; Kaempfer, 1979). Binding of Met-tRNA$_f$ to eIF-2 is inhibited competitively by mRNA (Kaempfer *et al.*, 1978*b*; Barrieux and Rosenfeld, 1978; Rosen *et al.*, 1981*a*; Chaudhuri *et al.*, 1981). Perhaps most convincing is the finding that binding of mRNA to purified eIF-2 preparations can be inhibited completely by competing amounts of Met-RNA$_f$, provided GTP is present (Rosen and Kaempfer, 1979). Thus, mRNA and Met-tRNA$_f$ are mutually exclusive in their binding to eIF-2, suggesting that, during initiation of translation, the interaction of a molecule of mRNA with eIF-2 on the 40 S ribosomal subunit could displace the previously bound Met-tRNA$_f$ from this factor. Rosen and Kaempfer (1979) have proposed that, during initiation, three processes may occur in one step: binding of mRNA to eIF-2, displacement of Met-tRNA$_f$ from eIF-2, and base-pairing between mRNA and Met-tRNA$_f$.

The elution of biologically active eIF-2 from mRNA-cellulose columns (Kaempfer *et al.*, 1978*a*) suggests that all three subunits of eIF-2 are found in the complex between eIF-2 and mRNA. Indeed, eIF-2 binds to RNA-Sepharose as a complex of three polypeptides (Vlasik *et al.*, 1980). By contrast, Barrieux and Rosenfeld (1978) concluded that binding of globin mRNA to eIF-2 induces the dissociation of the α- and γ-subunits. It now seems likely that the latter observation was an artifact occurring during their isoelectric focusing analysis.

5.1.2.   The Site in mRNA Recognized by eIF-2

Globin mRNA molecules lacking the terminal poly (A) tail or an additional 90 nucleotides from the 3′-untranslated region bind to eIF-2 as tightly as native globin mRNA, with an apparent dissociation constant of 5×10^{-9} M at 150 mM KCl (Kaempfer et al., 1979a).On the other hand, cap analogues inhibit binding of both mRNA and Met-tRNA$_f$ to eIF-2 (Kaempfer et al., 1978b). Although this could suggest that the cap in mRNA interacts with eIF-2, the genomic RNA species from mengovirus or STNV bind extremely well to eIF-2, in fact, even better than globin mRNA, yet they do not carry a cap structure (Kaempfer et al., 1978b, 1981; Rosen et al., 1982). This and the observation that eIF-2 prefers native globin mRNA by five orders of magnitude over cap analogues led to the suggestion that binding of eIF-2 to mRNA occurs primarily at an internal sequence (Kaempfer et al., 1978b).

5.1.2a.   Satellite Tobacco Necrosis Virus RNA

STNV RNA is particularly suitable for analyzing the binding site for eIF-2, because it is an efficient mRNA for translation (Leung et al., 1976; Herson et al., 1979), has a known sequence of 1,239 nucleotides encoding only a single protein (Ysebaert et al., 1980), lacks poly (A), and possesses an unmodified 5′ end (Wimmer et al., 1968; Horst et al., 1971) that can be labeled with polynucleotide kinase. RNA isolated from virions migrates, after 5′-end labeling, as an heterogeneous collection of fragments with only a minor amount of label in intact viral RNA. Such preparations contain approximately 35 different 5′-end sequences, as judged by fingerprinting after digestion with T1 or pancreatic RNase, attesting to the presence of many fragments originating from internal regions of the viral RNA molecule (Kaempfer et al., 1981). When this RNA mixture was offered to eIF-2 and RNA bound by eIF-2 was isolated and fingerprinted with either RNase, one major spot was observed, migrating precisely at the 5′ end of intact viral RNA. Sequence verification confirmed that eIF-2 binds selectively to STNV RNA fragments starting with the 5′ end of intact viral RNA (Kaempfer et al., 1981).

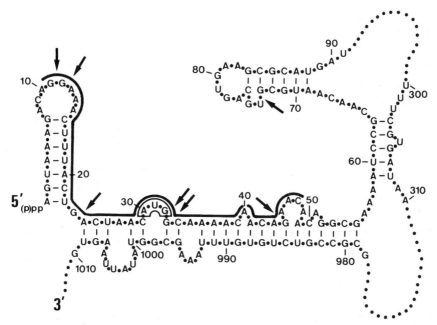

Fig. 1. Secondary structure model for the 5' end of STNV RNA. The model depicts stable secondary interactions, based on sequence (Ysebaert *et al.*, 1980) and prominent sites of ribonuclease T₁ cleavage (arrows; Kaempfer *et al.*, 1981). (Line) nucleotides protected by 40 S ribosomal subunits against nucleases (Browning *et al.*, 1980). For eIF-2-binding site, see text. From Kaempfer *et al.*, 1981.

To map the eIF-2 binding site more exactly, intact 5'-end-labeled STNV RNA was isolated and digested partially with RNase T1, to generate a nested set of labeled RNA fragments, all containing the 5' end of intact RNA and extending to various points within the molecule. Figure 1 depicts the 5'-terminal sequence of STNV RNA, including the unique AUG initiation codon located at positions 30–32. Arrows denote G residues sensitive to RNase T1 attack. eIF-2 does not bind the 32-nucleotide 5'-terminal fragment ending with the AUG codon, or shorter ones, but it does bind to the 44-nucleotide fragment or larger ones, and with the same specificity as to intact viral RNA (Kaempfer *et al.*, 1981). This places the 3'-proximal boundary of the eIF-2 binding site at or near nucleotide 44. Indeed, binding of eIF-2 to intact STNV RNA greatly increases the sensitivity of the RNA to cleavage by RNase T1 at nucleotide 44, or by nuclease P1 near position 60, attesting to a conformational change induced at these points by the binding of the initiation factor molecule. On the 5'-terminal side, eIF-2 shields positions 11, 12, 23, 32, and 33 against digestion (Fig. 1), placing the boundary at or before position 10 (Kaempfer *et al.*, 1981). Since the G residues at positions 2 and 7 are hydrogen bonded and thus resistant to nuclease attack (Leung *et al.*, 1979), it is not certain if the eIF-2 binding site extends to the physical 5′ end. The striking aspect of the eIF-2 binding site is, however, that it overlaps virtually completely with the binding site for 40 S ribosomal subunits (Browning *et al.*, 1980), depicted by the line in Fig. 1. Thus, eIF-2 by itself recognizes virtually the same nucleotide sequence that is protected by 40 S ribosomal subunits carrying eIF-2, Met-tRNA$_f$, and all other components needed for initiation of translation.

5.1.2b.   Mengovirus RNA

Mengovirus RNA has a length of about 7500 nucleotides and contains a poly (C) tract located within 500–700 nucleotides from the 5′ end (Perez-Bercoff and Gander, 1977). In the closely related FMDV RNA, the major initiation site for translation lies downstream from the poly (C) tract (Sangar *et al.*, 1980), while in poliovirus RNA, which lacks poly (C), translation starts at position 742 (Kitamura *et al.*, 1981). In 40 S or 80 S initiation complexes formed on mengovirus RNA in whole L cell or ascites cell lysates, four specific oligonucleotides, 15–28 bases long, are protected against RNase T1 attack (Perez-Bercoff and Kaempfer, 1982). These map into at least two widely separated domains in the viral RNA molecule at internal sites downstream from the poly (C) tract. When intact mengovirus RNA is offered to eIF-2 in the absence of other components for translation and the sequences protected by the initiation factor are isolated, three specific oligonucleotides are recovered out of this very long nucleotide sequence, and these are identical with three of the four protected in initiation complexes (Perez-Bercoff and Kaempfer, 1982). The absence of any additional nucleotide in detectable amounts shows that eIF-2 recognizes a highly specific site, or sites, in the mengovirus RNA molecule.

5.1.3.   Relationship between Binding Site Sequences for Ribosomes and for eIF-2

The striking resemblance between the binding site for eIF-2 and that for 40 S ribosomal subunits in STNV RNA (Kaempfer *et al.*,

1981) raised the possibility that, during initiation of translation, binding of a ribosome to mRNA may be guided to a significant extent by eIF-2. The findings with mengovirus RNA (Perez-Bercoff and Kaempfer, 1982) provide additional and independent support for this concept. The virtual identity of the binding sites in mengovirus RNA for ribosomes on one hand and for eIF-2 on the other reinforces the results with STNV RNA and point to a critical role for eIF-2 in recognition of mRNA by ribosomes.

The oligonucleotides in mengovirus RNA protected by eIF-2 alone or by complete initiation complexes map at several locations within the RNA molecule. Possibly, eIF-2 and ribosomes recognize a site composed of different regions held together by secondary or tertiary structure. Alternatively, there may be more than one binding site for eIF-2 and ribosomes. The finding that two N-formylmethionyl tryptic peptides can be detected when poliovirus (Celma and Ehrenfeld, 1975; Jense et al., 1978; Knauert and Ehrenfeld, 1979; Ehrenfeld and Brown, 1981) or mengovirus (Degener et al., 1983) RNA is translated in vitro tends to support the latter explanation. The specific binding of eIF-2 to internal sites in the mengovirus RNA molecule indicates that a free 5' end is not required for this interaction. The results with STNV RNA (Kaempfer et al., 1981) also show that if a 5' end is required for recognition by eIF-2, it is not sufficient per se, as the factor is able to select a specific RNA fragment from among a collection of fragments bearing over 30 different 5' ends.

5.1.4. Studies with Monovalent Anions

Translation of globin mRNA in intact or micrococcal nuclease-treated reticulocyte lysates (Pelham and Jackson, 1976) is inhibited by increasing concentrations of Cl^- or OAc^-, the former being more inhibitory (Fig. 2A). These anions primarily inhibit the translation of α-globin mRNA, resulting in a decrease in the α/β-globin synthetic ratio (Fig. 2B). The inhibition of α-globin mRNA translation by Cl^- is readily relieved by the addition of eIF-2 (Di Segni et al., 1979). Since Cl^- or OAc^- ions inhibit translation principally by affecting the binding of mRNA to 40 S subunits carrying Met-tRNA$_f$, but have little effect on the formation of 40 S/Met-tRNA$_f$ complexes (Weber et al., 1977), the ability of eIF-2 to reverse the translational inhibition by Cl^- suggests that this anion may inhibit the interaction between eIF-2 and mRNA during translation. Indeed, when the effect of Cl^-

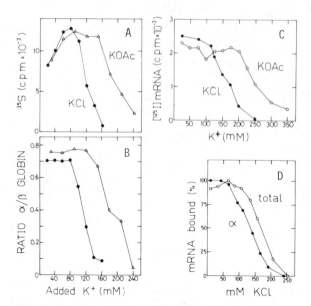

Fig. 2. (A) Effect of KCl and KOAc on translation of globin mRNA, (B) on the α/β-globin synthetic ratio, and (C) on complex formation between ^{125}I-labeled globin mRNA and eIF-2; (D) effect of KCl on complex formation between eIF-2 and labeled α-globin mRNA or total globin mRNA. In (A), cpm denote incorporation of [35S]methionine into protein. In (C), cpm denote mRNA retained on nitrocellulose membranes in the presence of a limiting amount of eIF-2. From Di Segni *et al.* (1979), with permission. Copyright 1979 American Chemical Society.

or OAc⁻ ions on complex formation between globin mRNA and eIF-2 is studied, a remarkably parallel inhibition is observed (Fig. 2C). Moreover, binding of purified α-globin mRNA to eIF-2 is more sensitive to salt than is binding of unfractionated globin mRNA (two-thirds of which consists of α-globin mRNA; Lodish, 1971; Fig. 2D), attesting to a greater affinity of β-globin mRNA for eIF-2 (Di Segni *et al.*, 1979).

The concept that a direct interaction between eIF-2 and mRNA occurs during initiation is raised by these observations and is further supported by studies of mRNA competition in translation (Section 6) and of the inhibitory action of double-stranded RNA on initiation (Section 7).

5.2. Other Initiation Factors

Besides eIF-2, the 40 S/Met-tRNA_f complex that binds to mRNA also contains all the polypeptides of eIF-3 and eIF-4C (see Section

4.4). Omission of eIF-4C causes only a marginal reduction in the binding of mRNA (Benne and Hershey, 1978). On the other hand, binding of mRNA is stimulated by eIF-3 and by two other initiation factors, eIF-4A and eIF-4B (Trachsel *et al.*, 1977; Benne and Hershey, 1978), and it requires the hydrolysis of ATP (Marcus, 1970; Trachsel *et al.*, 1977; Benne and Hershey, 1978).

Progress in understanding the roles of eIF-3, eIF-4A, and eIF-4B in binding of mRNA has been slow. As already mentioned, eIF-3 contains a complex mixture of polypeptides. eIF-4B has been partially purified as an 80,000 M_r polypeptide (Trachsel *et al.*, 1977; Benne and Hershey, 1978) but its purification has been problematic. Various activities have been ascribed to this factor, including cap recognition (Shafritz *et al.*, 1976), a role in the poliovirus-mediated shut-off of host protein synthesis (see Section 8.4; Rose *et al.*, 1978), and ATP-dependent binding of mRNA in the presence of eIF-4A (Grifo *et al.*, 1982). eIF-4A has been purified as a 46,000 M_r polypeptide and is mainly characterized by its ability to retain mRNA on filters in the presence of (impure) eIF-4B and ATP (Grifo *et al.*, 1982).

The importance of the cap structure in stabilizing the binding of mRNA to 40 S ribosomal subunits, as well as the inhibition of translation by cap analogues (see Section 3.2), both suggested that the cap is recognized by one or more proteins. In pinpointing the protein(s) involved, two lines of research converged. One of these was to study the crosslinking of proteins to the cap during initiation of translation (Sonenberg and Shatkin, 1977), while the other was to study the reversal of the initiation block observed in extracts of poliovirus-infected HeLa cells (Rose *et al.*, 1978). Initially, eIF-3 and eIF-4B were thought to be involved, but then a 24,000 M_r cap-binding protein (24-CBP) that copurifies with either factor was isolated (Sonenberg *et al.*, 1979; Bergmann *et al.*, 1979; Trachsel *et al.*, 1980). The 24-CBP binds specifically to the cap but does not restore the poliovirus-induced block; the latter activity resides in an 8–10 S complex of several proteins that includes 24 CBP and polypeptides of about 28,000, 46,000, 80,000, and 200,000 M_r, and has been termed CBP II (Tahara *et al.*, 1981; Sonenberg, 1981) or eIF-4F (Grifo *et al.*, 1983). The 46,000 M_r polypeptide, in turn, is probably identical with eIF-4A (Grifo *et al.*, 1983).

Curiously, eIF-4A, eIF-4B, and eIF-4F stimulate translation of the uncapped STNV RNA to the same extent as that of globin mRNA (Grifo *et al.*, 1983), indicating that, conceivably, they interact with other features in mRNA besides the cap. Indeed, translation of the noncapped poliovirus RNA is unusually dependent upon eIF-4A (Daniels-McQueen *et al.*, 1983). In this context, it may be noted that translation of STNV or the uncapped cowpea mosaic virus (CPMV) RNAs in the wheat germ cell-free system is sensitive to inhibition by cap analogues (Seal *et al.*, 1978). Binding of the uncapped mengovirus RNA to eIF-2, moreover, is also sensitive to inhibition by cap analogues (Kaempfer *et al.*, 1978*b*). These observations all suggest that recognition of the cap by initiation factors may be part of a more general interaction with the mRNA molecule.

Sonenberg (1981) observed that crosslinking of the larger polypeptides in CBP II (eIF-4F), but not of 24-CBP, to the cap requires ATP-Mg^{2+}; nonhydrolyzable analogues cannot substitute for ATP. Subsequent findings by Grifo *et al.* (1982, 1983) showed that retention of mRNA on filters by a combination of eIF-4A and eIF-4B containing eIF-4F, or crosslinking of eIF-4F to the cap, depends on ATP-Mg^{2+} and is inhibited by cap analogues. Sonenberg (1981) suggested that the cap-binding proteins may act to unwind the secondary structure of the 5′ end of capped eukaryotic mRNAs. Indeed, the extent to which a monoclonal antibody directed against the cap-binding proteins inhibits initiation complex formation can be correlated with the degree of secondary structure in mRNA (Sonenberg *et al.*, 1981), and a reduction in 5′-end secondary structure of mRNA reduces its dependence on cap-binding proteins for translation (Sonenberg *et al.*, 1982; Lee *et al.*, 1983). This fits with earlier findings that binding of ribosomes to inosine-substituted reovirus mRNA, which possesses less secondary structure, is less dependent on ATP than is binding to native mRNA (Morgan and Shatkin, 1980; Kozak, 1980*a,b*). An earlier interpretation of these results was that, during initiation, ribosomes scan the mRNA molecule from its 5′ end (Kozak, 1980*a*) and that this movement requires ATP hydrolysis (Kozak, 1980*b*). Jackson (1982) has found that efficient formation of 80 S initiation complexes on uncapped mRNA, EMC RNA, CPMV RNA, or poly (AUG), can occur in the absence of ATP, while binding of 40 S subunits to capped mRNA does require ATP and is inhibited by nonhydrolyzable ATP analogues. Moreover, if 40 S ribosomal subunits reach the initiation codon by scanning the mRNA from its 5′ end as suggested (Kozak, 1980*a*), one might expect to see queues of 40 S subunits on EMC RNA which is thought to possess a 5′-leader sequence of at least 500 nucleotides, yet only one ribosome is detectable on EMC RNA in conditions for initiation complex formation (Jackson, 1982). Clearly, more work is needed before the respective roles of ATP hydrolysis, cap recognition, unwinding of secondary structure in mRNA, and ribosome movement across mRNA molecules in initiation will be understood.

The data available as of summer 1983 indicate that the number of different polypeptide chains involved, besides eIF-2, in binding of mRNA may be more limited than would seem, as the various complexes or chains designated as eIF-3, eIF-4A, eIF-4B, eIF-4F (CPB-II), and 24-CBP possess extensive structural and functional relationships and thus may be different, and possibly artifactual, derivatives of a simpler set of proteins.

No sequence studies of the interaction of these proteins with mRNA have yet been reported, and specific binding to a region in mRNA remains to be demonstrated. On the basis of the available data, it is tempting to suggest that eIF-2 recognizes the sequence and conformation around the initiation site in mRNA, while the above-mentioned proteins may unwind secondary structure in mRNA and anchor it at the 5'-terminal cap structure.

6. TRANSLATIONAL CONTROL BY mRNA COMPETITION

The molar translation yield of a given mRNA, that is, the frequency of translation of this mRNA, is not a constant property but is subject to regulation. Experimentally, this is observed by a shift in the relative amounts of individual proteins synthesized as a function of (1) the rate of overall protein synthesis, and (2) the amounts and types of mRNA present.

Selective translation of certain mRNA species over other ones is often involved in the regulation of eukaryotic gene expression during growth, differentiation, or virus infection. The so-called discrimination of mRNA occurs mainly at the initiation step which involves the recognition of mRNA and its binding to the 40 S ribosomal subunit. These observations have raised the question whether there exist mRNA-specific factors that promote the translation of certain mRNA species, but not of other ones. So far, little evidence for such absolute specificity has been obtained. Thus, although Gette and Heywood (1979) suggest that eIF-3 from chick muscle can be separated into a core and discriminatory components that preferentially stimulate the translation of myosin mRNA, such preferential stimulation can be explained equally well in other ways, chiefly by different requirements, in terms of amount, for the same initiation factors by different mRNA species. Restated more directly, this implies that individual mRNA species possess different affinities for the standard initiation factors involved in the binding of mRNA. There is now good evidence that this concept is correct, that mRNA competition during translation does exist, and constitutes an essential aspect of translational control (Walden *et al.*, 1981; Kaempfer, 1982, 1984).

In order to identify initiation factors involved in mRNA competition, a number of studies have employed reconstituted cell-free systems (Golini *et al.*, 1976; Ray *et al.*, 1983). A major problem with such systems is that certain components may be present in excess, while others may be limiting or partially inactivated, precluding efficient initiation. However, meaningful studies of translation initiation frequency *in vitro* can only be done in systems where ribosomes can cycle rapidly and repeatedly over mRNA. To date, the only cell-free systems that translate mRNA at high efficiency are the reticulocyte lysate (see Jackson, 1982) and the micrococcal nuclease-treated reticulocyte lysate (Pelham and Jackson, 1976). The latter system offers several advantages over reconstituted cell-free systems. It responds to translational control signals (see below), it is capable of extensive and efficient initiation in conditions more likely to be representative of protein synthesis in intact cells, and except for mRNA, it contains all other components for translation in a proportion much closer to that of the intact cell.

6.1. mRNA Competition for eIF-2

6.1.1. α- and β-Globin mRNA

In reticulocytes, β-globin is synthesized on polysomes containing about 1.5 times as many ribosomes as those engaged in α-globin synthesis (Hunt *et al.*, 1968). An explanation of this observation could be either that the rate of initiation on each mRNA is identical, elongation being more rapid on α-globin mRNA, or alternatively, that the elongation rates are identical, but initiation of translation on β-globin mRNA is more frequent (Lodish, 1971). In the latter case, there must be more α-globin mRNA than β-globin mRNA in the reticulocyte, in order to account for the observed equimolar synthesis of α- and β-globin. Lodish (1971) showed that the second explanation is correct, by reducing the rate of elongation with drugs to the point where initiation no longer limited protein synthesis. Under such conditions, the α/β-globin synthetic ratio increased from 1.0 to 1.5, and poly-somes bearing nascent α- or β-globin contained the same number of ribosomes.

Any decrease in the rate of initiation of translation leads to a drop in the α/β-globin synthetic ratio, while any increase causes α-globin mRNA translation to rise more than translation of β-globin mRNA (Lodish, 1974). This is the result expected if one assumes that each mRNA has its own rate constant for binding to ribosomes at initiation; any nonspecific reduction in the rate of initiation at or be-fore binding of mRNA will then result in a preferential inhibition of the translation of the mRNA species with the lower rate constant, in this case, α-globin mRNA (Lodish, 1974).

Upon addition of increasing amounts of globin mRNA to mRNA-dependent *in vitro* translation systems, in particular, the micrococcal nuclease-treated reticulocyte lysate, there is about equimolar syn-thesis of α- and β-globin as long as the amount of mRNA does not exceed saturation level; but at higher levels of mRNA, the α/β syn-thetic ratio decreases drastically, to values as low as 0.1, even though overall translation remains constant (McKeehan, 1974; Kabat and Chappell, 1977; Di Segni *et al.*, 1979). This is the result expected if (1) α- and β-globin mRNAs compete with each other at initiation of translation, and (2) the competition favors β-globin mRNA. Consid-ering that the difference in initiation rate between α- and β-globin mRNA is only 1.5-fold in the intact reticulocyte lysate (Lodish, 1971), it can be amplified greatly by the application of competition pressure, that is, by addition of greater than saturating amount of mRNA. Used in this manner, the technique provides a sensitive and convenient tool for ordering mRNA species according to their ability to compete in translation ("initiation strength"; see also Section 6.1.3).

Di Segni *et al.* (1979) studied the effect of eIF-2 on translational competition between α- and β-globin mRNA. Addition of a constant amount of eIF-2 to translation mixtures containing increasing amounts of globin mRNA led to a rise in the α/β-globin synthetic ratio that was more pronounced, the lower the mRNA concentration. At low levels of mRNA, eIF-2 raised the α/β ratio to a value of about 1.5, the relative content of α- and β-globin mRNA in total globin mRNA (Lodish, 1971), as expected for a condition in which the mRNAs no longer compete. Yet, eIF-2 did not stimulate the number of initiations at any mRNA concentration (Di Segni *et al.*, 1979). Thus, the increase in α/β synthetic ratio observed in the presence of eIF-2 involves a stimulation of initiation on α-globin mRNA, concomitant with decreased initiation on β-globin mRNA, demonstrating that eIF-2 does not exert a specific effect on α-globin mRNA translation, but instead causes relief of competition. The finding that β-globin mRNA binds to eIF-2 with higher affinity than does α-globin mRNA (see Fig. 2D) supports the concept that translational competition between α- and β-globin mRNA is based on a direct competition for eIF-2 at the mRNA-binding step (Di Segni *et al.*, 1979).

6.1.2. Picornaviral and Globin mRNA

More quantitative evidence for a direct relationship between the affinity of a given mRNA for eIF-2 and its ability to compete in translation came from a study of the competition between globin mRNA and mengovirus RNA in the nuclease-treated reticulocyte lysate (Rosen *et al.*, 1982). In conditions where the total number of initiations remains constant, addition of mengovirus RNA leads to a drastic decrease in globin mRNA translation, accompanied by increasing synthesis of viral protein. Half-maximal inhibition of globin mRNA translation occurs when 35 molecules of globin mRNA are present for every molecule of mengovirus RNA. Assuming that equal proportions of these RNA species are translationally active, this means that a molecule of mengovirus RNA competes 35-fold more strongly in translation than does, on average, a molecule of globin mRNA (Rosen *et al.*, 1982).

In conditions where globin synthesis is greatly depressed by the presence of mengovirus RNA, addition of eIF-2 does not stimulate overall translation, yet restores globin synthesis to the level seen in the absence of competing viral RNA (Rosen *et al.*, 1982). In the absence of mengovirus RNA, on the other hand, globin synthesis is not stimulated by added eIF-2. Hence, addition of eIF-2 allows the more weakly competing, but more numerous, globin mRNA molecules to initiate translation at the expense of the more strongly competing, but less numerous, viral RNA molecules. The fact that eIF-2 acts to shift translation in favor of globin synthesis showed clearly that globin mRNA and mengovirus RNA compete for eIF-2, but did not eliminate the possibility that eIF-2 could act in a nonspecific manner, as by increasing the pool of 40 S/Met-tRNA$_f$ complexes. RNA-binding experiments, however, revealed that mengovirus RNA and globin mRNA compete directly for eIF-2, with an affinity ratio (30:1) that matches very closely with that observed in translation competition experiments (Rosen *et al.*, 1982). In good agreement with these findings, mengovirus RNA is about 30-fold more effective than globin mRNA in inhibiting ternary complex formation between eIF-2, Met-tRNA$_f$, and GTP (Rosen *et al.*, 1981*a*) and its binding to eIF-2 is markedly more salt resistant than is binding of globin mRNA (Rosen *et al.*, 1982). The high affinity of mengovirus RNA for eIF-2 involves specific RNA sequences (see Section 5.1.2b) and is not related simply to nucleotide length, for VSV negative-strand RNA, which is even longer, binds only very weakly and nonspecifically to eIF-2 and lacks the high-affinity binding site found in the far shorter VSV mRNA molecules (Kaempfer *et al.*, 1978*a*).

6.1.3. Cellular mRNA

One would predict that the more evenly different species of mRNA are expressed, the less extreme must be the differences in competing ability between them. Turning to the cellular mRNA population, one may ask if individual species of mRNA compete in translation, and if so, whether they differ in competing ability. While this question has been answered affirmatively for the mRNA pair that encodes α- and β-globin, far less work has been done on other and, in particular, less differentiated mRNA populations. Recently, competition pressure analysis was used to study this point for three liver mRNA species, encoding haemopexin, ferritin, and albumin (Kaempfer and Konijn, 1983). The application of competition pressure, through the addition of greater than saturating amounts of liver mRNA to a micrococcal nuclease-treated reticulocyte lysate, reveals distinct behavior of the three mRNA species, as judged by specific, quantitative immunoprecipitation of their respective translation products (Fig. 3A). At mRNA levels greatly in excess of saturation, a general inhibition of translation is often observed (Pelham and Jackson, 1976; Di Segni *et al.*, 1979), but at low and intermediate levels of mRNA, the proportion of total protein synthesized as albumin, ferritin, and haemopexin exhibits characteristic differences. Synthesis of haemopexin (hpx) declines well before that of ferritin (fer) and albumin (alb), while ferritin synthesis levels off earlier than albumin synthesis. The protein synthetic ratios plotted in Fig. 3B show that with increasing mRNA concentration, the alb/fer ratio increases 1.6-fold, the fer/hpx ratio 2.4-fold, and the alb/hpx ratio 3.8-fold before reaching essentially plateau value. These results can be interpreted

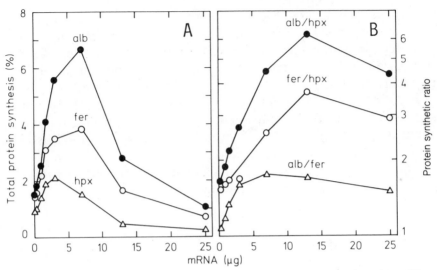

Fig. 3. (A) Synthesis of albumin, ferritin, and haemopexin as a function of liver mRNA concentration, (B) protein synthetic ratios calculated from the values in (A). From Kaempfer and Konijn (1983), with permission.

in analogy to the α/β-globin system to mean that the three mRNA species compete, and do so in the order hpx < fer < alb of ascending competing ability. By comparison to α- and β-globin (Di Segni *et al.*, 1979), the relative abilities of haemopexin, ferritin, and albumin mRNA to complete in translation seem to differ far less extensively. Yet, these small differences can be magnified considerably by the application of competition pressure *in vitro*; translation of the more effectively competing mRNA species is progressively increased at the expense of translation of more weakly competing ones. Thus, haemopexin, ferritin, and albumin mRNA could be rank-ordered in this respect, independent of their relative abundance in the total liver mRNA population and in spite of the fact that they were present in a complex mixture of mRNA species.

In conditions of competition, translation of ferritin mRNA responds more readily to added eIF-2 than does translation of albumin mRNA; the response of ferritin mRNA is that expected for an mRNA species that competes more weakly for eIF-2 (Kaempfer and Konijn, 1983).

The cases discussed in this and the preceding two sections are clear-cut examples of mRNA competition for a general initiation factor that results in the preferential translation of certain mRNA species

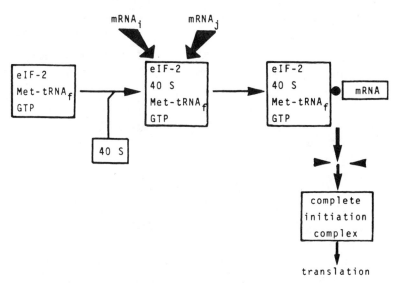

Fig. 4.   Proposed mechanism of mRNA competition for eIF-2 during initiation of protein synthesis (see text). From Kaempfer (1982), with permission.

over other ones. Addition of the initiation factor leads to preferential stimulation of translation of the weakly competing mRNA species, at the expense of the stronger ones. Clearly, such preferential stimulation does *not* mean that the factor is specific for the mRNA in question. Quite the opposite is true, the stronger-initiating mRNA has a higher affinity for the factor.

Studies on the role of eIF-2 in translational competition between α- and β-globin mRNA (Di Segni *et al.*, 1979) or between mengovirus RNA and globin mRNA (Rosen *et al.*, 1982) support the interpretation that mRNA competition is for eIF-2 molecules located in 40 S/Met-tRNA$_f$ complexes. An increase in the formation of these complexes will lead to relief of competition between two mRNAs by increasing the number of eIF-2 molecules available for binding to mRNA. For any pair of mRNA species i and j that differ in competing ability (Fig. 4), the relative frequency of their translation will not equal their molar ratio as long as the target of competition, eIF-2, is limiting. Thus, if mRNA$_i$ is the species with higher affinity and mRNA$_j$ is present in higher concentration (as is the case for the above-cited examples), then any increase in the concentration of eIF-2 molecules located in 40 S/eIF-2/Met-tRNA$_f$ complexes will result in a relatively greater stimulation of binding of the weaker species, mRNA$_j$, up to the point where binding of i and j is exactly according to their molar ratio and there no longer is competition. Quite independently of this competition, the actual rate of protein synthesis may be limited at a step beyond the interaction with eIF-2, as indicated by the wedges in Fig. 4. This is true for the examples cited above, as addition of eIF-2 did not lead to stimulation of translation at any mRNA concentration and hence did not limit translation. In such conditions, any increased translation of mRNA$_j$ must be concomitant with decreased translation of the stronger species, mRNA$_i$, as is indeed observed (Di Segni *et al.*, 1979; Rosen *et al.*, 1982).

In general, the relative translation rates of mRNA species will be determined (1) by their intrinsic affinity properties for components of the initiation step, (2) by their molar ratios, and (3) by the extent to which the target of competition (for instance, eIF-2) is limiting. This means that both the introduction of new mRNA species, as during differentiation or virus infection, and changes in initiation factor activity, as produced by heme deprivation, double-stranded RNA, or interferon (see Section 7), affect the relative expression of each mRNA in the cell.

6.2. mRNA Competition for Other Initiation Factors

Early work on competition between α- and β-globin mRNA (Kabat and Chappell, 1977) suggested an involvement of eIF-4B and eIF-4A, but neither direct binding of these factors to globin mRNA, nor the predicted greater affinity of β-globin mRNA were demonstrated. A role for eIF-4A and eIF-4B (considered at the time 50% pure) was also suggested by Golini *et al.* (1976) who studied translation of globin mRNA in the presence of EMC RNA in a cell-free system from mouse ascites cells. In that study, however, total translation was stimulated by addition of the factors, although globin mRNA translation was stimulated more than translation of EMC RNA. As mentioned in Section 6, any component that increases the rate of overall translation is expected to cause greater stimulation of translation of the weaker mRNA species, in this case, globin mRNA. Thus, the results did not show if globin mRNA and EMC RNA actually compete for eIF-4A or eIF-4B. Higher levels of the eIF-4B preparation considerably inhibited overall translation; however, globin mRNA translation increased, while EMC translation decreased (Golini *et al.*, 1976). On the basis of these observations, it was suggested that EMC RNA out-competes globin mRNA for this factor.

Indeed, EMC RNA was found to bind about eight-fold more tightly than globin mRNA to the eIF-4B preparation (Baglioni *et al.*, 1978).

The subsequent separation of eIF-4B into the 80,000 M_r eIF-4B and the complex of cap-binding proteins, CBP II (Tahara *et al.*, 1981), most likely identical with eIF-4F (Grifo *et al.*, 1983), and the apparent presence of eIF-4A in the eIF-4F (Grifo *et al.*, 1983), raised the question of which of these polypeptides was actually responsible for the earlier effects. To study this point, Ray *et al.* (1983) used a reconstituted cell-free system for translation from Krebs ascites cells, supplememted with crude ribosomal salt wash proteins (an unknown mixture of initiation factors and other components) and purified initiation factor preparations. A mixture of the reovirus mRNA species and globin mRNA was added as template, and the ability of increasing amounts of initiation factors to differentially stimulate translation of individual reovirus polypeptides was measured. Both eIF-4A and CBP II were found to possess this ability (Ray *et al.*, 1983). Unfortunately, it was not shown that the total number of initiations remained constant in conditions where the relative translation rate of individual proteins changed, so that, unlike in the studies of Kabat and Chappell (1977), Di Segni *et al.*, (1979), and Rosen *et al.* (1982), it cannot be concluded that the initiation factors actually relieved competition. Since Ray *et al.* (1983) used a reconstituted cell-free system, it could be that eIF-4A or CBP II were limiting and that they observed a nonspecific stimulation of protein synthesis. Ray *et al.* (1983) concluded that neither eIF-2 nor eIF-4B was able to stimulate translation differentially. This conclusion is questionable, since eIF-2 had a clear, though weaker, effect, even though it was added in 20- to 40-fold lower molar amounts than eIF-4A or CBP II; even fewer molecules of eIF-4B were present. Translating globin mRNA in a Krebs ascites cell-free system, Parets-Soler *et al.* (1981) observed that a fraction containing 28,000 and 50,000 M_r proteins, reminiscent of CBP II, differentially stimulates α-globin synthesis. Here, too, it is not clear if they observed true relief of translational competition.

Clearly, it will be necessary to demonstrate, first, relief of translational competition in an efficient and more native cell-free system, such as the micrococcal nuclease-treated reticulocyte lysate, and under conditions where the total number of initiation events does not change, and second, selective and specific binding of eIF-4A or CBP-II to individual mRNA species, before it can be concluded with certainty that these proteins can act as targets of mRNA competition.

It is interesting to note that the effects of eIF-4A and CBP II on differential translation of reovirus RNA species were comparable when uncapped reovirus RNA was substituted (Ray *et al.*, 1983). This result extends data reviewed in Section 5.2 indicating that these proteins interact with internal features of mRNA, even though CBP II and the eIF-4A apparently contained therein can be crosslinked readily to the cap structure (Tahara *et al.*, 1981; Sonenberg, 1981; Grifo *et al.*, 1983).

6.3. Further Examples of mRNA Competition

Studying the effect of low doses of cycloheximide on polysome size and translation yield (Lodish, 1971; see Section 6.1.1), Walden *et al.* (1981) showed that *in vivo*, the various reovirus mRNAs initiate translation less effectively than do most host mRNAs. This relationship is preserved *in vitro* (Brendler *et al.*, 1981). When protein synthesis in mouse myeloma cells is slowed by hypertonic salt concentrations in the medium (Nuss and Koch, 1976*a*) or by starvation or actinomycin D (Sonenshein and Brawerman, 1976*a*), synthesis of light and heavy immunoglobulin chains becomes predominant, indicating that they are encoded by particularly strong mRNA species. Indeed, these mRNAs are preferentially translated in a wheat germ cell-free system when competition pressure is applied by mRNA concentration, by KOAc (see Fig. 2), or by inhibitory amounts of poly (A) (Sonenshein and Brawerman, 1976*b*). Furthermore, a correspondence between initiation efficiency *in vivo* and *in vitro* was suggested by Wieringa *et al.* (1981) for the liver mRNAs encoding very low-density lipoprotein II, serum albumin, and vitellogenin. They observed that, in whole cells, the packing density of ribosomes per nucleotide decreases, while *in vitro*, the sensitivity of translation to cap analogues or aurin tricarboxylic acid increases in this order for the three mRNA species.

On the other hand, Asselbergs *et al.* (1980) observed translational competition between globin mRNA, lens crystallin mRNAs, and turnip yellow mosaic virus RNA in the micrococcal nuclease-treated reticulocyte lysate, but not after microinjection into *Xenopus* oocytes (Asselbergs *et al.*, 1979). The reason for this discrepancy is not clear, as Laskey *et al.* (1977) have shown that in oocytes, a component of the translation machinery, and not mRNA, limits the rate of protein synthesis.

Ignotz *et al.* (1981), studying the selective inhibition of ribosomal protein synthesis in chick embryo fibroblasts deprived of insulin, concluded that it is due to less efficient initiation of translation on some ribosomal protein mRNAs, and suggested that these mRNAs compete less effectively in translation. Likewise, synthesis of proinsulin appears to occur on a weak mRNA template, and increases preferentially when overall translation is stimulated in whole cells by glucose (Lomedico and Saunders, 1977).

That the mRNAs of many viruses, including polio and VSV, are preferentially translated in intact, infected cells has been demonstrated ingeniously by their relative resistance to hypertonic shock (NaCl), a treatment that inhibits initiation of translation (Nuss *et al.*, 1975; Nuss and Koch, 1976*b*). On the other hand, Lodish and Porter (1980) could find no evidence for preferential translation of VSV mRNAs over typical cellular mRNAs in infected cells. Preferential translation over host mRNAs has been shown for EMC virus RNA *in vivo* (Jen *et al.*, 1978) and *in vitro* (Lawrence and Thach, 1974; Golini *et al.*, 1976; Svitkin *et al.*, 1978), as well as for mengovirus RNA *in vitro* (Abreu and Lucas-Lenard, 1976; Hackett *et al.*, 1978*a,b*; Rosen *et al.*, 1982).

In the reticulocyte lysate, translation of the G and N mRNAs of VSV proceeds with the same initiation efficiency, but initiation on G mRNA is more sensitive to inhibition (Lodish and Froshauer, 1977). Indeed, in infected cells, synthesis of G protein is correspondingly more sensitive to hypertonic shock than is synthesis of N protein (Nuss and Koch, 1976*b*). In the wheat germ system, STNV RNA is an order of magnitude more effectively translated than globin mRNA, although the use of a plant-derived cell-free system may have limited the effectiveness of the globin mRNA; on the other hand, STNV RNA competes more effectively in that system than a series of other plant viral mRNAs, with alfalfa mosaic virus coat protein mRNA (AMV-RNA 4) being the next strongest competitor (Herson *et al.*, 1979). Interestingly, in that study, several viral RNAs, such as AMV-RNA 3, competed less effectively than globin mRNA.

Chroboczek *et al.* (1980), studying translation of plant viral RNA in the wheat germ system, found that, in the presence of mRNA from developing cotyledons of *Vicia faba*, translation of nonstructural BMV mRNAs is arrested, while synthesis of BMV coat protein is only slightly inhibited. Similarly, synthesis of nonstructural polypeptides on tobacco mosaic virus RNA is arrested when *Vicia* mRNA is added. It seems that mRNA for seed storage proteins is able to out-compete mRNAs for nonstructural viral polypeptides that are needed for virus replication. Chroboczek *et al.* proposed that this property may explain the immunity of plant seed against viral infections.

The thesis seems warranted, therefore, that mRNA competition in translation is an ubiquitous phenomenon that influences gene expression during normal cell growth, as well as during differentiation or virus infection.

7. TRANSLATIONAL CONTROL BY REGULATION OF eIF-2 ACTIVITY

Mechanisms that influence the activity of initiation factors should have a profound effect on translation, particularly if they affect initiation factors that are involved in the binding and recognition of mRNA. It is now clear that a multitude of conditions affects the activity of eIF-2; these will be reviewed in this section. An additional case of modulation of initiation factor activity, concerning the virus-induced inactivation of cap-binding proteins, will be described in Section 8.4.

7.1. Heme Deprivation

Protein synthesis in reticulocyte lysates is regulated by the availability of heme, as it is in whole cells (Zucker and Schulman, 1968; Adamson *et al.*, 1968). In the absence of added heme, initiation of translation continues abated for about 5–10 min, but then stops abruptly. During this period, the level of 40 S/Met-tRNA$_f$ complexes declines (Legon *et al.*, 1973). The lesion in initiation can be restored by the addition of an initiation factor (Kaempfer and Kaufman, 1972) identified later as eIF-2 (Kaempfer, 1974*b*; Clemens, 1976). The absence of heme leads apparently to the inactivation of eIF-2.

Incubation of lysates in the absence of heme leads to the activation of an inhibitor of initiation that is able to depress initiation in lysates containing heme (Gross and Rabinovitz, 1972*a,b*). These investigators identified two general types of inhibitor: a reversible one that causes only transient inhibition, and an irreversible one that causes near-total cessation of translation. The reversible form appears after brief incubation of reticulocyte postribosomal supernatant, while the irreversible one is formed after longer incubation or in the presence of *N*-ethylmaleimide (NEM). This irreversible form is designated as the heme-controlled repressor (HCR; reviewed by Jackson, 1982; Austin and Kay, 1982). All three forms are believed to arise from the same proinhibitor (Gross, 1974). The irreversible or NEM-activated HCR forms are more stable than the reversible repressor, but the latter is most likely responsible for the observed translational control by heme. It is generally assumed that the reversible repressor acts like HCR, which has been studied in far greater detail.

HCR copurifies with a cAMP-independent protein kinase that specifically phosphorylates the α-subunit of eIF-2 (see Section 5.1; Levin *et al.*, 1976; Kramer *et al.*, 1976; Gross and Mendelewski, 1977), most likely at a single site. The kinase, a protein of about 90,000 M_r, readily phosphorylates itself (Gross and Mendelewski, 1978). Farrell *et al.* (1977) demonstrated that phosphorylation of eIF-2 is linked to inhibition of translation in the absence of heme and showed that ATP hydrolysis is needed to establish inhibition. This finding suggested that phosphorylation of eIF-2 might be the cause of its inactivation. However, phosphorylated eIF-2 is apparently as active in ternary complex formation with Met-tRNA$_f$ and GTP, in binding of Met-tRNA$_f$ to 40 S subunits, and in stimulating translation as is native eIF-2 (Trachsel and Staehelin, 1978; Safer *et al.*, 1977; Bonne *et al.*, 1980). Indeed, phosphorylated and nonphosphorylated eIF-2 are equally able to restore translation in heme-deficient reticulocyte lysates (Safer *et al.*, 1977; Benne *et al.*, 1980). In all these assays, however, eIF-2 molecules function stoichiometrically rather than catalytically, and large amounts of the factor are required for activity. This led to the suggestion that phosphorylation might impair the catalytic recycling of eIF-2 between rounds of initiation (Cherbas and London, 1976), although the opposite interpretation, that phosphorylation has no effect on recycling, was also offered (Benne *et al.*, 1980).

The isolation of an activity that is not eIF-2, yet restores translation in heme-deficient lysates brought the solution: this activity turned out to be one of the cofactors involved in eIF-2 function, namely, eIF-2B which catalyzes GTP/GDP exchange (De Haro and Ochoa, 1978; Ranu and London, 1979; Amesz *et al.*, 1979; Siekierka *et al.*, 1981; Austin and Kay, 1982). Indeed, Siekierka *et al.* (1982) showed that phosphorylation of the α-subunit of eIF-2, but not of the β-subunit, specifically blocks the ability of eIF-2B to exchange GDP for GTP in eIF-2. Clemens *et al.* (1982), moreover, were able to show that HCR prevents GTP/GDP exchange in eIF-2, in a reaction that requires ATP hydrolysis. These findings were confirmed by Matts *et al.* (1983). They explained the long-puzzling observation that high concentrations of GTP (1–2 mM) prevent the decline in protein synthesis occurring in heme-deficient lysates and that the effect of GTP is counteracted by ATP (Balkow *et al.*, 1975; Ernst *et al.*, 1976). Previously, it seemed reasonable to explain this phenomenon primarily in terms of an effect of GTP on the eIF-2 kinase, although there was no direct evidence to support this contention (Ernst *et al.*, 1976; Jackson, 1982). Now, however, it seems equally possible that GTP exerts its effect on eIF-2 itself. Apparently, high concentrations of GTP can also effect GDP/GTP exchange in eIF-2 (Pain and Clemens, 1983) when the normal, eIF-2B-dependent mechanism is no longer operational.

The picture that emerges from these studies thus is that, in the absence of heme, an inactive eIF-2 kinase is converted to an active form that phosphorylates the α-subunit of eIF-2. The phosphorylated factor is normally active in initiation of translation, but the released eIF-2/GDP complex is unable to undergo GDP/GTP exchange that is catalyzed by eIF-2B. Consequently, recycling of eIF-2 is prevented, and initiation comes to a halt.

An important but yet unexplained observation is the lack of a direct relationship between the extent of inhibition of translation and the extent of phosphorylation of eIF-2. Thus, almost complete inhibition of translation can be observed when no more than 30% of the eIF-2 pool is phosphorylated (Leroux and London, 1982; Safer *et al.*, 1977; Amesz *et al.*, 1979). There are several possible explanations for this discrepancy. One hypothesis is that eIF-2B is present in limiting amounts and that phosphorylation of a minority of eIF-2 molecules is sufficient to tie up the entire eIF-2B pool, leaving none available to recycle as yet unphosphorylated eIF-2. Indeed, Matts *et al.* (1983) have found that phosphorylated eIF-2 forms a complex with eIF-2B that is not readily dissociable. Another possibility is that the phosphorylation of eIF-2 is not the primary cause for inhibition but occurs only when normal functioning of eIF-2 is prevented by other means. Evidence in favor of such an explanation is available for the very similar inhibitory effect on translation exerted by dsRNA (Rosen *et al.*, 1981a; see Section 7.2). Jagus and Safer (1981a,b) have reported that there are conditions in which eIF-2 is inactivated, but where there is no apparent change in its state of phosphorylation (see Section 7.4.1).

7.2.   Double-Stranded RNA

One of the clearest instances of translational control that can be studied *in vitro* is the block in initiation of translation in reticulocyte lysates observed in the presence of dsRNA (Ehrenfeld and Hunt, 1971). Initiation of translation becomes inhibited with kinetics virtually identical to those seen in the absence of heme (Hunter *et al.*, 1975) and as in that case, the block involves inactivation of an initiation factor (Kaempfer and Kaufman, 1973) identified as eIF-2 (Kaempfer, 1974*b*; Clemens *et al.*, 1975); addition to this factor is sufficient to overcome the dsRNA-mediated inhibition (Kaempfer, 1974*b*).

The presence of dsRNA in lysates leads to phosphorylation of eIF-2 (Farrell *et al.*, 1977) at the same site in the α-subunit that is phosphorylated by HCR (Ernst *et al.*, 1980). The protein kinase responsible for this phosphorylation is activated by dsRNA and is antigenically distinct from HCR (Petryshyn *et al.*, 1979). This dsRNA-activated inhibitor of translation is ribosome-associated, while HCR is found in the postribosomal supernatant (Farrell *et al.*, 1977). The purified dsRNA-activated kinase undergoes autophosphorylation upon activation, yielding a labeled 67,000 M_r component (Levin *et al.*, 1981).

In the presence of dsRNA, 40 S/Met-tRNA$_f$ complexes disappear (Darnbrough *et al.*, 1972) while eIF-2 becomes phosphorylated (Farrell *et al.*, 1977). It is thus tempting to assume that, as in the case of heme deprivation, failure of phosphorylated eIF-2 to recycle at initiation is at the basis of the translational inhibition elicited by dsRNA. Two types of evidence indicate, however, that the phosphorylation of eIF-2 is the result of a more complex series of events. The first is that low concentrations (0.1–100 ng/ml) of dsRNA effectively inhibit initiation of translation in reticulocyte lysates, while by contrast, high concentrations (10–20 μg/ml) do not (Hunter *et al.*, 1975). Indeed, at high concentrations of dsRNA, eIF-2 does not become phosphorylated (Farrell *et al.*, 1977). Second, according to the kinase model in its simplest form, dsRNA, by inducing the phosphorylation of eIF-2 and preventing the recycling of eIF-2, abolishes binding of Met-tRNA$_f$ to 40 S ribosomal subunits and, hence, subsequent binding of mRNA and initiation of translation cannot take place. If that were correct, then one would predict that dsRNA should inhibit translation in general, irrespective of the type of mRNA being translated. Rosen *et al.* (1981*a*), however, discovered that this is not the case and demonstrated, instead, mRNA specificity in the inhibitory action of dsRNA on translation. These apparent paradoxes are considered in the next section.

7.2.1. The mRNA-Specific Action of Double-Stranded RNA

Translation of mengovirus or Coxsackie virus RNA in the mRNA-dependent reticulocyte lysate is resistant to inhibition by dsRNA in conditions where translation of globin or ascites tumor cell mRNA is sensitive (Rosen *et al.*, 1981*a*). The inhibition of globin mRNA translation by dsRNA can be reversed completely by the addition of eIF-2. The unabated translation of mengovirus RNA in the presence of dsRNA is dependent on continued initiation; not only is translation of mengovirus RNA as sensitive as that of globin mRNA to the specific inhibitors of initiation, pactamycin and aurin tricarboxylate, but unlike for globin mRNA, when mengovirus RNA is the template, the number of initiation events is not decreased by the presence of dsRNA. The resistance of mengovirus RNA translation to dsRNA is not caused by a lesser dependence on initiation factors, but by a failure of dsRNA to establish inhibition when mengovirus RNA is used as messenger. Rosen *et al.* (1981*a*) found that the nature of the mRNA being translated is a critical factor in the formation of a dsRNA-activated inhibitor of translation. Mengovirus or Coxsackie virus RNA, they showed, prevents the formation of dsRNA-activated inhibitor, while globin mRNA does not. Yet, once it is allowed to form, the inhibitor is as effective in blocking translation of mengovirus RNA as it is in blocking translation of globin mRNA, when assayed in the presence of non-inhibitory, high concentrations of dsRNA.

During translation of globin mRNA in the template-dependent lysate, the presence of dsRNA stimulates the phosphorylation of the α-subunit of eIF-2, as well as a 67,000 M_r band (the kinase) that is characteristic of the state of translational inhibition (Farrell *et al.*, 1977). By contrast, when mengovirus RNA is used as template, the dsRNA-dependent phosphorylation of either polypeptide is significantly depressed (Rosen *et al.*, 1981*a*). Moreover, in a ribosomal system containing eIF-2, dsRNA, and ATP, mengovirus RNA inhibits the phosphorylation of the α-subunit of eIF-2 and of the 67,000 M_r polypeptide, apparently in a competitive manner. Globin mRNA does not possess this property.

Direct RNA binding competition studies reveal that dsRNA competes with mRNA for eIF-2, binding this factor more strongly than globin mRNA, but more weakly than mengovirus RNA (Rosen *et al.*, 1981*a*). A molecule of mengovirus RNA, it will be recalled (see Section 6.1.2), exhibits a 30-fold greater affinity for eIF-2 than does a molecule of globin mRNA (Rosen *et al.*, 1982). The progressively greater affinities of globin mRNA, dsRNA, and mengovirus RNA for eIF-2 are also reflected by the ability of these RNA species to competitively inhibit the binding of Met-tRNA$_f$ to eIF-2 (Rosen *et al.*, 1981*a*; Section 5.1.1).

These findings show mRNA specificity in the inhibitory action of dsRNA on translation. The correlation between the results of translation, inhibitor formation, phosphorylation, and binding competition experiments suggests that the affinity of a given mRNA species for eIF-2 is important in determining the sensitivity of its translation to dsRNA. The observations support the concept that the rate-determining event in the establishment of inhibition of translation by dsRNA involves competition between dsRNA and mRNA, with inhibitor formation, phosphorylation, and inactivation of eIF-2 depending on the outcome of this competition. The data on mRNA specificity in the action of dsRNA on translation of Rosen *et al.* (1981*a*) are wholly consistent with the concept that a direct interaction between mRNA and eIF-2 occurs during translation, as indicated already by translational competition and RNA-affinity studies (Sections 5 and 6).

Rosen *et al.* (1981*a*) considered two possible explanations for their findings. The first is that the rate-determining step in the action of dsRNA involves a competition between mRNA and dsRNA for a proinhibitor. The proinhibitor would exhibit a preference for mengovirus RNA, but activation would require its binding to dsRNA. While this model satisfactorily explains the observation that the nature of the mRNA being translated is critical for the formation of the dsRNA-activated inhibitor, rather than for its action, it fails to explain the observed correlation between the results of translation and phosphorylation experiments with the RNA-affinity properties of eIF-2. The essential feature of the second model considered is that the rate-determining step in the inactivation of eIF-2 by dsRNA involves a direct competition of mRNA and dsRNA for eIF-2 (Rosen *et al.*, 1981*a*). Binding of eIF-2 to mRNA will lead to protein synthesis, regenerating eIF-2 for further rounds of competition. The sequence of events leading to phosphorylation and inactivation of eIF-2, then, would take place once dsRNA binds to eIF-2. Accordingly, during protein synthesis in the presence of dsRNA, the relative affinities of

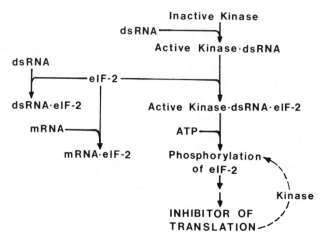

Fig. 5. Model for the mechanism of action of dsRNA on translation. From Rosen *et al.* (1981*a*), with permission. Copyright 1981 American Chemical Society.

mRNA and dsRNA for eIF-2, and their effective concentrations, will determine whether eIF-2 binds to mRNA or dsRNA and, hence, whether translation or inhibition will ensue.

Hunter *et al.* (1975) and Farrell *et al.* (1977) found that formation of the dsRNA-activated inhibitor is ATP-dependent and does not occur if high concentrations of dsRNA are present, but that these levels of dsRNA do not impair the action of the inhibitor. To reconcile this hitherto paradoxical observation with, first, the finding of increased eIF-2 kinase activity in the presence of dsRNA and a correlation between phosphorylation of eIF-2 and translational inhibition (Farrell *et al.*, 1977), and second, the mRNA specificity of dsRNA action (Rosen *et al.*, 1981*a*); the latter suggested the following hypothesis (see Fig. 5). An inactive eIF-2 kinase is activated by forming a complex with dsRNA. Before this activated dsRNA-kinase complex can phosphorylate eIF-2, this initiation factor must bind to the dsRNA molecule in the complex, forming a ternary complex consisting of kinase, dsRNA, and eIF-2. mRNA can compete with the kinase-bound dsRNA for eIF-2, and when mRNA binds to eIF-2 (in 40 S complexes; see Section 6.1.3), the above ternary complex will not be formed. In the presence of high concentrations of dsRNA, eIF-2 will tend to bind to free dsRNA molecules rather than to those coupled with the kinase, and ternary complex formation also will not occur. Once the ternary complex is formed, ATP-dependent phosphorylation of eIF-2 takes place. The phosphorylation of eIF-2 leads to the formation of a translational inhibitor. Since this inhibitor is active in the presence of high concentrations of dsRNA, it may act by making other eIF-2 molecules substrates for an eIF-2-kinase that does not depend on dsRNA, although other explanations are possible. This point is treated more extensively in Section 7.5.

7.3. Interferon

Translation is more sensitive to inhibition by dsRNA in extracts from interferon-treated cells than in those from control cells (Kerr *et al.*, 1974). As in reticulocyte lysates, dsRNA causes eIF-2 to be phosphorylated to a significant extent in interferon-treated cell extracts, but much less in extracts of uninfected cells (Lebleu *et al.*, 1976; Roberts *et al.*, 1976; Zilberstein *et al.*, 1978). The translational block observed in the presence of dsRNA in extracts of interferon-treated mouse L cells can be overcome completely by the addition of eIF-2 (Kaempfer *et al.*, 1979*b*)

A dsRNA-activated eIF-2 kinase that resembles the dsRNA-dependent kinase present in reticulocyte lysates (Farrell *et al.*, 1977) has been purified from interferon-treated L and ascites cells (Sen *et al.*, 1978; Kimchi *et al.*, 1979; Samuel, 1979). The level of this kinase increases between three- and ten-fold after treatment with interferon; interferons α, β, and γ all induce the enzyme (Hovanessian *et al.*, 1980; Lengyel, 1982*a,b*). Indeed, addition of dsRNA to intact, interferon-treated cells induces the phosphorylation of endogenous eIF-2, while addition to control cells does not (Gupta, 1979).

These observations can account for the greater sensitivity of translation to dsRNA in extracts from interferon-treated cells. Apparently, interferon treatment induces a rise in the level of dsRNA-dependent eIF-2-kinase, but since the enzyme is inactive in the absence of sufficient dsRNA, protein synthesis in interferon treated, uninfected cells continues normally. During infection, however, virus-generated dsRNA potentiates the kinase and the resulting extensive phosphorylation of eIF-2 leads to a general inhibition of initiation of translation. As noted for reticulocyte lysates, in extracts of interferon-treated cells, too, there is no good correlation between the extent of phosphorylation of eIF-2 and inhibition of translation (Jacobsen *et al.*, 1983).

In addition to a direct effect on eIF-2 activity, interferons induce other activities that affect translation less immediately. These include

the induction of an ATP- and dsRNA-dependent (2'-5')oligo (A) synthetase, a phosphodiesterase that degrades (2'-5')oligo (A), and a (2'-
5')oligo (A)-activated endoribonuclease. This ribonuclease attacks
mRNA as well as rRNA (Wreschner *et al.*, 1981). Some tRNAs are
also inactivated by interferon treatment of cells. In addition, methylation of 5' ends of newly synthesized viral RNA molecules is impaired. These phenomena have recently been reviewed in detail by
Lengyel (1982*a,b*). The importance of the (2'-5')oligo (A)-activated
ribonuclease for the action of interferon *in vivo* is apparent from the
finding that, in a clone of mouse cells that lack the enzyme, interferon
is able to induce the antiviral state against certain viruses, but not
against EMC (Epstein *et al.*, 1981).

One question that deserves attention is the fact that, at least in
the case of some virus–cell systems, viral RNA translation is inhibited
preferentially over host mRNA translation, for example, in reovirus-
infected L cells (Gupta *et al.*, 1974). An interesting possibility was
raised by Nilsen and Baglioni (1979). They showed that, in extracts
of interferon-treated cells, VSV mRNA hybridized with poly (U) at
its poly (A) tail or EMC RNA hybridized with poly (I) at its poly (C)
tract are more rapidly degraded than the corresponding control
mRNAs. They proposed that, in infected, interferon-treated cells,
activation of the endoribonuclease takes place near the replicative
intermediate of RNA viruses, because the dsRNA moiety therein promotes the formation of (2'-5')oligo (A) in its vicinity. As a result, the
viral mRNA portion in the replicative intermediate may be more sensitive to degradation than host mRNA.

7.4. Other Conditions that Affect eIF-2 Activity

An inhibition of initiation very similar in characteristics to that
observed in the absence of heme or in the presence of dsRNA can
be seen if a reticulocyte lysate is preincubated at high temperature
(42–44°C) (Mizuno, 1977; Bonanou-Tzedaki *et al.*, 1978) or is subjected to high hydrostatic pressure (Jackson, 1982). Both treatments
are now believed to cause activation of the HCR (Mizuno, 1977; Ernst
et al., 1982; Jackson, 1982).

Most of the information on the regulation of eIF-2 activity comes
from studies of reticulocyte lysates. It appears, however, that very
similar, if not identical mechanisms operate in a variety of mammalian
cell types. Initiation of translation in a cell-free system from HeLa

cells exhibits characteristics that strikingly resemble those of the reticulocyte lysate, such as the need for heme in continued initiation, the formation of an HCR-like inhibitor, and the ability of 2 mM GTP to largely replace the need for heme (Weber *et al.*, 1975). Cell-free extracts from resting lymphocytes have a reduced ability to form 40 S initiation complexes, compared to extracts from mitogen-stimulated lymphocytes, and this difference can be overcome by addition of eIF-2 (Kay *et al.*, 1979). Another example concerning nucleated mammalian cells, amino acid starvation, is reviewed in Section 7.4.2.

7.4.1. Sugar Phosphates and Reducing Agents

A link between the activity of eIF-2 in translation and the energy metabolism of the cell was first suggested by findings showing that initiation in lysates from reticulocytes that have been incubated anaerobically, or under other conditions that lead to depletion of the intracellular ATP pool, is inhibited even in the presence of hemin, but can be reactivated by adding fructose-1,6-diphosphate together with dithiothreitol, or by adding glucose-6-phosphate and NADP (Giloh and Mager, 1975; Giloh *et al.*, 1975; Ernst *et al.*, 1978*a*). A need for sugar phosphates and reducing agents was also perceived in reticulocyte lysates subjected to gel filtration (Hunt, 1976). The need for reducing power to maintain protein synthesis had already been revealed by the fact that oxidized glutathione (GSSG) inhibits initiation in a manner closely resembling that seen in the absence of heme (Kosower *et al.*, 1972). In the presence of GSSG, activation of the heme-controlled kinase, HCR, can be observed (Ernst *et al.*, 1978*b*), while a reversible form of this kinase is readily inactivated in the presence of dithiothreitol (Gross, 1978). Moreover, in gel-filtered lysates incubated without reducing agents, increased phosphorylation of the α-subunit of eIF-2 can be observed (Jackson *et al.*, 1983). Jagus and Safer (1981*a*), however, could find no evidence for increased phosphorylation of eIF-2 under the same conditions, and concluded that one or more critical sulfhydryl groups in eIF-2 itself, required for the activity or catalytic utilization of this factor, becomes oxidized (Jagus and Safer, 1981*b*). In this context, it has long been known that treatment with NEM abolishes the ability of eIF-2 to bind Met-tRNA$_f$ in the presence of GTP, but does not affect the mRNA-binding activity of eIF-2 (Kaempfer, 1974*b*).

The fact that, as in heme-deficient reticulocyte lysates (Balkow *et al.*, 1975), translation in gel-filtered lysates can be restored by high levels (2 mM) of GTP (Ernst *et al.*, 1976; Jagus and Safer, 1981*a*), suggests that gel filtration impairs the ability of eIF-2 to interact with GTP. Most likely, this is because phosphorylation of the α-subunit of eIF-2 prevents GTP/GDP exchange by eIF-2B, as in the case of heme deprivation (see Sections 7.1 and 7.5).

At first, it seemed as if the need for sugar phosphates and reducing agents might be related, as inhibition of translation in gel-filtered lysates, in lysates from cells subjected to anaerobiosis, or in lysates incubated with GSSG can be prevented by dithiothreitol or by glucose-6-phosphate and NADP that generate NADPH with the aid of glucose-6-phosphate dehydrogenase. However, a role for sugar phosphates *per se*, with only a minor role for NADPH, was suggested by several studies (Ernst *et al.*, 1978*b*; Lenz *et al.*, 1978; West *et al.*, 1979). Indeed, sugar phosphates, but not NADPH, enhance binding of Met-tRNA$_f$ to 40 S ribosomal subunits (Lenz *et al.*, 1978; West *et al.*, 1979). Hunt *et al.* (1983) solved the problem of dissociating the production of NADPH from the presence of sugar phosphates by passing a gel-filtered lysate over 2′,5′ ADP-Sepharose. This treatment revealed a requirement for NADPH, thioredoxin, and thioredoxin reductase for the maintenance of high protein synthesis activity in reticulocyte lysates. The sugar phosphate requirement can be met by glucose-6-phosphate, 2-deoxy-glucose-6-phosphate, and fructose-1,6-diphosphate, but not by 6-phosphogluconate. The reducing power requirement can be met by dithiothreitol or by an NADPH-generating system in conjunction with a functional thioredoxin reductase/thioredoxin system (Hunt *et al.*, 1983).

While, as just mentioned, it is not yet clear if the reducing power is needed to prevent activation of an eIF-2 kinase, to maintain eIF-2 in an active form, or both, Jackson *et al.* (1983) concluded that sugar phosphates do not act by affecting the phosphorylation of eIF-2. The role of sugar phosphates is still unresolved, but they clearly act cooperatively with reducing agents in preventing the decreased binding of Met-tRNA$_f$ to 40 S subunits (Jackson, 1982; Jackson *et al.*, 1983). Finally, the observation that the presence of 1 mM ATP and 0.13 mM GTP during gel filtration of a reticulocyte lysate prevents the impairment of eIF-2 activity (Jagus and Safer, 1981*a*) may reflect the need for eIF-2 to be bound to GTP, but may also indicate a more complex involvement of these nucleotides.

7.4.2. Amino Acid Starvation

Protein synthesis in mammalian cell cultures is closely regulated by the supply of essential nutrients (van Venrooij *et al.*, 1972; Austin and Clemens, 1981). In such cells, deprivation of amino acids, glucose, or serum results in a reduction in translation that occurs immediately in cells deprived of an essential amino acid and within 1 hr in cells deprived of glucose (van Venrooij *et al.*, 1972). Refeeding amino acid-deprived cells with the missing amino acid results in full recovery of the rate of translation within minutes, and the recovery is seen even if RNA synthesis is blocked (van Venrooij *et al.*, 1972; Sonenshein and Brawerman, 1977; Pain *et al.*, 1980). In extracts of amino acid-starved cells, initiation is inhibited, but it can be restored completely by the addition of eIF-2 (Pain *et al.*, 1980). Although eIF-2 is phosphorylated in extracts of both fed and starved cells, there is no noticeable increase in phosphorylation in starvation (Austin and Clemens, 1981). Here, too, the extent of phosphorylation of eIF-2 does not correlate well with the extent of translational inhibition, but again this may be explained, as suggested in Section 7.1, by irreversible binding of a limiting eIF-2 recycling factor. An inhibitor of initiation that resembles HCR in activity has been detected in extracts from ascites tumor cells, and its level is slightly higher in amino acid-starved cells; heme stimulates translation in extracts of fed cell extracts, but not in those from starved cells (Austin and Clemens, 1981).

7.5. Nature of the Translational Inhibitor

With the more recent insight into the control of eIF-2 activity by heme (Section 7.1), it becomes interesting to consider the possibility that the translational inhibitor in the hypothesis of Rosen *et al.* (1981*a*; Fig. 5), concerning the closely related action of dsRNA, may be phosphorylated eIF-2 itself, which, by tying up eIF-2B (Matts *et al.*, 1983), prevents the recycling of other, active eIF-2 molecules. This would fit rather well with the observation, already mentioned in Section 7.1, that complete inhibition of translation can be seen when only 20–30% of eIF-2 is phosphorylated. Active but nonrecycling eIF-2 molecules, in turn, might become prone to phosphorylation by the heme-controlled eIF-2 kinase, thus amplifying the extent of inhibition.

The apparently paradoxical observation that high concentrations of GTP not only restore translation in gel-filtered reticulocyte lysates (Jagus and Safer, 1981*a*), in lysates deprived of heme (Balkow *et al.*, 1975; Ernst *et al.*, 1976), incubated in the presence of dsRNA (Ernst *et al.*, 1976), or subjected to high temperature (Mizuno, 1977), as well as in HeLa cell extracts (Weber *et al.*, 1977), but actually block *formation* of the heme-controlled inhibitor (Balkow et al., 1975) is in agreement with this concept. It was previously assumed that GTP may somehow prevent activation of various eIF-2 kinases. As stated in Section 7.1, however, an excess of GTP may promote GTP/GDP exchange in eIF-2, even if it cannot be catalyzed by eIF-2B. Conceivably, such exchange may enable phosphorylated eIF-2 molecules to continue to recycle in initiation, and consequently, to leave eIF-2B molecules available for normal recycling of nonphosphorylated eIF-2 molecules. The latter, in turn, may be less prone to the action of eIF-2 kinases as long as they are recycling actively.

Should this as yet unproven view be correct, two types of translational inhibitor should be clearly distinguished. The first comprises the various kinases that phosphorylate the α-subunit of eIF-2 and are controlled by heme or activated by dsRNA or other conditions (Section 7.4). These kinases repress translation because they cause eIF-2 to be phosphorylated. The second type of inhibitor is phosphorylated eIF-2 itself, which, by competing with active eIF-2 for the GDP/GTP exchange factor, eIF-2B, depletes the pool of recycling eIF-2B as just suggested and prevents recycling of eIF-2. The observation that phosphorylated eIF-2 is able to stimulate translation in heme-deficient lysates (Safer *et al.*, 1977; Benne *et al.*, 1980) is not at variance with this view, because the observed effect concerns noncatalytic functioning of eIF-2.

While the final explanation of the phenomena connected with translational control by dsRNA and by heme deprivation may turn out different from that considered here, it is clear from the observations discussed in Section 7.2.1 that the steps leading to phosphorylation of eIF-2 may be under more subtle and complex control than first thought, even though the effect of phosphorylation on eIF-2 activity has now been clarified.

8. TRANSLATIONAL REGULATION BY OTHER MEANS

This section is devoted to several cases of translational control where specific initiation factor(s), if involved, have not yet been identified, or where more than one mechanism of control is activated, as in the case of virus-induced shut-off of host translation.

8.1. Translational Repression

In the cell, messenger RNA does not exist as free molecules, but appears to be complexed with protein in messenger ribonucleoprotein (mRNP) particles. Although the procedure of cell breakage and extraction can yield artifactual RNP complexes (Girard and Baltimore, 1966), there is now good evidence that mRNP particles do exist in the cell and play an important role in translational control. For a more detailed review of mRNP structure and function, see Hershey (1982*b*) and Schmid *et al.* (1982).

Gross *et al.* (1964) and Spirin and coworkers (Spirin, 1969) were the first to observe that unfertilized eggs contain mRNPs that are inactive in translation, and Spirin termed them "informosomes." Within minutes of fertilization, the rate of translation in the sea urchin egg rises dramatically, increasing about 25-fold within an hour (Regier and Kafatos, 1977). Before fertilization, less than 1% of the egg's ribosomes is found in polysomes, but the percentage increases 60-fold thereafter (Humphreys, 1971). The increase in rate of protein synthesis occurs in the absence of new mRNA synthesis, for it is also seen in enucleated or actinomycin D-treated embryos (Gross *et al.*, 1964; Davidson, 1976). The increase in translation is not accompanied by a change in the rate of initiation, as the packing density of ribosomes on mRNA does not change significantly upon fertilization (Humphreys, 1969, 1971). Although an increase of up to two-fold in elongation rate can be observed (Brandis and Raff, 1979), this is insufficient to account for the far more dramatic rise in the rate of translation. Gross *et al.* (1973) found that the unfertilized egg contains amounts of mRNA activity comparable to those of the embryo, when mRNA is phenol-extracted and translated *in vitro*; in the egg, however, mRNA is complexed with protein. These results indicate that masking of potentially active mRNA by proteins in the unfertilized egg, and mobilization of stored mRNA upon fertilization, control the rate of translation in early embryogenesis (Woodland and Wilt, 1980). According to Brandhorst (1976), fertilization does not result in the expression of a novel set of proteins, but in a strong overall increase in translation of the same set expressed before fertilization.

In the surf clam, however, there is compelling evidence that fertilization results in the selective activation of translation of specific mRNA species (Rosenthal *et al.*, 1980). mRNA extracted before and after fertilization gave identical patterns of protein synthesized upon translation in the RNA-dependent reticulocyte lysate, while homogenates from fertilized eggs showed the appearance and extensive synthesis of novel proteins, and decreased synthesis of others, when compared to homogenates from oocytes. Clearly, the translational repression active in the clam oocyte acts in an mRNA-specific manner. Rather similar results were obtained in studies with starfish oocytes (Rosenthal *et al.*, 1982) and with *Drosophila* oocytes and young embryos (Mermod *et al.*, 1980).

In muscle development, a number of myofibrillar proteins appears at specific stages of differentiation. There is evidence that expression of myosin heavy chain mRNA may involve its synthesis and storage as mRNP before the mRNA is mobilized into polysomes at a specific stage (Buckingham *et al.*, 1974; Heywood and Rich, 1968; Jain and Sarkar, 1979), but the data are not yet as convincing as in the previous examples.

In cells growing under steady-state conditions, up to 50% of the mRNA may be in mRNP, depending on the rate of growth. In HeLa cells, pulse-chase studies support the view that mRNP are not intermediate forms of mRNA that later enter polysomes, and indeed, the RNA in the mRNP compartment turns over some four-fold more rapidly (Spohr *et al.*, 1970; Mauron and Spohr, 1978). The mRNA population in mRNPs and in polysomes differs in complexity, both in ascites tumor cells (McMullen *et al.*, 1979; Kinniburgh *et al.*, 1979) and in avian erythroblasts (Maundrell *et al.*, 1979). The mRNPs in the latter system are translationally inactive *in vitro* (Civelli *et al.*, 1980). In rabbit reticulocytes, α-globin mRNA is found in free mRNPs, while β-globin mRNA is not (Gianni *et al.*, 1972), suggesting a possible relationship between initiation strength and the formation of mRNPs in that system. However, such a relationship, if it exists, cannot account for the observations in the surf clam of Rosenthal *et al.* (1980).

The nature of the repressor proteins that prevent the translation of mRNA in mRNA particles has not yet been resolved in spite of much diligent effort, especially by Scherrer *et al.* in the avian globin synthesis system (Vincent *et al.*, 1981; reviewed by Hershey, 1982*b*). Thus, today, it is not yet clear how masking and unmasking is accomplished at the protein level. It is particularly difficult to determine if a given mRNP complex is relevant to the intact cell, or an artifact formed during or after its extraction from the cell. The finding that the heat shock response in *Xenopus* oocytes involves reversible masking of mRNA (Bienz and Gurdon, 1982) may be useful to analyse this question.

Another example of mRNA masking is suggested by the iron-induced synthesis of ferritin. In reticulocytes of the bullfrog tadpole, synthesis of ferritin increases 40- to 50-fold in the presence of iron, while the level of ferritin mRNA, translatable after extraction in the wheat germ system, does not change perceptibly (Shull and Theil, 1982). It has been suggested that liver ferritin protein molecules not complexed with iron may bind to their own mRNA templates and act as translational repressor (Zähringer *et al.*, 1976), but evidence for this model has so far been lacking.

A number of reports describe low-molecular weight RNAs that inhibit mRNA translation, usually in a nonspecific manner, but their relevance to translational control remains to be demonstrated (see Austin and Kay, 1982).

Cervera *et al.* (1981) have found that cytoskeleton preparations of HeLa cells contain active mRNA, while free mRNA not found in polysomes is absent. Conceivably, proteins in mRNP particles may act to prevent the attachment of mRNA to the cytoskeleton.

As opposed to the above-reviewed studies, which all point to a negative role for proteins in free mRNP particles, polysome-derived mRNP particles contain proteins that are essential for translation. In a Krebs ascites tumor cell-free system from which RNA-binding proteins have been removed, at least in part, by passage over RNA-Sepharose, polysome-derived mRNPs are translated, while deproteinized mRNA is not (Schmid *et al.*, 1982). The polysomal mRNPs contain fewer proteins than free mRNPs, and are as active as deproteinized mRNA in whole cell-free systems for translation (reviewed by Hershey, 1982*b*).

8.2. Heat Shock

In a large variety of organisms, heat shock causes a general inhibition of translation of mRNA from previously active genes, accompanied with extensive synthesis of a small set of heat shock proteins (see Schlesinger *et al.*, 1982). Thus, in *Drosophila* cells, raising the temperature from 25 to 36°C results in cessation of transcription of previously expressed genes and activation of transcription of heat shock mRNAs. Translation of all pre-existing mRNA species ceases, but they are not degraded, and their translation resumes if the temperature is returned to 25°C (Lindquist, 1981). Lysates from normal cells will translate normal and heat shock mRNA equally well, but lysates from heat-shocked cells will translate only heat shock mRNA (Storti *et al.*, 1980; Krüger and Benecke, 1981). Apparently, heat shock induces a reversible change in the translational machinery that allows the efficient translation of only heat shock mRNA because it differs structurally from normal RNA, in a manner that is not yet understood.

A remarkable variety of mechanisms has been proposed to explain the heat shock response in different cells. Lindquist *et al.* (1982) and Di Domenico *et al.* (1982) suggest that, in *Drosophila* and yeast cells, heat shock proteins repress their own synthesis. The response in *Xenopus* oocytes differs from that in somatic cells in that only translational control appears to be involved; heat shock mRNA is masked and inactive in translation at normal temperatures, but is unmasked by high temperature (Bienz and Gurdon, 1982). At that temperature, translation of normal mRNA is inhibited by another mechanism. Lowering the temperature results in reactivation of normal mRNA and remasking of heat shock mRNA.

In *Drosophila* and HeLa cells, a specific lesion in the elongation process was first thought to be responsible for the inhibition of translation of normal mRNA (Scott and Pardue, 1981; Thomas and Mathews, 1982), but it is now apparent that both initiation and elongation are reduced 15- to 30-fold in rate (Ballinger and Pardue, 1983). The nature of these lesions is not yet clear.

In reticulocyte lysates, on the other hand, it is the activation of the eIF-2 kinase that seems primarily responsible for the observed inhibition, here at initiation of translation (Mizuno, 1977; Bonanou-Tzedaki *et al.*, 1978; Ernst *et al.*, 1982).

8.3. Phosphorylation of Ribosomal Protein

In many systems, the degree of phosphorylation of ribosomal protein S6 shows a strong positive correlation with the rate of initiation of protein synthesis (Gressner and Wool, 1974; Martini and Kruppa, 1979; Ballinger and Hunt, 1981; Thomas *et al.*, 1980*b*; Glover, 1982). However, increased S6 phosphorylation can also be observed in cases where protein synthesis does not occur, as when isolated ribosomes are exposed to insulin receptor and insulin (Rosen *et al.*, 1981*b*) or in whole cells during cycloheximide treatment (Thomas *et al.*, 1980*b*). On the other hand, a function for S6 phosphorylation in translation is suggested by the findings that complete or partial inhibition of S6 phosphorylation by methylxanthines leads to a corresponding inhibition of protein synthesis (Thomas *et al.*, 1980*b*). Indeed, 40 S ribosomal subunits containing the most highly phosphorylated forms of S6 have an apparent selective advantage in entering polysomes (Duncan and McConkey, 1982; Thomas *et al.*, 1982).

8.4. Virus-Induced Shut-Off of Host Translation

Many cytopathogenic viruses induce a decline in the rate of host mRNA translation upon infection. Since this shut-off of host translation is accompanied by extensive synthesis of viral protein, and since the decline in host protein synthesis is not concomitant with degradation or inactivation of host mRNA, it is clear that translational control mechanisms are involved in this phenomenon. A number of explanations have been invoked to explain shut-off by various viruses. Here, a few of these will be examined that are related directly to the material already reviewed above, but for a more extensive treatment, surveys by Koch *et al.* (1982) and elsewhere in this volume should be consulted.

One prominent mechanism that probably operates in many virus infections is mRNA competition. The molecular basis for such competition has been reviewed in Section 6. It is useful to consider that competition at initiation of translation is effective not only in cases of viruses producing mRNA that is of unusually high competitive strength, such as the picornaviruses, EMC, mengo, or STNV (Section 6). In principle, even viral mRNA of weak or moderate ability to compete in translation will lead to a reduction in cellular protein synthesis, as long as it is synthesized in large amounts during infection. This is a consequence of the fact that initiation is the rate-limiting step in protein synthesis. A good example of such a virus is VSV (Lodish and Porter, 1980). After infection, the total amount of translatable mRNA in the cell increases about threefold. Although the VSV mRNA is initiated with approximately the same efficiency as host mRNA, the large excess of viral RNA leads to translational competition at the expense of host mRNA and a commensurate decline in host protein synthesis.

A mechanism that may be restricted more specifically to certain viruses is the inactivation of cap-binding proteins accompanying, for example, the shut-off of host protein synthesis by poliovirus (see Chapter by E. Ehrenfeld in this volume). Lee and Sonenberg (1982) have demonstrated that the ability of five polypeptides, associated with the CBP II complex, to be crosslinked to the cap is reduced in extracts of poliovirus-infected cells. This reduction correlates with the inability of initiation factor preparations from infected cells to restore translation of capped mRNAs in extracts of poliovirus-infected cells. Indeed, initiation factor preparations from poliovirus-infected cells have the ability to inactivate cap-binding proteins and to impair the restoring activity of initiation factors from uninfected cells (Brown and Ehrenfeld, 1980; Lee and Sonenberg, 1982). It is generally believed that poliovirus, by inactivating cap-binding proteins, removes the ability of host mRNA to compete in translation, thereby allowing translation of the uncapped viral RNA, which is otherwise a relatively ineffective template, unusually dependent on eIF-4A (Daniels-McQueen et al., 1983). If that is correct, then the cardioviruses, such as EMC and mengovirus, may depend far less on inactivation of cap-binding proteins, as their RNAs compete exceedingly well in translation (Section 6).

A related change occurs during the late stages of replication of reovirus in L cells. The early reovirus mRNA is capped and is translated in competition with host mRNA. Late in infection, however, the viral mRNA is uncapped, and it cannot be translated in extracts of uninfected cells, even though it is an efficient template for translation in extracts of infected cells (Skup et al., 1981). Apparently, reovirus induces a gradual modification in the cap-binding proteins, such that late in infection, translation of host mRNA and early viral mRNA comes to a halt, permitting only the translation of uncapped viral mRNA.

In cells infected with Semliki Forest virus, synthesis of early, nonstructural viral proteins is encoded by the genomic 42 S viral RNA. Late in infection, synthesis of host and early viral proteins ceases, although 42 S RNA and host mRNA persist, and a subgenomic 26 S viral RNA, comprising the 3'-proximal third of the 42 S RNA molecule, serves as efficient template for the synthesis of structural virus proteins (see Kääriäinen and Söderlund, 1978). Although the late mRNA is capped in this case, infected cells become deficient in cap-binding proteins in such a manner that only 26 S viral mRNA is translated (Van Steeg et al., 1981). Apparently, the 26 S viral mRNA has a very high ability to compete in translation when competition is sharpened by a decrease in cap-binding proteins.

Several host–virus interactions at the level of translational control may involve eIF-2. The ability of a strong viral RNA template, such as mengovirus RNA, to out-compete host mRNA for eIF-2 (Rosen et al., 1982) will lead to the selective translation of viral mRNA concomitant with a displacement of host mRNA from the ribosomes.

Even though in mengovirus-infected cells, the physical intactness of host mRNA as a whole does not change perceptibly (Colby *et al.,* 1974), host mRNA, once displaced, may become more prone to limited nuclease attack at sites essential for translation.

A second mechanism involving eIF-2 may operate during infection: resistance of viral RNA translation to inhibition by dsRNA, coupled with sensitivity of host mRNA translation. Early in infection, dsRNA begins to be formed as a result of viral RNA replication. Initially, the level of dsRNA is low, but it increases during infection. A role for dsRNA in shut-off of host protein synthesis was first suggested by Hunt and Ehrenfeld (1971) who showed that cytoplasm from poliovirus-infected cells inhibits initiation of translation in reticulocyte lysates, and identified the inhibitor as dsRNA (Ehrenfeld and Hunt, 1971). A role for dsRNA in shut-off was rejected later when a differential effect of dsRNA on host and viral mRNA translation could not be demonstrated in cell-free systems capable of only limited initiation (Robertson and Mathews, 1973; Celma and Ehrenfeld, 1974). By using a reticulocyte lysate system able to initiate translation with an efficiency resembling that of the intact cell, however, Rosen *et al.* (1981*a*) could show that translation of mengovirus RNA is, in fact, completely resistant to inhibition by concentrations of dsRNA that totally inhibit the translation of globin or ascites tumor cell mRNA. As reviewed in Section 7.2, dsRNA causes the inactivation of eIF-2, but fails to do so when mengovirus RNA is used as template. The most likely explanation of this observation is suggested by the finding that dsRNA competes with mRNA for eIF-2, binding it more weakly than mengovirus RNA, but more strongly than globin mRNA (Rosen *et al.,* 1981*a*). The high affinity of mengovirus RNA for eIF-2 may thus serve to permit continued viral mRNA translation even in the presence of amounts of virus-generated dsRNA that are sufficient to inhibit host protein synthesis.

The salt optimum for translation of viral RNA appears to be higher than that for host mRNA, for exposure of infected cells to hypertonic salt concentrations leads to a block in initiation of host mRNA translation, while still permitting translation of viral RNA (Nuss *et al.,* 1975). Indeed, picornavirus-infected cells contain higher salt concentrations than do uninfected cells (Carrasco and Smith, 1976; Egberts *et al.,* 1977), but the change may be too gradual to account for shut-off of host protein synthesis. The observation that complexes between mengovirus RNA and eIF-2 can still form at salt concentrations at which complexes between globin mRNA and eIF-

2 are no longer stable (Rosen *et al.*, 1982) is consistent with the differential effect of salt on protein synthesis in virus-infected cells.

The high affinity of mengovirus RNA for eIF-2 could thus contribute in a number of ways to the transition from host to viral protein synthesis. After infection, when viral RNA begins to accumulate, favorable competition for eIF-2 will lead to progressive displacement of host mRNA from polysomes. Viral dsRNA formed during replication is expected to cause inactivation of some eIF-2 molecules, acting thereby to decrease the pool size of eIF-2 and to sharpen the competition between host and viral mRNA for eIF-2, to the advantage of the stronger template, viral RNA. Likewise, an increase in intracellular salt concentrations will cause a preferential reduction in the binding of host mRNA to eIF-2. These effects will further increase the advantage of the stronger viral mRNA and enhance the shift from cellular to viral protein synthesis. Later in infection, when more dsRNA has accumulated, viral mRNA is the predominant template in translation and protects the residual eIF-2 molecules against inactivation by dsRNA; viral proteins continue to be synthesized at a high rate. Should the viral RNA be able to compete more favorably for other initiation factors as well, then this will enhance the above course of events.

Yet another type of regulation is suggested by the finding that vaccinia virus particles or cores inhibit initiation of translation in reticulocyte lysates (Person *et al.*, 1980; Cooper and Moss, 1978). Vaccinia virus RNA transcripts, generated in infected cells or *in vitro* by viral cores, inhibit in such lysates or in wheat germ extracts the translation of globin mRNA or mRNA from HeLa cells and even the translation of EMC RNA, yet they do not inhibit translation of vaccinia viral mRNA (Coppola and Bablanian, 1983). Since the vaccinia transcripts are not translated particularly efficiently in these experiments, favorable competition by these RNA species is unlikely to account for the results. The inhibitory activity appears to consist of RNA, but it is not dsRNA. The nature of this inhibitory RNA, as well as the mechanism by which it apparently discriminates against nonvaccinia mRNA species, remain to be characterized.

9. PARAMETERS DETERMINING TRANSLATIONAL EFFICIENCY OF AN mRNA SPECIES

One of the outstanding questions that remains to be clarified is the very wide range in initiation efficiency encountered among eukaryotic mRNA species. That this range is at least 100-fold follows from the observations that mengovirus RNA competes 35-fold more effectively at initiation than globin mRNA (Rosen *et al.*, 1982), while globin mRNA is a more effective template than albumin mRNA which, in turn, competes significantly better than haemopexin mRNA in translation (Kaempfer and Konijn, 1983). Although in the latter cases the relative initiation efficiencies have not been quantitated precisely, the 100-fold range may well be a conservative estimate.

The role of the 5'-leader sequence in determining initiation efficiency is difficult to determine on the basis of presently available data. This point is well illustrated by the five mRNA species of VSV. *In vivo*, these mRNAs are apparently translated with identical efficiencies of initiation (Villareal *et al.*, 1976), yet they possess 5' leaders that are unrelated in sequence and vary in length from 10–41 nucleotides (Rose, 1980).

This variation in 5' ends in the face of identical initiation efficiency suggests that it may not be the leader sequence *per se* that determines initiation strength, but the interaction of this sequence with other, internal parts of the mRNA molecule. Since the coding sequence in an mRNA molecule is dictated by the individual protein encoded, and since the 3'-untranslated sequence is also highly variable (Section 3.5), any stable interaction between 5' leader and internal sequences is only possible if the leader sequence is especially tailored to allow such a fit. If this view is correct, then an important determinant for the efficiency of initiation is the structure generated by this interaction around the initiation codon.

The fact that denaturation of eukaryotic mRNA does not lead to binding of ribosomes at internal sites in mRNA (Kozak, 1980*a*; Collins *et al.*, 1982; Zagorska *et al.*, 1982) is not in conflict with this concept. It merely indicates that structure in mRNA is not important for determining *where* initiation occurs. As mentioned in Section 3.1, this point should be separated clearly from the question of *how often* initiation takes place. Two mRNA species can vary widely in initiation efficiency even if, as is generally the case, initiation occurs at the first AUG codon in both. It is here that the contribution of structure may be of essence.

The absence of 5'-proximal sequences in both eIF-2 and ribosome-protected segments of mengovirus RNA (Perez-Bercoff and Kaempfer, 1982) supports the concept that the RNA of this virus has evolved a highly efficient mechanism of initiation that bypasses the need for either a 5' end or a 5'-terminal cap structure. Conceivably, this is because a structured site, highly favorable for initiation, is generated within this RNA molecule. Indeed, the related EMC RNA possesses an unusual primary and secondary structure at the start of the coding region, and this highly structured site is resistant to nuclease treatment (Smith, 1975; Porter *et al.*, 1975). The existence of a structure possessing very high affinity for eIF-2 (and possibly for other components of initiation) would obviate the need for the additional stabilization imparted by binding at the cap.

Yet, in STNV RNA, which is also an efficiently competing mRNA species lacking a 5'-terminal cap (Wimmer *et al.*, 1968; Horst *et al.*, 1971), the eIF-2- and ribosome-binding sites lie close to the 5' end (Browning *et al.*, 1980; Kaempfer *et al.*, 1981). High efficiency of initiation may, in this case, result from a suitable structure at the initiation site, coupled with proximity to the 5' end (Fig. 1). As the affinity properties of the structure generated around the initiation codon become less favorable, proximity to the 5' end may start to provide a contribution. For even less favorable initiation sites, the additional contribution of the 5'-terminal cap becomes increasingly important, as evidenced for all cellular and many viral mRNA species. Apparently, eIF-2 recognizes in different mRNA species a common conformation existing around the initiation codon, but differing in subtle ways that are important for determining individual binding affinities.

A regulatory role for the cap structure in translational control in intact mammalian cells is particularly well illustrated by findings of Cordell *et al.* (1982). In normal pancreatic tissue, the two rat insulin genes are expressed about equally, but in tumor tissue, one of these genes is expressed ten-fold less, in spite of the fact that equivalent amounts of mRNA are produced. Insulin mRNA from the underexpressed gene is ten-fold less active in *in vitro* translation, but this defect can be repaired by treatment with vaccinia virus capping extract.

The accessibility of the AUG initiation codon to factors and ribosomes may well be influenced by secondary structure. Diminished accessibility has been suggested for α-globin mRNA as compared to β-globin mRNA (Pavlakis *et al.*, 1980), but the generality of this concept as a determinant of initiation efficiency remains to be documented.

ACKNOWLEDGMENTS

I thank Mrs. E. Herskovics for typing the manuscript. Research of the author was supported by the National Academy of Sciences of Israel, the Gesellschaft fuer Strahlen- und Umweltforschung (Muenchen), and the National Council for Research and Development of Israel.

10. REFERENCES

Abreu, S. L., and Lucas-Lenard, J., 1976, Cellular protein synthesis shutoff by mengovirus: Translation of nonviral and viral mRNAs in extracts from uninfected and infected Ehrlich ascites tumor cells, *J. Virol.* **18:**182.

Adamson, S. D., Herbert, E., and Godchaux, W., 1968, Factors affecting the rate of protein synthesis in lysate systems from reticulocytes, *Arch. Biochem. Biophys.* **125:**671.

Amesz, H., Goumans, H., Haubrich-Morree, T., Voorma, H. O., and Benne, R., 1979, Purification and characterisation of a protein factor that reverses the inhibition of protein synthesis by the heme-regulated translational inhibitor in rabbit reticulocyte lysates, *Eur. J. Biochem.* **98:**513.

Anderson, C. W., Lewis, J. B., Atkins, J. F., and Gesteland, R. F., 1974, Cell-free synthesis of adenovirus 2 proteins programmed by fractionated messenger RNA: A comparison of polypeptide products and messenger RNA lengths, *Proc. Natl. Acad. Sci. USA* **71:**2756.

Asselbergs, F. A. M., van Venrooij, W. J., and Bloemendal, H., 1979, Messenger RNA competition in living *Xenopus* oocytes, *Eur. J. Biochem.* **94:**249.

Asselbergs, F. A. M., Meulenberg, E., van Venrooij, W. J., and Bloemendal, H., 1980. Preferential translation of mRNAs in an mRNA-dependent reticulocyte lysate, *Eur. J. Biochem.* **109:**159.

Austin, S. A., and Clemens, M. J., 1981, The effects of haem on translational control of protein synthesis in cell-free extracts from fed and lysine-deprived Ehrlich ascites cells, *Eur. J. Biochem.* **117:**601.

Austin, S. A., and Kay, J. E., 1982, Translational regulation of protein synthesis in eukaryotes, *Essays Biochem.* **18:**79.

Baglioni, C., Simili, M., and Shafritz, D. A., 1978, Initiation activity of EMC virus RNA, binding to eIF-4B and shut-off of host cell protein synthesis, *Nature* **275:**240.

Balkow, K., Hunt, T., and Jackson, R. J., 1975, Control of protein synthesis in reticulocyte lysates: The effect of nucleotide triphosphates on formation of the translational repressor. *Biochem. Biophys. Res. Commun.* **67:**366.

Ballinger, D. G., and Hunt, T., 1981, Fertilization of sea urchin eggs is accompanied by 40 S ribosomal subunit phosphorylation, *Dev. Biol.* **87:**277.

Ballinger, D. G., and Pardue, M. L., 1983, The control of protein synthesis during heat shock in *Drosophila* cells involves altered polypeptide chain elongation rates, *Cell* **33:**103.

Banerjee, A. K., 1980, 5′-Terminal cap structure in eucaryotic messenger ribonucleic acids, *Microbiol. Rev.* **44:**175.

Baralle, F. E., and Brownlee, G. G., 1978, AUG is the only recognised signal sequence in the 5′ non-coding regions of eukaryotic mRNA, *Nature* **274:**84.

Barrieux, A., and Rosenfeld, M. G., 1977, Characterization of GTP-dependent Met-tRNA$_f$ binding protein, *J. Biol. Chem.* **252:**3843.

Barrieux, A., and Rosenfeld, M. G., 1978, mRNA-induced dissociation of initiation factor 2, *J. Biol. Chem.* **253:**6311.

Benne, R., and Hershey, J. W. B., 1976, Purification and characterization of initiation factor IF-E3 from rabbit reticulocytes, *Proc. Natl. Acad. Sci. USA* **73**:3005.

Benne, R., and Hershey, J. W. B., 1978, The mechanism of action of protein synthesis initiation factors from rabbit reticulocytes, *J. Biol. Chem.* **253**:3078.

Benne, R., Wong, C., Luedi, M., and Hershey, J. W. B., 1976, Purification and characterization of initiation factor IF-E2 from rabbit reticulocytes, *J. Biol. Chem.* **251**:7675.

Benne, R., Salimans, M., Goumans, H., Amesz, H., and Voorma, H. O., 1980, Regulation of protein synthesis in rabbit reticulocyte lysates. Phosphorylation of eIF-2 does not inhibit its capacity to recycle, *Eur. J. Biochem.* **104**:501.

Bergmann, I. E., and Brawerman, G., 1977, Control of breakdown of polyadenylate sequence in mammalian polyribosomes: Role of poly (adenylic acid)–protein interactions, *Biochemistry* **16**:259.

Bergmann, J. E., Trachsel, H., Sonenberg, N., Shatkin, A. J., and Lodish, H. F., 1979, Characterization of rabbit reticulocyte factor that stimulates the translation of mRNA lacking 5'-terminal 7-methylguanosine, *J. Biol. Chem.* **254**:1440.

Bienz, M., and Gurdon, J. B., 1982, The heat shock response in *Xenopus* oocytes is controlled at the translational level, *Cell* **29**:811.

Bonanou-Tzedaki, S. A., Smith, K. E., Sheeran, B. A., and Arnstein, H. R. V., 1978, Reduced formation of initiation complexes between met-tRNA$_f$ and 40s ribosomal subunits in rabbit reticulocyte lysates at elevated temperature, *Eur. J. Biochem.* **84**:601.

Bos, J. L., Polder, L. J., Bernards, R., Schrier, P. I., van der Elsen, P. J., van der Eb, A. J., and van Ormondt, H., 1981, The 2.2 kb Elb mRNA of human Ad12 and Ad5 codes for two tumor antigens starting at different AUG triplets, *Cell* **27**:121.

Brandhorst, B. P., 1976, Two-dimensional gel patterns of protein synthesis before and after fertilization of sea urchin eggs, *Dev. Biol.* **52**:310.

Brandis, J. W., and Raff, R. A., 1979, Elevation of protein synthesis is a complex response to fertilization, *Nature* **278**:467.

Brawerman, G., 1981, The role of the poly(A) sequence in mammalian messenger RNA, *Crit. Rev. Biochem.* **10**:1.

Breathnach, R., and Chambon, P., 1981, Organization and expression of eucaryotic split genes coding for proteins, *Annu. Rev. Biochem.* **50**:349.

Brendler, T., Godefroy-Colburn, T., Yu, S., and Thach, R. E., 1981, The role of mRNA competition in regulating translation. Comparisons of *in vitro* and *in vivo* results, *J. Biol. Chem.* **256**:11755.

Brown, B., and Ehrenfeld, E., 1980, Initiation factor preparations from poliovirus infected cells restrict translation in reticulocyte lysates, *Virology* **103**:327.

Browning, K. S., Leung, D. W., and Clark, J. M., Jr., 1980, Protection of satellite tobacco necrosis virus ribonucleic acid by wheat germ 40 S and 80 S ribosomes, *Biochemistry* **19**:2276.

Buckingham, M. E., Caput, D., Cohen, A., Whalen, R. G., and Gros, F., 1974, The synthesis and stability of cytoplasmic messenger RNA during myoblast differentiation in culture, *Proc. Natl. Acad. Sci. USA* **71**:1466.

Cancedda, R., Villa-Komaroff, L., Lodish, H. F., and Schlesinger, M., 1975, Initiation sites for translation of Sindbis virus 42S and 26S messenger RNAs, *Cell* **6**:215.

Carrasco, L., and Smith, A. E., 1976, Sodium ions and the shut-off of host cell protein synthesis by picornaviruses, *Nature* **264**:807.

Castel, A., Kraal, B., Konieczny, A., and Bosch, L., 1979, Translation by *Escherichia coli* ribosomes of alfalfa mosaic virus RNA 4 can be initiated at two sites on the monocistronic message, *Eur. J. Biochem.* **101**:123.

Celma, M. L., and Ehrenfeld, E., 1974, Translation of poliovirus RNA *in vitro*: Detection of two different initiation sites, *J. Mol. Biol.* **98**:761.

Cervera, M., Dryfus, G., and Penman, S., 1981, Messenger RNA is translated when associated with the cytoskeletal framework in normal and VSV-infected HeLa cells, *Cell* **23**:113.

Chaudhuri, A., Stringer, E. A., Valenzuela, D., and Maitra, U., 1981, Characterization of eIF-2 containing two polypeptide chains of M_r = 48,000 and 38,000, *J. Biol. Chem.* **256**:3988.

Checkley, J. W., Cooley, L., and Ravel, J. M., 1981, Characterization of eIF-3 from wheat germ, *J. Biol. Chem.* **256**:1582.

Chen, Y., Woodley, C., Bose, K., and Gupta, N. K., 1972, Protein synthesis in rabbit reticulocytes: Characteristics of a Met-tRNA$_f^{Met}$ binding factor, *Biochem. Biophys. Res. Commun.* **48**:1.

Cherbas, L., and London, I. M., 1976, On the mechanism of delayed inhibition of protein synthesis in heme-deficient rabbit reticulocyte lysates, *Proc. Natl. Acad. Sci. USA* **73**:3506.

Chroboczek, J., Witt, M., Ostrowka, K., Bassuner, R., Puchel, M., and Zagorski, W., 1980, Seed transmissibility of plant viruses may be modulated by competition between viral and cellular messengers. A proposal, *Plant Sci. Lett.* **19**:263.

Civelli, O., Vincent, A., Maundrell, K., Buri, J-F., and Scherrer, K., 1980. The translational repression of globin mRNA in free cytoplasmic ribonucleoprotein particles, *Eur. J. Biochem.* **107**:577.

Clark, B. F. C., and Marcker, K. A., 1966, The role of *N*-formyl-methionyl-sRNA in protein biosynthesis, *J. Mol. Biol.* **17**:394.

Clemens, M. J., 1976, Functional relationships between reticulocyte polypeptide chain initiation factor (IF-MP) and the translational inhibitor involved in regulation of protein synthesis by haemin, *Eur. J. Biochem.* **66**:413.

Clemens, M. J., Safer, B., Merrick, W. C., Anderson, W. F., and London, I. M., 1975, Inhibition of protein synthesis in rabbit reticulocyte lysates by double stranded RNA and oxidized glutathione: Indirect mode of action on polypeptide chain initiation, *Proc. Natl. Acad. Sci. USA* **72**:1286.

Clemens, M. J., Pain, V. M., Wong, S-T., and Henshaw, E. C., 1982, Phosphorylation inhibits guanine nucleotide exchange on eukaryotic initiation factor 2, *Nature* **296**:93.

Colby, D. S., Finnerty, V., and Lucas-Lenard, J., 1974, Fate of mRNA of L-cells infected with mengovirus, *J. Virol.* **13**:858.

Collins, P., Fuller, F., Marcus, P., Hightower, L., and Ball, L. A., 1982, Synthesis and processing of Sindbis virus nonstructural proteins *in vitro*, *Virology* **118**:363.

Cooper, J. A., and Moss, B., 1978, Transcription of vaccinia virus mRNA coupled to translation *in vitro*, *Virology* **88**:149.

Coppola, G., and Bablanian, R., 1983, Discriminatory inhibition of protein synthesis in cell-free systems by vaccinia transcripts, *Proc. Natl. Acad. Sci. USA* **80**:75.

Cordell, B., Diamond, D., Smith, S., Punter, J., Schone, H., and Goodman, H. M., 1982, Disproportionate expression of the two nonallelic rat insulin genes in a pancreatic tumor is due to translational control, *Cell* **31**:531.

Daniels-McQueen, S., Detjen, B., Grifo, J. A., Merrick, W. C., and Thach, R. E., 1983, Unusual requirements for optimum translation of polio viral RNA *in vitro, J. Biol. Chem.* **258**:7195.

Darnbrough, C., Hunt, T., and Jackson, R. J., 1972, A complex between Met-tRNA$_f$ and native 40 S subunits in reticulocyte lysates and its disappearance during incubation with double-stranded RNA, *Biochem. Biophys. Res. Commun.* **48**:1556.

Darnbrough, C. H., Legon, S., Hunt, T., and Jackson, R. J., 1973, Initiation of protein synthesis: Evidence for messenger RNA independent binding of methionyl-transfer RNA to the 40 S ribosomal subunit, *J. Mol. Biol.* **76**:379.

Dasgupta, A., Das, A., Roy, R., Ralston, R., Majumdar, A., and Gupta, N. K., 1978, Protein synthesis in rabbit reticulocytes. XXI. Purification and properties of a protein factor (Co-EIF-1) which stimulates Met-tRNA$_f$ binding to EIF-1, *J. Biol. Chem.* **253**:6054.

Davidson, E. H., 1976, *Gene Activity in Early Development,* Academic Press, New York.

Degener, A. M., Pagnotti, P., Facchini, J., and Perez-Bercoff, R., 1983, Genomic RNA of mengovirus. Translation of its two cistrons in lysates of interferon-treated cells, *J. Virol.* **45**:889.

De Haro, C., and Ochoa, S., 1978, Mode of action of the hemin-controlled inhibitor of protein synthesis: Studies with factors from rabbit reticulocytes, *Proc. Natl. Acad. Sci. USA* **75**:2713.

De Haro, C., and Ochoa, S., 1979, Further studies on the mode of action of the hemin-controlled translational inhibitor: Stimulating protein acts at level of binary complex formation, *Proc. Natl. Acad. Sci. USA* **76**:2163.

De Wachter, R., 1979, Do eukaryotic mRNA 5' noncoding sequences base-pair with the 18 S ribosomal RNA 3' terminus? *Nucl. Acids Res.* **7**:2045.

Di Domenico, B., Bugaisky, G., and Lindquist, S., 1982, The heat shock response is self-regulated at both the transcriptional and post-transcriptional levels, *Cell* **31**:593.

Di Segni, G., Rosen, H., and Kaempfer, R., 1979, Competition between α- and β-globin messenger ribonucleic acids for eukaryotic initiation factor 2, *Biochemistry* **18**:2847.

Duncan, R., and McConkey, E. H., 1982, Preferential utilization of phosphorylated 40-S ribosomal subunits during initiation complex formation, *Eur. J. Biochem.* **123**:535.

Efstratiadis, A., Kafatos, F. C., and Maniatis, T., 1977, The primary structure of β-globin mRNA as determined from cloned DNA, *Cell* **10**:571.

Efstratiadis, A., Posakony, J. W., Maniatis, T., Lawn, R. M., O'Connell, C., Spritz, R. A., DeRiel, J. K., Forget, B. G., Weissman, S. M., Slightom, J. L., Blechl, A. E., Smithies, O., Baralle, F., Shoulders, C. C., and Proudfoot, N. J., 1980, The structure and evolution of the human β-globin gene family, *Cell* **21**:653.

Egberts, E., Hackett, P., and Traub, P., 1977, Alteration of the intracellular energetic and ionic conditions by mengovirus infection of Ehrlich ascites tumor cells and its influence on protein synthesis in the midphase of infection, *J. Virol.* **22**:591.

Ehrenfeld, E., and Brown, D., 1981, Stability of poliovirus RNA in cell-free translation systems utilizing two initiation sites, *J. Biol. Chem.* **256**:2656.

Ehrenfeld, E., and Hunt, T., 1971, Double-stranded poliovirus RNA inhibits initiation of protein synthesis by reticulocyte lysates, *Proc. Natl. Acad. Sci. USA* **68**:1075.

Epstein, D. A., Czarniecki, C. W., Jacobsen, H., Friedman, R. M., and Panet, A., 1981, A mouse cell line, which is unprotected by interferon against lytic virus infection, lacks ribonuclease F activity, *Eur. J. Biochem.* **118**:9.

Ernst, V., Levin, D. H., Ranu, R. S., and London, I. M., 1976, Control of protein synthesis in reticulocyte lysates: Effects of cyclic AMP, ATP and GTP on inhibitions induced by heme-deficiency, dsRNA and a reticulocyte translational inhibitor. *Proc. Natl. Acad. Sci. USA* **73**:1112.

Ernst, V., Levin, D. H., and London, I. M., 1978a, Evidence that glucose 6-phosphate regulates protein synthesis initiation in reticulocyte lysates, *J. Biol. Chem.* **253**:7163.

Ernst, V., Levin, D. H., and London, I. M., 1978b, Inhibition of protein synthesis initiation by oxidized glutathione, *Proc. Natl. Acad. Sci. USA* **75**:4110.

Ernst, V., Levin, D. H., Leroux, A., and London, I. M., 1980, Site-specific phosphorylation of the α-subunit of eukaryotic initiation factor eIF-2 by the heme-regulated and double-stranded RNA-activated eIF-2α kinases from rabbit reticulocyte lysates, *Proc. Natl. Acad. Sci. USA* **77**:1286.

Ernst, V., Zukofsky-Baum, E., and Reddy, P., 1982, Heat shock, protein phosphorylation and the control of translation in rabbit reticulocytes, reticulocyte lysate, and HeLa cells, in: *Heat Shock, from Bacteria to Man* (M. J. Schlesinger, M. Ashburner, and A. Tissières, eds.), pp. 215–225, Cold Spring Harbor Laboratory, Cold Spring Harbor, New York

Falvey, A., and Staehelin, T., 1970, Structure and function of mammalian ribosomes. II. Exchange of ribosomal subunits at various stages of *in vitro* polypeptide synthesis, *J. Mol. Biol.* **53**:21.

Farrell, P. J., Balkow, K., Hunt, T., Jackson, R. J., and Trachsel, H., 1977, Phosphorylation of initiation factor eIF-2 and the control of reticulocyte protein synthesis, *Cell* **11**:187.

Fellner, P., 1979, General organization and structure of the picornavirus genome, in: *The Molecular Biology of Picornaviruses* (R. Perez-Bercoff, ed.), pp. 25–47, Plenum Publishing, New York.

Furuichi, Y., LaFiandra, A., and Shatkin, A. J., 1977, 5'-Terminal structure and mRNA stability, *Nature* **266**:235.

Gette, W. R., and Heywood, S. M., 1979, Translation of myosin heavy chain mRNA in an eIF-3- and messenger-dependent muscle cell-free system, *J. Biol. Chem.* **254**:9879.

Ghosh-Dastidar, P., Gilbin, D., Yaghmai, B., Das, A., Das, H. K., Parkhurst, H. K., and Gupta, N. K., 1980, A study of the mechanism of interreaction of fluorescently labeled Co-eIF-2A with eIF-2 using fluorescence polarization, *J. Biol. Chem.* **255**:3826.

Gianni, A. M., Giglioni, B., Ottolenghi, S., Comi, P., and Guidotti, G. G., 1972, Globin α-chain synthesis directed by "supernatant" 10S RNA from rabbit reticulocytes, *Nature New Biol.* **240**:183.

Giloh, H., and Mager, J., 1975, Inhibition of peptide chain initiation in lysates from ATP-depleted cells. I. Stages in the evolution of the lesion and its reversal by thiol compounds, cyclic AMP or purine derivatives and phosphorylated sugars, *Biochim. Biophys. Acta* **414**:293.

Giloh, H., Schochat, L., and Mager, J., 1975, Inhibition of peptide chain initiation in lysates from ATP-depleted cells, *Biochim. Biophys. Acta* **414**:309.

Girard, M., and Baltimore, D., 1966, The effect of HeLa cell cytoplasm on the rate of sedimentation of RNA, *Proc. Natl. Acad. Sci. USA* **56**:999.

Glanville, N., Ranki, M., Morser, J., Kääriäinen, L., and Smith, A. E., 1976, Initiation of translation directed by 42 S and 26 S RNAs from Semliki Forest virus *in vitro, Proc. Natl. Acad. Sci. USA* **73**:3059.

Glover, C. V. C., 1982, Heat shock induces rapid dephosphorylation of a ribosomal protein in *Drosophila, Proc. Natl. Acad. Sci. USA* **79**:1781.

Glover, J. F., and Wilson, T. M. A., 1982, Efficient translation of the coat protein cistron of tobacco mosaic virus in a cell-free system from *Escherichia coli, Eur. J. Biochem.* **122**:485.

Golini, F., Thach, S. S., Birge, C. H., Safer, B., Merrick, W. C., and Thach, R. E., 1976, Competition between cellular and viral mRNAs *in vitro* is regulated by a messenger discriminatory factor, *Proc. Natl. Acad. Sci. USA* **73**:3040.

Gressner, A. M., and Wool, I. G., 1974, The phosphorylation of liver ribosomal proteins *in vivo, J. Biol. Chem.* **249**:6917.

Grifo, J. A., Tahara, S. M., Leis, J. P., Morgan, M. A., Shatkin, A. J., and Merrick, W. C., 1982, Characterization of eukaryotic initiation factor 4A, a protein involved in ATP-dependent binding of globin mRNA, *J. Biol. Chem.* **257**:5246.

Grifo, J. A., Tahara, S. M., Morgan, M. A., Shatkin, A. J., and Merrick, W. C., 1983, New initiation factor activity required for globin mRNA translation, *J. Biol. Chem.* **258**:5804.

Gross, K. W., Jacobs-Lorena, M., Baglioni, C., and Gross, P. R., 1973, Cell-free translation of maternal messenger RNA from sea urchin eggs, *Proc. Natl. Acad. Sci. USA* **70**:2614.

Gross, M., 1974, Control of globin synthesis by hemin: An intermediate form of the translational repressor in rabbit reticulocyte lysates, *Biochim. Biophys. Acta* **366**:319.

Gross, M., 1978, Regulation of protein synthesis by hemin: Effect of dithiothreitol on the formation and activity of the hemin controlled translational repressor, *Biochim. Biophys. Acta* **520**:642.

Gross, M., and Mendelewski, J., 1977, Additional evidence that the hemin controlled repressor from rabbit reticulocytes is a protein kinase, *Biochem. Biophys. Res. Commun.* **74**:559.

Gross, M., and Mendelewski, J., 1978, Control of protein synthesis by hemin. An association between the formation of the hemin-controlled translational repressor and the phosphorylation of a 100,000 molecular weight protein, *Biochim. Biophys. Acta* **520**:650.

Gross, M., and Rabinovitz, M., 1972a, Control of globin synthesis in cell-free preparations of reticulocytes by formation of a translational repressor that is inactivated by haemin, *Proc. Natl. Acad. Sci. USA* **69**:1565.

Gross, M., and Rabinovitz, M., 1972b, Control of globin synthesis by hemin: Factors influencing formation of an inhibitor of globin chain initiation in reticulocyte lysates, *Biochim. Biophys. Acta* **287**:340.

Gross, P. R., Malkin, J. L., and Moyer, W. A., 1964, Templates for the first proteins of embryonic development, *Proc. Natl. Acad. Sci. USA* **51**:407.

Grunstein, M., and Schedl, P., 1976, Isolation and sequence analysis of sea urchin (*Lytechinus pictus*) histone H4 messenger RNA, *J. Mol. Biol.* **104**:323.

Gupta, N. K., 1982, Regulation of eIF-2 activity and initiation of protein synthesis in mammalian cells, in: *Protein Biosynthesis in Eukaryotes* (R. Perez-Bercoff, ed.), pp. 157–166, Plenum Publishing, New York.

Gupta, S. L., 1979, Specific protein phosphorylation in interferon-treated uninfected and virus infected mouse L929 cells: Enhancement by double-stranded RNA, *J. Virol.* **29**:301.

Gupta, S. L., Sopori, M. L., and Lengyel, P., 1974, Release of the inhibition of mRNA translation in extracts of interferon-treated Ehrlich ascites tumor cells by added tRNA, *Biochem. Biophys. Res. Commun.* **57**:763.

Guthrie, C., and Nomura, M., 1968, Initiation of protein synthesis: A critical test of the 30 S subunit model, *Nature* **219**:232.

Hackett, P. B., Egberts, E., and Traub, P., 1978a, Selective translation of mengovirus RNA over host mRNA in homologous, fractionated, cell-free translational systems from Ehrlich-ascites tumor cells, *Eur. J. Biochem.* **83**:353.

Hackett, P. B., Egberts, E., and Traub, P., 1978b, Translation of ascites and mengovirus RNA in fractionated cell-free systems from uninfected and mengovirus-infected Ehrlich-ascites tumor cells, *Eur. J. Biochem.* **83**:341.

Hagenbüchle, O., Santer, M., Steitz, J. A., and Mans, R. J., 1978, Conservation of the primary structure at the 3' end of 18 S rRNA from eucaryotic cells, *Cell* **13**:551.

Heindell, H. C., Liu, A., Paddock, G. V., Studnicker, G. M., and Salser, W., 1978, The primary structure of rabbit α-globin mRNA, *Cell* **15**:43.

Hershey, J. W. B., 1982a, The initiation factors, in: *Protein Biosynthesis in Eukaryotes* (R. Perez-Bercoff, ed.), pp. 97–117, Plenum Publishing, New York.

Hershey, J. W. B., 1982b, Messenger ribonucleoprotein particles, in: *Protein Biosynthesis in Eukaryotes* (R. Perez-Bercoff, ed.), pp. 157–166, Plenum Publishing, New York.

Herson, D., Schmidt, A., Seal, S., Marcus, A., and van Vloten-Doting, L., 1979, Competitive mRNA translation in an *in vitro* system from wheat germ, *J. Biol. Chem.* **254**:8245.

Heywood, S. M., and Rich, A., 1968, *In vitro* synthesis of native myosin, actin, and tropomyosin from embryonic chick polyribosomes, *Proc. Natl. Acad. Sci. USA* **59**:590.

Hickey, E. D., Weber, L. A., and Baglioni, C., 1976, Inhibition of initiation of protein synthesis by 7-methylguanosine-5'-monophosphate, *Proc. Natl. Acad. Sci. USA* **73**:19.

Horst, H., Fraenkel-Conrat, H., and Mandeles, S., 1971, Terminal heterogeneity at both ends of the satellite tobacco necrosis virus ribonucleic acid, *Biochemistry* **10**:4748.

Hovanessian, A. G., Meurs, E., Aujean, O., Vaquero, C., Stefanos, S. and Falcoff, E., 1980, Antiviral response and induction of specific proteins in cells treated with immune (Type II) interferon analogous to that from viral interferon (Type I)-treated cells, *Virology* **104**:195.

Howard, G., Adamson, S., and Herbert, E., 1970, Subunit recycling during translation in a reticulocyte cell-free system, *J. Biol. Chem.* **245**:6237.

Humphreys, T., 1969, Efficiency of translation of messenger-RNA before and after fertilization in sea urchins, *Dev. Biol.* **20**:435.

Humphreys, T., 1971, Measurements of messenger RNA entering polysomes upon fertilization of sea urchin eggs, *Dev. Biol.* **26**:201.

Hunt, T., 1974, The control of globin synthesis in rabbit reticulocytes, *Ann. N.Y. Acad. Sci.* **241**:223.

Hunt, T., 1976, Control of globin synthesis, *Br. Med. Bull.* **32**:257.

Hunt, T., and Ehrenfeld, E., 1971, Cytoplasm from poliovirus infected HeLa cells inhibits cell-free haemoglobin synthesis, *Nature New Biol.* **230**:91.

Hunt, T., Hunter, T., and Munro, A., 1968, Control of haemoglobin synthesis: A difference in the size of the polysomes making α and β chains, *Nature* **220**:481.

Hunt, T., Herbert, P., Campbell, E. A., Delidakis, C., and Jackson, R. J., 1983, The use of affinity chromatography on 2'5' ADP-sepharose reveals a requirement for NADPH, thioredoxin and thioredoxin reductase for the maintenance of high protein synthesis activity in rabbit reticulocyte lysates, *Eur. J. Biochem.* **131**:303.

Hunter, T., Hunt, T., Jackson, R. J., and Robertson, H. D., 1975, The characteristics of inhibition of protein synthesis by double-stranded ribonucleic acid in reticulocyte lysates, *J. Biol. Chem.* **250**:409.

Hunter, A. R., Hunt, T., Knowland, J., and Zimmern, D., 1976, Messenger RNA for the coat protein of tobacco mosaic virus, *Nature* **260**:759.

Ignotz, G. G., Hokari, S., DePhilip, R. M., Tsukada, K., and Lieberman, I., 1981, Lodish model and regulation of ribosomal protein synthesis by insulin-deficient chick embryo fibroblasts, *Biochemistry* **20**:2550.

Jackson, R. J., 1982, The cytoplasmic control of protein synthesis, in: *Protein Biosynthesis in Eukaryotes* (R. Perez-Bercoff, ed.), pp. 363–418, Plenum Publishing, New York.

Jackson, R. J., Herbert, P., Campbell, E. A., and Hunt, T., 1983, The roles of sugar phosphates and thiol-reducing systems in the control of reticulocyte protein synthesis, *Eur. J. Biochem.* **131**:313.

Jacobsen, H., Epstein, D. A., Friedmann, R. A., Safer, B., and Torrence, P. F., 1983, Double-stranded RNA-dependent phosphorylation of protein P1 and eukaryotic initiation factor 2α does not correlate with protein synthesis inhibition in a cell-free system from interferon-treated mouse L cells, *Proc. Natl. Acad. Sci. USA* **80**:41.

Jacobson, M. F., and Baltimore, D., 1968, Polypeptide cleavages in the formation of poliovirus proteins, *Proc. Natl. Acad. Sci. USA* **61**:77.

Jagus, R., and Safer, B., 1981a, Activity of eukaryotic initiation factor 2 is modified by processes distinct from phosphorylation. Activities of eukaryotic initiation factor 2 and eukaryotic initiation factor 2α kinase in lysate gel filtered under different conditions, *J. Biol. Chem.* **256**:1317.

Jagus, R., and Safer, B., 1981b, Activity of eukaryotic initiation factor 2 is modified by processes distinct from phosphorylation. Activity of eukaryotic initiation factor 2 in lysate modified by oxidation-reduction state of its sulfhydryl groups, *J. Biol. Chem.* **256**:1324.

Jain, S. K., and Sarkar, S., 1979, Poly(riboadenylate)-containing messenger ribonucleoprotein particles of chick embryonic muscles, *Biochemistry* **18**:745.

Jay, G., Nomura, S., Anderson, C. W., and Khoury, G., 1981, Identification of the SV40 agnogene product: A DNA binding protein, *Nature* **291**:346.

Jen, G., Birge, C. H., and Thach, R. E., 1978, Comparison of initiation rates of encephalomyocarditis virus and host protein synthesis in infected cells, *J. Virol.* **27**:640.

Jense, H., Knauert, F., and Ehrenfeld, E., 1978, Two initiation sites for translation of poliovirus *in vitro*: Comparison of LSc and Moloney strains, *J. Virol.* **28**:387.

Jones, R. L., Sadnik, I., Thompson, H. A., and Moldave, K., 1980, Studies on native ribosomal subunits from rat liver, *Arch. Biochem. Biophys.* **199**:277.

Kääriäinen, L., and Söderlund, H., 1978, Structure and replication of α-viruses, *Curr. Top. Microbiol. Immunol.* **82:**15.

Kabat, D., and Chappell, M. R., 1977, Competition between globin messenger ribonucleic acids for a discriminating initiation factor, *J. Biol. Chem.* **252:**2684.

Kaempfer, R., 1968, Ribosomal subunit exchange during protein synthesis, *Proc. Natl. Acad. Sci. USA* **61:**106.

Kaempfer, R., 1969, Ribosomal subunit exchange in the cytoplasm of a eukaryote, *Nature* **222:**950.

Kaempfer, R., 1970, Dissociation of ribosomes on polypeptide chain termination and origin of single ribosomes, *Nature* **228:**534.

Kaempfer, R., 1971, Control of single ribosome formation by an initiation factor for protein synthesis, *Proc. Natl. Acad. Sci. USA* **68:**2458.

Kaempfer, R., 1972, Initiation factor IF-3: A specific inhibitor of ribosomal subunit association, *J. Mol. Biol.* **71:**583.

Kaempfer, R., 1974a, The ribosome cycle, in: *Ribosomes* (M. Nomura, A. Tissières, and P. Lengyel, eds.), pp. 679–704, Cold Spring Harbor Laboratory, Cold Spring Harbor, New York.

Kaempfer, R., 1974b, Identification and RNA binding properties of an initiation factor capable of relieving translational inhibition induced by heme deprivation or double-stranded RNA, *Biochem. Biophys. Res. Commun.* **61:**591.

Kaempfer, R., 1979, Purification of initiation factor eIF-2 by RNA-affinity chromatography, in: *Methods in Enzymology*, Vol. 60, Part G (L. Grossman and K. Moldave, eds.), pp. 247–255, Academic Press, New York.

Kaempfer, R., 1982, Messenger RNA Competition, in: *Protein Biosynthesis in Eukaryotes* (R. Perez-Bercoff, ed.), pp. 441–458, Plenum Publishing, New York.

Kaempfer, R., 1984, Differential gene expression by messenger RNA competition for eukaryotic initiation factor 2, in: *Protein Synthesis: Translation and Post-Translational Events* (A. K. Abraham, T. S. Eikhom, and I. F. Pryme, eds.), pp. 57–76, The Humana Press, Clifton, New Jersey.

Kaempfer, R., and Kaufman, J., 1972, Translational control of hemoglobin synthesis by an initiation factor required for recycling of ribosomes and for their binding to messenger RNA, *Proc. Natl. Acad. Sci. USA* **69:**3317.

Kaempfer, R., and Kaufman, J., 1973, Inhibition of cellular protein synthesis by double-stranded RNA: Inactivation of an initiation factor, *Proc. Natl. Acad. Sci. USA* **70:**1222.

Kaempfer, R., and Konijn, A. M., 1983, Translational competition by mRNA species encoding albumin, ferritin, haemopexin and globin, *Eur. J. Biochem.* **131:**545.

Kaempfer, R., Meselson, M., and Raskas, H., 1968, Cyclic dissociation into stable subunits and reformation of ribosomes during bacterial growth, *J. Mol. Biol.* **31:**277.

Kaempfer, R., Hollender, R., Abrams, W. R., and Israeli, R., 1978a, Specific binding of messenger RNA and methionyl-tRNA$_f^{Met}$ by the same initiation factor for eukaryotic protein synthesis, *Proc. Natl. Acad. Sci. USA* **75:**209.

Kaempfer, R., Rosen, H., and Israeli, R., 1978b, Translational control: Recognition of the methylated 5' end and an internal sequence in eukaryotic mRNA by the initiation factor that binds methionyl-tRNA$_f^{Met}$, *Proc. Natl. Acad. Sci. USA* **75:**650.

Kaempfer, R., Hollender, R., Soreq, H., and Nudel, U., 1979a, Recognition of messenger RNA in eukaryotic protein synthesis: Equilibrium studies of the interaction between messenger RNA and the initiation factor that binds methionyl-tRNA$_f$, *Eur. J. Biochem.* **94**:591.

Kaempfer, R., Israeli, R., Rosen, H., Knoller, S., Zilberstein, A., Schmidt, A., and Revel, M., 1979*b*, Reversal of the interferon-induced block of protein synthesis by purified preparations of eucaryotic initiation factor 2, *Virology* **99**:170.

Kaempfer, R., Van Emmelo, J., and Fiers, W., 1981, Specific binding of eukaryotic initiation factor 2 to satellite tobacco necrosis virus RNA at a 5'-terminal sequence comprising the ribosome binding site, *Proc. Natl. Acad. Sci. USA* **78**:1542.

Kaempfer, R., Rosen, H., Di Segni, G., and Knoller, S., 1984, Structural feature of picornavirus RNA involved in pathogenesis: A very high affinity binding site for a messenger RNA-recognizing protein, in: *Developments in Molecular Virology, Vol. 6, Mechanisms of Viral Pathogenesis* (A. Kohn and P. Fuchs, eds.), pp. 180–200, Martinus Nijhoff, The Hague.

Kakidani, H., Furutani, Y., Takahashi, H., Noda, M., Morimoto, Y., Hirose, T., Asai, M., Inayama, S., Nakanishi, S., and Numa, S., 1982, Cloning and sequence analysis of cDNA for porcine β-neo-endorphin/dynorphin precursor, *Nature* **298**:245.

Kay, J. E., Wallace, D. M., Benzie, C. R., and Jagus, R., 1979, Regulation of protein synthesis during lymphocyte activation by phytohaemagglutinin, in: *Cell Biology and Immunology of Leukocyte Function* (M. R. Quastel, ed.), pp. 107–114, Academic Press, New York.

Kaziro, Y., 1978, The role of guanosine 5'-triphosphate in polypeptide chain elongation, *Biochim. Biophys. Acta* **505**:95.

Kelley, D., Coleclough, C., and Perry, R. P., 1982, Functional significance and evolutionary development of the 5'-terminal regions of immunoglobulin variable-region genes, *Cell* **29**:681.

Kerr, I. M., Brown, R. E., and Ball, L. A., 1974, Increased sensitivity of cell free protein synthesis to double stranded RNA after interferon treatment, *Nature* **250**:57.

Kimchi, A., Zilberstein, A., Schmidt, A., Shulman, L., and Revel, M., 1979, The interferon induced protein kinase PK-i from mouse L cells, *J. Biol. Chem.* **254**:9846.

Kinniburgh, A., McMullen, M. D., and Martin, T. E., 1979, Distribution of cytoplasmic poly(A)$^+$ RNA sequences in free mRNP and polysomes of mouse ascites cells, *J. Mol. Biol.* **132**:695.

Kitamura, N., Semler, B., Rothberg, P., Larsen, G., Adler, C., Dorner, A., Emini, E., Hanecak, R., Lee, J., van der Werf, S., Anderson, C. W., and Wimmer, E., 1981, Primary structure, gene organization and polypeptide expression of poliovirus RNA, *Nature* **291**:547.

Klein, W. H., Nolan, C., Lazar, J. M., and Clark, J. M., Jr., 1972, Translation of satellite tobacco necrosis virus RNA: Characterization of *in vitro* procaryotic and eucaryotic translation products, *Biochemistry* **11**:2009.

Knauert, F., and Ehrenfeld, E., 1979, Translation of poliovirus RNA *in vitro*: Studies on *N*-formylmethionine-labeled polypeptides initiated in cell-free extracts prepared from poliovirus infected HeLa cells, *Virology* **93**:537.

Koch, F., Koch, G., and Kruppa, J., 1982, Virus-induced shut-off of host specific protein synthesis, in: *Protein Biosynthesis in Eukaryotes* (R. Perez-Bercoff, ed.), pp. 339–361, Plenum Publishing, New York.

Konienczny, A., and Safer, B., 1983, Purification of the eukaryotic initiation factor 2-eukaryotic initiation factor 2B complex and characterization of its guanine nucleotide exchange activity during protein synthesis initiation, *J. Biol. Chem.* **258**:3402.

Kosower, N. S., Vanderhoff, G. A., and Kosower, E. M., 1972, Glutathione. VIII. The effects of glutathione disulfide on initiation of protein synthesis, *Biochim. Biophys. Acta* **272**:623.

Kozak, M., 1978, How do eucaryotic ribosomes select initiation regions in messenger RNA? *Cell* **15**:1109.

Kozak, M., 1980*a*, Influence of mRNA secondary structure on binding and migration of 40 S ribosomal subunits, *Cell* **19**:79.

Kozak, M., 1980*b*, Role of ATP in binding and migration of 40 S ribosomal subunits, *Cell* **22**:459.

Kozak, M., 1981*a*, Mechanism of mRNA recognition by eukaryotic ribosomes during initiation of protein synthesis, *Curr. Top. Microbiol. Immunol.* **93**:81.

Kozak, M., 1981*b*, Possible role of flanking nucleotides in recognition of the AUG initiator codon by eukaryotic ribosomes, *Nucl. Acids Res.* **9**:5233.

Kozak, M., 1983, Comparison of initiation of protein synthesis in procaryotes, eucaryotes, and organelles, *Microbiol. Rev.* **47**:1.

Kramer, G., Cimadevilla, M., and Hardesty, B., 1976, Specificity of the protein kinase activity associated with the hemin-controlled repressor of rabbit reticulocytes, *Proc. Natl. Acad. Sci. USA* **73**:3078.

Kronenberg, H., Roberts, B., and Efstratiadis, A., 1979, The 3' noncoding region of β-globin mRNA is not essential for *in vitro* translation, *Nucl. Acids Res.* **6**:153.

Krüger, C., and Benecke, B-J., 1981, *In vitro* translation of *Drosophila* heat-shock and non-heat-shock mRNAs in heterologous and homologous cell-free systems, *Cell* **23**:595.

Laskey, R. A., Mills, A. D., Gurdon, J. B., and Partington, G. A., 1977, Protein synthesis in oocytes of *Xenopus laevis* is not regulated by the supply of mRNA, *Cell* **11**:345.

Lawrence, C., and Thach, R. E., 1974, Encephalomyocarditis virus infection of mouse plasmacytoma cells. I. Inhibition of cellular protein synthesis, *J. Virol.* **14**:598.

Lebleu, B., Sen, G. C., Shaila, S., Cabrer, B., and Lengyel, P., 1976, Interferon, double-stranded RNA and protein phosphorylation, *Proc. Natl. Acad. Sci. USA* **73**:3107.

Lee, K. A. W., and Sonenberg, N., 1982, Inactivation of cap-binding proteins accompanies the shut-off of host protein synthesis by poliovirus, *Proc. Natl. Acad. Sci. USA* **79**:3447.

Lee, K. A. W., Guertin, D., and Sonenberg, N., 1983, mRNA secondary structure as a determinant in cap recognition and initiation complex formation, *J. Biol. Chem.* **258**:707.

Legon, S., Jackson, R. J., and Hunt, T., 1973, Control of protein synthesis in reticulocyte lysates by haemin, *Nature New Biol.* **241**:150.

Lehtovaara, P., Söderlund, H., Keränen, S., Pettersson, R., and Kääriäinen, L., 1982, Extreme ends of the genome are conserved and rearranged in the defective interfering RNAs of Semliki Forest virus, *J. Mol. Biol.* **156**:731.

Lengyel, P., 1982*a*, Biochemistry of interferons and their actions, *Annu. Rev. Biochem.* **51**:251.

Lengyel, P., 1982*b*, Interferon action: Control of RNA processing, translation and degradation, in: *Protein Biosynthesis in Eukaryotes* (R. Perez-Bercoff, ed.), pp. 459–483, Plenum Publishing, New York.

Lenz, J. R., Chatterjee, G. E., Maroney, P. A., and Baglioni, C., 1978, Phosphorylated sugars stimulate protein synthesis and Met-tRNA$_f$ binding activity in extracts of mammalian cells, *Biochemistry* **17**:80.

Leroux, A., and London, I. M., 1982, Regulation of protein synthesis by phosphorylation of eukaryotic initiation factor 2α in intact reticulocytes and reticulocyte lysates, *Proc. Natl. Acad. Sci. USA* **79**:2147.

Leung, D. W., Gilbert, C. W., Smith, R. E., Sasavage, N. L., and Clark, J. M., Jr., 1976, Translation of satellite tobacco necrosis virus ribonucleic acid by an *in vitro* system from wheat germ, *Biochemistry* **15**:4943.

Leung, D. W., Browning, K. S., Heckmann, J. E., RajBhandary, U. L., and Clark, J. M., Jr., 1979, Nucleotide sequence of the 5′ terminus of satellite tobacco necrosis virus ribonucleic acid, *Biochemistry* **18**:1361.

Levin, D. H., Kyner, D., and Acs, G., 1973, Protein initiation in eukaryotes: Formation and function of a ternary complex composed of a partially purified ribosomal factor, methionyl transfer RNA, and guanosine triphosphate, *Proc. Natl. Acad. Sci. USA* **70**:41.

Levin, D. H., Ranu, R., Ernst, V., and London, I. M., 1976, Regulation of protein synthesis in reticulocyte lysates: Phosphorylation of the methionyl-tRNA$_f$ binding factor by the protein kinase activity of the translational inhibitor isolated from heme-deficient lysates, *Proc. Natl. Acad. Sci. USA* **73**:3112.

Levin, D. H., Petryshyn, R., and London, I. M., 1981, Characterisation of purified double-stranded RNA-activated eIF-2 kinase from rabbit reticulocytes, *J. Biol. Chem.* **256**:7638.

Lin, S., and Riggs, A. D., 1972, Lac repressor binding to non-operator DNA: Detailed studies and a comparison of equilibrium and rate competition methods, *J. Mol. Biol.* **72**:671.

Lin, S., and Riggs, A. D., 1975*a*, The general affinity of *lac* repressor for *E. coli* DNA: Implications for gene regulation in procaryotes and eucaryotes, *Cell* **4**:107.

Lin, S., and Riggs, A. D., 1975*b*, A comparison of *lac* repressor binding to operator and nonoperator DNA, *Biochem. Biophys. Res. Commun.* **62**:704.

Lindquist, S., 1981, Regulation of protein synthesis during heat shock, *Nature* **293**:311.

Lindquist, S., DiDomenico, B., Bugaisky, G., Kurtz, S., Petko, L., and Sonoda, S., 1982, Regulation of heat-shock response in *Drosophila* and yeast, in: *Heat Shock, from Bacteria to Man* (M. J. Schlesinger, M. Ashburner, and A. Tissières, eds.), pp. 167–175, Cold Spring Harbor Laboratory, Cold Spring Harbor, New York.

Lloyd, M. A., Osbourne, J. C., Safer, B., Powell, G. M., and Merrick, W. C., 1980, Characteristics of eukaryotic initiation factor 2 and its subunits, *J. Biol. Chem.* **255**:1189.

Lodish, H. F., 1971, Alpha and beta globin mRNA: Different amounts and rates of initiation of translation, *J. Biol. Chem.* **246**:7131.

Lodish, H. F., 1974, Model for the regulation of mRNA translation applied to haemoglobin synthesis, *Nature* **251**:385.

Lodish, H. F., 1976, Translational control of protein synthesis, *Annu. Rev. Biochem.* **45**:39.

Lodish, H. F., and Froshauer, S., 1977, Rates of initiation of protein synthesis by two purified species of vesicular stomatitis virus messenger RNA, *J. Biol. Chem.* **252**:8804.

Lodish, H. F., and Jacobsen, M., 1972, Regulation of hemoglobin synthesis. Equal rates of translation and termination of α and β globin chains, *J. Biol. Chem.* **247**:3622.

Lodish, H. F., and Porter, M., 1980, Translational control of protein synthesis after infection by vesicular stomatitis virus, *J. Virol.* **36**:719.

Lodish, H. F., and Rose, J. K., 1977, Relative importance of 7-methylguanosine in ribosome binding and translation of vesicular stomatitis virus mRNAs in wheat germ and reticulocyte cell-free systems, *J. Biol. Chem.* **252**:1181.

Lomedico, P. T., and Saunders, G. F., 1977, Cell-free modulation of proinsulin synthesis, *Science* **198**:620.

Maitra, U., Stringer, E. A., and Chaudhuri, A., 1982, Initiation factors in protein biosynthesis, *Annu. Rev. Biochem.* **51**:869.

Majumdar, A., Roy, R., Das, A., Dasgupta, A., and Gupta, N. K., 1977, Protein synthesis in rabbit reticulocytes XIX. eIF-2 promotes dissociation of met-tRNA$_f$. eIF-1. GTP complex and met-tRNA$_f$ binding to 40s ribosomes, *Biochem. Biophys. Res. Commun.* **78**:161.

Marcus, A., 1970, Tobacco mosaic virus ribonucleic acid-dependent amino acid incorporation in a wheat embryo system, *J. Biol. Chem.* **245**:955.

Martin, S., Zimmer, E., Davidson, W., Wilson, A., and Kan, Y. W., 1981, The untranslated regions of β-globin mRNA evolve at a functional rate in higher primates, *Cell* **25**:737.

Martini, O. H. W., and Kruppa, J., 1979, Ribosomal phosphorylation of mouse myeloma cells, *Eur. J. Biochem.* **95**:349.

Matts, R. L., Levin, D. H., and London, I. M., 1983, Effect of phosphorylation of the α-subunit of eukaryotic initiation factor 2 on the function of reversing factor in the initiation of protein synthesis, *Proc. Natl. Acad. Sci. USA* **80**:2559.

Maundrell, K., Maxwell, E. S., Civelli, O., Vincent, A., Goldenberg, S., Buri, J-F., Imaizumi-Scherrer, M-T., and Scherrer, K., 1979, Messenger ribonucleoprotein complexes in avian erythroblasts: Carriers of post-transcriptional regulation? *Mol. Biol. Rep.* **5**:43.

Mauron, A., and Spohr, G., 1978, Kinetics of synthesis of cytoplasmic messenger-like RNA not associated with ribosomes in HeLa cells, *Eur. J. Biochem.* **82**:619.

McKeehan, W. L., 1974, Regulation of hemoglobin synthesis: Effect of concentration of messenger ribonucleic acid, ribosome subunits, initiation factors, and salts on ratio of α and β chains synthesized *in vitro*, *J. Biol. Chem.* **249**:6517.

McMullen, M. D., Shaw, P. H., and Martin, T. E., 1979, Characterization of poly(A)$^+$ RNA in free mRNP and polysomes of mouse Taper ascites cells, *J. Mol. Biol.* **132**:679.

Mermod, J. J., Schatz, G., and Crippa, M., 1980, Specific control of messenger RNA translation in *Drosophila* oocytes and embryos, *Dev. Biol.* **75**:177.

Meyer, L. J., Brown-Leudi, M., Corbett, S., Tolan, D. R., and Hershey, J. W. B., 1981, The purification and characterization of multiple forms of protein synthesis eukaryotic initiation factors 2, 3 and 5 from rabbit reticulocytes, *J. Biol. Chem.* **256**:351.

Miyata, T., Yasunaga, T., and Nishida, T., 1980, Nucleotide sequence divergence and functional constraint in mRNA evolution, *Proc. Natl. Acad. Sci. USA* **77**:7328.

Mizuno, S., 1977, Temperature sensitivity of protein synthesis initiation. Inactivation of a ribosomal factor by an inhibitor formed at elevated temperatures, *Arch. Biochem. Biophys.* **179**:289.

Morgan, M. A., and Shatkin, A. J., 1980, Initiation of reovirus transcription by ITP and properties of m^7 I-capped, inosine-substituted mRNAs, *Biochemistry* **19**:5960.

Nilsen, T. W., and Baglioni, C., 1979, Mechanism for discrimination between viral and host mRNA in interferon-treated cells, *Proc. Natl. Acad. Sci. USA* **76**:2600.

Nudel, U., Soreq, H., Littauer, U. Z., Marbaix, G., Huez, G., Leclercq, M., Hubert, E., and Chantrenne, H., 1976, Globin mRNA species containing poly(A) segments of different lengths, their functional stability in *Xenopus* oocytes, *Eur. J. Biochem.* **64**:115.

Nuss, D. L., and Koch, G., 1976a, Variation in the relative synthesis of immunoglobulin G and non-immunoglobulin G proteins in cultured MPC-11 cells with changes in the overall rate of polypeptide chain initiation and elongation, *J. Mol. Biol.* **102**:601.

Nuss, D. L., and Koch, G., 1976b, Differential inhibition of vesicular stomatitis virus polypeptide synthesis by hypertonic initiation block, *J. Virol.* **17**:283.

Nuss, D. L., Oppermann, H., and Koch, G., 1975, Selective blockage of initiation of host protein synthesis in RNA-virus infected cells, *Proc. Natl. Acad. Sci. USA* **72**:1258.

Nygard, O., Westermann, P., and Hultin, T., 1980, Met-tRNA$_f^{Met}$ is located in close proximity to the β subunit of eIF-2 in the eucaryotic initiation complex, eIF-2. Met-tRNA$_f^{Met}$. GDPCP, *FEBS Lett.* **113**:125.

Ochoa, S., and de Haro, C., 1979, Regulation of protein synthesis in eukaryotes, *Annu. Rev. Biochem.* **48**:549.

Pain, V. M., and Clemens, M. J., 1983, Assembly and breakdown of mammalian protein synthesis initiation complexes: Regulation by guanine nucleotides and by phosphorylation of initiation factor eIF-2, *Biochemistry* **22**:726.

Pain, V. M., Lewis, J. A., Huvos, P., Henshaw, E. C., and Clemens, M. J., 1980, The effects of amino acid starvation on regulation of polypeptide chain initiation in Ehrlich ascites tumor cells, *J. Biol. Chem.* **255**:1486.

Palmiter, R. D., 1972, Regulation of protein synthesis in chick oviduct, *J. Biol. Chem.* **247**:6770.

Palmiter, R. D., 1974, Differential rates of initiation on conalbumin and ovalbumin messenger ribonucleic acid in reticulocyte lysates, *J. Biol. Chem.* **249**:6779.

Parets-Soler, A., Reibel, L., and Schapira, G., 1981, Differential stimulation of α- and β-globin mRNA translation by M_r 50,000 and 28,000 polypeptide-containing fractions isolated from reticulocyte polysomes, *FEBS Lett.* **136**:259.

Pavlakis, G. N., Lockard, R. E., Vamvakopoulos, N., Rieser, L., RajBhandary, V. L., and Vournakis, J. N., 1980, Secondary structure of mouse and rabbit α- and β-globin mRNAs: Differential accessibility of α and β initiator AUG codons towards nucleases, *Cell* **19**:91.

Pelham, H. R. B., and Jackson, R. J., 1976, An efficient mRNA-dependent translation system from reticulocyte lysates, *Eur. J. Biochem.* **67**:247.

Perez-Bercoff, R. (ed.), 1982, *Protein Biosynthesis in Eukaryotes*, Plenum Publishing, New York.

Perez-Bercoff, R., and Gander, M., 1977, The genomic RNA of mengovirus. I: Location of the poly(C) tract, *Virology* **80**:426.

Perez-Bercoff, R., and Kaempfer, R., 1982, Genomic RNA of Mengovirus: Recognition of common features by ribosomes and eukaryotic initiation factor 2, *J. Virol.* **41**:30.

Person, A., Ben-Hamida, F., and Beaud, G., 1980, Inhibition of 40 S-Met-tRNA$_f^{Met}$ ribosomal initiation complex formation by vaccinia virus, *Nature* **287**:355.

Peterson, D. T., Merrick, W. C., and Safer, B., 1979, Binding and release of radiolabeled eukaryotic initiation factors 2 and 3 during 80 S initiation complex formation, *J. Biol. Chem.* **254**:2509.

Petryshyn, R., Trachsel, H., and London, I. M., 1979, Regulation of protein synthesis in reticulocyte lysates: Immune serum inhibits heme-regulated protein kinase activity and differentiates heme-regulated protein kinase from double-stranded RNA induced protein kinase, *Proc. Natl. Acad. Sci. USA* **76**:1575.

Porter, A. G., Frisby, D. P., Carey, N. H., and Fellner, P., 1975, Nucleotide sequence studies on picornavirus RNAs, in: *In Vitro Transcription and Translation of Viral Genomes* (A. L. Haenni and G. Beaud, eds.), pp. 169–176, INSERM, Paris.

Ranu, R. S., and London, I. M., 1979, Regulation of protein synthesis in rabbit reticulocyte lysates: Additional initiation factor required for formation of ternary complex (eIF-2-GTP-met.tRNA$_f$) and demonstration of the inhibitory effect of heme-regulated protein kinase. *Proc. Natl. Acad. Sci. USA* **76**:1079.

Ray, B. K., Brendler, T. G., Adya, S., Daniels-McQueen, S., Miller, J. K., Hershey, J. W. B., Grifo, J. A., Merrick, W. C., and Thach, R. E., 1983, Role of mRNA competition in regulating translation: Further characterization of mRNA discriminatory factors, *Proc. Natl. Acad. Sci. USA* **80**:663.

Regier, J. C., and Kafatos, F. C., 1977, Absolute rates of protein synthesis in sea urchins with specific activity measurements of radioactive leucine and leucyl-tRNA, *Dev. Biol.* **57**:270.

Revel, M., and Groner, Y., 1978, Post-transcriptional and translational control of gene expression in eukaryotes, *Annu. Rev. Biochem.* **47**:1079.

Roberts, W. K., Hovanessian, A., Brown, R. E., Clemens, M. J., and Kerr, I. M., 1976, Interferon mediated protein kinase and low-molecular weight inhibitor of protein synthesis, *Nature* **264**:477.

Robertson, H. D., and Mathews, M. B., 1973, Double-stranded RNA as an inhibitor of protein synthesis and as a substrate for a nuclease in extracts of Krebs II ascites cells, *Proc. Natl. Acad. Sci. USA* **70**:225.

Rose, J. K., 1980, Complete intergenic and flanking gene sequences from the genome of vesicular stomatitis virus, *Cell* **19**:415.

Rose, J. K., Trachsel, H., Leong, K., and Baltimore, D., 1978, Inhibition of translation by poliovirus: Inactivation of a specific initiation factor, *Proc. Natl. Acad. Sci. USA* **73**:2732.

Rosen, H., and Kaempfer, R., 1979, Mutually exclusive binding of messenger RNA and initiator methionyl transfer RNA to eukaryotic initiator factor 2, *Biochem. Biophys. Res. Commun.* **91**:449.

Rosen, H., Knoller, S., and Kaempfer, R., 1981*a*, Messenger RNA specificity in the inhibition of eukaryotic translation by double-stranded RNA, *Biochemistry* **20**:3011.

Rosen, O. M., Rubin, C. S., Cobb, M. H., and Smith, C. J., 1981*b*, Insulin stimulates the phosphorylation of ribosomal protein S6 in a cell-free system derived from 3T3-L1 adipocytes, *J. Biol. Chem.* **256**:3630.

Rosen, H., Di Segni, G., and Kaempfer, R., 1982, Translational control by messenger RNA competition for eukaryotic initiation factor 2, *J. Biol. Chem.* **257**:946.

Rosenthal, E. T., Hunt, T., and Ruderman, J. V., 1980, Selective translation of mRNA controls the pattern of protein synthesis during early development of the surf clam *Spisula solidissima, Cell* **20**:487.

Rosenthal, E. T., Brandhorst, B. P., and Ruderman, J. V., 1982, Translationally mediated changes in patterns of protein synthesis during maturation of starfish oocytes, *Dev. Biol.* **91**:215.

Roy, R., Ghosh-Dastidar, P., Das, A., Yaghmai, B., and Gupta, N. K., 1981, Protein synthesis in rabbit reticulocytes. Co-eIF-2A reverses mRNA inhibition of ternary complex (Met-tRNA$_f$.eIF-2.GTP) formation by eIF-2, *J. Biol. Chem.* **256**:4719.

Russell, D. W., and Spremulli, L. L., 1979, Purification and characterization of ribosome dissociation factor (eIF-6) from wheat germ, *J. Biol. Chem.* **254**:8796.

Russell, D. W., and Spremulli, L. L., 1980, Mechanism of action of the wheat germ ribosomal dissociation factor: Interaction with the 60 S subunit, *Arch. Biochem. Biophys.* **201**:518.

Sabol, S., and Ochoa, S., 1971, Ribosomal binding of labelled initiation factor F3, *Nature New Biol.* **234**:233.

Sacerdot, C., Fayat, G., Dessen, P., Springer, M., Plumridge, J., Grunberg-Manago, M., and Blanquet, S., 1982, Sequence of a 1.26 kb DNA fragment containing the structural gene for *E. coli* initiation factor IF3: presence of an AUU initiator codon, *EMBO J.* **1**:311.

Safer, B., Peterson, D., and Merrick, W. C., 1977, The effect of hemin controlled repressor on initiation factor functions during sequential formation of the 80s initiation complex, in: *Translation of Natural and Synthetic Polynucleotides* (A. B. Legocki, ed.), pp. 24–31, Poznan Agricultural University Press, Poznan, Poland.

Samuel, C. E., 1979, Mechanism of interferon action: Phosphorylation of protein synthesis initiation factor eIF-2 in interferon treated human cells by a ribosome associated kinase processing site specificity similar to hemin-regulated rabbit reticulocyte kinase, *Proc. Natl. Acad. Sci. USA* **76**:600.

Sangar, D. V., Black, D. N., Rowlands, D. J., Harris, T. J. R., and Brown, F., 1980, Location of the initiation site for protein synthesis on foot-and-mouth disease virus RNA by *in vitro* translation of defined fragments of the RNA, *J. Virol.* **33**:59.

Sargan, D. R., Gregory, S. P., and Butterworth, P., 1982, A possible novel interaction between the 3'-end of 18 S ribosomal RNA and the 5'-leader sequence of many eukaryotic messenger RNAs, *FEBS Lett.* **147**:133.

Schlesinger, M. J., Ashburner, M., and Tissières, A. (eds.), 1982, *Heat Shock, from Bacteria to Man*, Cold Spring Harbor Laboratory, Cold Spring Harbor, New York.

Schmid, H.-P., Köhler, K., and Setyono, B., 1982, Possible involvement of messenger RNA-associated proteins in protein synthesis, *J. Cell Biol.* **93**:893.

Schreier, M. H., and Staehelin, T., 1973, Initiation of eukaryotic protein synthesis: (Met-tRNA$_f$.40 S ribosome) initiation complex catalysed by purified initiation factors in the absence of mRNA, *Nature New Biol.* **242**:35.

Schreier, M. H., Erni, B., and Staehelin, T., 1977, Initiation of mammalian protein synthesis. I. Purification and characterization of seven initiation factors, *J. Mol. Biol.* **116**:727.

Scott, M. P., and Pardue, M. L., 1981, Translational control in lysates of *Drosophila melanogaster* cells, *Proc. Natl. Acad. Sci. USA* **78**:3353.

Seal, S. N., Schmidt, A., Tomaszewski, M., and Marcus, A., 1978, Inhibition of mRNA translation by the cap analogue, 7-methylguanosine-5'-phosphate, *Biochem. Biophys. Res. Commun.* **82:**553.

Sen, G. C., Taira, H., and Lengyel, P., 1978, Interferon, double-stranded RNA and protein phosphorylation. Characteristics of a double-stranded RNA-activated protein kinase system partially purified from interferon-treated Ehrlich ascites tumor cells, *J. Biol. Chem.* **253:**5915.

Setzer, D. R., McGrogan, M., Nunberg, J., and Schimke, R. T., 1980, Size heterogeneity in the 3' end of dihydrofolate reductase messenger RNAs in mouse cells, *Cell* **22:**361.

Shafritz, D., Weinstein, J., Safer, B., Merrick, W. C., Weber, L., Hickey, E., and Baglioni, C., 1976, Evidence for role of m⁷G-phosphate group in recognition of eukaryotic mRNA by initiation factor IF-M3, *Nature* **261:**291.

Shatkin, A. J., 1976, Capping of eucaryotic mRNAs, *Cell* **9:**645.

Shatkin, A. J., 1982, A closer look at the 5'-end of mRNA in relation to initiation, in: *Protein Biosynthesis in Eukaryotes* (R. Perez-Bercoff, ed.), pp. 199–221, Plenum Publishing, New York.

Shaw, M. W., Choppin, P. W., and Lamb, R. A., 1983, A previously unrecognized influenza B virus glycoprotein from a bicistronic mRNA that also encodes the viral neuraminidase, *Proc. Natl. Acad. Sci. USA* **80:**4879.

Shih, D. S., and Kaesberg, P., 1973, Translation of brome mosaic viral RNA in a cell-free system derived from wheat embryo, *Proc. Natl. Acad. Sci. USA* **70:**1799.

Shine, J., and Dalgarno, L., 1974, The 3'-terminal sequence of *E. coli* 16 S ribosomal RNA: Complementarity to nonsense triplets and ribosome binding sites, *Proc. Natl. Acad. Sci. USA* **71:**1342–1346.

Shull, G. E., and Theil, E. C., 1982, Translational control of ferritin synthesis by iron in embryonic reticulocytes of the bullfrog, *J. Biol. Chem.* **257:**14187.

Siekierka, J., Mitsui, K. I., and Ochoa, S., 1981, Mode of action of the heme controlled translational inhibitor: Relationship of eukaryotic initiation factor 2-stimulating protein to translation restoring factor, *Proc. Natl. Acad. Sci. USA* **78:**220.

Siekierka, J., Mauser, L., and Ochoa, S., 1982, Mechanism of polypeptide chain initiation in eukaryotes and its control by phosphorylation of the α subunit of initiation factor 2, *Proc. Natl. Acad. Sci. USA* **79:**2537.

Siekierka, J., Manne, V., Mauser, L., and Ochoa, S., 1983, Polypeptide chain initiation in eukaryotes: Reversibility of the ternary complex-forming reaction, *Proc. Natl. Acad. Sci. USA* **80:**1232.

Sippel, A. E., Stavrianopoulos, J. G., Schutz, G., and Feigelson, P., 1974, Translational properties of rabbit globin mRNA after specific removal of poly(A) with ribonuclease H, *Proc. Natl. Acad. Sci. USA* **71:**3143.

Skup, D., Zarbl, H., and Millward, S., 1981, Regulation of translation in L-cells infected with reovirus, *J. Mol. Biol.* **151:**35.

Smith, A. E., 1975, Control of translation of animal virus messenger RNA, in: *Control Processes in Virus Multiplication*, (D. C. Burke and W. C. Russell, eds.), pp. 183–223, 25th Symp. Soc. Gen. Microbiol., Cambridge University Press, Cambridge.

Smith, A. E., Kamen, R., Mangel, W., Shure, H., and Wheeler, T., 1976, Location of the sequences coding for capsid proteins VP1 and VP2 on polyoma virus DNA, *Cell* **9:**481.

Sonenberg, N., 1981, ATP/Mg^{2+}-dependent cross-linking of cap binding proteins to the 5′ end of eukaryotic mRNA, *Nucl. Acids Res.* **9**:1643.

Sonenberg, N., and Shatkin, A. J., 1977, Reovirus RNA can be covalently cross-linked via the 5′-cap to proteins in initiation complexes, *Proc. Natl. Acad. Sci. USA* **74**:4288.

Sonenberg, N., Rupprecht, K. M., Hecht, S. M., and Shatkin, A. J., 1979, Eukaryotic mRNA cap binding protein: Purification by affinity chromatography on Sepharose-coupled m^{7}GDP, *Proc. Natl. Acad. Sci. USA* **76**:4345.

Sonenberg, N., Guertin, D., Cleveland, D., and Trachsel, H., 1981, Probing the function of the eukaryotic 5′ cap structure by using a monoclonal antibody directed against cap-binding proteins, *Cell* **27**:563.

Sonenberg, N., Guertin, D., and Lee, K. A. W., 1982, Capped mRNAs with reduced secondary structure can function in extracts of poliovirus-infected cells, *Mol. Cell Biol.* **2**:1633.

Sonenshein, G. E., and Brawerman, G., 1976a, Regulation of immunoglobulin synthesis in mouse myeloma cells, *Biochemistry* **15**:5497.

Sonenshein, G. E., and Brawerman, G., 1976b, Differential translation of mouse myeloma messenger RNAs in a wheat-germ cell-free system, *Biochemistry* **15**:5501.

Sonenshein, G. E., and Brawerman, G., 1977, Entry of mRNA into polyribosomes during recovery from starvation in mouse sarcoma 180 cells, *Eur. J. Biochem.* **73**:307.

Soreq, H., Nudel, U., Salomon, R., Revel, M., and Littauer, U. Z., 1974, *In vitro* translation of polyadenylic acid-free rabbit globin mRNA, *J. Mol. Biol.* **88**:233.

Soreq, H., Sagar, A., and Sehgal, P., 1981, Translational activity and functional stability of human fibroblast β$_1$ and β$_2$ interferon mRNAs lacking 3′-terminal RNA sequences, *Proc. Natl. Acad. Sci. USA* **78**:1741.

Spirin, A. S., 1969, Informosomes, *Eur. J. Biochem.* **10**:20.

Spohr, G., Granboulan, N., Morel, C., and Scherrer, K., 1970, Messenger RNA in HeLa cells: An investigation of free and polyribosome-bound cytoplasmic messenger ribonucleoprotein particles by kinetic labelling and electron microscopy, *Eur. J. Biochem.* **17**:296.

Storti, R. V., Scott, M. P., Rich, A., and Pardue, M. L., 1980, Translational control of protein synthesis in response to heat shock in *Drosophila melanogaster* cells, *Cell* **22**:825.

Summers, D. F., and Maizel, J. V., 1968, Evidence for large precursor proteins in poliovirus synthesis, *Proc. Natl. Acad. Sci. USA* **59**:966.

Svitkin, Y. V., Ginevskaya, V. A., Ugarova, T. Y., and Agol, V. I., 1978, A cell-free model of the encephalomyocarditis virus-induced inhibition of host cell protein synthesis, *Virology* **87**:199.

Tahara, S. M., Morgan, M. A., and Shatkin, A. J., 1981, Two forms of purified m^{7}G-cap binding protein with different effects on capped mRNA translation in extracts of uninfected and poliovirus infected HeLa cells, *J. Biol. Chem.* **256**:7691.

Talkington, C. A., and Leder, P., 1982, Rescuing the *in vitro* function of a globin pseudogene promoter, *Nature* **298**:192.

Thach, R. E., Sundararajan, T. A., Dewey, K., Brown, J. C., and Doty, P., 1966, Translation of synthetic messenger RNA, *Cold Spring Harbor Symp. Quant. Biol.* **31**:85.

Thomas, G. P., and Mathews, M. B., 1982, Control of polypeptide chain elongation in the stress response: A novel translational control, in: *Heat Shock, from Bacteria to Man* (M. J. Schlesinger, M. Ashburner, and A. Tissieres, eds.), pp. 207–213, Cold Spring Harbor Laboratory, Cold Spring Harbor, New York.

Thomas, A., Goumans, H., Voorma, H. O., and Benne, R., 1980*a*, The mechanism of action of eukaryotic initiation factor 4C in protein synthesis, *Eur. J. Biochem.* **107**:39.

Thomas, G., Siegmann, M., Kubler, A-M., Gordo, J., and Jimenez de Asua, L., 1980*b*, Regulation of 40 S ribosomal protein S6 phosphorylation in Swiss mouse 3T3 cells, *Cell* **19**:1015.

Thomas, G., Martin Perez, J., Siegmann, M., and Otto, A., 1982, The effect of serum, EGF, PGF2α and insulin on S6 phosphorylation and the initiation of protein and DNA synthesis, *Cell* **30**:235.

Trachsel, H., and Staehelin, T., 1978, Binding and release of eucaryotic initiation factor eIF 2 and GTP during protein synthesis initiation, *Proc. Natl. Acad. Sci. USA* **75**:204.

Trachsel, H., and Staehelin, T., 1979, Initiation of mammalian protein synthesis. The multiple functions of the initiation factor eIF-3, *Biochim. Biophys. Acta* **565**:305.

Trachsel, H., Erni, B., Schreier, M., and Staehelin, T., 1977, Initiation of mammalian protein synthesis. The assembly of the initiation complex with purified initiation factors, *J. Mol. Biol.* **116**:755.

Trachsel, H., Erni, B., Schreier, M. H., Braun, L., and Staehelin, T., 1979, Purification of seven protein synthesis initiation factors from Krebs II ascites cells, *Biochim. Biophys. Acta* **561**:484.

Trachsel, H., Sonenberg, N., Shatkin, A. J., Rose, J. K., Leong, K., Bergmann, J. E., Gordon, J., and Baltimore, D., 1980, Purification of a factor that restores translation of VSV mRNA in extracts from poliovirus-infected HeLa cells, *Proc. Natl. Acad. Sci. USA* **77**:770.

Valenzuela, P., Chaudhuri, A., and Maitra, U., 1982, Eukaryotic ribosomal subunit anti-association activity of calf liver is contained in a single polypeptide chain protein of M_r = 25,500 (eukaryotic initiation factor 6), *J. Biol. Chem.* **257**:7712.

Van Steeg, M., Van Grinsven, M., Van Mansfeld, F., Voorma, H. O., and Benne, R., 1981, Initiation of protein synthesis in neuroblastoma cells infected by Semliki Forest virus, *FEBS Lett.* **129**:62.

Van Venrooij, W. J. W., Henshaw, E. C., and Hirsh, C. A., 1972, Effects of deprival of glucose or individual amino acids on polyribosome distribution and rate of protein synthesis in cultured mammalian cells, *Biochim. Biophys. Acta* **259**:127.

Villareal, L. P., Breindl, M., and Holland, J. J., 1976, Determination of molar ratios of vesicular stomatitis virus induced RNA species in BHK_{21} cells, *Biochemistry* **15**:1663.

Vincent, A., Goldenberg, S., Standart, N., Civelli, O., Imaizumi-Scherrer, T., Maundrell, K., and Scherrer, K., 1981, Potential role of mRNP proteins in cytoplasmic control of gene expression in duck erythroblasts, *Mol. Biol. Rep.* **7**:71.

Vlasik, T., Domogatsky, S., Bezlepkina, T., and Ovchinnikov, L., 1980, RNA-binding activity of eukaryotic initiation factors of translation, *FEBS Lett.* **116**:8.

Von Hippel, P., Revzin, A., Gross, C. A., and Wang, A. C., 1974, Non-specific DNA binding of genome regulating proteins as biological control mechanism. The *lac* operon: Equilibrium aspects, *Proc. Natl. Acad. Sci. USA* **71**:4808.

Walden, W. E., Godefroy-Colburn, T., and Thach, R. E., 1981, The role of mRNA competition in regulating translation. Demonstration of competition *in vivo, J. Biol. Chem.* **256:**11739.

Walton, G. M., and Gill, G. N., 1976, Regulation of ternary protein synthesis initiation complex formation by the adenylate energy charge, *Biochim. Biophys. Acta* **418:**195.

Weber, L. A., Feman, E. R., and Baglioni, C., 1975, A cell-free system from HeLa cells active in initiation of protein synthesis, *Biochemistry* **14:**5315.

Weber, L. A., Hickey, E. D., Maroney, P. A., and Baglioni, C., 1977, Inhibition of protein synthesis by Cl⁻, *J. Biol. Chem.* **252:**4007.

Weber, L., Hickey, E., and Baglioni, C., 1978, Influence of potassium salt concentration and temperature on inhibition of mRNA translation by 7-methylguanosine 5′-monophosphate, *J. Biol. Chem.* **253:**178.

Weiss, S. R., Varmus, H. E., and Bishop, J. M., 1977, The size and genetic composition of virus-specific RNAs in the cytoplasm of cells producing avian sarcoma-leukosis viruses, *Cell* **12:**983.

Wengler, G., Wengler, G., and Gross, H. J., 1979, Replicative form of Semliki Forest virus RNA contains an unpaired guanosine, *Nature* **282:**754.

West, D. K., Lenz, J. R., and Baglioni, C., 1979, Stimulation of protein synthesis and Met tRNA_f binding by phosphorylated sugars: Studies on their mechanism of action, *Biochemistry* **18:**624.

Wieringa, B., van der Zwaag, J., Mulder, J., Ab, G., and Gruber, M., 1981, Translation *in vivo* and *in vitro* of mRNAs coding for vitellogenin, serum albumin and very-low density lipoprotein II from chicken liver. A difference in translational efficiency, *Eur. J. Biochem.* **114:**635.

Wimmer, E., Chang, A. Y., Clark, J. M., Jr., and Reichmann, M. E., 1968, Sequence studies of satellite tobacco necrosis virus RNA: Isolation and characterization of a 5′-terminal trinucleotide, *J. Mol. Biol.* **38:**59.

Wodnar-Filipowicz, A., Szczesna, E., Zan-Kowalczewska, M., Muthukrishnan, S., Szybiak, U., Legocki, A., and Filipowicz, W., 1978, 5′-Terminal 7-methylguanosine and mRNA function, *Eur. J. Biochem.* **92:**69.

Woodland, H. R., and Wilt, F. H., 1980, The stability and translation of sea urchin histone messenger RNA molecules injected into *Xenopus laevis* eggs and developing embryos, *Dev. Biol.* **75:**214.

Wreschner, D. H., McCauley, J. W., Skehel, J. J., and Kerr, I. M., 1981, Interferon action: Sequence specificity of the ppp(A2′p)_nA-dependent ribonuclease, *Nature* **289:**414.

Yoo, O. J., Powell, C. T., and Agarwal, K. L., 1982, Molecular cloning and nucleotide sequence of full-length cDNA coding for porcine gastrin, *Proc. Natl. Acad. Sci. USA* **79:**1049.

Ysebaert, M., van Emmelo, J., and Fiers, W., 1980, Total nucleotide sequence of a nearly full-size DNA copy of satellite tobacco necrosis virus RNA, *J. Mol. Biol.* **143:**273.

Zagorska, L., Chroboczek, J., Klita, S., and Szafranski, P., 1982, Effect of secondary structure of mRNA on the formation of initiation complexes with prokaryotic and eukaryotic ribosomes, *Eur. J. Biochem.* **122:**265.

Zähringer, J., Baliga, B. S., and Munro, H. N., 1976, Subcellular distribution of total poly(A)-containing RNA and ferritin mRNA in the cytoplasm of rat liver, *Biochem. Biophys. Res. Commun.* **68:**1088.

Zilberstein, A., Kimchi, A., Schmidt, A., and Revel, M., 1978, Isolation of two interferon induced translational inhibitors: A protein kinase and an oligo-isoadenylate synthetase, *Proc. Natl. Acad. Sci. USA* **75:**4734.

Zingsheim, H. P., Geisler, N., Weber, K., and Mayer, F., 1977, Complexes of *Escherichia coli lac*-repressor with non-operator DNA revealed by electron microscopy: Two repressor molecules can share the same segment of DNA, *J. Mol. Biol.* **115:**565.

Zucker, W. V., and Schulman, H. M., 1968, Stimulation of globin chain initiation by haemin in the reticulocyte cell free system, *Proc. Natl. Acad. Sci. USA* **59:**582.

CHAPTER 4

Picornavirus Inhibition of Host Cell Protein Synthesis

Ellie Ehrenfeld

Departments of Biochemistry
and Cellular, Viral, and Molecular Biology
University of Utah School of Medicine
Salt Lake City, Utah 84132

1. INTRODUCTION AND SCOPE

Infection of cultured cells with many lytic viruses results in a marked decrease in the rate of cellular protein synthesis. Usually, this decrease is accompanied by increasing rates of viral protein synthesis, marked cytopathic effects, and ultimately cell death. In most cases, it is not known whether the "shut-off" of host cell protein synthesis results from an active process induced by the virus evolved for that (or some other) purpose, or whether it is merely a passive result of another viral function, such as production of large quantities of viral mRNA which compete effectively with their cellular counterparts. In the case of poliovirus, however, three types of studies suggested that the former, active type of mechanism was at work. Kinetic analysis of the rate of protein synthesis in cells synchronously infected with high multiplicities of virus showed that cellular protein synthesis could be virtually completely inhibited prior to the synthesis of significant quantities of viral RNA and protein (Summers *et al.*, 1965). In addition, infection in the presence of 1–3 mM guanidine, which prevents detectable replication of viral RNA, nevertheless results in viral inhibition of host cell protein synthesis (Holland, 1964;

Bablanian *et al.*, 1965, Penman and Summers, 1965). Last, infection with a temperature-sensitive mutant of poliovirus that synthesizes no single-stranded RNA at restrictive temperature nevertheless induces normal inhibition of cellular protein synthesis (Hewlett *et al.*, 1982). All of these results argue against a competition between cellular and viral mRNAs for cellular components as an explanation for the selective inhibition of cellular protein synthesis. A large body of experimental work on this subject has been performed with poliovirus-infected cells, and consequently, the major focus of this review is on the inhibiton of protein synthesis by poliovirus.

For many years, it has been assumed that all processes related to virus infection observed with one picornavirus would be common to all other picornaviruses, since the only major differences among them were thought to reside in the structure of capsid proteins which determined target cell and species specificity, and physicochemical properties such as density, stability, antigenicity, etc. Details of intracellular replication mechanisms were assumed to be uniformly applicable. Studies of the inhibition of host cell protein synthesis by encephalomyocarditis (EMC) virus and mengovirus have been conducted, and until recently, these results have been included with the data on poliovirus-induced shut-off. One recent study, however, demonstrated an apparent difference between events related to cellular protein synthesis inhibition in the same cells infected with poliovirus or with EMC virus (Jen *et al.*, 1980). Thus, the results of studies of picornaviruses other than poliovirus will be summarized in a separate section below. In addition, it appears that the effects of virus infection on host cell protein synthesis may also be a function of the cell type used (Otto and Lucas-Lenard, 1980; Jen and Thach, 1982). Throughout the following discussion, therefore, it should be remembered that the effects of poliovirus (or other picornavirus) infection on cellular protein synthesis during a natural infection of specific target cells have never been observed or analyzed.

2. PROPERTIES OF INHIBITION OF HOST CELL PROTEIN SYNTHESIS BY POLIOVIRUS

2.1. General Description

The first descriptions of the rapid and progressive decline in the rate of protein synthesis in poliovirus-infected cells were reported 20

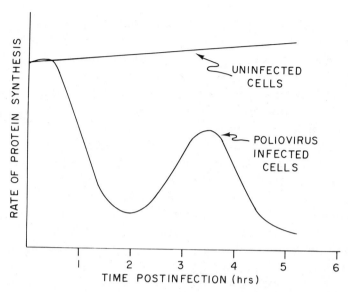

Fig. 1. Rates of protein synthesis in poliovirus-infected and mock-infected HeLa cells. The curves shown were generated from numerous experiments in which HeLa cells were incubated for 15 min with radiolabeled amino acids at various times postinfection or mock infection with poliovirus. The cells were harvested immediately following the labeling period, and incorporation of amino acids into macromolecular protein was determined by scintillation spectroscopy of TCA-precipitable material.

years ago when it was observed that the incorporation of radioactive amino acids into protein was drastically reduced at early times after infection (Zimmerman *et al.*, 1963; Holland and Peterson, 1964). Analysis of the sedimentation properties of polyribosomes in infected cells showed a progressive disaggregation of ribosomes, followed by the subsequent formation of larger, virus-specific polysome structures which were consistent with the larger size of viral mRNA, compared with the cellular mRNA population (Penman *et al.*, 1963). Figure 1 shows a typical curve of the rates of protein synthesis in HeLa cells infected with poliovirus at a multiplicity of 100 pfu/cell. No effect is observed for the first 30 min after infection, presumably the time required for adsorption, penetration, and uncoating. The rate of protein synthesis then declines rapidly, reaching a minimum at approximately 2–2.5 hr postinfection. At this time, almost no polyribosomes are detectable. A second burst of protein synthesis then occurs, which represents the synthesis of viral proteins exclusively (Summers *et al.*, 1965). The rate of inhibition is apparently a function of the multiplicity of infection (Holland and Peterson, 1964; Borgert *et al.*, 1971); however, more detailed measurements indicate that the slope of the shut-off curve is constant, but an increasing lag occurs before the onset of inhibition at lower multiplicities (Helentjaris and Ehrenfeld, 1977).

During the period of polysome disaggregation, the specific activity of nascent polypeptide chains and the polypeptide elongation rates on the fraction of remaining polyribosomes remains normal (Summers and Maizel, 1967; Leibowitz and Penman, 1971). These analyses suggested that the cause of the declining rates of protein synthesis was a decreased association of cellular mRNA with ribosomes, resulting in a block at the initiation step of protein synthesis (Willems and Penman, 1966; Leibowitz and Penman, 1971). Direct measurements of initiation complex formation demonstrated that 80 S complexes containing ribosomes, mRNA, and met-tRNA$_f^{met}$ failed to form in infected cells during the shut-off period, and that the step in protein synthesis inhibited by virus infection therefore preceded the entry of components into the 80 S initiation complex (Ehrenfeld and Manis, 1979). An *in vitro* system, which is thought to mimic the selective inhibition seen *in vivo*, indicated that mRNA failed to bind to 40 S ribosomal subunits, and that the inhibited step therefore preceded the 60 S junction reaction (Brown and Ehrenfeld, 1980).

2.2. Requirement for Expression of Viral Genome

Soon after the realization that poliovirus infection markedly interfered with host cell protein synthesis, and preliminary studies of the cause of the inhibition were conducted, efforts were directed toward determining the mediator of the inhibitory event. Several investigators showed that the shut-off of host protein synthesis occurred or was only slightly affected when cells were infected in the presence of 1 mM guanidine, which prevents detectable virus replication (Holland, 1964; Bablanian et al., 1965; Penman and Summers, 1965). This demonstrated that inhibition could be effected by the initial infecting virus. However, Penman and Summers (1965) and Borgert et al. (1971), using puromycin or cycloheximide at the time of virus infection, found that a period of protein synthesis was required after removal of the inhibitor before the inhibition of cell protein synthesis could be manifested, and thus suggested that inhibition required translation of the infecting viral genome. Holland (1964), on the other hand, concluded that puromycin or fluorophenylalanine failed to suppress or delay the shut-off effect. It should be noted that, in the latter study, virus replication was allowed to proceed for 1–1.5 hr before addition of the drugs, so that some translation of the parental viral genomes may have occurred. In support of this interpretation, it is likely that translation of input viral RNA and synthesis of at least some viral RNA also occurs in the presence of guanidine (Noble and Levintow, 1970).

The above-mentioned inhibitor studies seem best interpreted as demonstrating that expression of the viral genome is required for inhibition of cellular protein synthesis. Inactivation of the infecting viral RNA with ultraviolet irradiation (Penman and Summers, 1965; Bablanian, 1972; Helentjaris and Ehrenfeld, 1977), with proflavine (Holland, 1964) or with hydroxylamine (Borgert et al., 1971) completely abolished the virus' ability to induce shut-off and, thus, the requirement of a functional viral genome for the inhibition of host cell protein synthesis is imperative. Inactivation of the host cell shut-off function follows single-hit kinetics, and involves damage to the infecting viral RNA rather than to capsid proteins. The target size for inactivation appears to be the entire genome (Borgert et al., 1971; Helentjaris and Ehrenfeld, 1977). The results demonstrate that the complete inhibition of host cell protein synthesis can be accomplished by one infectious viral genome per cell.

The results of studies of other drugs or conditions restrictive to poliovirus replication are all consistent with the conclusion that a functional, infecting viral genome must enter the cell and be expressed, although not necessarily replicated, in order for inhibition of cellular protein synthesis to occur. Thus, agents which inactivate the infecting genome (UV, proflavine, hydroxylamine) or which prevent virus uncoating or very early events (thiopyrimidine derivative, S-7; LaColla et al., 1972) abolish shut-off. Agents which prevent viral gene expression (general inhibitors of protein synthesis) also prevent shut-off; whereas, agents which permit expression of the infecting viral genome but interfere with replication (guanidine, nonpermissive temperature for virus mutants which are temperature-sensitive for RNA synthesis) do not prevent inhibition of cellular protein synthesis. On this basis, the polio antiviral drug, D-penicillamine, which does not prevent shut-off (Merryman et al., 1974) can be assigned a site of action subsequent to translation of the infecting genome.

2.3. Poliovirus Inhibition of Other Viral Protein Synthesis

The results of early studies of the kinetics and characteristics of poliovirus-induced inhibition of cellular protein synthesis were subsequently used to strong advantage to allow the analysis of the synthesis of virus-specific proteins, in the absence of any background of cellular protein synthesis. Conditions were optimized to achieve virtually complete shut-off of host protein synthesis (infection in the presence of actinomycin D (Penman *et al.*, 1963) and guanidine to prevent virus replication but allow shut-off), with subsequent reversal of the guanidine and radiolabeling of viral proteins (Summers *et al.*, 1965). It was hoped by some investigators that the ability to eliminate host cell background with replicating poliovirus could be applied to the analysis of other viral proteins. By superinfection of the cells with other viruses, those products which were normally obscured by host cell synthesis could be studied (Bablanian and Russell, 1974). However, it quickly became apparent that the mechanism utilized by poliovirus to selectively discriminate against translation of cellular mRNAs also worked against translation of many other viral mRNAs. Replication of herpes simplex virus (HSV; Saxton and Stevens, 1972) and vesicular stomatitis virus (VSV; Doyle and Holland, 1972; Ehrenfeld and Lund, 1977) was completely restricted by preinfection or superinfection with poliovirus, and the interference in both cases was shown to be a specific inhibition of protein synthesis, resulting in polysome disaggregation. No effect was seen on HSV or VSV RNA or DNA synthesis, and expression, but no replication of the poliovirus genome was required for restriction to occur. Thus, poliovirus inhibition of other viral protein synthesis appears to occur by the same mechanism as inhibition of host cell translation. Poliovirus interference with the growth of Newcastle Disease Virus (Ito *et al.*, 1968) and of adenovirus (Bablanian and Russell, 1974) have been reported, and is also likely due to inhibition of viral protein synthesis, although the site of interference was not localized.

Since normal yields of poliovirus are obtained from the dually infected cells, the inhibition of HSV, VSV, or cellular translation represents a selectivity in mRNA recognition. Other picornaviruses, however, are immune to the mRNA disciminatory process and are insensitive to inhibition by poliovirus infection. Mengovirus (McCormick and Penman, 1968) and EMC virus (Detjen *et al.*, 1981) both replicate normally in the presence of replicating or nonreplicating poliovirus, although host cell protein synthesis is sharply inhibited during the same dual infection. Similarly, guanidine-resistant strains of poliovirus grow normally in cells preinfected with guanidine-sensitive virus in the presence of guanidine (Cords and Holland, 1964).

There is one reported exception to the ability of poliovirus to inhibit all protein synthesis other than that resulting from translation of picornavirus mRNA. Viral protein synthesis in SV5-infected rhesus monkey kidney cells was significantly enhanced following superinfection with poliovirus in the presence of guanidine (Choppin and Holmes, 1967). Cellular protein synthesis was inhibited in the dually infected cells, and this apparent discrimination was interpreted as providing a functional distinction between SV5 messenger and cellular messenger RNA. Further studies of this interaction have not been pursued, but should be investigated in light of current models for the mechanism of poliovirus-induced inhibition (see below).

2.4. Integrity of Untranslated mRNA

The kinetics and properties of polysome disaggregation in infected cells and the failure of ribosomes to form initiation complexes with cellular mRNAs provided important clues as to the nature of the defect in the protein synthesis apparatus induced by poliovirus. Shortly after infection, cellular mRNAs no longer associate with ribosomes to initiate new rounds of polypeptide synthesis. Since the ribosomes, tRNAs, and other components are subsequently utilized for virus-specific translation and thus are unlikely to be altered, it was reasoned that either the host cell mRNA was modified so as to be unable to bind ribosomes, or the "binding machinery," collectively known as initiation factors, was modified so as to inactivate one or more steps in the binding reaction. The cellular mRNA which was no longer translated in infected cells was examined first, and found to be both structurally and functionally intact. The average size of cellular mRNAs was unchanged compared with uninfected cells (Willems and Penman, 1966; Leibowitz and Penman, 1971; Fernandez-Munoz and Darnell, 1976), and it contained a normal sequence of poly(A) at the 3' end (Koschel, 1974). The 5'-terminal methylated cap structures remained intact, and no other base modifications could be detected (Fernandez-Munoz and Darnell, 1976). Cellular mRNA extracted from cytoplasmic extracts of infected cells was capable of stimulating protein synthesis in vitro (Kaufmann et al., 1976). In the case of VSV-infected cells which were superinfected and consequently shut off with poliovirus, VSV mRNA synthesis continued at normal rates after superinfection, resulting in the accumulation of large amounts of intact, polyadenylated, capped, and methylated mRNA, but this mRNA did not associate with ribosomes and was not translated. However, when extracted from superinfected cells, the VSV mRNA was efficiently translated *in vitro* to produce *bona fide* VSV proteins (Ehrenfeld and Lund, 1977). The untranslated VSV mRNA accumulated in the cytoplasm of the poliovirus superinfected cells associated with cellular proteins in the form of messenger ribonucleoprotein particles (mRNPs). These mRNPs were also able to direct the translation of VSV polypeptides *in vitro*, indicating that no specific "blocking factor" was complexed with the RNA so as to sequester it from ribosome binding (Jones and Ehrenfeld, 1983). Thus, the failure of mRNA to initiate translation in poliovirus-infected cells could not be attributed to any modification or inactivation of the mRNA, and the explanation for its failure to be translated was sought elsewhere.

3. PROPOSED MODELS FOR INHIBITION

3.1. Double-Stranded RNA

The demonstration that expression of a functional viral genome was required, albeit at low levels, to elicit inhibition of initiation of translation of cellular mRNAs led several investigators to search for a virus-specific product which could be implicated in the shut-off reaction.

One approach was to prepare extracts of poliovirus-infected HeLa cells and to assay for an inhibitor of initiation by addition to a reticulocyte lysate translation reaction. The reticulocyte lysate had been well-characterized as a system capable of efficient initiation *in vitro*. A potent inhibitor of initiation was indeed found in the cytoplasm of poliovirus-infected cells, which was not present in similar preparations from uninfected HeLa cells (Hunt and Ehrenfeld, 1971), and the inhibitor was subsequently identified as double-stranded poliovirus RNA (Ehrenfeld and Hunt, 1971). In order to qualify as a specific inhibitor of host cell protein synthesis, however, it was necessary to demonstrate that viral protein synthesis was immune to the inhibitory effects of the putative shut-off factor. When double-stranded RNA was put to this test, no specificity between cellular and viral protein synthesis could be demonstrated (Celma and Ehrenfeld, 1974); rather, all protein synthesis was inhibited equally. In addition, concentrations of viral double-stranded RNA required to achieve inhibition in HeLa cell extracts were not attained during the period of host cell shut-off, but rather correlated with the later times of declining viral protein synthesis at the end of the infectious cycle. Subsequent studies of the mechanism of double-stranded RNA-induced inhibition of translational initiation in reticulocyte lysates demonstrated the existence of a dsRNA-activated protein kinase which phosphorylates the small subunit of eIF-2 (Farrell *et al.*, 1977). Since this initiation factor functions at a step prior to mRNA binding, it is not surprising that no specificity between cellular and viral mRNAs occurred. Thus, although this approach elucidated a potent and interesting inhibitor of protein synthesis, it failed to reveal the mediator of poliovirus-specific inhibition of host cell protein synthesis.

3.2. Viral Capsid Proteins

A second model to account for the observed inhibition of cellular mRNA translation developed from a genetic analysis of temperature-sensitive mutants of poliovirus. These mutants were screened for their ability to repress protein synthesis after infection in the presence of guanidine at restrictive temperature (Steiner-Pryor and Cooper, 1973). Six mutants were identified which showed reduced inhibition of amino acid incorporation into protein at a single time point (4–5 hr postinfection with guanidine) at the restrictive, compared with the permissive, temperature. All six of these mutants carried defects in the genes for capsid proteins, and it was concluded that repression of host cell protein synthesis was dependent upon the configuration of some product of the coat protein gene(s). The finding of capsid proteins (VP0, VP1, VP3) apparently associated with the small ribosomal subunit and other ribosomal structures was presented as biochemical evidence to support this model (Wright and Cooper, 1974), which proposed that a complex of capsid proteins combined with the ribosomal subunit and thereby prevented its association with cellular (but presumably not viral) mRNA. Several questions can be raised about the genetic experiments and the assay used to determine the ability of various mutants to inhibit protein synthesis under various conditions. At higher multiplicities of infection, all of the mutants did repress protein synthesis at both temperatures, although to variable extents. Measurements at a single time point may have revealed differences in the rate of inhibition as opposed to the absolute ability to inhibit. Mutations in the capsid protein gene(s) may have altered the rate of initial translation of the infecting genome and thus affected the kinetics of shut-off. The described association of capsid proteins with ribosomal structures was never shown to have had any functional consequences for the translational machinery. In addition, the model implies a stoichiometric rather than catalytic function of the putative capsid protein inhibitor, and thus fails to account for the observed shut-off that occurs in the presence of guanidine, when significant quantities of capsid protein do not accumulate. It is relevant that defective polioviruses containing large deletions in the coat protein genes nevertheless inhibit protein synthesis in a fashion similar to that of wild-type poliovirus (Cole and Baltimore, 1973). Another study reported that poliovirus capsids, in the form of either intact or disrupted virions, inhibited protein synthesis in reticulocyte lysates (Racevskis *et al.*, 1976), but no test of specificity for inhibition of cellular over viral translation was performed. In contrast, other measurements of inhibition *in vitro* by purified virus failed to demonstrate any effect on translation (Hunt and Ehrenfeld, 1971). Since no evidence has been presented to the contrary, it is possible that the observed inhibition in the former study was due to a general inhibition of all protein synthesis in the *in vitro* system caused by unidentified contaminants.

3.3. Inherent Translational Efficiencies

An alternative type of explanation for the specific discrimination against host cell protein synthesis in poliovirus-infected cells stemmed from the observation that initiation of protein synthesis could be selectively inhibited in HeLa cells and in poliovirus-infected HeLa cells by increasing the osmolarity of the growth medium (Saborio *et al.*, 1974). The inhibition was independent of the solute used to increase the osmolarity. However, virus-directed protein synthesis was observed to be relatively more resistant to inhibition by hypertonic medium than was cellular protein synthesis, a fact which was interpreted as indicating that initiation of viral RNA translation was intrinsically more efficient than that of cellular mRNA (Nuss *et al.*, 1975). These workers, therefore, proposed that the virus-specific or virus-induced factor involved in suppression of host protein synthesis could function by indiscriminantly lowering the rate of peptide chain initiation. Under such conditions, translation of viral mRNA, when it was synthesized, could occur due to its inherently strong affinity for ribosomes, whereas, cellular mRNA would fail to be translated. The mechanism by which a general reduction in the rate of initiation of protein synthesis might occur was not defined. However, the model requires that poliovirus RNA be a more efficient mRNA than those which remain untranslated in the infected cell. Unfortunately, other than the increased resistance of poliovirus protein synthesis to hypertonic medium, no direct measurements of ribosome or initiation factor binding affinities of poliovirus mRNA were available. Subsequent measurements of overall translation efficiencies *in vitro* showed polio RNA to be a relatively "poor" mRNA (Shih *et al.*, 1978), unable to compete with VSV mRNAs or other mRNAs whose translation is inhibited after poliovirus infection (Rose *et al.*, 1978; Brown and Ehrenfeld, 1980). It is, of course, possible that these *in vitro* measurements are not true reflections of *in vivo* reactions.

The suggestion that viral mRNA is an inherently efficient mRNA and can effectively compete with cellular mRNAs for some limiting component of the protein-synthesizing machinery has acquired experimental support for picornaviruses other than polioviruses. A direct competition model for shut-off of host cell translation has been proposed, specifically, to describe the shut-off induced by the cardioviruses, EMC, and mengo (see Section 7, below). However, these models do not apply to poliovirus-induced inhibition because, as noted above, cessation of protein synthesis in poliovirus-infected cells occurs before detectable viral RNA is synthesized and shut-off does occur after infection in the presence of guanidine or with a mutant virus temperature-sensitive for RNA synthesis at restrictive temperature.

3.4. Membrane Alterations and Intracellular Ionic Modifications

The idea that infection produced a general change in the intracellular environment which affected initiation complex formation in infected cells was extended by Carrasco and Smith (1976), who noted that alterations in the cell membranes of infected cells had been described which could lead to disruptions in permeability and, therefore, alter the ionic gradient between the inside and outside of the cell. Since numerous cell processes were known to be markedly affected by specific ion concentrations, they proposed that a viral protein might directly affect the cell membrane so as to cause an alteration in the intracellular ionic environment, particularly an entry of Na^+

into the cytoplasm, which inhibited cellular protein synthesis (Carrasco, 1977). The new ionic conditions would additionally favor translation of viral mRNA as well as eliminate the competition for cellular components. Support for this model was presented in the form of experiments which demonstrated that a 30 mM increase in Na^+ concentration in an *in vitro* translation system inhibited translation of globin mRNA or total mouse ascites cell polyadenylated RNA, whereas, the translation of polio RNA (or EMC RNA) was stimulated (Carrasco and Smith, 1976). Although alterations in the permeability of poliovirus-infected cells undoubtedly do occur, direct measurements of Rb^+ uptake, Na^+ accumulation, Na^+-K^+ ATPase activity, and other monovalent cation metabolic changes in infected cells showed that the changes occur later in infection, subsequent to the virus-induced inhibition of host cell protein synthesis (Nair *et al.*, 1979; Nair, 1981; Lacal and Carrasco, 1983). In addition, all of the ionic changes measured were prevented by guanidine, which does not affect inhibition of cellular protein synthesis. Thus, the modification of monovalent ion concentration in poliovirus-infected cells is not involved in the early shut-off events. Similar conclusions were drawn by analysis of the ion dependence of *in vitro* translation in extracts from poliovirus-infected cells (Bossart and Bienz, 1981).

3.5. Inactivation of Initiation Factors

The above sections have described several different mechanisms which have been proposed and explored during the last decade to explain the selective inhibition of host cell protein synthesis in poliovirus-infected cells. Admittedly, this author's bias has presented each mechanism as a straw man, requiring the reader to await what is perceived at this time to be the correct explanation for this aspect of the regulation of protein synthesis in poliovirus-infected cells. The favored model will be discussed in this and subsequent sections. It is important to state, however, that there is no convincing evidence that other picornaviruses are necessarily similar to poliovirus in the mechanism(s) utilized for host protein synthesis inhibition and that the mechanisms described above, as well as others, cannot all be dismissed in every case of picorna virus-induced protein synthesis inhibition. Thus, the data for other picornaviruses will be reviewed separately.

Currently, the most preferred mechanism for poliovirus-induced inhibition of cellular protein synthesis is that a factor or factors required for the initiation of translation of cellular mRNAs becomes specifically inactivated early after infection (Ehrenfeld, 1982). This hypothesis grew from an analysis of protein synthesis carried out *in vitro* by extracts prepared from poliovirus-infected cells. These unfractionated cell extracts, containing endogenous mRNA, faithfully reproduce the inhibition of cellular protein synthesis, despite the continued presence of cellular mRNA in such extracts. When extracts were prepared at early times (2 hr) after infection, when shut-off had occurred but before appreciable viral synthesis had begun, there were virtually no initiation events detectable *in vitro*. Extracts prepared at later times (4 hr) after infection showed resumed initiation activity, but totally under the direction of viral mRNA (Celma and Ehrenfeld, 1974; Kaufmann *et al.*, 1976; Bossart and Bienz, 1981). Fractionation of these extracts, as well as extracts from uninfected cells, into a soluble fraction, a ribosomal salt wash, and salt-washed ribosomes which still carried endogenous mRNA, enabled Kaufmann *et al.* (1976) to reconstitute a protein-synthesizing system and to localize the inactive component(s). They demonstrated that the ribosomal salt wash from poliovirus-infected cells, which contains protein synthesis initiation factors, was unable to stimulate translation by ribosomes carrying endogenous mRNA from uninfected cells and, thus, the crude initiation factor preparation was responsible for the failure of infected cells to translate cellular mRNA. These same preparations of initiation factors readily stimulate initiation of translation of endogenous poliovirus mRNA (Helentjaris and Ehrenfeld, 1978) and, thus, were shown to display the mRNA discrimination exhibited by infected cells *in vivo*. Control experiments with ribosomal salt wash preparations and translation systems from VSV-infected cells demonstrated the specificity of these assays and showed that initiation factors from poliovirus-infected cells failed to support synthesis of VSV proteins *in vitro*, whereas, initiation factors from VSV-infected cells were fully active for translation of both cellular and viral mRNAs (Brown *et al.*, 1980). Since several initiation factors have been shown to bind mRNAs and some are known to be required for binding of mRNA to ribosomal subunits, the hypothesis that one (or more) initiation factors might be specifically inactivated in poliovirus-infected cells so as to eliminate initiation complex formation with host cell mRNA appeared reasonable.

4. INITIATION FACTORS

4.1. Summary of Eukaryotic Initiation Factors

Initiation of protein synthesis in mammalian cells proceeds by a complex process whereby the assembly of mRNA, the ribosome, and initiator met-tRNA$_f$ into an initiation complex is catalyzed by a group of proteins called initiation factors. By definition, these proteins are not required for polypeptide chain elongation. A detailed discussion of this process, and the role of individual initiation factors can be found in this volume (Kaempfer, 1984), or in other recent reviews (Benne and Hershey, 1978; Jagus *et al.*, 1981) which also contain pertinent references. Nine initiation factors have been highly purified from rabbit reticulocytes, and still other factors have been described which may serve auxiliary functions, or indeed may qualify as initiation factors in their own right. A very brief (and oversimplifed) description of the role of the nine, characterized initiation factors is listed below, along with the step in initiation complex formation in which each participates:

Step 1. eIF-2 forms a ternary complex with met-tRNA$_f$ and GTP.

Step 2. eIF-3 and eIF-4C react with the 40 S ribosomal subunit and allow binding of the ternary complex.

Step 3. eIF-4A, eI-F4B, and eIF-4F function to promote binding of mRNA to the ribosomal subunit carrying initiator met-tRNA. eIF-3 is essential for this step, and eIF-1 and eIF-4C appear to stimulate binding.

Step 4. eIF-5 catalyzes the junction of the 60 S ribosomal subunit to form an 80 S initiation complex.

Step 5. eIF-4D stimulates the reaction of met-tRNA$_f$ with puromycin.

4.2. Involvement of Initiation Factors in Poliovirus-Induced Inhibition of Cellular Protein Synthesis: Discovery of Cap-Binding Protein

The experiments with poliovirus-infected cell extracts and ribosomal salt washes focused attention on those initiation factors which were involved with steps surrounding the recognition and binding of mRNA to the ribosome. Two different approaches were applied to identify the factor(s) which was responsible for the failure of infected cell extracts to initiate translation of cellular mRNA. Rose *et al.* (1978) prepared extracts from poliovirus-infected cells in which the endogenous viral mRNA was hydrolyzed by treatment with micrococcal nuclease. These extracts translated polio RNA nearly as well as control extracts, but were completely inactive in translating the VSV mRNA, again showing that the *in vitro* assay reproduces the selective translation inhibition seen *in vivo*. Purified (or partially purified) initiation factors from rabbit reticulocytes were then singly added to the infected cell extract to determine whether VSV translation could be restored, and it was demonstrated that preparations of eIF-4B did allow synthesis of VSV proteins. These workers initially concluded that eIF-4B was inactivated in poliovirus-infected cells. A different laboratory attempted to purify individual initiation factors directly from uninfected and from poliovirus-infected HeLa cells and then to test their ability to function in the translation of endogenous cellular mRNAs by a fractionated cell extract (Helentjaris *et al.*, 1979). Crude ribosomal salt wash was fractionally precipitated with ammonium sulfate to yield an A cut (0–40% saturation) and a B cut (40–70% saturation), which were then reconstituted in various combinations to localize the defective component. The B cut (known to contain eIF-1, 2, 4A, 4C, 4D, and 5) was active, regardless of the source of cells from which it was prepared; whereas, the A cut (containing eIF-3 and 4B) from infected cells was totally inactive. Separation of eIF-3 from eIF-4B and analysis of the individual factor activities showed that eIF-4B from infected cells retained normal activity, but eIF-3 was inactive.

The resolution of this apparent discrepancy was provided by an independent discovery by Sonenberg *et al.* (1978) that a small polypeptide (M_r 24,000) which specifically bound the m^7G cap structure at the 5′ end of capped mRNAs, was present in preparations of both eIF-3 and eIF-4B purified from rabbit reticulocytes. The name cap-binding protein (CBP) was given to this polypeptide. The activity which restored VSV mRNA translation by poliovirus-infected cell extracts, and which had previously been ascribed to eIF-4B was subsequently shown to be due to CBP that had copurified with eIF-4B (Trachsel *et al.*, 1980). Subsequent comparisons of eIF-4B from uninfected and poliovirus-infected HeLa cells showed no differences in the number of molecules per cell, molecular size, or extent of covalent modification of this initiation factor (Duncan *et al.*, 1983). Similarly, the eIF-3 preparations isolated from poliovirus-infected cells which were inactive *in vitro*, likely contained both eIF-3 and CBP, since the two proteins have been shown to cosediment in sucrose gradients containing low salt; indeed, an affinity column of eIF-3 coupled to Sepharose was used at one time to purify the CBP (Trachsel *et al.*, 1980).

The implication of these findings is that poliovirus infection causes an inactivation of CBP and, thereby, inhibits translation of capped, cellular mRNAs. This idea was inherently attractive, since poliovirus mRNA is not capped, but rather terminates with a 5' pUp (Hewlett *et al.*, 1976; Nomoto *et al.*, 1976), and this structural difference between viral and cellular mRNAs provided a possible basis for the discrimination in translation displayed by infected cells and cell extracts. Indeed, the 24,000-dalton CBP, purified by affinity chromatography on m^7GDP-Sepharose columns (Sonenberg *et al.*, 1979b) was shown to stimulate translation of capped (Sindbis virus, reovirus, and globin) mRNAs in uninfected HeLa cell extracts, but had no stimulatory effect on translation of naturally uncapped (EMC and satellite tobacco necrosis virus) mRNAs (Sonenberg *et al.*, 1980). The importance of the cap group in facilitating translation of capped mRNAs has been reviewed (Banerjee, 1980) and the ability of cap analogues (m^7GMP, m^7GDP, or m^7GTP) to inhibit translation of capped mRNAs in a competitive fashion has been well documented. Furthermore, it is known that the step in protein synthesis initiation in which the cap group functions is during binding of the mRNA to the small ribosomal subunit, prior to its conversion to an 80 S initiation complex. This is consistent with the step shown to be inhibited after poliovirus infection.

5. CAP RECOGNITION ACTIVITY

5.1. The 24,000-Dalton Cap-Binding Protein

The initial identification of a cap-binding protein resulted from experiments by Sonenberg and Shatkin (1977), who developed a crosslinking assay to detect polypeptides which were in physical proximity to the 5' ends of capped mRNAs. Reovirus mRNA was synthesized containing ^{3}H-label in the methyl groups of the 5' cap. The mRNA was oxidized with $NaIO_4$ so as to convert the 2',3'-*cis*-diol of the 5'-terminal m^7G to a reactive dialdehyde, and the oxidized mRNA was then incubated with ribosomal salt wash from rabbit reticulocytes or from Ehrlich ascites cells (Sonenberg *et al.*, 1978). The products were then reduced with $NaBH_3CN$, resulting in covalent linkage between the mRNA 5' termini and any amino groups of neighboring proteins. The bulk of the mRNA was disgested with ribonuclease and the proteins to which the ^3H-label had been transferred were displayed on polyacrylamide gels. The specificity of the interaction was ascertained by crosslinking in the presence of cap analogues which were known to inhibit mRNA binding to ribosomes. Analysis of crude preparations of initiation factors yielded a single polypeptide, M_r 24,000, which was sensitive to competition by cap analogues. A similar polypeptide with identical crosslinking characteristics was also demonstrated in HeLa cells (Hansen and Ehrenfeld, 1981; Lee and Sonenberg, 1982), and will hereafter be referred to as 24-CBP. The 24-CBP was found in purified preparations of eIF-3 and, to a lesser extent, in eIF-4B. Several other polypeptides in eIF-3 were crosslinked but not inhibited by cap analogues. Interestingly, different degrees of crosslinking to these eIF-3 polypeptides were observed with different mRNAs (Sonenberg et al., 1979a). Thus, although the 24-CBP appeared to be the sole cap-specific binding protein, eIF-3 may interact with the CBP near the site of cap attachment at the time of mRNA binding to the ribosome.

An alternative assay for detecting cap-binding protein activity was utilized by Hellman et al. (1982), who looked for proteins capable of reversing the ability of cap analogues to inhibit translation in a rabbit reticulocyte lysate. The basis of this assay is that cap analogues inhibit protein synthesis by binding to a receptor (CBP) which normally recognizes the capped end of mRNA. Fractionation of the reticulocyte ribosomal salt wash led to the purification of a 24,000-dalton protein which appeared identical to the one which reacted in the crosslinking assay.

5.2. ATP/Mg^{2+}-Dependent Cap-Binding Proteins

The conditions of the crosslinking assay described above were modified to examine the effect of ATP and Mg^{2+} on the crosslinking of reticulocyte initiation factors to mRNA caps (Sonenberg, 1981; Sonenberg et al., 1981). In addition to the 24-CBP, several other polypeptides having molecular weights of 28,000, 50,000, and 80,000 daltons were shown to specifically crosslink to the oxidized terminus of reovirus mRNA. Specific crosslinking of the latter polypeptides was totally dependent upon ATP and Mg^{2+}, in contrast to crosslinking of the 24-CBP. The binding of mRNA to the small ribosomal subunit initiation complex has long been known to require ATP, although the precise mechanism of ATP involvement has not been elucidated. In addition, a monoclonal antibody prevented the crosslinking of all four polypeptides to mRNA caps, suggesting the presence of a common epitope in all four proteins. This raised the possibility of a functional (or artifactual) proteolytic cleavage in the generation of 24-CBP. A major difficulty in interpreting these results, however, is the uncertainty regarding what polypeptides the antibody was directed against. Immunoaffinity chromatography of a crude initiation factor preparation yielded a 50K polypeptide as the major antigen retained by the monoclonal antibody, and this 50K polypeptide was shown to share common tryptic peptides with the 24-CBP (Sonenberg et al., 1981). However, it now seems unlikely that this 50K protein is the same as that which shows ATP-dependent crosslinking to oxidized mRNA (see below). In addition, crosslinking of the higher molecular weight proteins was not tested in the absence of 24-CBP, thus, the possibility remains that only the 24-CBP recognizes and binds cap groups (without requiring ATP-Mg^{2+}), while the other proteins then bind in the vicinity of the complex in an ATP-dependent step. If the antibody were directed against the CBP and prevented its interaction with the cap group, then subsequent crosslinking of the other polypeptides would be inhibited. Competition by cap analogues in the crosslinking would occur similarly for all four proteins.

A possible clue to the identities of the proteins which crosslink in an ATP/Mg^{2+}-dependent fashion was provided by experiments designed to characterize the initiation factor, eIF-4A (Grifo et al., 1982). This factor has a molecular weight described variously as 44,000–50,000 daltons. It was shown that eIF-4A could bind mRNA (measured by retention on nitrocellulose filters) in a reaction that required ATP as well as eIF-4B ($M_r = 80,000$) which contained CBP. The binding reaction was inhibited by m^{7}GMP. Importantly, when oxidized ^{3}H-cap-labeled reovirus mRNA was incubated with various combinations of eIF-4A, eIF-4B, CBP, and ATP/Mg^{2+} in a crosslinking assay, both initiation factors demonstrated an ATP-dependent crosslinking that was cap specific. No crosslinking was observed with eIF-4A alone, and only limited crosslinking of eIF-4B occurred without eIF-4A. Thus, in the presence of CBP, both eIF-4A and eIF-4B bind to the 5′ terminus of mRNA in an ATP-dependent and cap-dependent reaction and the 80K and 50K polypeptides described by Sonenberg were likely eIF-4B and eIF-4A, respectively. Indeed the identity of the 50K protein with eIF-4A has been conclusively demonstrated, since anti-eIF-4A monoclonal antibody immunoprecipitates the crosslinked 50K polypeptide (Edery *et al.*, 1983).

5.3. Cap-Binding Protein Complex: eIF-4F

Although the 24-CBP has been purified to apparent homogeneity (Sonenberg *et al.*, 1981; Hellman *et al.*, 1982), numerous studies indicated that the CBP also associated with other proteins to form a high molecular weight complex. Hellman *et al.* (1982) noted that the CBP exhibited a significant degree of heterogeneity during various steps in its purification. In addition, at least one of the functional activities ascribed to the CBP, the ability to restore translation of capped mRNAs in a poliovirus-infected cell extract, was reported to be extremely labile in highly purified preparations (Trachsel *et al.*, 1980), suggesting that other proteins might be involved in this activity. Indeed, subsequent reports showed no restoring activity by free 24-CBP alone (Tahara *et al.*, 1981).

Tahara *et al.* (1981) was the first to clearly demonstrate that 24-CBP was present in two differently sedimenting forms in reticulocyte ribosomal salt wash. In these studies, it was important to dissociate CBP from eIF-3 by treatment with 0.5 M KCl and then to eliminate eIF-3 by velocity sedimentation prior to further purification of CBP on a m^{7}GDP-Sepharose affinity column. This protocol prevented reassociation of CBP with eIF-3. The CBP was identified by crosslinking to oxidized, cap-labeled mRNA, and was found to resolve into a rapidly-sedimenting (7–10 S) complex, containing 24-CBP and several higher molecular weight proteins, and a slowly-sedimenting (< 6 S) form comprised primarily of the 24-CBP. Both forms could stimulate translation of capped (but not uncapped) mRNAs in uninfected HeLa cell extracts, but only the large complex restored translated of capped, Sindbis virus mRNA in polio-infected cell extracts. The large complex of CBP plus other polypeptides was named CBP II. Hansen *et al.* (1982*a*) also described a rapidly-sedimenting and slowly-sedimenting form of CBP isolated from the ribosomal salt wash of HeLa cells. However, in this report, CBP was not purified from eIF-3, so the polypeptides present in the large CBP complex could not be analyzed. Significantly, poliovirus-infected cells contained no detectable large CBP complex, although the slowly-sedimenting form was present. In addition, slowly-sedimenting CBP was found in the soluble

(200,000 g supernatant) fraction of both uninfected and poliovirous-infected HeLa cell extracts (Hansen *et al.*, 1982*b*). Neither the S-200 nor the ribosomal salt wash from poliovirus-infected cells contained restoring activity for poliovirus-infected cell extracts.

The distribution of 24-CBP in large and small complexes in the total cell has never been quantitated. This is partly because most investigators use only a fraction of the cell or a fraction of the ribosomal salt wash as starting material, and also because the cross-linking assay most often used to detect CBP is not necessarily a quantitative measure of the amount of CBP present. Furthermore, it has not been determined whether the 24-CBP present in various forms is truly identical, or whether 24-CBP in the large complex, for example, has undergone some modification. The apparent molecular weight of 24-CBP in all forms is the same, as judged by mobility on SDS–polyacrylamide gels, and 24-CBP in all forms appears capable of crosslinking to oxidized mRNA caps. Further comparisons of charge properties or peptide maps have not yet been reported. The functional activity of restoration of the ability of poliovirus-infected cell extracts to translate capped mRNAs appears to require the large CBP complex. It is not known whether free 24-CBP (CBP I) or a smaller form of a CBP complex has any function in protein synthesis, although it does stimulate translation of capped mRNAs *in vitro*, and some investigators refer to 24-CBP as eIF-4E.

Purification of the large CBP complex has been achieved in two laboratories. One group (Grifo *et al.*, 1983) obtained an apparently distinct, multisubunit protein complex from crude eIF-4B preparations. The protocol utilized dissociation of the complex from eIF-3 by treatment with high salt and separation from eIF-4B on an ultragel AC-A34 column in high salt, followed by further purification. The isolated complex contained four major polypeptides with molecular weights described as 24,000, 46,000, 73,000, and 200,000. The complex was active in restoring capped mRNA translation in polio-infected cell extracts, and appeared similar to the previously described CBP II. In fact, since the 73,000-dalton polypeptide was not present in CBP II (Tahara *et al.*, 1981), the authors suggested that this polypeptide was not part of the complex required for restoring activity. The 24,000-dalton component in both preparations crosslinked specifically to the capped 5' end of oxidized mRNA, confirming its identity with CBP I. Furthermore, the complex was required in a factor-dependent translation system for maximal stimulation of globin synthesis, in addition to, and independent of eIF-3, eIF-4A, and eIF4B.

The authors concluded that the CBP-containing complex therefore constituted a new initiation factor activity (termed eIF-4F) with a unique functional role, which had presumably previously been undetected because it contaminated preparations of eIF-4B and/or eIF-3. eIF-4F was shown to promote maximal mRNA binding to the 40 S ribosomal subunit, in addition to eIF-2, eIF-3, eIF-4A, and eIF-4B. Curiously, all factor requirements, including those for eIF-4F were the same for translation of globin mRNA and for STNV RNA, a naturally uncapped mRNA. A possible role of eIF-4F for translation of the latter RNA was not proposed.

The 46,000-dalton component of eIF-4F was identified in the same report as being identical to eIF-4A, which was purified independently. Both proteins had the same apparent molecular weight and net charge, as indicated by their mobilities in two-dimensional gels. In addition, eIF-4F could replace eIF-4A in the eIF-4B and ATP-dependent crosslinking to oxidized mRNA.

A second group of investigators (Edery *et al.*, 1983) purified a multisubunit protein complex which contained three major polypeptides, of apparent molecular weights, 24,000, 50,000, and 210,000. The complex had restoring activity for polio-infected HeLa cell extracts; thus, it is functionally analogous to CBP II or eIF-4F. They also identified the 50K polypeptide as eIF-4A by immunoprecipitation with anti-eIF-4A monoclonal antibody, and further demonstrated similarity of tryptic peptide maps between the two proteins. An interesting observation was that, although the majority of peptides were clearly common to both polypeptides, one consistent difference in the peptide maps was evident. Since the majority of eIF-4A in ribosomal salt wash purified independently from the CBP complex and only a small proportion appears to be associated with CBP, the difference in tryptic peptides may reflect a modification of eIF-4A that contributes to its distribution between the CBP complex and the free form. The eIF-4A in the CBP complex crosslinks in an ATP-dependent fashion to oxidized mRNA cap groups in the presence but not in the absence of added eIF-4B. Thus, the CBP complex contains the activity required for crosslinking of both eIF-4A and eIF4B, and the crosslinking of eIF-4a in the complex requires eIF-4B. The findings in this report suggest that eIF-4B can interact with the CBP complex, but is not an integral part of it. The detection of a 73,000-dalton polypeptide in the complex isolated by Grifo *et al.* (1983) suggests that it might represent associated eIF-4B (reported $M_r = 75,000–80,000$), but characterization of this component has not yet been performed.

5.4. Cap-Binding Proteins in Poliovirus-Infected Cells

The fact that cell extracts derived from poliovirus-infected cells fail to translate capped mRNAs (Rose *et al.*, 1978; Sonenberg *et al.*, 1982; Hansen *et al.*, 1982*a*; Jones and Ehrenfeld, 1983) and that translation in these extracts can be restored by the CBP-containing complex (CBP II or eIF-4F) purified from rabbit reticulocytes (Tahara *et al.*, 1981; Grifo *et al.*, 1983; Edery *et al.*, 1983) firmed the hypothesis that the CBP complex was somehow inactivated in poliovirus-infected cells. Initiation factor preparations from poliovirus-infected cells contain no restoring activity (Hansen and Ehrenfeld, 1981; Lee and Sonenberg, 1982). Hansen *et al.* (1982*a*) established the fact that the rapidly-sedimenting CBP-containing complex was absent or dissociated in poliovirus-infected cells, although free CBP (CBP I) could still be crosslinked to oxidized mRNA (Hansen and Ehrenfeld, 1981; Hansen *et al.*, 1982*a,b*; Lee and Sonenberg, 1982). Direct demonstration that the CBP complex, isolated from poliovirus-infected HeLa cells, was inactive was recently achieved (Etchison *et al.*, submitted) after the development of assays designed to measure the separate activities of eIF-3 and CBP complex. By this means, it was shown that eIF-3 measured in the presence of purified CBP complex (Edery *et al.*, 1983) is equally active in infected and uninfected cells, whereas, no CBP complex activity was detected.

What, then, is the biochemical reason for the inactivation of the CBP complex and the structural dissociation of the polypeptides? The 24-CBP appears to remain in infected cells, able to recognize and bind to mRNA cap groups, but no detailed structural analyses of the protein from either uninfected or infected cells have been performed (see above). The 50,000-dalton subunit represents only a subset of the cell's eIF-4A population, and may contain a covalent modification, as evidenced by a small difference in tryptic peptide maps between the polypeptide isolated from the CBP complex and the majority of eIF-4A isolated independently (Edery *et al.*, 1983). Thus, although total cellular eIF-4A has been reported to be both structurally (Duncan *et al.*, 1983) and functionally (Etchison *et al.*, submitted) identical in poliovirus-infected HeLa cells compared with uninfected cells, the possibility still remains that virus infection affects the subset of eIF-4A in the CBP complex.

The large polypeptide in the CBP complex, p220, is perhaps the most interesting of all. Little is known about this polypeptide, except its described presence in purified preparations of CBP complex. However, antiserum which was prepared against eIF-3 preparations containing CBP complex was able to detect the p220 antigen in lysates from uninfected HeLa cells, but this antigen was absent from polio-infected cells (Etchison *et al.*, 1982). Instead, antigenetically-related polypeptides of 100,000–130,000 daltons, presumed degradation products of p220, were detected in the infected cell lysates. The time course for degradation of p220 correlated with that for inhibition of cellular protein synthesis *in vivo*. Affinity-purified antibodies against p220 also reacted with the 220,000-dalton component of the purified CBP complex. The authors suggested that specific degradation of p220 may be the cause of the dissociation and inactivation of the CBP complex in poliovirus-infected cells.

Lee and Sonenberg (1982) analyzed the crosslinking pattern of cap-binding proteins in crude initiation factor preparations from poliovirus-infected and uninfected HeLa cells in the presence of ATP/ Mg^{2+}. As discussed above, these conditions yield specific crosslinking of 80, 50, 32, and 28K polypeptides, in addition to the 24-CBP whose crosslinking is independent of ATP/Mg^{2+}. This study demonstrated a complete loss of specific crosslinking of all higher molecular weight CBPs, as well as a significant reduction in the specific crosslinking of the 24-CBP, in initiation factor preparations from infected cells. In light of the subsequent clarification of the identity of at least some of these polypeptides (see above, this section), as well as the demonstrated dependence of their crosslinking upon functional 24-CBP, most likely in the form of CBP complex, the absence of ATP-dependent crosslinking in poliovirus-infected cells is not surprising. Very likely, functional binding of the CBP complex (which is absent in polio-infected cells) to the cap group of mRNA is required for the other proteins to bind. Impairment of cap recognition ability in the infected cell results in failure of the other cap-binding proteins to be crosslinked.

5.5. Function of the Cap-Binding Proteins

The function of the cap group on mRNAs and presumably, therefore, of the cap recognition proteins has been probed by the construction of mRNAs with altered structures, as well as by analysis of mRNA function in poliovirus-infected cell extracts which lack cap-binding activity. Morgan and Shatkin (1980) and Kozak (1980*a,b*) prepared reovirus mRNAs with reduced secondary structure by substituting inosine for guanosine nucleotides, thereby preventing G:C base-pairing, or by reacting native reovirus mRNA with bisulfite to convert C to U. These mRNAs were able to bind wheat germ ribosomes and form initiation complexes in a relatively cap-independent and ATP-independent manner. Naturally, uncapped mRNAs such as EMC virus RNA or cowpea mosaic virus RNA are less dependent upon ATP for initiation complex formation than capped mRNAs (Jackson, 1982). In addition, a monoclonal antibody with anti-CBP activity was shown to inhibit ribosome binding to native reovirus mRNA but did not inhibit binding to inosine-substituted mRNA (Sonenberg et al., 1981). The requirement for a cap, therefore, appears to be reduced by unfolding of the mRNA, and these results were taken as support of the hypothesis that cap-binding proteins are involved in an ATP-dependent melting of 5' mRNA secondary structure to facilitate ribosome attachment (Sonenberg, 1981). Presumably, the model calls for binding of the cap group via 24-CBP in eIF-4F, the CBP complex, with subsequent hydrolysis of ATP, allowing attachment of eIF-4A and eIF-4B. The energy released would be absorbed to destabilize the secondary structure at the 5' end of the mRNA, ultimately facilitating ribosome attachment. Unfolded (inosine-substituted) mRNAs would not require cap recognition and ATP-dependent destabilization for ribosome attachment. In this model, the CBPs would serve as RNA unwinding or melting proteins. Interestingly, the requirement for CBP complex for translation of globin mRNA in an initiation factor-dependent, fractionated translation system is abolished under low salt conditions, under which (among other things) the mRNA is presumably relatively unfolded (D. Etchison, unpublished observations).

Lee et al. (1983) attempted to directly measure the effect of mRNA secondary structure on its reaction with cap-binding proteins. The 50,000- and 80,000-dalton, ATP/Mg^{2+}-dependent CBPs in crude rabbit reticulocyte ribosomal salt wash showed the usual ATP-dependence for crosslinking to the oxidized cap structure of native reovirus mRNA. However, when inosine-substituted mRNA was used, specific crosslinking of these polypeptides occurred in the absence of ATP/Mg^{2+}. These results were contradicted in a subsequent report (Tahara et al., 1983) which demonstrated that ATP was required for crosslinking purified eIF-4A and 4B to oxidized, inosine-substituted mRNA as well as for authentic ribosome binding by the denatured RNA. The reason for the discrepant results by these two laboratories has not yet been elucidated. Perhaps small amounts of ATP were present in the crude ribosomal salt wash utilized by Lee *et al.* (1983); perhaps differences in salt concentration and/or other factors affected the degree of secondary structure remaining in the inosine-substituted mRNA. In any event, the role of ATP in initiation of eukaryotic protein synthesis is still not defined. It may be required for binding of initiation factors, for correct binding of ribosomal subunits, and/or for migration of the ribosomal subunit to the initiating AUG. A recent report describes experiments which demonstrate the eIF-4A is the component that physically interacts with ATP during protein synthesis initiation (Seal *et al.*, 1983). Direct proof of whether or how the binding of CBP affects the 5'-terminal structure of mRNA must await direct physical measurements. Results of such studies are not yet available.

A second approach to understanding the function of the CBP complex has been the analysis of events occurring in a poliovirus-infected cell extract, on the assumption that these extracts represent a system which is specifically impaired in cap recognition function. Again, inosine-substituted reovirus mRNA was prepared to provide an mRNA with reduced secondary structure. This "denatured" mRNA was able to form initiation complexes in a polio-infected cell extract, in contrast to the inability of native reovirus mRNA to form such complexes (Sonenberg *et al.*, 1982). It is unfortunate that actual translation of inosine-substituted mRNA could not be measured (presumably due to misreading of I-containing codons), to confirm the legitimacy of these initiation complexes (see Tahara *et al.*, 1983). Furthermore, a capped mRNA from alfalfa mosaic virus which is predicted by computer analysis to be devoid of stable secondary structure by virtue of an A + U-rich 5'-leader region, was translated with reasonable efficiency in the polio-infected cell extract. Translation of this RNA had previously been shown to be resistant to inhibition by cap analogues, and to a monoclonal antibody with anti-CBP activity (Sonenberg *et al.*, 1981). Thus, translation of the alfalfa mosaic virus RNA appears to be relatively cap independent, perhaps related to 5'-end secondary structures.

A possibly related observation is that polio-infected cell extracts which do not translate capped mRNAs under reaction conditions optimal for translation by uninfected extracts will perform translation if the salt concentration is reduced (Hansen and Ehrenfeld, unpublished observations). Several investigators have demonstrated a lower salt optimum for translation of chemically or enzymatically uncapped mRNAs (Brown *et al.*, 1982), and have reported that the importance of cap groups in stimulating the interaction between mRNA and ribosome decreases with decreasing ionic strength (Weber *et al.*, 1977). Their observations have generally been interpreted to mean that unfolding of the mRNA at low ionic strength bypasses the requirement for the cap group, and presumably also the requirement for a functional cap recognition activity.

At present, the function of the cap group and of the cap-binding proteins is suggested only from indirect experimental evidence, and is still a matter of much speculation. Fortunately, several well-qualified laboratories are directing major efforts toward this problem, and we can anticipate a much improved understanding in the near future.

6. POLIOVIRAL MEDIATOR OF HOST CELL SHUT-OFF

As is evident from the above discussion, the precise biochemical lesion induced by virus infection which leads to disruption of the cap-binding protein complex and loss of function of cap-binding activity has not yet been defined. Whatever the nature of this lesion, however, it must be effected or induced by some viral gene product. Rose *et al.* (1978) found that the ability of uninfected HeLa cell extracts to translate VSV mRNA was abolished by preincubation of the extract for 30 or 60 min with an extract from infected cells. The *in vitro* inactivation was restored by subsequent addition of eIF-4B which contained CBP. This experiment suggests that an activity which causes a slow inactivation of the cap recognition complex is present in poliovirus-infected cells, and that preformed, inactive cap-binding complex does not actively inhibit translation. Brown and Ehrenfeld (1980) described an activity in ribosomal salt wash from infected cells which restricted the ability of rabbit reticulocyte lysates to translate VSV mRNA. These inactivated lysates retained their ability to translate polio RNA, but failed to translate several capped mRNAs. The inactivating activity was localized to the 0–40% ammonium sulfate fractional precipitate of the ribosomal salt wash. In addition, both the restoring activity and the crosslinking ability of initiation factors from uninfected cells were impaired by preincubation with ribosomal salt wash from polio-infected cells (Lee and Sonenberg, 1982).

All of these data indicate that a catalytic activity is present in poliovirus-infected cells which can inactivate the function of the cap recognition complex. Each of these observations represents a potential assay for the purification of the inactivator, but this has not yet been accomplished. It is a bit confusing that the viral (or virus-induced) inactivator localizes to the same cell fraction (0–40% ammonium sulfate cut of the ribosomal salt wash) as the cap-binding protein complex itself, which is presumed to be the substrate for the inactivator. Examination of the viral proteins that fractionate with the inactivating activity in the ribosomal salt wash revealed a disappointingly complete spectrum of viral proteins, with no evident enrichment for one or more viral products (unpublished observations). The finding that one component of the putative cap-binding protein complex, p220, is cleaved in infected cells has led to the attractive but as yet unproved idea that the viral protease may be responsible for the cleavage of p220 and, therefore, for the inactivation of the CBP complex. Since infected cells show no gross changes in the sizes of the general protein population, the viral protease clearly does not indiscriminantly hydrolyze proteins. In fact, the cleavages catalyzed by the polio protease on viral protein substrates are known to be extremely specific. Thus, if p220 is a substrate for this interesting protease, it is likely a significant reaction. Etchison *et al.* (1982) reported that incubation of an uninfected HeLa cell ribosomal cell wash, which contained intact p220, with ribosomal salt wash from polio-infected cells for 15 or 30 min at 37°C, resulted in the disappearance of p220 from the uninfected cell sample. Incubation of the uninfected cell ribosomal salt wash alone had no detectable effect on the integrity of p220. Similarly, incubation of purified cap-binding protein complex (p220, eIF-4A, and 24-CBP) which was radiolabeled *in vitro* by reductive methylation with [^{14}C]formaldehyde, showed degradation of p220 after incubation with infected cell extracts (Lee and Sonenberg, personal communication). No effect on eIF-4A or 24-CBP was observed. The degradation of p220 was inhibited by several protease inhibitors (TPCK, PMSF, NEM, iodoacetamide) as well as by Zn^{2+} reagents which have been used to inhibit processing of poliovirus proteins *in vivo*. If cleavage of p220 is the cause of inactivation of the CBP complex, and if the viral protease is the cause of p220 cleavage, then a consistent explanation for the mechanism of host cell shutoff by this virus will be reached.

Virus-specific phosphorylation of ribosomal proteins and/or proteins associated with ribosome preparations has been described (Tershak, 1978; James and Tershak, 1981) and thought to be a possible regulatory reaction involved in inhibition of cellular protein synthesis. Although no data are available to relate the phosphorylation events with the regulation of protein synthesis, it is interesting that one of the virus-specific phosphorylations affected a 24,000-dalton polypeptide. Phosphorylation of the 24-CBP has not been reported.

At the present time, it is assumed that a viral gene product either acts directly or indirectly to cause inactivation of the cap recognition system in HeLa cells. However, direct identification of the mediator of the inactivation has not been demonstrated.

7. OTHER PICORNAVIRUSES

7.1. Differences from Poliovirus

The idea that poliovirus infection results in the inactivation of an initiation factor required for the translation of capped mRNAs and, thus, effects a transition from host to viral mRNA translation, accommodates both the effects seen *in vivo* during poliovirus infection and the observations made during studies of translation by infected cell extracts *in vitro*. Similar analyses of cells or extracts of cells infected with EMC or mengovirus, however, suggested important differences from polio infections, and revealed no comparable mRNA-specific changes in the host translational machinery. Several groups of investigators, studying either EMC or mengovirus infection of Krebs II, mouse plasmacytoma, or Ehrlich ascites tumor cells, found that extracts prepared from infected and uninfected cells were equally active in the translation of both cellular and viral mRNAs (Svitkin *et al.*, 1974; Lawrence and Thach, 1974; Abreu and Lucas-Lenard, 1976; Hackett *et al.*, 1978a). In some cases, infected cell extracts translated both viral and cellular mRNAs less actively than their uninfected counterparts, but this was likely due to the presence of small amounts of viral double-stranded RNA in the infected cell preparations, since the magnitude of the effect appeared to correlate with the sensitivity of a given cell extract to double-stranded RNA. The important point is that there is no evidence for a selective inhibition of host mRNA translation in infected cell extracts, and these extracts appear to posses all of the factors and functional components required for translation of cellular mRNAs.

The observed differences between the results obtained by investigators of protein synthesis regulation by the cardioviruses and those studying poliovirus remained unreconciled for several years, and were largely ignored by each other. This was likely due to a firm belief held by almost all picornavirologists that the basic replication strategies, including this specific interaction with the host cell, must be uniform among all members of the virus group, although Svitkin *et al.* (1978) pointed out that the properties of EMC and poliovirus-infected cells suggested that different mechanisms were operating to effect a preferential synthesis of viral over cellular proteins. To their credit, Jen *et al.* (1980) performed a direct comparison of the characteristics of host cell protein synthesis inhibition and of the initiation factor activity derived from the same HeLa cells infected with either EMC or polio. Ribosomal salt washes were prepared from uninfected and from EMC- and polio-infected HeLa cells, and these were tested for the ability to support translation of capped (globin) and uncapped (EMC) mRNAs in a fractionated HeLa cell translation system. As previously described, the ribosomal salt wash from polio-infected cells showed a marked preferential reduction in activity for globin mRNA translation compared with EMC RNA translation. In contrast, EMC-infected cell ribosomal salt wash showed no selective activity between mRNAs. Both virus-infected cell ribosomal salt wash preparations were less active than uninfected cell ribosomal salt wash, possibly again due to the presence of small amounts of viral double-stranded RNA.

The same study drew attention to the differences in kinetics of host cell shut-off in the two virus-infected cells, documenting the rapid inhibition of cellular protein synthesis by polio infection which occurred prior to and during the time that viral protein synthesis was increasing, and which demonstrated the selectivity of the inhibition. EMC protein synthesis, in the same cells, was not preceded by or accompanied by an abrupt inhibition of cellular protein synthesis. Rather, concurrent synthesis of cellular and viral protein was seen throughout the cycle with the ratio of viral to total protein synthesis steadily increasing with time.

7.2. mRNA Competition

Since extracts of infected and uninfected cells translated both viral and cellular mRNAs with equal efficiencies, other explanations for the preferential synthesis of viral mRNA were sought. Simultaneous addition of viral and cellular mRNAs to either the plasmacytoma (Lawrence and Thach, 1974) or Ehrlich ascites (Abreu and Lucas-Lenard, 1976) cell-free systems resulted in suppression of translation of cellular mRNAs, whereas, viral RNA was translated preferentially. At subsaturating concentrations of mRNAs, both types of templates could be translated simultaneously, although nonviral mRNA translation was selectively reduced. At saturating concentrations, complete suppression of translation of cellular mRNA occurred. Such competition experiments led to the suggestion that EMC or mengoviral RNA had a high affinity for some factor, possibly an initiation factor, which was limiting in the cell and necessary for translation of cellular mRNA.

Since relatively high concentrations of viral mRNA were required to exclude cellular protein synthesis *in vitro*, Svitkin *et al.* (1978) investigated whether the relative mRNA concentrations present in infected cells could in fact explain the pattern of protein synthesis observed *in vivo*. The natural mixture of polyadenylated cellular and viral RNA was isolated from EMC-infected Krebs-2 cells, and this mixture was translated *in vitro*. Viral RNA constituted no more than 20% of the total RNA isolated by their procedures, yet predominantly viral proteins were synthesized. Thus, the relative mRNA concentrations that approximated those present in the infected cell caused preferential translation of viral rather than cellular mRNAs. These experiments supported the proposed model of preferential viral mRNA translation due to effective competition for some limiting component of protein synthesis. This model was originally proposed as a general scheme for protein synthesis regulation by Lodish (1974). A similar study of competitive RNA translation was performed in a fractionated protein synthesis system from mengovirus-infected Ehrlich ascites tumor cells. These experiments showed that the fraction from infected cells responsible for preferential translation of viral RNA over cellular mRNA was predominantly the ribosomal salt wash (Hackett *et al.*, 1978b). Preliminary attempts to identify the putative limiting factor(s) were conducted by Golini *et al.* (1976), who translated mixtures of EMC and globin mRNA in a fractionated translation system from mouse plasmacytoma cells and demonstrated that the competitive suppression of globin mRNA translation by EMC RNA could be relieved by the addition of a partially purified preparation of eIF-4B (formerly called IF-M3 by some laboratories) from rabbit reticulocytes. Independent measurements of RNA binding to eIF-4B by retention on nitrocellulose filters demonstrated that EMC RNA has a greater affinity for this initiation factor than globin mRNA or other capped mRNAs (Baglioni *et al.*, 1978). The filter binding assay utilized by these workers, however, would have been unable to distinguish the activity of a contaminating protein in their eIF-4B preparation. The lack of purity of initiation factor preparations available at that time and the paucity of information about initiation factor activities make the assignment of eIF-4B as the limiting component tentative at best.

Despite an incomplete understanding of the basis for mRNA competition in EMC or mengovirus-infected cells, there appeared to be general satisfaction with the competition model as an explanation for regulation of protein synthesis in cardiovirus-infected cells. The reason for the lack of a common strategy utilized by poliovirus and EMC or mengo was puzzling, although it was suggested that, due to inherent RNA sequence and/or structure properties, poliovirus RNA suffered from a relatively poor translation initiation activity and, therefore, was required to evolve a mechanism to eliminate competition by host mRNAs, which took the form of a cap-specific protein synthesis inhibition (Jen *et al.*, 1980). Indeed, poliovirus RNA functions as a significantly less efficient mRNA *in vitro* than does EMC RNA (Shih *et al.*, 1978). EMC RNA, on the other hand, appears to be an inherently strong initiator both *in vivo* (Jen *et al.*, 1978) and *in vitro* and, thus, modification of the host's translational machinery may be unnecessary.

7.3. Host Cell-Specific Interactions

A complication to this perhaps simplistic reasoning arose with the observation that EMC infection of mouse L cells resulted in a rapid inhibition of host cell translation, prior to the utilization of viral mRNA to synthesize viral proteins (Jen and Thach, 1982). These kinetics differed from those observed in mouse plasmacytoma, Krebs, or Ehrlich ascites cells, and were not consistent with direct competition of viral with cellular mRNA to preferentially yield viral protein synthesis. Different responses in protein synthesis patterns by different cell types was also reported for mengovirus (Otto and Lucas-Lenard, 1980).

Infected L cell extracts showed no reduction in initiation activity on cellular or globin mRNA *in vitro*. Fractionation of these extracts, however, revealed a marked increase in a component which stimulated translation of capped (but not uncapped EMC) mRNAs in the soluble (200,000 $\times$ g supernatant) cell fraction, compared with uninfected cells (Jen and Thach, 1982). This suggested that EMC infection of L cells resulted in a subcellular redistribution of at least some protein synthesis initiation factors. The nature of the released component was not further characterized, but a relationship to the cap-binding protein was implied. Thus, while no apparent inactivation of specific initiation factors can be detected in EMC-infected L cells, an effect on the cap recognition activity may be occurring, which could result from a virus-induced event similar to that which occurs in poliovirus-infected cells.

7.4.   Role of Alterations in Ionic Environment

None of the studies of EMC or mengovirus effects on cellular protein synthesis exclude a possible causal relationship between these effects and alterations in cell membrane integrity which lead to changes in intracellular ionic conditions. Lacal and Carrasco (1982) showed a positive correlation in time between the reduction in L cell protein synthesis following EMC virus infection and an increase in Rb^+ uptake, changes in membrane potential and modification in membrane permeability. Such changes did not correlate in time with host cell shut-off in poliovirus-infected HeLa cells. Other parameters, such as ATP leakage from infected cells occurred at later times after infection, and might be responsible for the eventual decline in viral protein synthesis (Egberts *et al.*, 1977; Lacal and Carrasco, 1982). A striking observation demonstrated that cellular protein synthesis in EMC-infected HeLa cells could be restored *in vivo* by reducing the concentration of monovalent cations in the extracellular medium (Alonso and Carrasco, 1981). The hypotonic medium, on the other hand, irreversibly prevented viral protein synthesis. It is true that the bulk of viral protein synthesis, in EMC or in polio-infected cells, occurs under ionic conditions which have been altered due to membrane alterations induced by infection. The role of these altered intracellular conditions in mRNA competition for initiation factors or even in possible redistribution or altered associations between initiation factors remains to be elucidated.

7.5.   Unstudied Picornaviruses

Virtually no studies of host cell protein synthesis inhibition have been conducted for other picornaviruses such as rhinovirus, foot-and-mouth disease virus, Theiler's virus, coxsackie, echo, or hepatitis A

virus. The latter is perhaps the most recent addition to the picornavirus group, and is clearly different in its interaction with its host cell,
since it produces no measurable cytopathology and appears not to
affect cellular protein synthesis at all.

8.   OTHER VIRUSES THAT MAY UTILIZE MECHANISMS SIMILAR TO POLIOVIRUS

Many lytic viruses, other than picornaviruses, markedly inhibit
host cell protein synthesis during the course of the infectious cycle.
None have been investigated to the same extent as poliovirus with
respect to the mechanism of this function. However, there are a few
preliminary studies which might be interpreted as indications of similar effects on initiation factor activity.

8.1.   Reovirus

Extracts from reovirus-infected L cells efficiently translate uncapped (5′ pGp . . .) reovirus or globin mRNA and translate the
capped species only poorly (Skup and Millward, 1980a). This is in
contrast to uninfected extracts, which, as usual, show a marked preference for capped over uncapped mRNA. Translation of either
mRNA is insensitive to inhibition by cap analogues in the infected
cell extract. Further work from the same laboratory (Zarbl et al.,
1980; Skup and Millward, 1980b) showed that progeny subviral particles isolated from infected cells synthesized uncapped mRNAs in
vitro, as opposed to the synthesis of capped RNAs produced by cores
isolated from mature virions. Uncapped mRNA production is apparently due to the presence of "masked" capping enzymes (guanyl
transferase and methylase) in the intracellular particles. Product
mRNAs from these particles fail to translate in the cap-dependent,
uninfected L cell extract, but are translated well by reovirus-infected
cell extracts. These results suggested that reovirus infection of L cells
induces a gradual transition from the host, cap-dependent translational machinery to a cap-independent translational mechanism, and
this transition is paralleled by the appearance in infected cells of uncapped viral mRNAs. Indeed, viral mRNAs isolated from polysomes
of infected cells showed a transition from capped 5′ termini (produced
by parental, infecting virus particles) to uncapped 5′ termini (Skup

et al., 1981). The mechanism of the change in translational specificity has not been defined, nor has the cap-binding protein complex been demonstrated to be involved in the discrimination between mRNAs. An analogy to the poliovirus system appears obvious, although several important differences exist. First, uncapped reovirus mRNA is not translated efficiently in unaffected cell lysates, whereas poliovirus RNA is. Secondly, reovirus-infected cell extracts translate uncapped reovirus mRNA's, but do not preferentially translate other uncapped mRNAs. Thus, the translational regulation observed in reovirus-infected L cells may involve a reovirus mRNA-specific mechanism rather than a general cap-specific discrimination.

A complication to the reovirus story was presented in a report from a different laboratory (Detjen *et al.*, 1982), whose analysis of protein synthesis in extracts from reovirus-infected SC-1 cells (a mouse fibroblast line) yielded findings contradictory to the previous studies. These authors found no evidence of transition from a cap-dependent to cap-independent translation in infected cells, since both extracts translated capped globin mRNA with equal efficiencies, and the infected cell extracts displayed no increase in the ability to translate uncapped globin mRNA. These analyses were performed with SC-1 cell extracts, but the same laboratory also examined L cells, and again found no changes in specificity, although the L cell extracts, were generally less active. The inhibition of host cell translation by reovirus is greater in L cells than in SC-1 cells, and it is interesting that this increased shut-off does not confer a significant growth advantage of reovirus in L cells. Detjen *et al.* (1982) concluded that their results were consistent with their previously published model (Walden *et al.*, 1981) for regulation of translation in reovirus-infected SC-1 cells by competition of host and viral mRNAs for a component of the unaltered, host protein-synthesizing apparatus. Resolution of the different findings in these two laboratories will likely require a careful comparison of the systems and methodologies used.

Both of the above analyses were conducted on cells infected with the Dearing strain of reovirus, type 3. Sharpe and Fields (1982) showed that type 2 reovirus inhibits cellular protein synthesis more effectively than type 3. By isolating recombinant viruses containing various combinations of double-stranded RNA segments derived from both strains of reovirus, they demonstrated that the S4 RNA segment, which encodes the major outer capsid protein of the virion, is responsible for the ability of type 2 reovirus to inhibit L cell macromolecular synthesis. Inactivation of type 2 reovirus by ultraviolet irradiation abolishes its ability to inhibit protein synthesis, suggesting that viral gene expression or replication is required to mediate the inhibition. No information is available regarding the mechanism of this inhibition.

8.2. Semliki Forest Virus

Cells infected with Semliki Forest Virus (SFV) synthesize only viral proteins, despite the continued presence of cellular mRNAs. Van Steeg *et al.* (1981a) examined initiation factor preparations from infected and uninfected neuroblastoma cells in a reconstituted protein-synthesizing system, and found a pronounced loss of activity of the crude ribosomal salt wash from infected cells for translation of either early SFV RNA or neuroblastoma polyadenylated mRNA templates. The infected cell initiation factor preparation was nearly fully active for translation of EMC virus RNA or late SFV mRNA. Although late SFV mRNA is capped, the synthesis of late proteins in infected cell lysates was insensitive to inhibition by cap analogues. These results suggested that virus infection resulted in a shift from cap-dependent to cap-independent protein synthesis. Initiation factors from infected cells could support translation of early SFV and host mRNAs if purified 24-CBP or purified eIF-4B were added. The authors argued against contamination of eIF-4B with CBP as an explanation for the former's restoring ability, although, admittedly, this is difficult to prove conclusively.

From these data, it is tempting to conclude that the inactivation of cap-dependent translation which seems to occur in poliovirus-infected HeLa cells, and perhaps in reovirus-infected L cells, also occurs in neuroblastoma cells infected with SFV. Although the relevance of a redirection of the protein-synthesizing system towards noncapped mRNAs is obvious for uncapped poliovirus or for uncapped reovirus mRNAs, the rationale for SFV-infected cells is obscured because SFV mRNAs, when isolated from infected cells, appear to be capped (Pettersson *et al.*, 1980). In addition, further analyses of initiation factors from SFV-infected cells suggest that the mechanisms operative in the shift in specificity might be different from the biochemical lesion(s) induced by poliovirus.

The first major difference emerged when both eIF-4B and 24-CBP were partially purified from SFV-infected neuroblastoma cells and compared with uninfected cell factors for activity in translation assays. No significant differences between infected and uninfected factors could be demonstrated. The authors concluded that the factors themselves were not inactivated in infected cells, but rather that an inhibitory component specifically blocks their action. This component may have been present in the infected cell ribosomal salt wash but was lost upon purification of eIF-4B and CBP. Furthermore, the partially purified eIF-4B and CBP from both infected cells and uninfected cells were able to restore activity equally to the infected cell crude ribosomal salt wash for translation of host mRNAs. These results are in contrast to those obtained with factors purified from poliovirus-infected cells, none of which have ever demonstrated restoring activity. Translation of late SFV mRNA, which occurs in the presence of the inhibited eIF-4B and CBP factors from infected cells, was shown to have a decreased requirement for both eIF-4B and CBP. This resulted in relatively active translation of late viral mRNA in spite of the reduced activity of these factors (Van Steeg et al., 1981b).

Efforts to identify the inhibitory component in infected cell ribosomal salt wash yielded a protein of M_r 33,000 daltons, which comigrated with the SFV capsid protein on SDS–polyacrylamide gels and which reacted with antibodies raised against viral capsid protein (H. Van Steeg, personal communication). This protein, purified from infected cell ribosomal salt wash, selectively inhibited translation in vitro of host and early viral mRNA, but had no effect on translation of late viral mRNA or EMC virus RNA. The mechanism by which SFV capsid protein interferes with translation initiation of some mRNAs is not understood, nor is it clear how SFV late mRNA eludes the inhibition, or what role cap recognition plays in this scheme. However, the overall strategies followed by SFV and poliovirus may be quite similar, despite differences in underlying mechanisms.

8.3.   Frog Virus 3

Frog virus 3 is a linear double-stranded DNA-containing virus belonging to the family Iridoviridae. Viral protein synthesis during the infectious cycle is regulated at both transcriptional and posttranscriptional levels, e.g., early mRNA transcripts continue to be synthesized and remain present late in infection, but they are not translated. Late mRNAs are translated very poorly in vitro in extracts from rabbit reticulocytes, wheat germ, or cultured BHK cells, although early mRNAs are translated efficiently (Raghow and Granoff,

1983). Infected cell extracts, on the other hand, translate both late and early mRNAs equally well. Translation of late mRNAs in uninfected cell extracts was markedly increased by the addition of crude preparations of initiation factors from infected cells (but not from uninfected cells), and this specific stimulation was due to enhanced binding of the mRNA to ribosomes to form initiation complexes. Unlike the case for poliovirus or for Semliki Forest virus, the regulatory activity in Frog virus 3-infected cells appears to be a virus-induced, positive factor, present only in infected cell ribosomal salt wash, which is required for translation of late viral mRNAs. This is in contrast to the negative effect of poliovirus on the cell's initiation factors, rendering them unable to translate capped mRNAs. The 5' termini of Frog virus 3 early and late mRNAs appear to be similar and capped, and no specific involvement of cap recognition or bypass has been examined as a basis for the discrimination. Interestingly, infected cell extracts also show a reduced ability to translate globin or cellular mRNA. At present, there is insufficient information about this system to speculate on its similarity or difference from poliovirus.

9. CONCLUDING REMARKS

The inhibition of cellular protein synthesis following virus infection is a widespread occurrence. As different types of viruses are examined, a variety of mechanisms are being implicated to account for the inhibition, including structural components of infecting virions that appear "toxic" to the cell, specific alterations of the cell's translational machinery, mRNA competition, etc. It is curious that even for poliovirus, whose ability to restrict its host's translation has been observed and studied for many years, it is not known whether this ability represents an essential function. No mutants have been isolated that are defective only in this function for which the consequences for virus replication can be determined. In those few cases where the kinetics or extent of host cell shut-off appears to vary in different host cells, no correlation with yield or "success" of infection has been documented. It, therefore, remains possible that the inhibition of cellular protein synthesis is an accidental event, resulting from the required utilization of a cellular translation component for some other aspect of virus replication. There are currently no data to weigh the likelihoods that host cell shut-off is either essential or dispensable for efficient virus replication, although teleogically, it would seem to work in the virus's favor.

The best studied case of viral inhibition of host cell protein synthesis is poliovirus infection of HeLa cells. Infected cells, and extracts prepared from these cells, are impaired in their ability to bind capped mRNAs to ribosomes, and thus fail to initiate translation. At least one multisubunit initiation factor (eIF-4F) which contains a documented cap-binding protein, and which is required for translation of most mRNAs, is functionally inactive and structurally disrupted following infection. The biochemical lesion underlying the dissociation of this factor is not yet understood, nor has its precise mechanism of action in uninfected cells been determined. Expression, but not extensive replication, of the infecting viral genome is required for inhibition to occur, but the molecular mediator of the inactivation remains to be identified. The fact that viral protein synthesis occurs under conditions of inhibition of cellular protein synthesis demands that viral mRNA initiate translation differently from its host cell. The discrimination apparently involves a lack of dependence on the cap structure, but additionally requires other sequence and/or structural features which permit initiation by a cap-independent mechanism (Brown *et al.*, 1982). Virtually nothing is known about how viral mRNA associates with ribosomes to enable it to bypass those steps in the usual initiation process which are inhibited following infection.

Although the biochemistry of translational initiation is steadily being unraveled, and the block imposed by poliovirus in HeLa cells is simultaneously becoming clarified, other picornaviruses may interact with other cell types in different ways. Infection of at least some cells with the cardioviruses, EMC or mengovirus, does not appear to produce the same initiation factor inactivation as does poliovirus in HeLa cells, and the regulation of protein synthesis in such cells is not well understood. The majority of picornaviruses, in natural host tissue or in cultured cells, have not been studied at all. Thus, despite a long-standing interest in the phenomenon of virus-induced interference with host cell protein synthesis, many questions remain to be answered.

10. REFERENCES

Abreu, S. L., and Lucas-Lenard, J., 1976, Cellular protein synthesis shut-off by mengovirus: Translation of nonviral and viral mRNAs in extracts from uninfected and infected Ehrlich ascites tumor cells, *J. Virol.* **18**:182.

Alonso, M. A., and Carrasco, L., 1981, Reversion by hypotonic medium of the shutoff of protein synthesis induced by encephalomyocarditis virus, *J. Virol.* **37**:535.

Bablanian, R., 1972, Depression of macromolecular synthesis in cells infected with guanidine-dependent poliovirus under restrictive conditions, *Virology* **47**:255.

Bablanian, R., and Russell, W. C., 1974, Adenovirus polypeptide synthesis in the presence of non-replicating poliovirus, *J. Gen. Virol.* **24**:261.

Bablanian, R., Eggers, H. J., and Tamm, I., 1965, Studies on the mechanism of poliovirus-induced cell damage. I. The relation between poliovirus-induced metabolic and morphologic alterations in cultured cells, *Virology* **26**:100.

Baglioni, C., Simili, M., and Shafritz, D. A., 1978, Initiation activity of EMC virus RNA, binding to initiation factor eIF-4B and shut-off of host cell protein synthesis, *Nature* **275**:240.

Banerjee, A. K., 1980, 5'-Terminal cap structure in eukaryotic messenger ribonucleic acids, *Microbiol. Rev.* **44**:175.

Benne, R., and Hershey, J. W. B., 1978, The mechanism of action of protein synthesis initiation factors from rabbit reticulocytes, *J. Biol. Chem.* **253**:3078.

Borgert, K., Koschel, K., Tauber, H., and Wecker, E., 1971, Effect of hydroxylamine on early functions of poliovirus, *J. Virol.* **8**:1.

Bossart, W., and Bienz, R., 1981, Regulation of protein synthesis in Hep-2 cells and their cytoplasmic extracts after poliovirus infection, *Virology* **111**:555.

Brown, B., and Ehrenfeld, E., 1980, Initiation factor preparations from poliovirus-infected cells restrict translation in reticulocyte lysates, *Virology* **103**:327.

Brown, D., Hansen, J., and Ehrenfeld, E., 1980, Specificity of initiation factor preparations from poliovirus-infected cells, *J. Virol.* **34**:573.

Brown, D., Jones, C. L., Brown, B. A., and Ehrenfeld, E., 1982, Translation of capped and uncapped VSV mRNAs in the presence of initiation factors from poliovirus-infected cells, *Virology* **123**:60.

Carrasco, L., 1977, The inhibition of cell functions after viral infection. A proposed general mechanism, *FEBS Lett.* **76**:11.

Carrasco, L., and Smith, A. E., 1976, Sodium ions and the shut-off of host cell protein synthesis by picornaviruses, *Nature* **264**:807.

Celma, M. L., and Ehrenfeld, E., 1974, Effect of poliovirus double-stranded RNA on viral and host-cell protein synthesis, *Proc. Natl. Acad. Sci. USA* **71**:2440.

Choppin, P. W., and Holmes, R. V., 1967, Replication of SV5 RNA and the effects of superinfection with poliovirus, *Virology* **33**:442.

Cole, C., and Baltimore, D., 1973, Defective interfering particles of poliovirus II. Nature of the defect, *J. Mol. Biol.* **76**:325.

Cords, C. E., and Holland, J. J., 1964, Interference between enteroviruses and conditions effecting its reversal, *Virology* **22**:226.

Detjen, B. M., Jen, G., and Thach, R. E., 1981, Encephalomyocarditis viral RNA can be translated under conditions of polio-induced translation shut-off *in vivo*, *J. Virol.* **38**:777.

Detjen, B. M., Walden, W. E., and Thach, R. E., 1982, Translational specificity in reovirus-infected mouse fibroblasts, *J. Biol. Chem.* **257**:9855.

Doyle, S., and Holland, J., 1972, Virus-induced interference in heterologously infected HeLa cells, *J. Virol.* **9**:22.

Duncan, R., Etchison, D., and Hershey, J. W. B., 1983, Protein synthesis eukaryotic initiation factors 4A and 4B are not altered by poliovirus infection of HeLa cells, *J. Biol. Chem.* **258**:7236.

Edery, I., Humbelin, M., Darveau, A., Lee, K., Milburn, S., Hershey, J., Trachsel, H., and Sonenberg, N., 1983, Involvement of eukaryotic initiation factor 4A in the cap recognition process, *J. Biol. Chem.* **258**:11398.

Egberts, E., Hackett, P., and Traub, P., 1977, Alteration of the intracellular energetic and ionic conditions by mengovirus infection of Ehrlich ascites tumor cells and its influence on protein synthesis in the midphase of infection, *J. Virol.* **22**:591.

Ehrenfeld, E., 1982, Poliovirus-induced inhibition of host cell protein synthesis, *Cell* **28**:435.

Ehrenfeld, E., and Hunt, T., 1971, Double-stranded poliovirus RNA inhibits initiation of protein synthesis by reticulocyte lysates, *Proc. Natl. Acad. Sci. USA* **68**:1075.

Ehrenfeld, E., and Lund, H., 1977, Untranslated vesicular stomatitis virus messenger RNA after poliovirus infection, *Virology* **80**:297.

Ehrenfeld, E., and Manis, S., 1979, Inhibition of 80S initiation complex formation by infection with poliovirus, *J. Gen. Virol.* **43**:441.

Etchison, D., Milburn, S. C., Edery, I., Sonenberg, N., and Hershey, J. W. B., 1982, Inhibition of HeLa cell protein synthesis following poliovirus infection correlates with the proteolysis of a 220,000 dalton polypeptide associated with eukaryotic initiation factor 3 and a cap binding protein complex, *J. Biol. Chem.* **257**:14806.

Etchison, D., Edery, I., Sonenberg, N., Hansen, J., Ehrenfeld, E., Milburn, S., and Hershey, J. W. B., Poliovirus infection of HeLa cells inhibits the activity of a cap binding protein complex in the translation of globin messenger RNA, submitted.

Farrell, P. J., Balkow, K., Hunt, T., Jackson, R. J., and Trachsel, H., 1977, Phosphorylation of initiation factor eIF-2 and the control of reticulocyte protein synthesis, *Cell* **11**:187.

Fernandez-Munoz, R., and Darnell, J. E., 1976, Structural difference between the 5′ termini of viral and cellular mRNA in poliovirus-infected cells: Possible basis for the inhibition of host protein synthesis, *J. Virol.* **18**:719.

Golini, F., Thach, S. S., Birge, C. H., Safer, B., Merrick, W. C., and Thach, R. E., 1976, Competition between cellular and viral mRNAs *in vitro* is regulated by a messenger discriminatory initiation factor, *Proc. Natl. Acad. Sci. USA* **73**:3040.

Grifo, J. A., Tahara, S. M., Leis, J. P., Morgan, M. A., Shatkin, A. J., and Merrick, W., 1982, Characterization of eukaryotic initiation factor 4A, a protein involved in ATP-dependent binding of globin mRNA, *J. Biol. Chem.* **257**:5246.

Grifo, J. A., Tahara, S. M., Morgan, M. A., Shatkin, A. J., and Merrick, W. C., 1983, New initiation factor activity required for globin mRNA translation, *J. Biol. Chem.* **258**:5804.

Hackett, P. B., Egberts, E., and Traub, P., 1978a, Translation of ascites and mengovirus RNA in fractionated cell-free systems from uninfected and mengovirus-infected Ehrlich ascites tumor cells, *Eur. J. Biochem.* **83**:341.

Hackett, P. B., Egberts, E., and Traub, P., 1978b, Selective translation of mengovirus RNA over host mRNA in homologous, fractionated, cell-free translational systems from Ehrlich ascites tumor cells, *Eur. J. Biochem.* **83**:353.

Hansen, J. L., and Ehrenfeld, E., 1981, Presence of the cap-binding protein in initiation factor preparations from poliovirus-infected HeLa cells, *J. Virol.* **38**:438.

Hansen, J. L., Etchison, D., Hershey, J. W. B., and Ehrenfeld, E., 1982a, Association of cap-binding protein with eIF-3 in initiation factor preparations from uninfected and poliovirus-infected cells, *J. Virol.* **42**:200.

Hansen, J. L., Etchison, D. O., Hershey, J. W. B., and Ehrenfeld, E., 1982*b*, Localization of cap-binding protein in subcellular fractions of HeLa cells, *Mol. Cell. Biol.* **2**:1639.

Helentjaris, T., and Ehrenfeld, E., 1977, Inhibition of host cell protein synthesis by UV-inactivated poliovirus, *J. Virol.* **21**:259.

Helentjaris, T., and Ehrenfeld, E., 1978, Control of protein synthesis in extracts from poliovirus-infected cells I. mRNA discrimination by crude initiation factors, *J. Virol.* **26**:510.

Helentjaris, T., Ehrenfeld, E., Brown-Luedi, M. L., and Hershey, J. W. B., 1979, Alterations in initiation factor activity from poliovirus-infected HeLa cells, *J. Biol. Chem.* **254**:10973.

Hellman, G. M., Chu, L-Y., and Rhoades, R. E., 1982, A polypeptide which reverses cap analogue inhibition of cell-free protein synthesis, *J. Biol. Chem.* **257**:4056.

Hewlett, M. J., Rose, J. K., and Baltimore, D., 1976, 5′ Terminal structure of poliovirus polyribosomal RNA is pUp, *Proc. Natl. Acad. Sci. USA* **73**:327.

Hewlett, M. J., Axelrod, J. H., Antorino, N., and Field, R., 1982, Isolation and preliminary characterization of temperature-sensitive mutants of poliovirus type 1, *J. Virol.* **41**:1089.

Holland, J. J., 1964, Inhibition of host cell macromolecular synthesis by high multiplicities of poliovirus under conditions preventing virus synthesis, *J. Mol. Biol.* **8**:574.

Holland, J. J., and Peterson, J. A., 1964, Nucleic acid and protein synthesis during poliovirus infection of human cells, *J. Mol. Biol.* **8**:556.

Hunt, T., and Ehrenfeld, E., 1971, Cytoplasmic from poliovirus-infected HeLa cells inhibits cell-free haemoglobin synthesis, *Nature New Biol.* **230**:91.

Ito, Y., Okazaki, H., and Ishida, N., 1968, Growth inhibition of Newcastle Disease virus upon superinfection of poliovirus in the presence of guanidine, *J. Virol.* **2**:645.

Jackson, R. J., 1982, The Control of initiation of protein synthesis in reticulocyte lysates, in: *Protein Biosynthesis in Eukaryotes* (R. Perez-Bercoff, ed.), pp. 362–418, Plenum Press, New York.

Jagus, R., Anderson, W. F., and Safer, B., 1981, The regulation of initiation of mammalian protein synthesis, in: *Progress in Nucleic Acid Research and Molecular Biology*, Vol. 25, (W. E. Cohn, and E. Volkin, eds.), pp. 127–185, Academic Press, New York.

James, L. A., and Tershak, D. R., 1981, Protein phosphorylations in poliovirus-infected cells, *Can. J. Microbiol.* **27**:28.

Jen, G., and Thach, R. E., 1982, Inhibition of host translation in encephalomyocarditis virus-infected L cells: A novel mechanism, *J. Virol.* **43**:250.

Jen, G., Birge, C. H., and Thach, R. E., 1978, Comparison of initiation rates of encephalomyocarditis virus and host protein synthesis in infected cells, *J. Virol.* **27**:640.

Jen, G., Detjen, B. M., and Thach, R. E., 1980, Shut-off of HeLa cell protein synthesis by encephalomyocarditis virus and poliovirus: A comparative study, *J. Virol.* **35**:150.

Jones, C., and Ehrenfeld, E., 1983, The effect of poliovirus infection on the translation *in vitro* of VSV messenger ribonucleoprotein particles, *Virology* **129**:415.

Kaempfer, R., 1984, Regulation of eukaryotic translation, in: *Comprehensive Virology*, Vol. 19, (H. Fraenkel-Conrat and R. E. Wagner, eds.), Plenum Press, New York.

Kaufmann, Y., Goldstein, E., and Penman, S., 1976, Poliovirus-induced inhibition of polypeptide initiation *in vitro* on native polyribosomes, *Proc. Natl. Acad. Sci. USA* **73**:1834.

Koschel, K., 1974, Poliovirus infection and poly (A) sequences of cytoplasmic cellular RNA, *J. Virol.* **13**:1061.

Kozak, M., 1980*a*, Influence of mRNA secondary structure on binding and migration of 40S ribosomal subunits, *Cell* **19**:79.

Kozak, M., 1980*b*, Role of ATP in binding and migration of 40S ribosomal subunits, *Cell* **22**:459.

Lacal, J. C., and Carrasco, L., 1982, Relationship between membrane integrity and the inhibition of host translation in virus-infected mammalian cells, *Eur. J. Biochem.* **127**:359.

Lacal, J. C., and Carrasco, L., 1983, Modification of membrane permeability in poliovirus-infected HeLa cells: Effect of guanidine, *J. Gen. Virol.* **64**:787.

LaColla, P., Marcialis, M. A., Mereu, G. P., and Loddo, B., 1972, Specific inhibition of poliovirus-induced blockade of cell protein synthesis by a thiopyrimidine derivative, *J. Gen. Virol.* **17**:13.

Lawrence, C., and Thach, R. E., 1974, Encephalomyocarditis virus infection of mouse plasmacytoma cells. I. Inhibition of cellular protein synthesis, *J. Virol.* **14**:598.

Lee, K. A. W., and Sonenberg, N., 1982, Inactivation of cap-binding protein accompanies the shut-off of host protein synthesis by poliovirus, *Proc. Natl. Acad. Sci. USA* **79**:3447.

Lee, L. A. W., Guertin, D., and Sonenberg, N., 1983, mRNA secondary structure as a determinant in cap recognition and initiation complex formation, *J. Biol. Chem.* **258**:707.

Leibowitz, R., and Penman, S., 1971, Regulation of protein synthesis in HeLa cells. III. Inhibition during poliovirus infection, *J. Virol.* **8**:661.

Lodish, H. F., 1974, Model for the regulation of mRNA translation applied to haemoglobin synthesis, *Nature* **251**:385.

McCormick, W., and Penman, S., 1968, Replication of mengovirus in HeLa cells preinfected with non-replicating poliovirus, *J. Virol.* **2**:859.

Merryman, P., Jaffe, I., and Ehrenfeld, E., 1974, Effect of D-penicillamine on poliovirus replication in HeLa cells, *J. Virol.* **13**:881.

Morgan, M. A., and Shatkin, A. J., 1980, Initiation of reovirus transcription by inosine 5'-triphosphate and properties of 7-methyl inosine-capped, inosine-substituted messenger RNA, *Biochemistry* **19**:5960.

Nair, C., 1981, Monovalent cation metabolism and cytopathic effects of poliovirus-infected HeLa cells, *J. Virol.* **37**:268.

Nair, C. N., Stowers, J. W., and Singfield, B., 1979, Guanidine-sensitive Na$^+$ accumulation by poliovirus-infected HeLa cells, *J. Virol.* **31**:184.

Noble, J., and Levintow, L., 1970, Dynamics of poliovirus-specific RNA synthesis and the effects of inhibitors of virus replication, *Virology* **40**:634.

Nomoto, A., Lee, Y. F., and Wimmer, E., 1976, The 5' end of poliovirus mRNA is not capped with m^{7}G(5')ppp(5')Np, *Proc. Natl. Acad. Sci. USA* **73**:375.

Nuss, D. L., Oppermann, H., and Koch, G., 1975, Selective blockage of initiation of host protein synthesis in RNA virus-infected cells, *Proc. Natl. Acad. Sci. USA* **72**:1258.

Otto, M. J., and Lucas-Lenard, J., 1980, The influence of the host cell on the inhibition of viral protein synthesis in cells doubly-infected with vesicular stomatitis virus and mengovirus, *J. Gen. Virol.* **50**:293.

Penman, S., and Summers, D., 1965, Effects on host cell metabolism following synchronous infection with poliovirus, *Virology* **27**:614.

Penman, S., Scherrer, K., Becker, Y., and Darnell, J. E., 1963, Polyribosomes in normal and poliovirus-infected HeLa cells and their relationship to messenger RNA, *Proc. Natl. Acad. Sci. USA* **49**:654.

Pettersson, R. F., Soderlund, H., and Kaariainen, L., 1980, The nucleotide sequences of the 5'-terminal T_1 oligonucleotides of Semliki Forest Virus 42S and 26S RNAs are different, *Eur. J. Biochem.* **105**:435.

Racevskis, J., Kerwar, S. S., and Koch, G., 1976, Inhibition of protein synthesis in reticulocyte lysates by poliovirus, *J. Gen. Virol.* **31**:135.

Raghow, R., and Granoff, A., 1983, Cell-free translation of frog virus 3 messenger RNAs: Initiation factors from infected cells discriminate between early and late viral mRNAs. *J. Biol. Chem.* **258**:571.

Rose, J. K., Trachsel, M., Leong, K., and Baltimore, D., 1978, Inhibition of translation by poliovirus inactivation of a specific initiation factor, *Proc. Natl. Acad. Sci. USA* **75**:2732.

Saborio, J. L., Pong, S. S., and Koch, G., 1974, Selective and reversible inhibition of initiation of protein synthesis in mammalian cells, *J. Mol. Biol.* **85**:195.

Saxton, R. E., and Stevens, J. C., 1972, Restriction of herpes simplex virus replication by poliovirus: A selective inhibition of viral translation, *Virology* **48**:207.

Seal, S. N., Schmidt, A., and Marcus, A., 1983, Eukaryotic initiation factor 4A is the component that interacts with ATP in protein chain initiation, *Proc. Natl. Acad. Sci. USA* **80**:6562.

Sharpe, A. H., and Fields, B. N., 1982, Reovirus inhibition of cellular RNA and protein synthesis: Role of the S4 gene, *Virology* **122**:381.

Shih, D. S., Shih, C. T., Kew, O., Pallansch, M., Rueckert, R., and Kaesberg, P., 1978, Cell-free synthesis and processing of the proteins of poliovirus, *Proc. Natl. Acad. Sci. USA* **75**:5807.

Skup, D., and Millward, S., 1980a, Reovirus-induced modification of cap-dependent translation in infected L cells, *Proc. Natl. Acad. Sci. USA* **77**:152.

Skup, D., and Millward, S., 1980b, mRNA capping enzymes are masked in reovirus progeny subviral particles, *J. Virol.* **34**:490.

Skup, D., Zarbl, H., and Millward, S., 1981, Regulation of translation in L-cells infected with reovirus, *J. Mol. Biol.* **151**:35.

Sonenberg, N., 1981, ATP/Mg^{++}-dependent crosslinking of cap binding proteins to the 5' end of eukaryotic mRNA, *Nucl. Acids Res.* **9**:1643.

Sonenberg, N., and Shatkin, A. J., 1977, Reovirus mRNA can be covalently cross-linked via the 5' cap to protein in initiation complexes, *Proc. Natl. Acad. Sci. USA* **74**:4288.

Sonenberg, N., Morgan, M., Merrick, W., and Shatkin, A. J., 1978, A polypeptide in eukaryotic initiation factors that crosslinks specifically to the 5'-terminal cap in mRNA, *Proc. Natl. Acad. Sci. USA* **75**:4843.

Sonenberg, N., Morgan, M. A., Testa, D., Collonno, R. J., and Shatkin, A. J., 1979a, Interaction of a limited set of proteins with different mRNAs and protection of 5'-caps against pyrophosphatase digestion in initiation complexes, *Nucl. Acids Res.* **7**:15.

Sonenberg, N., Ruprecht, K., Hecht, S., and Shatkin, A., 1979*b*, Eukaryotic mRNA cap binding protein: Purification by affinity chromatography on sepharose-coupled m^7 GDP, *Proc. Natl. Acad. Sci. USA* **76**:4345.

Sonenberg, N., Trachsel, H., Hecht, S., and Shatkin, A., 1980, Differential stimulation of capped mRNA translation *in vitro* by cap binding protein, *Nature* **285**:331.

Sonenberg, N., Guertin, D., Cleveland, D., and Trachsel, H., 1981, Probing the structure of the eukaryotic 5' cap structure using a monoclonal antibody directed against cap binding proteins, *Cell* **27**:563.

Sonenberg, N., Guertin, D., and Lee, K., 1982, Capped mRNAs with reduced secondary structure can function in extracts from poliovirus-infected cells, *Mol. Cell. Biol.* **2**:1633.

Steiner-Pryor, A., and Cooper, P. D., 1973, Temperature-sensitive poliovirus mutants defective in repression of host protein synthesis are also defective in structural protein, *J. Gen. Virol.* **21**:215.

Summers, D. F., and Maizel, J. V., 1967, Disaggregation of HeLa cell polysomes after infection with poliovirus, *Virology* **31**:550.

Summers, D. F., Maizel, J. V., and Darnell, J. E., 1965, Evidence for virus-specific noncapsid proteins in poliovirus-infected cells, *Proc. Natl. Acad. Sci. USA* **54**:505.

Svitkin, Y. V., Agarova, T. Y., Ginevskaya, V. A., Kalinina, N. O., Scarlat, I. V., and Agol, V. I., 1974, Efficiency of translation of viral and cellular mRNAs in extracts from cells infected with encephalomyocarditis virus, *Intervirology* **4**:214.

Svitkin, Y. V., Ginevskaya, V. A., Ugarova, T. Y., and Agol, V. I., 1978, A cell-free model of the encephalomyocarditis virus-induced inhibition of host cell protein synthesis, *Virology* **87**:199.

Tahara, S. M., Morgan, M. A., and Shatkin, A. J., 1981, Two forms of purified m^{7}G-cap binding protein with different effects on capped mRNA translation in extracts of uninfected and poliovirus-infected HeLa cells, *J. Biol. Chem.* **256**:7691.

Tahara, S. M., Morgan, M. A., and Shatkin, A. J., 1983, Binding of inosine-substituted mRNA to reticulocyte ribosomes and eukaryotic initiation factors 4A and 4B requires ATP, *J. Biol. Chem.* **258**:11350.

Tershak, D. A., 1978, Protein kinase activity of polysome–ribosome preparations from poliovirus-infected cells, *Biochem. Biophys. Res. Commun.* **80**:283.

Trachsel, H., Sonenberg, N., Shatkin, A. J., Rose, J. K., Leong, K., Bergman, J. E., Gordon, J., and Baltimore, D., 1980, Purification of a factor that restores translation of VSV mRNA in extracts from poliovirus-infected HeLa cells, *Proc. Natl. Sci. Acad. Sci. USA* **77**:770.

Van Steeg, H., Thomas, A., Verbeck, S., Kasperaitis, M., Voorma, H. O., and Benne, R., 1981*a*, Shut-off of neuroblastoma cell protein synthesis by Semliki Forest Virus: Loss of ability of crude initiation factors to recognize early Semliki Forest Virus and host mRNAs, *J. Virol.* **38**:728.

Van Steeg, H., Van Grinsven, M., Van Mansfield, F., Voorma, H. O., and Benne, R., 1981*b*, Initiation of protein synthesis in neuroblastoma cells infected by Semliki Forest Virus: A decreased requirement of late viral mRNA for eIF-4B and cap binding protein, *FEBS Lett.* **129**:62.

Walden, W., Godfrey-Colburn, T., and Thach, R. E., 1981, The role of mRNA competition in regulating translation. I. Demonstration of competition *in vivo*, *J. Biol. Chem.* **256**:11739.

Weber, L. A., Hickey, E. D., Nuss, D. L., and Baglioni, C., 1977, 5'-Terminal 7-methylguanosine and mRNA function: Influence of potassium concentration on translation *in vitro, Proc. Natl. Acad. Sci. USA* **72:**318.

Willems, M., and Penman, S., 1966, The mechanism of host cell protein synthesis inhibition by poliovirus, *Virology* **30:**355.

Wright, P. J., and Cooper, P. D., 1974, Poliovirus proteins associated with ribosomal structures in infected cells, *Virology* **59:**1.

Zarbl, H., Skup, D., and Millward, S., 1980, Reovirus progeny subviral particles synthesize uncapped mRNA, *J. Virol.* **34:**497.

Zimmerman, E. F., Heeter, M., and Darnell, J. E., 1963, RNA synthesis in poliovirus-infected cells, *Virology* **19:**400.

Rhabdovirus Cytopathology: Effects on Cellular Macromolecular Synthesis

Robert R. Wagner and James R. Thomas

Department of Microbiology
The University of Virginia School of Medicine
Charlottesville, Virginia 22908

John J. McGowan

Department of Microbiology
The Uniformed Services University of the Health Sciences
Bethesda, Maryland 20014

1. INTRODUCTION

Viruses belonging to the family *Rhabdoviridae* are widely distributed throughout the animal and plant kingdoms and cause severe diseases in mammals, fish, and plants. The economic impact of the rhabdovirus diseases of plants and fish are quite considerable. The major diseases of mammals are vesicular stomatitis of cattle and swine with man as in incidental host and rabies, which affects domestic and wild mammals and is an ancient scourge of man. The rhabdoviruses generally cause acute infections but can persist under certain conditions for long periods (Wagner *et al.*, 1963; Youngner and Preble, 1980; Holland *et al.*, 1980). Rabies and possibly other rhabdoviruses of mammals have a particular predilection for infection of the central nervous system (Murphy, 1977). Two extremely virulent viruses of man, Marburg and Ebola viruses, which cause almost uniformly fatal infections, are similar in many respects to rhabdoviruses but may eventually be classified in a separate family (Kiley *et al.* 1982).

Most of the numerous mammalian rhabdoviruses can be classified in two genera: *Vesiculovirus*, the prototype of which is vesicular stomatitis virus (VSV), and *Lyssavirus*, the prototype of which is rabies virus. The plant rhabdoviruses undoubtedly comprise at least one more genus, as do the rhabdoviruses that infect fish (Matthews, 1982). However, it is characteristic of rhabdoviruses, particularly the *Vesiculovirus* group, that a single member can infect animals of many species, ranging from vertebrates to invertebrates. The varios VSV species are frequently isolated from insects, which are often vectors for transmission of infection, and these rhabdoviruses were originally considered to be arboviruses. Vertebrate and invertebrate cells serve as susceptible hosts to somewhat varying degrees. VSV can grow quite well at either 20°C or 37°C; in fact, 31°C is frequently used as the optimal temperature to produce virus progeny in highest titer. Fish rhabdoviruses, such as the virus of hemorrhagic septicemia and infectious hematopoetic necrosis virus, will replicate well only at the usual poikilothermic temperature of the host fish and not at higher temperature, probably because of the heat lability of the virion polymerase (McAllister and Wagner, 1975).

The International Committee on Taxonomy of Viruses (Matthews, 1982) provided a working definition of a rhabdovirus as a rod-shaped negative-strand virus surrounded by a lipoprotein envelope which encloses a helical nucleocapsid with a nonsegmented single strand of RNA that cannot serve as messenger but is complementary to the viral messenger RNAs. All rhabdoviruses examined carefully to date contain an RNA-dependent RNA polymerase coded by the viral genome. Reviews of the biology, genetics, and chemistry of rhabdoviruses are available in earlier volumes of *Comprehensive Virology* (Wagner, 1975; Pringle, 1977) and a series of three volumes edited by Bishop (1979).

Most of the research in this field has been done with the prototype vesicular stomatitis virus because of its rapid growth to high titer in a wide variety of cell types and relative ease for purifying large amounts of homogeneous virus particles. VSV rapidly kills many host cells and even more rapidly shuts off cellular macromolecular synthesis (Weck and Wagner, 1978). On the other hand, infection of cells with rabies virus results in only a delayed cytopathic effect and disease probably results late from an immunological response to persistent virus in the host (Murphy, 1977). In this chapter, we direct most of our attention to VSV as a model system for studying cytopathogenicity and inhibition of cellular macromolecular synthesis by a negative-strand RNA virus. It should be made clear at the outset that all rhabdoviruses do not behave in the same way. Moreover, we must keep in mind that there is only circumstantial, but no direct, evidence that inhibition by VSV of cellular macromolecular synthesis is *the* cause of cell death later in infection.

2. PROPERTIES OF RHABDOVIRUSES: A BRIEF SURVEY

This short summary of the chemistry and biology of rhabdoviruses is intended to provide certain background information required to interpret the mechanisms by which rhabdoviruses perturb cell functions. These comments are not intended to serve as a comprehensive analysis of the structure and function of rhabdoviruses; other reviews and current literature must be consulted for in-depth principles of rhabdovirology. Unless indicated otherwise, all properties described here refer to those of the Indiana serotype of vesicular stomatitis virus (VSV-Indiana), which serves as the prototypic model of all rhabdoviruses.

2.1 Structure–Function Relationships

2.1.1 Virion Structural Components

VSV, probably like all other rhabdoviruses is composed of ~74% protein, ~20% lipid, ~3% carbohydrate, and ~3% RNA (Wagner, 1975). The virion is bullet shaped and is assembled from two components, the nucleocapsid and the membrane. The membrand contains ~50% lipid and two protein: the integral glycoprotein (G), which protrudes externally, and the peripheral matrix (M) protein, which lines the inner bilayer of the membrane (Patzer *et al.*, 1979; Zakowski and Wagner, 1980). Based on clones cDNA sequences of their mRNAs, the Indiana serotype G protein is composed of 511 amino acides and is glycosylated at two separate asparagine residues, whereas, the M protein consists of 229 amino acids and is not glycosylated (Rose and Gallione, 1981). The G protein is the major antigenic determinant responsible for type specificity and gives rise to neutralizing antibody (Kelley *et al.*, 1972; Volk *et al.*, 1982). The M protein appears to serve as the "glue" that attaches the nucleocapsid to the membrane; the M protein is quite basic ($pI \simeq 9.1$) and inhibits transcription by binding to the nucleocapsid (Carroll and Wagner, 1979). The basic M protein also binds to acidic phospholipid head-groups, by which means it appear to attach the nucleocapsid–M protein complex to phosphatidylserine residues that line the inner surface of the virion membrane (Zakowski *et al.*, 1981; Wiener *et al.*, 1983).

The nucleocapsid of VSV contains a single strand of RNA, $M_r \simeq 3.68 \times 10^6$ and ~11,162 nucleotides (Manfred Schubert, personal communication), closely associated with ~1,500 copies of the major structural N protein plus ~50 copies each of two minor proteins, L and NS. As determined by cloned cDNA sequences of their mRNAs, the N protein contains 422 amino acids and the NS protein contains 222 amino acids (Gallione *et al.*, 1981); the L protein has not been sequenced but its molecular weight is roughly 190,000, representing about one-half the coding potential of the VSV genome. Both the L and NS proteins comprise the RNA polymerase and are required for transcription of the VSV genome negative-strand RNA, which functions only when encapsidated with N protein (Emerson and Wagner, 1972, 1973; Emerson and Yu, 1975).

2.1.2. Cycle of Infection

2.1.2a. Adsorption

Adsorption of VSV, and presumably other rhabdoviruses as well, is an inefficient process. The ratio of physical particles to infectious units (based on plaque-forming units) is rarely less than 5 to 1. The viral attachment organ is the glycoprotein spike, removal of which reduces infectivity $>10^5$-fold; partitioning isolated glycoprotein into the membrane of spikeless virions can restore infectivity ~100-fold (Bishop *et al.*, 1975). The concept that the terminal sialic acid on the carbohydrate chain(s) of VSV G protein was responsible for efficient adsorption (Schloemer and Wagner, 1975) has been refuted (Cartwright and Brown, 1977). Nucleocapsids free of envelope are infectious at low efficiency but only if they contain intact transcriptase activity in the form of L and NS proteins (Bishop *et al.*, 1974).

2.1.2b. Penetration

Penetration of host cells by VSV follows adsorption by a mechanism that has long been in dispute. Whereas adsorption can occur at 4°C, penetration is an energy-dependent event and requires physiological temperatures. Electron microscopic studies of VSV penetration resulted in conflicting results. Heine and Schnaitman (1971) presented evidence that the membrane of VSV can fuse with the surface cytoplasmic membrane, thereafter discharging the nucleocapsid into the cytoplasm. Simpson *et al.* (1969), on the other hand, could find only evidence for entry of intact VS virions in phagocytic vesicles. In an exhaustive series of studies, Helenius and his colleagues have come to the conclusion that enveloped viruses adsorb to cell surfaces at the site of histocompatibility antigens and coated pits; ingestion of the adsorbed virus then appears to occur by endocytosis of coated vesicles, a process that applies to VSV (Matlin *et al.*, 1982) and other enveloped viruses, particularly the well-studied Semliki Forest virus (Helenius and Marsh, 1982). After endocytosis, the coated vesicle fuses in a succession of events at pH <6 with lysosomes, thus resulting in release of the nucleocapsid (Marsh *et al.*, 1983). Certain aspects of this series of events have not been confirmed by other investigators. Oldstone *et al.* (1980) reported that cells lacking H-2 or HLA histocompatibility antigen were readily penetratable by Semliki Forest virus. The findings that inhibition of endocytosis by cytochalasin B did not influence infection with Sindbis virus or vesicular stomatitis virus, and that chloraquine did not block ingestion of virus particles but greatly reduced their yields, led McCoombs *et al.* (1981) to conclude "that endocytosis is not essential for the infection of cultured cells by Sindbis virus or vesicular stomatitis virus." Nevertheless, it seems likely that the Helenius hypothesis is largely true and that VSV and other enveloped viruses can enter cells by endocytosis and are uncoated (demembraned) by reaction of endocytic vesicles with lysosomes at low pH. Another possibility is that VSV and other enveloped viruses can use the alternative pathways of endocytosis or cytoplasmic membrane fusion to deposit the membrane-stripped nucleocapsid into the host cell cytoplasm. In either case, the virus membrane is removed by fusion with surface or internal cell membranes.

2.1.2c. Transcription

Transcription is the first viral metabolic event in cells penetrated by infectious VSV nucleocapsids. The viral transcriptase functions in the presence of actinomycin D and inhibitors of protein synthesis, such as cycloheximide; this has been called primary transcription, which takes place on input (parental) nucleocapsids, rather than on progeny nucleocapside (Huang and Manders, 1972; Perrault and Holland, 1972; Flamand and Bishop, 1973). The early literature on *in vivo* VSV transcription is reviewed by Wagner (1975). It is now known that transcription begins at the extreme 3′ end of the minus-strand genome by synthesis of a plus-strand (complementary) 47-nucleotide leader RNA (Colonno and Banerjee, 1977). Transcription then proceeds by sequential synthesis in decreasing molar ratios of mRNAs for the N, NS, M, G, and L proteins (Villareal *et al.*, 1976). Each of these mRNAs is capped and polyadenylated (Abraham *et al.*, 1975; Banerjee and Rhodes, 1976). Much of the mRNA and the plus-strand leader is encapsidated to form mRNPs (Rosen *et al.*, 1982) and 18 S leader (Blumberg and Kolakofsky, 1981).

2.1.2d.   Translation

Translation of each mRNA, but not the leader RNA sequence, proceeds immediately after transcription. All five VSV proteins are synthesized throughout the cycle of infection but the amount of each protein synthesized is roughly in decreasing order for the N, NS, M, G, and L proteins (Wagner *et al.*, 1970; Mudd and Summers, 1970; Hsu *et al.*, 1979). The G protein is synthesized from mRNA on endoplasmic reticulum membrane-associated polyribosomes by means of a signal sequence, step-wise glycosylation, and migration of vesicles to the cytoplasmic membrane for insertion of the processed, fully glycosylated protein (Knipe *et al.*, 1977; Rothman *et al.*, 1980). The other four VSV proteins are synthesized from monocistronic mRNAs on cytoplasmic polyribosomes (Morrison and Lodish, 1975).

2.1.2e.   Replication

Replication of VSV RNA appears to be coupled with translation of the nucleocapsid proteins N, L, and NS (Wertz and Levine, 1973; Wertz, 1980; Rubio *et al.*, 1980). The plus-strand RNA is apparently synthesized by read-through transcription of the entire genome and is completely encapsidated by the N protein as well as by small amounts of polymerase proteins L and NS (Wertz, 1980; Davis and Wertz, 1982). This plus-strand RNA serves as template for replication of progeny negative-strand RNA, fully encapsidated with N, L, and

NS proteins. Complete replication of full-length plus- and minus-strand nucleocapsids has been achieved in a cell-free coupled transcription–translation reaction in the presence of a required cellular factor(s) (Hill *et al.*, 1981; Patton *et al.*, 1983). In fact, Wertz (1983) has been able to replicate the RNA of defective-interfering VSV nucleocapsids in a cell-free system coupled with VSV mRNA translation but free of infectious full-length nucleocapsids.

2.1.2f. Assembly

Assembly of mature VS virions apparently begins by association of the independently synthesized M protein with the progeny nucleocapsid (Wilson and Lenard, 1981). Then, it is believed that the nucleocapsid–M protein complex migrates to a region of the cytoplasmic membrane, which contains newly inserted but randomly distributed VSV glycoprotein (Wagner *et al.*, 1971). The nucleocapsid is presumably next enveloped in the G protein-converted cell membrane leading to budding and release of fully formed and infectious VS virions. The evidence for these assembly/budding steps is largely circumstantial and is far from complete, but recent data by Jacobs and Penhoet (1982) indicate that viral nucleocapsid–M protein complexes promote lateral diffusion of G protein in the plane of the membrane bilayer to those regions destined for nucleocapsid envelopment and budding.

2.2. Reproduction Strategies of Rhabdoviruses

Insight into the pathogenetic potentials of rhabdoviruses clearly requires an understanding of their physiological functions. Using vesicular stomatitis virus as the model system, only three physiological functions, all concerned with viral reproduction, have been definitively identified as viral in origin: transcription, translation, and replication. All three processes can be carried out in cell-free systems, but only viral transcription is completely independent of cellular components. Translation, of course, requires preformed cellular ribosomes and energy sources. *In vitro* replication of the VSV genome requires unidentified cellular factor(s) (Hill *et al.*, 1981; Patton *et al.*, 1983). The only VSV genome-specified products, made in infected cells, are five proteins and various species of RNA including: (1) five messengers, (2) plus- and minus-strand leader sequences; (3) two full-length RNAs complementary and anticomplementary to the parental viral genome, (4) various defective-interfering particle RNAs, (5) possibly double-stranded RNA transcriptive and replicative intermediates, and (6) a series of oligonucleotides, varying in length from 11 to >50 nucleotides, representing the 5' termini of leader or messenger RNAs (Pinney and Emerson, 1982; Iverson and Rose, 1982). All of these VSV-specified products are possible candidates as inhibitors of host cell macromolecular synthesis. This section briefly examines the strategies thought to be used by VSV, the model rhabdovirus, to synthesize these RNAs and proteins that could be involved in its cytopathology.

2.2.1. Transcription

The deproteinized RNA of VSV, unlike poliovirus RNA, is not infectious (Huang and Wagner, 1966b). Baltimore et al. (1970) first described the virion-associated RNA-dependent RNA polymerase and worked out the basic conditions for VSV transcription in vitro. The enzymology of the VSV transcriptase was described by Emerson and Wagner (1972, 1973) with the absolute requirements for N protein-encapsidated genome RNA serving as a template for reconstitution with both the L and NS proteins; L and NS proteins, from homologous or heterologous virions, are both required to initiate the transcriptase reaction (Emerson and Yu, 1975; Mellon and Emerson, 1978). Under these conditions, the nucleocapsid is infectious (Bishop et al., 1974) and codes for all five VSV monocistronic messenger RNAs in vitro (Bishop et al., 1974) and in vivo (Banerjee et al., 1977). The gene order of the five VSV cistrons was worked out by Ball and White (1976) and by Abraham and Banerjee (1976a) as 3'-N,NS,M,G,L-5', which also represents the decreasing molar ratios in which the five mRNAs are synthesized based on their proximity to the 3'-genome terminus. Colonno and Banerjee (1977) described a 47-nucleotide leader RNA sequence which, unlike the five mRNAs, is neither capped, polyadenylated, nor translated, but initiates the transcription process and is made in larger molar amounts than any mRNA. All five VSV mRNAs are capped and polyadenylated but no one, to date, has identified capping or adenylating enzymes (Abraham and Banerjee, 1976b). Another VSV protein which regulates transcription is the matrix (M) protein, which binds to nucleocapsids (Wilson and Lenard, 1981) and inhibits transcription by $\pm 80\%$ (Clinton et al., 1979; Carroll and Wagner, 1978, 1979).

The evidence is quite clear from UV-mapping data and other studies (Ball and White, 1976; Abraham and Banerjee, 1976a) that transcription of the VSV genome begins at a single site at the extreme 3' end and proceeds in a linear sequence to synthesize complementary RNAs in the order 3'-leader-N-NS-M-G-L. Major problems have arisen in identifying a unifying hypothesis that will account for two alternative pathways in transcription: (1) synthesis of separate leader and five capped and polyadenylated mRNAs, and (2) read-through of the minus-strand template to make a full-length complementary plus-strand (Ball and Wertz, 1981). Two models were originally postulated to conform with the UV-mapping data: (1) sequential synthesis along the entire length of the genome with cleavage and processing to make capped and polyadenylated mRNAs (Colonno and Banerjee, 1976) and (2) a stop–start model (Banerjee et al., 1977), in which the polymerase moves down the genome, initiating and terminating at specific intergenomic signals (Herman et al., 1980). The first model could be readily discarded for complete lack of evidence for a messenger RNA precursor in vitro or in vivo which could undergo cleavage and processing. A third model, which Iverson and Rose (1982) called the simultaneous initiation model, was proposed by Testa et al. (1980). The basis for this latter model was the finding from the onset of in vitro transcription of three oligonucleotides, in addition to the leader, the RNA sequences of which are homologous to the 5' termini of N and NS mRNA but had a 5' triphosphate and no cap. Similar oligonucleotides, 11–14 bases in length and identical to the 5' terminus of the N gene, were also described by Pinney and Emerson (1982). These findings led to the hypothesis that the VSV transcripts are initiated internally at the origin of each cistron by the virion RNA polymerase.

Two separate sets of experiments appear to refute the hypothesis of simultaneous internal initiation of VSV transcripts. By studying kinetics of mRNA synthesis analyzed by specific RNase T_1 oligonucleotides complementary to the N, NS, and M genes, Iverson and Rose (1982) obtained results "inconsistent with a model of vesicular stomatitis virus transcription involving simultaneous initiation and presynthesis of leader RNAs 30–70 nucleotides long for each mRNA." Emerson (1982), in very different experiments, found by reconstitution of VSV template and RNA polymerase in the absence of UTP and GTP that only leader gene products were synthesized and the 5'-terminal mRNA oligonucleotides were detected only after transcription of full-length leader was permitted. She came to the conclusion that the VSV "polymerase does not enter the genome independently at each gene, but each polymerase begins transcription at the 3' end of the genome, and reaches internal genes only by sequentially transcribing the 3' preceding sequences." It would appear, therefore, at this stage of our knowledge, that "transcription of vesicular stomatitis virus mRNAs is due to obligatory entrance of all polymerases at the leader gene, and suggests that the transcriptase and replicase may recognize the same promoter" (Emerson, 1982). The nature of that promoter is yet to be determined.

2.2.2. Translation

The mRNAs of VSV appear to be translated in a manner quite similar to other eukaryotic mRNAs (Breindl and Holland, 1975). The VSV mRNAs are polyadenylated and are capped *in vitro* as well as *in vivo* by incorporating the methyl group of S-adenosyl-L-methionine during synthesis by the virion-associated polymerase (Rhodes *et al.*, 1974). If methylation and capping are inhibited by the analogue S-adenosylhomocysteine, the uncapped VSV mRNAs are poorly translated in a wheat-germ system (Both *et al.*, 1975). However, Rose and Lodish (1976) found that removal of the 5'-terminal 7-methyguanosine did not greatly reduce the translational activity of VSV mRNAs in a reticulocyte–lysate cell-free system. These findings raise the question of the roles played by capping or other processing of VSV mRNAs as a potential inhibitors of cellular protein synthesis. In addition, certain VSV proteins are phosphorylated to varying degrees by a kinase, probably of host cell origin (Imblum and Wagner, 1974). The degree of phosphorylation of two species of VSV NS protein appears to determine its role in VSV transcriptase activity (Kingsford and Emerson, 1980). These and other properties of VSV proteins and their specific mRNAs are potentially involved in competition for translation of cellular proteins. Although different VSV mRNAs appear to be equally effective in binding to ribosomes and for initiation and elongation of polypeptides, the N-protein mRNA appears to be more efficient in translation than the G-protein mRNA under conditions of inhibition by high-salt concentrations or by aurintricarboxylate, suggesting difference requirements for factors that initiate translation (Lodish and Froshauer, 1977). Lodish and Porter (1980) have reported that VSV mRNAs in general out-compete cellular mRNAs for binding to ribosomes although they do not differ significantly in initiation of translation in reticulocyte lysates. These authors concluded that this competitive advantage of VSV mRNAs for ribosomes explains their capacity to reduce cellular protein synthesis but, as will be indicated in a later section, evidence by other techniques controverts such a conclusion (Dunigan and Lucas-Lenard, 1983; Schnitzlein *et al.*, 1983).

2.2.3. Replication

Unlike transcription, replication of VSV requires coupled translation (Huang and Manders, 1972; Wertz and Levine, 1973; Davis and Wertz, 1982) and a host factor(s) (Hill *et al.*, 1981; Patton *et al.*, 1983). In all liklihood, the same endogenous polymerase functions for replication as well as transcription. Although all full-length plus-strand and negative-strand RNA molecules and some messengers and leader are encapsidated with N protein (Blumberg and Kolakofsky, 1981), partially double-stranded replicative intermediates can be found in infected cells (Wertz, 1978); as described in a later section, these replicative intermediates may be involved in inhibiting cellular protein synthesis (Thomas and Wagner, 1982). The viral N protein is thought to modulate transcription and replication by its ability to bind to nascent leader RNA, thus promoting read-through of the termination signals as the full-length RNA is assembled into nucleocapsids (Schubert *et al.*, 1982). Experiments by Blumberg *et al.* (1981) have led to the hypothesis that interaction of VSV leader RNA and nucleocapsid protein may control VSV genome replication.

It is becoming lncreasingly evident that the VSV polymerase serves the dual purpose of messenger transcription and replication of the entire VSV genome. The apparent site of entry of the polymerase is the 3′ terminus of the VSV genome (Emerson, 1982), which codes for the plus-strand leader RNA. It has been proposed that the 5′ terminus of the leader RNA serves as the nucleation site for binding of newly synthesized N protein to form the ribonucleocapsid, thus regulating the switch from transcription to replication (Blumberg *et al.*, 1981, 1983). The La protein of systemic lupus erythematosus has been found complexed with both the plus-strand leader RNA (Kurilla and Keene, 1984) as well as the minus-strand leader (Wilusz *et al.*, 1983), and has been proposed as a host cell factor that may control the level of N protein which ostensibly affects the switch from VSV transcription to replication.

Assuming that the leader gene is the site of entry for the VSV polymerases, it has been proposed that the VSV leader RNA sequence regulates the switch from transcription to replication (Blumberg *et al.*, 1981, 1983). This led Kolakofsky and co-workers to speculate that the switch involves the leader sequence and the specific concentration of N protein. Wilusz *et al.* (1984) have proposed that formation of an La-protein-leader RNA complex may serve as an *in vivo* attenuator of transcription to ensure adequate levels of viral protein accumulation before the switch to replication takes place. Evidence for displacement of La protein by viral N protein on the plus-strand leader RNA is consistent with this possibility (Kurilla and Keene, 1984).

The leader sequence is thought to contain the nucleation site for nucleocapsid assembly within the first 18–20 nucleotides of both the wild type and DI leader RNA. Of the five strains of VSV examined, there is a remarkable homology in the first 18 nucleotides with the consensus sequence 3′-UGCUUN-UNNUNNUUUGU-5′ (Giorgi *et al.*, 1983). The nucleocapsid assembly signal is thought to be a five times repeated A residue in every third position at the 5′ end of the leader RNA (Giorgi *et al.*, 1983). This repeated pattern in the proposed encapsidation signal was pointed out by Blumberg *et al.* (personal communication) to be analogous to the encapsidation signal for tobacco mosaic virus. Perhaps, as suggested by Rose (1980), these genes may have a common ancestry.

The importance of the leader sequence in the shut-off of host-cell macromolecular synthesis will be discussed in detail later. We wish to emphasize here two aspects of the VSV leader RNA pertinent to the shut-off of host cellular macromolecular synthesis. First, the leader sequence contains nucleotide sequences which are essential for controlling VSV transcription, replication, and encapsidation of the virus. Second, there already is some preliminary evidence that VSV contains sequences analogous to other eukaryotic genes. Given the conservation of sequences essential in the transcription of eukaryotic genes, it is tempting to speculate that the VSV leader RNA does have sequences which are essential for viral RNA synthesis which are also essential for cellular transcription.

2.3. Rhabdovirus Genetics

2.3.1. Virus Species and Strain Variation

The rhabdoviruses are closely related morphologically by protein composition and, even to some extent, by base sequence homology. The two serotypes of VSV, New Jersey and Indiana, exhibit 50% amino acid identity of their type-specific glycoproteins, and about 20% identity was found between the glycoproteins of two separate genera, VSV-New Jersey and rabies virus (Gallione and Rose, 1983). However, there appear to be considerable differences in the replication of rabies and vesicular stomatitis viruses. VSV grows readily and produces almost a normal yield of progeny virus in enucleated cells (Follett *et al.*, 1974), whereas, rabies virus undergoes an abortive infection in enucleated cells by producing virus-directed mRNA and proteins but no infectious virions (Wiktor and Koprowski, 1974). Also, unlike VSV, rabies virus does not inhibit cellular protein synthesis in infected BHK-21 cells (Matsumoto, 1974), although the ERA rabies strain will selectively suppress protein synthesis in BHK-21 cells infected at a high multiplicity under hypertonic conditions (Madore and England, 1975). It should also be noted that different serotypes and strains of VSV vary somewhat in their capacity to inhibit cellular macromolecular synthesis. The Indiana serotype inhibits cellular RNA synthesis in MPC-11 cells more efficiently than does the New Jersey serotype (Grinnell and Wagner, 1983). In fact, Lodish and Porter (1981) found that the San Juan strain of the Indiana serotype of VSV was more effective in inhibiting cellular protein synthesis than was the Glasgow strain of the same serotype. It is clearly inadvisable to compare results of experiments performed with different serotypes or strains and, perhaps, even with different passage levels of the same strain.

2.3.2. Conditional Lethal Temperature-Sensitive Mutants

There is a vast literature on temperature-sensitive and other mutants of vesicular stomatitis viruses, a field pioneered and reviewed in this series by Pringle (1977). The temperature-sensitive (*ts*) mutants of VSV serotypes fall into at least five groups as determined by cross-complementation; these mutants arise spontaneously or by exposure to various mutagens. No attempt will be made to review the field here but, for purposes of this chapter, it should be stated that each complementation group can be tentatively identified phenotypically with a specific protein of the Indiana serotype as follows: I-L protein transcriptase (Hunt *et al.*, 1976), II-NS protein, III-M protein, IV-N protein, and V-G protein (see Pringle, 1977). The vast majority of all *ts* mutants fall into complementation group I. As expected, no *ts* mutants have been described for the leader sequence gene but leader RNA synthesis is probably restricted in group I mutants. Temperature-sensitive mutants generally do not cause acute encephalitis in mice as do wild-type VS viruses because they are usually restricted at temperatures 37–40°C (Wagner, 1974). However, some but not other mutants in complementation group III were found to be neurovirulent for mice (Rabinowitz *et al.*, 1981) and were capable of causing extensive morphological changes in cultured neuroblastoma N-18 cells (Dille *et al.*, 1981); these effects appear to occur fairly late in infection and were not equated with effects on cellular macromolecular synthesis. As will be described later, cytopathogenic effects and inhibition of cellular macromolecular synthesis do not occur at nonpermissive temperature in cells infected with certain group I mutants restricted in transcription (Marcus and Sekellick, 1975; McAllister and Wagner, 1976; McGowan and Wagner, 1981).

Certain mutants of VSV are conditionally restricted in HeLa or HEp-2 cells (Simpson and Obijeski, 1974; Obijeski and Simpson, 1974) or chick embryo cells (Pringle, 1978); these are referred to as host-restricted (*hr*) mutants. Such *hr* mutants are also temperature-sensitive (*ts*) and are often, but not always, restricted in RNA synthesis (RNA⁻); the lesion is probably in the L gene (the polymerase gene). Nonconditional *hr* mutants, which failed to multiply in chick embryo cells at 31°C or 39°C, isolated by Pringle (1978), were also restricted in polymerase activity in the restricted host. These *hr* mutants suggest that host factors can be significant in VSV polymerase expression and, hence, may be important as determinants of viral cytopathogenicity and viral effects on cellular macromolecular synthesis. An interesting mutant derived from the HR wild-type VSV that has not been adequately studied is T1026, which is RNA⁺ but allowed "essentially normal DNA synthesis and division" in cells infected at restrictive temperature (Farmilo and Stanners, 1972). This mutant is purported to have a genetic function, termed P, which inhibits initiation of translation (Stanners *et al.*, 1977), a postulate questioned by Lodish and Porter (1981).

2.3.3. Nonconditional Mutants

Less well-studied VSV mutants have been identified by their reduced cytopathogenicity, usually manifested as a small plaque. One such small-plaque mutant, first identified by Wagner *et al.* (1963), was found to grow to lower titer than the wild-type parent and to establish persistent infection more readily, but had a very high reversion frequency.

A more stable and more interesting small-plaque mutant (S_2) was selected by Youngner and Wertz (1968) from a large-plaque (L_1) variant of VSV. This S_2 mutant was found to be a more efficient inhibitor of protein synthesis in L cells (Wertz and Youngner, 1972) and also synthesized a disproportionately large amount of 12 S–15 S mRNA than did the L_1 wild type (Wertz and Levine, 1973). Despite this excess production of mRNA by the S_2 VSV mutant, much less viral protein was synthesized in infected cells when compared to the wild type (Davis and Wertz, 1980). Mutants like this provide important probes for studying the relationship of viral RNA and protein synthesis to the cytopathic and cell-inhibitory functions of VSV.

2.3.4. Defective–Interfering (DI) Virus Particles

Stocks of most animal viruses contain defective particles in which varying amounts of the viral genome are deleted and many of these interfere with replication of the homologous parental virus (see reveiw in this series by Huang and Baltimore, 1977). Among the best studied of these DI particles are those of vesicular stomatitis virus, partially because they occur in great abundance and frequently contaminate preparations of wild-type ("standard") VSV. By definition, DI particles cannot replicate and require *wt* virus as helper, as was evident in the original description by Cooper and Bellett (1959). Owing to their much smaller size, DI particles of VSV (originally called T particles) are relatively easy to separate from standard infectious VSV (bullet-shaped or B virions) by rate zonal centrifugation (Huang *et al.*, 1966) and can be assayed by their capacity to inhibit VSV replication (Huang and Wagner, 1966*a*). VSV DI particles comprise two main groups: (1) those in which the distal 5' half of the genome, most of the L gene, is deleted; almost all of these arise from the heat-resistant VSV variant and are designated HR-LT (Leamnson and Reichmann, 1974), and (2) those in which the proximal 50–90% of the 3' end of the VSV genome is deleted, representing a deletion of all genes except the 5' portions of the L gene. The HR-LT 3'-DI particles retain all but the L gene and can readily transcribe plus-strand leader and the N, NS, M, and G mRNAs but cannot replicate without *wt* helper (Colonno *et al.*, 1977; Marcus *et al.*, 1977). The glycoprotein mRNA transcribed in cells infected with HR-LT can be elongated by reading through the stop signal between the G and L genes, allowing transcription of the undeleted portion of the L gene (Herman, 1983). The truly defective 5'-DI particles, on the other hand, have no capacity to transcribe anything but a unique DI leader sequence (Emerson *et al.*, 1977) despite the fact that the virion contains an active polymerase (Emerson and Wagner, 1972).

The genetic make-up, origin, and the characteristic properties of 5'-DI particles of rhabdoviruses and other negative-strand viruses have been reviewed by Lazzarini *et al* (1981). In order to replicate, the 5'-DI particles are thought to require competent initiation sites at their 3' terminus, equivalent to those of full-length standard virus, as well as sequences at the 5' end compatible with initiation of encapsidation. In these respects, the 3'- to 5'-terminal oligonucleotides are presumably derived by copy choice from the parental virus genome, and are also similar in their 3' terminus to that of the parental genome (minus-strand) and anti-genome (plus-strand). At least four classes of DI particles can be identified on the basis of their genome structure: (1) a true simple deletion mutant in which the 5' half, the L protein gene, is deleted, i.e., DI-LT-3' which has been sequenced (Epstein *et al.*, 1980), (2) DI-5' panhandles containing information from the 5' half of the VSV parental genome but terminating at the 3' end with a short complementary sequence which self-anneals to form a panhandle (Schubert *et al.*, 1978, (3) the "snapback" or "hairpin" DI particle contains genetic information derived from the 5' end of the parental genome but, instead of 3' sequences, it forms an exact complement of the 5' end to give an almost perfect duplex when deproteinized (Schubert and Lazzarini, 1981*a*), and (4) prototype DI-LT$_2$ appears to have compound initiation sites at the 3' end derived from both the 3' and 5' ends of the parental genome, including a complete leader gene and N gene sequences, and was derived from

DI-HR-LT (Keene *et al.*, 1981). Lazzarini *et al.* (1981) present clear models of how these DI particles originate and replicate; this pioneering laboratory has also been investigating the recombinational events that generate DI particles such as DI-LT by use of cDNA clones of different portions of the VSV genome (Yang and Lazzarini, 1983). Moreover, these studies provide excellent model systems for probing the transcriptive and replicative strategies of VSV and other negative-strand viruses.

2.3.5. Mapping and Sequencing

Certain concepts about rhabdovirus genomes could be deduced by kinetic analysis of transcription and translation, but *ts* mutant phenotypes for gene assignment, and by defective-interfering particles. However, the gene sequence of VSV has been established primarily by the UV-inactivation studies of Ball and White (1976) and Abraham and Banerjee (1976*a*). These and subsequent studies, based on UV-target size and sequential transcription, provided the standard linear map of VSV genes as 3'-leader-N-NS-M-G-L-5'. Each of these cistrons shows increasing resistance to UV inactivation ordered from 5' to 3', terminating in the extreme UV resistance of the 47-nucleotide leader gene (Colonno and Banerjee, 1977). However, the real breakthrough in rhabdovirus genetics came about from the availability of cDNA clones of four VSV messengers. Complete cDNA and protein sequences are now available for the N, NS, M, and G cistrons of the Indiana serotype (Rose and Gallione, 1981; Gallione *et al.*, 1981) as well as the glycoprotein genes of VSV-New Jersey (Gallione and Rose, 1983) and of rabies virus (Rose *et al.*, 1982). Intergenic sequences at the N-NS gene junctions have been reported by McGeoch and Dolan (1979), and those between the NS, M, G, and L genes by Rose (1980). These studies and others (McGeoch *et al.*, 1980) also provide initiation and termination codons for N mRNA of both Indiana and New Jersey serotypes. The region specifying polyadenylation of the L gene (Schubert and Lazzarini, 1981*b*) and other VSV genes have also been identified. Many of the VSV genes are now cloned in vectors that allow expression in transfected cells of VSV products such as the N protein (Sprague *et al.*, 1982) and the G protein (Rose, 1982).

3. CELLULAR RESPONSES TO RHABDOVIRUS INFECTION

3.1. Cytopathology

A general survey of cytopathic effects of rhabdoviruses is provided in the first chapter of this volume (Wagner, 1984) and will only be briefly summarized here. Certain rhabdoviruses, particularly VSV, have a very wide host range, from insects to mammals, and are quite virulent for mammals. Rhabdoviruses of plants and fish probably have a narrow host range, wherease, rhabdoviruses of the rabies genus are widespread among mammals but are much less acutely pathogenic than are vesicular stomatitis viruses.

The review by Bablanian (1975) provides an excellent description of the cytopathology caused by VSV. Clearly, there are two distinct types of cytopathic effects resulting from VSV infection. (1) a rapid cellular response at high multiplicity, characterized by cell rounding by 1 hr postinfection (Baxt and Bablanian, 1976a), which was once thought not to require active viral synthetic functions and was dubbed cytotoxic, and (2) a slower response that appears to require active VSV replication, usually accompanied by release of progeny virions from the infected cell. Quite characteristic of the rapid cellular response is early inhibition of cellular RNA, DNA, and protein synthesis (McGowan and Wagner, 1981). These early events were once thought to be due to input parental virion structural components because they were noted at high multiplicity and with nonreplicating DI particles and UV-inactivated wild-type virus, as well as in the presence of cycloheximide (Baxt and Bablanian, 1976b). It is now quite clear that very high multiplicities of VSV DI-5' particles, when *completely* free of standard wild-type VSV, do not kill cells (Marcus and Sekellick, 1974) and do not inhibit cellular RNA synthesis (Weck and Wagner, 1979a). By the same token, UV irradiation at moderate doses does not eliminate all biological activity of VSV, which can still retain the capacity to kill cells (Marvaldi *et al.*, 1977) and to inhibit cellular nucleic acid (McGowan and Wagner, 1981) and protein synthesis (Marvaldi *et al.*, 1978). This subject will be treated in greater detail in later sections of this chapter. Suffice it to say that the most compelling data suggest that certain viral gene functions must retain their activity in order to kill cells, In fact, it seems quite clear from genetic studies by Marcus and his colleagues (1975, 1977) that VSV can kill cells only if the infecting virion retains a certain degree of transcriptase activity. Moreover, only a single standard virus particle, but not

VSV DI-5′ particles, is sufficient to kill a cell (Marcus and Sekellick, 1976).

3.2. Variations in Host Cell Susceptibility

It has long been known that cells vary greatly in their susceptibility to viral infection, even when no differences can be demonstrated in their surface receptors for virus adsorption. Moreover, the same virus clone can inhibit macromolecular synthesis in one cell type to a greater extent than another, even though virus yields may not differ significantly. For example, Baxt and Bablanian (1976*b*) showed that VSV inhibits nucleic acid synthesis in BHK-21 cells more readily than it does in LLC-MK2 cells. Weck and Wagner (1978) also reported that MPC-11 mouse myeloma cells were more susceptible to VSV shut-off of cellular RNA synthesis than were BHK-21 or mouse L cells.

Cell differentiation may play a role in cellular susceptibility to viral infection as illustrated by the finding that VSV replication is restricted in one human lymphoblastoid cell line but not in another or in HeLa cells (Nowakowski *et al.*, 1973). Robertson and Wagner (1981) also found that HeLa cells and L cells are more permissive for VSV growth than an end-stage myeloma cell line, MPC-11, which, in turn, is more permissive than the Abelson virus-transformed 18–81 pre-B cell that does not secrete immunoglobulin. The converse relationship was found when we tested these same cells for the capacity of VSV to shut-off cellular RNA synthesis; the susceptibility of these cells to VSV inhibition of cellular RNA synthesis could be ranked in the order 18–81 > MPC-11 > L cells > HeLa cells (Robertson and Wagner, 1981). This difference in susceptibility to inhibition of cellular macromolecular synthesis does not appear to be related to the specific proteins produced by differentiated cells. Myeloma cells infected with VSV exhibit far less inhibition of the synthesis of immunoglobulin than of other cellular proteins (Nuss and Koch, 1976*b*). Similarly, globin synthesis in differentiated Friend erythroleukemia cells is more resistant to shut-off by VSV infection than are other proteins of the same cell (Nishioka and Silverstein, 1978). Cellular factors of unknown nature were also found to determine the ability of mengovirus to inhibit the synthesis of VSV protein or of VSV to inhibit the synthesis of mengovirus proteins in doubly infected HeLa, CHO, or L-929 cells (Otto and Lucas-Lenard, 1980). VSV also appears to inhibit uptake of uridine in infected chick embryo cells (Genty, 1975) but not in other cells (Genty and Berreur, 1975; Weck and Wagner, 1978). Quite obviously, the factors that control cellular susceptibility to infection with VSV, or other viruses for that matter, remain unclear.

4. PROTEIN SYNTHESIS INHIBITION BY VESICULAR STOMATITIS VIRUS

4.1. Definition of the Problem

Among the rhabdoviruses, only vesicular stomatitis virus has been studied to any extent for its capacity to shut off cellular protein synthesis. The vast majority of these studies have been performed with one or another of the strains of the Indiana serotype of VSV, although some studies have been carried out with the New Jersey serotype (Yaoi *et al.*, 1970). Although almost all the studies described in this chapter are concerned with VSV-Indiana virus without identifying the strain, it is wise to keep in mind that the San Juan strain, the one generally used, has been found to inhibit cellular protein synthesis more efficiently than the Glasgow strain (Lodish and Porter, 1981). Early investigators considered multiplicity of infection to be an important factor in the degree and rate of protein synthesis inhibition in VSV-infected cells (reviewed by Bablanian, 1975). At VSV multiplicities of 100 PFU/cell, protein synthesis in rabbit kidney cells was inhibited by >80% by 2 hr postinfection but a longer time was required for comparable inhibition at lower multiplicities (Yamazaki and Wagner, 1970; Wagner *et al.*, 1970). Wertz and Youngner (1972) found that host cell determinants may play an important role in VSV multiplicity-dependent inhibition of cellular protein synthesis, as judged by different responses of chick embryo fibroblasts and mouse L cells.

The phenomenon of protein synthesis inhibition in VSV-infected cells has been well described for various other virus–host systems. The characteristics of inhibition appear to depend in part on the host cell studied and on the multiplicity of infection. From pulse-labeling experiments, Mudd and Summera (1970) reported a 90% inhibition of total protein synthesis at 4 hr postinfection in HeLa cells by the Indiana serotype of VSV. These studies were done on HeLa cells grown in suspension infected at a multiplicity of infection of 85. Only a 40%

reduction in total protein synthesis was observed by 4 hr when the multiplicity of infection was decreased to 10. At an MOI of 50, Otto and Lucas-Lenard (1980) noted a 70% inhibition at 4 hr in HeLa cells grown in monolayers. At high multiplicity of infection (200), VSV was noted to inhibit total protein synthesis in LLC-MK2 cells by 80% at 4 hr postinfection, but BHK-21 cells were inhibited by only 45% (Baxt and Bablanian, 1976*b*). At an MOI of 50, VSV inhibited HeLa cell and CHO cell total protein synthesis by 70% by 4 hr postinfection, but L-929 cells were inhibited by only 45% (Otto and Lucas-Lenard, 1980).

Pulse-labeling of cells followed by measurement of total isotope incorporation at various time postinfection cannot differentiate between an effect on cell-specific vs. viral-specific protein synthesis. In an effort to determine to what extent cell-specific protein synthesis was being affected, McAllister and Wagner (1976) compared, by SDS–polyacrylamide gel electrophoresis, the synthesis of specific cellular proteins from uninfected and infected L cells and determined that VSV inhibited the synthesis of cellular proteins by about 80% at an MOI of 10 by 5 hr postinfection. In a similar manner, Lodish and Porter (1980) reported a 35% inhibition of BHK cell total protein synthesis by 4 hr postinfection at an MOI of 10, but noted that cell-specific proteins were being synthesized at a rate that was only 25% of that in uninfected cells.

Similar studies on multiplicity-dependent inhibition of cellular protein (and RNA) synthesis led a number of investigators to hypothesize that structural components of input parental virions were the factors that shut off cellular macromolecular synthesis. Huang and Wagner (1965) reported that nonreplicating VSV inhibits cellular RNA synthesis, and Baxt and Bablanian (1976*b*) found that "nontranscribable" VSV DI and UV-irradiated standard virions were as effective as fully infectious VSV in inhibiting cellular protein synthesis. These early studies were, of course, performed before a clear knowledge of the effects of different doses of UV irradiation on VSV transcription or the significance of contaminating infectious B (standard) virions present in DI-particle preparations.

The most likely candidate for inhibition of protein synthesis by structural components of input VSV is the glycoprotein; therefore, it is of interest that enormous amounts of isolated glycoprotein did not inhibit protein synthesis in BHK-21F cells but did inhibit cellular RNA and DNA synthesis (McSharry and Choppin, 1978). In an attempt to resolve the question of input virion components as factors contributing to inhibition of host cell macromolecular synthesis, Wertz and Youngner (1972) hypothesized that "two mechanisms may be involved in the inhibition of host protein synthesis by VSV: (1) an initial multiplicity-dependent and ultraviolet-insensitive inhibition, and (2) a progressive ultraviolet-sensitive inhibition." Despite considerable evidence now available and presented below to favor newly synthesize viral products as the major or sole mechanism for inhibition of cellular protein synthesis, the role of input virion toxic components has not been ruled out completely, even though enormous amounts of isolated VSV glycoprotein were found to have no effect on cellular protein synthesis (McSharry and Choppin, 1978). Despite considerable earlier evidence for a toxic component, much of it from his own data, great credit should be given to Bablanian (1975) for stating: "On the whole, these data suggest that a viral component could be the cause of host inhibitions, but they do not rule out the possibility that a product of the viral RNA polymerase may be responsible. The RNA polymerase of the virion could still be functional after UV irradiation."

Another problem that arises is whether VSV inhibition of cellular protein synthesis is a direct effect of the virus or is secondary to its inhibition of cellular RNA synthesis. Without much data, Bablanian (1975) came up with the wise observation that "the speed with which protein synthesis is inhibited makes it seem unlikely that it is related to the virus-induced inhibition of host RNA synthesis." In a recent study, McGowan and Wagner (1981) were able to demonstrate by pulse-labeling with [^{3}H]leucine, [^{3}H]uridine, and [^{3}H]thymidine that inhibition of protein, RNA, and DNA all occurred very early and comcomitantly after VSV infection of MPC-11 myeloma cells (Fig. 1). Moreover, suppression of cellular RNA and DNA synthesis was maximal at 2 hr postinfection but total protein synthesis inhibition was much less and took longer. Of considerable importance is the finding that VSV infection does not result in degradation of host cell mRNA (Nishioka and Silverstein, 1978; Lodish and Porter, 1980; Weck and Wagner, 1978).

4.2. Host Cell and Environmental Factors

The extent and rate of protein synthesis inhibition by VSV also appears to be host cell dependent. Wertz and Youngner (1970) noted considerable differences in susceptibility of chick embryo fibroblasts

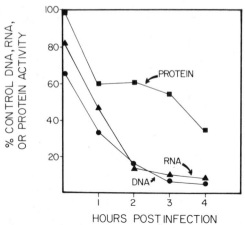

Fig. 1. Comparative inhibition of DNA, RNA, and protein syntheses in synchronized MPC-11 cells infected with VSV (MOI ≈10). Cells pulsed for 10 min at 1-hr intervals with [³H]thymidine (2 μCi/ml), ³H-labeled L-amino acids (5 μCi/ml), or [³H]uridine (2 μCi/ml). ³H-labeled amino acid incorporation into uninfected cell protein remained constant (1×10^3 cpm per 4×10^5 cells) relative to the range of [³H]thymidine incorporation into cellular DNA (3×10^2 cpm at time zero to 1×10^4 cpm per 4×10^5 cells by 4 hr after mock infection). [³H]uridine incorporation into uninfected cells ranged from 1×10^3 to 5×10^3 cpm per 4×10^5 cells. Data from McGowan and Wagner (1981).

and mouse L cells. In contrast to linear inhibition of DNA and RNA synthesis, protein synthesis inhibition in MPC-11 cells was not linear after 1 hr and was not maximal until 4 hr postinfection (McGowan and Wagner, 1981). In mouse L cells, VSV inhibition of protein synthesis was relatively linear for 4 hr and reached lower levels than in MPC-11 cells, despite a concommitant switch from host cell-specific to virus-specific protein synthesis (McAllister and Wagner, 1976). BHK-21 cells appear to be the least susceptible to VSV inhibition of cellular protein synthesis which reached levels of 65% that of uninfected control cells at 4 hr postinfection with the Glasgow strain of VSV-Indiana (Lodish and Porter, 1980) and somewhat lower with the San Juan strain of VSV-Indiana (Lodish and Porter, 1981). The host also appears to play a role in inhibition by VSV of mengovirus protein synthesis and, conversely, the capacity of mengovirus to inhibit VSV protein synthesis in three different cell types doubly infected with both viruses (Otto and Lucas-Lenard, 1980).

Two other factors that appear to control cell response to protein synthesis inhibition by VSV are ionic environment and co-infection with another virus, each of which is related to preferential translation of competing messenger RNAs. It is of peripheral interest that a hypertonic environment of VSV-infected cells inhibits synthesis of the membrane-associated viral G and M proteins to a greater extent than the nucleocapsid proteins (Nuss and Koch, 1976a). Hypertonic environment appears to favor translation of poliovirus and VSV mRNA in preference to host cell mRNA (Nuss *et al.*, 1975). It has been hypothesized that viral infection alters membrane permeability leading to increased intracellular K^+ concentration, in which environment viral mRNA is translated preferentially over K^+-susceptible cellular mRNA. However, this hypothesis has been challenged by Francoeur and Stanners (1978) who found no alteration in K^+ concentration or transport in VSV-infected L cells until late in infection, whereas, protein synthesis inhibition occurred very early.

Superinfection with poliovirus has a striking effect on VSV-infected HeLa-S_3 cells; the evidence in these experiments was quite convincing that preformed VSV mRNA was not translated following infection with poliovirus (Ehrenfeld and Lund, 1977; Trachsel *et al.*, 1980). Of course, the effect of VSV on cellular protein synthesis could not be determined in this experiment because poliovirus itself drastically inhibits cellular protein synthesis. As mentioned above, the ascendency of the mRNAs of two viruses may also depend on the type of cell which is doubly infected with the two competing viruses (Otto and Lucas-Lenard, 1980).

Only limited studies have been performed on the variable capacity of VSV to inhibit synthesis of specific proteins in differentiated or undifferentiated cells. In a very early study, it was found that wild-type VSV, which itself is a relatively poor inducer of interferon synthesis in certain cells (Wertz and Youngner, 1970; Sekellick and Marcus, 1979), rapidly shuts off interferon synthesis induced by Newcastle disease virus probably by inhibiting synthesis of interferon mRNA (Wagner and Huang, 1966; Wertz and Youngner, 1970). In the differentiated mouse plasmocytoma MPC-11 cells infected with VSV, translation of nonimmunoglobulin mRNAs was suppressed to a much greater extent than was translation of immunoglobulin mRNA; 41% inhibition in the synthesis of non-IgG proteins was coincident with a 28% inhibition for heavy chain and a 7% inhibition for light chain synthesis (Nuss and Koch, 1976*b*). Very similar resistance of IgG heavy and light chain synthesis was noted in the protein synthesis inhibitory effect of hypertonic medium in the same MPC-11 cell system (Nuss and Koch, 1976*c*). VSV infection of another differentiated cell line, Friend erythroleukemia, revealed that synthesis of globin was completely resistant to inhibition despite the fact that VSV reduced total cell protein synthesis; in contrast, herpes simplex virus drastically suppressed globin synthesis, ostensibly by degradation of globin mRNA (Nishioka and Silverstein, 1978). These two sets of experiments suggest that specialized proteins synthesized by end-stage differentiated cells may be more resistance to the protein synthesis inhibitory activity of VSV than are the usual protein synthesized by all cells.

Specific proteins synthesized by undifferentiated cells appear to be quite susceptible to inhibition by VSV infection. We have found that actin synthesis by L cells following VSV infection is inhibited to about the same extent as most other cell proteins (Thomas and Wagner, unpublished experiments). In fact, McAllister and Wagner (1976) did not detect, by polyacrylamide gel electrophoresis, any specific proteins in undifferentiated L cells that were resistant to the protein synthesis inhibitory activity of VSV. By the same token, all the ribosomal proteins synthesized by L cells were inhibited to about the same extent after VSV infection (Marvaldi and Lucas-Lenard, 1977); in fact, VSV infection had an even greater inhibitory effect on incorporation of ribosomal proteins into newly assembled ribosomes (Jaye *et al.*, 1980).

4.3. Viral Properties as Candidates for Inhibiting Cell Protein Synthesis

The capacity of VSV to shut off host cell protein synthesis could be due either to structural components of the input virion or to newly synthesized and/or assembled products. Evidence presented above indicates that parental virion structural components are unlikely to play a major role in inhibiting cellular protein synthesis or killing of the host cell (Marcus and Sekellick, 1974). This leaves a newly synthesized product(s) as the likely culprit(s) for inhibiting protein synthesis of the VSV-infected host cell. The only products coded by the viral genome are RNA and proteins, since lipids and carbohydrates are host derived (McSharry and Wagner, 1971a; McSharry and Wagner, 1971b). Identification of a newly synthesized viral inhibitor must rely on biological and chemical methods, both of which are reviewed here. The biological methods are based essentially on genetics and physical inactivation of the viral genome. The most reliable chemical techniques for identifying the inhibitor(s) depend largely on cell-free translation methodology.

The major products of VSV transcription are complementary RNA molecules, either full-length templates to replicate progeny negative strands, or monocistronic messengers for each of the five viral genes preceded by a plus-strand uncapped and nonpolyadenylated leader that has no messenger function (Abraham and Banerjee, 1976b; Colonno and Banerjee, 1976). The major controversy concerning VSV inhibition of cellular protein synthesis is whether the phenomenon is due to more efficient translatability of viral mRNA competing successfully for ribosomal binding sites with endogenous cellular mRNAs or whether the viral products (messengers, leader, proteins, or nucleocapsids) directly block cellular mRNA translation by selected perturbation of ribosomal functions required for translation of cellular mRNAs but not viral mRNAs. We shall also address this question.

4.4. VSV Genetics and Protein Synthesis Inhibition

The three types of genetic variants used to probe the biological properties of VSV that lead to inhibition of host cell protein synthesis are deletion mutants (defective-interfering virus), temperature-sensitive (*ts*) mutants, and nonconditional mutants.

4.4.1. Defective–Interfering VSV

A number of investigators had reported in earlier studies that 5'-DI particles of VSV, in which one-half or more of the 3' end of the genome is deleted, were fully capable of inhibiting cellular protein synthesis (Baxt and Bablanian, 1976b). It seems likely that this apparent effect of DI particles on protein synthesis inhibition is due to difficult-to-detect contamination with the helper B virions (standard infection virus). Schnitzlein *et al.* (1983) detected little or no inhibition of BHK cell protein synthesis by various types of highly purified 5'-DI particles, even at high multiplicities of infection. These DI particles fully retained capacity to inhibit replication of infectious B particles and, as will be discussed later, the 5'-DI particles did not affect the capacity of infectious B virions to inhibit cellular protein synthesis (Schnitzlein *et al.*, 1983). It seems safe to conclude that 5'-DI particles of VSV contain no structural components or biological activity that will inhibit cellular protein synthesis.

4.4.2. Temperature-Sensitive Mutants

McAllister and Wagner (1976) compared the synthesis of cellular proteins, differentiated from VSV proteins by polyacrylamide gel electrophoresis, in L cells infected with *wt* VSV and various *ts* mutants. Two group I mutants restricted in transcription (Hunt *et al.*, 1976) and, particularly *ts* G114(I), completely failed to inhibit cellular protein synthesis at nonpermissive temperature. In sharp contrast, a group IV RNA⁻ mutant and a group II RNA$^\pm$ mutant, neither of which is restricted in primary transcription, shut off cellular protein synthesis at nonpermissive temperatures as well as did the wild-type VSV (McAllister and Wagner, 1976). Marvaldi *et al.* (1977), using the same and some additional *ts* mutants, came to similar conclusions; they found that several group I *ts* mutants failed to inhibit cellular protein synthesis but the leaky mutant *ts*05(I) did, even at restrictive temperature. Somewhat in contrast to the results of McAllister and Wagner (1976), Marvaldi *et al.* (1977) reported that one *ts* mutant in complementation group IV did inhibit protein synthesis at restrictive temperature, but another group IV mutant, *ts*G41(IV), and a group II mutant *ts*G22(II), failed to inhibit cellular protein synthesis under restrictive conditions. This latter result was not confirmed by Schnitzlein *et al.* (1983) who found that replication mutant *ts*G22(II) did inhibit protein synthesis in BHK-21 cells infected at nonpermissive temperature. There is general agreement among all these investigators that VSV transcription is the minimal event required for inhibition of host cell protein synthesis. However, none of these experiments rule out synthesis of VSV proteins as potential inhibitors of cell protein synthesis; in fact, Marvaldi *et al.* (1977) were inclined to favor implication of VSV mRNAs and perhaps proteins N and NS based on their genetic analysis.

4.4.3. Nonconditional Mutants

A small plaque (S$_2$) mutant of VSV, which is more efficient in inhibition of host cell protein synthesis than wild-type virus (Wertz and Youngner, 1970; 1972) has been characterized (Davis and Wertz, 1980). This mutant makes two or three times the amount of viral mRNA in chick embryo cells than does wild-type virus, but synthesizes only about half as much viral protein. The mRNA transcribed by this mutant appears to be identical to wild-type mRNA in size, extent of polyadenylation at 3′ termini, and extent of capping of 5′ termini. It would appear, however, that the S$_2$ mutant transcripts are associated with smaller polysomes than are the wild-type transcripts. Furthermore, this reduction in polysome size appears to be the result of a decreased efficiency in protein synthesis initiation and not the result of an increased rate of elongation, or competition among a limited number of ribosomes for an excess of mRNA.

Stanners *et al.* (1977) have described a *ts* mutant derived from the VSV HR strain which contains a second non-*ts* mutation in the viral polymerase gene, (which they call ''P'') purported to be responsible for inhibition of cellular protein synthesis. This mutant appeared to inhibit initiation of translation in some cells but not others. This mutant (T1026) and its revertant R1 have interesting properties for probing VS viral effects on cellular protein synthesis, but Lodish and Porter (1977) have strongly questioned the validity of a specific genetic ''P'' function that phenotypically expresses inhibition of cellular protein synthesis.

4.5. Inhibitors of VSV Functions that Affect Protein Synthesis Inhibition

A variety of chemical and physical agents have been used in an attempt to dissect and identify the VSV function(s) that inhibit cellular protein synthesis.

4.5.1. Inhibitors of Viral Protein Synthesis

Yamazaki and Wagner (1970) found that pretreatment of primary rabbit kidney cells with highly purified rabbit interferon greatly reduced VSV replication and inhibited synthesis of virus-specific proteins; despite this inhibition of viral protein synthesis, interferon did not prevent the switch-off by VSV of cellular protein synthesis. This observation was interpreted, in retrospect erroneously, to be due to ''an input virion component rather than a newly synthesized viral product.'' It was originally thought that primary transcription of VSV is inhibited in interferon-treated cells (Marcus *et al.*, 1971; Manders *et al.*, 1972), but it now seems clear that only secondary transcription and not primary transcription is inhibited in interferon-pretreated cells (Repik *et al.*, 1974). There is now quite convincing evidence that the interferon-induced inhibitor of VSV replication functions at the translational level (Sen, 1982). These circumstantial data point to the hypothesis that a VSV transcription product, alone or in association with another factor, is the inhibitor of cellular protein synthesis.

Use of the general protein synthesis inhibitor, cycloheximide, led to the hypothesis that *de novo* VSV protein synthesis is required for inhibiting cellular protein synthesis because, when cycloheximide was removed from infected cells, viral protein synthesis preceded suppression of cellular protein synthesis inhibition (Wertz and Youngner, 1972). In our experience, experiments such as these are difficult to interpret because cycloheximide has many side effects, such as inhibition of cellular RNA synthesis (Weck and Wagner, unpublished data).

4.5.2. Ultraviolet Irradiation and Heat

VS virions inactivated by ultraviolet irradiation or heat have been tested for their capacity to inhibit cellular protein synthesis. Baxt and Bablanian (1976*b*) have shown that heat-inactivated VSV-Indiana does not shut off cellular protein synthesis but UV-irradiated VSV does. Earlier studies by Yaoi *et al.* (1970) revealed that VSV-New Jersey inactivated by UV light also inhibited protein synthesis, as well as RNA and DNA synthesis, in chick embryo fibroblasts. Similar studies on UV-irradiated VSV and cellular RNA synthesis had been reported by Huang and Wagner (1965). Interpretation of these earlier studies is somewhat obscured by difficulties in quantitating the UV dose and by wide variations in multiplicity of infection. Later studies by Marvaldi *et al.* (1978) were performed when knowledge of the VSV genome and its susceptibility to UV irradiation were much better understood.

Because VSV is sequentially transcribed (Ball and White, 1976; Abraham and Banerjee, 1976*a*), UV irradiation of the virus will prevent expression of any genes that are downstream from a pyrimidine dimer induced by the irradiation. Since the entire genome is required for the successful replication of the virus (infectivity), the UV-target size of this function can be taken as the entire genome. By comparing the relative doses of UV irradiation required to produce a given loss of infectivity with that required to give the same loss in protein synthesis inhibition, a target size for protein synthesis inhibition can be determined which represents that portion of the genome required to affect protein synthesis inhibition. This has been done by Marvaldi *et al.* (1978) for VSV infection of mouse L cells at MOIs of 10 and 100. Their results indicate that transcription of one-fifth of the VSV genome, the N gene, and possibly the NS gene, is required for in-

hibition of host cell protein synthesis. These results have been confirmed in our laboratory by dose-responses to UV irradiation of VSV
used to infect mouse L cells (Thomas, Carroll, and Wagner, unpublished results). Similarly, others have determined the target sizes required for nucleic acid synthesis inhibition by vesicular stomatitis
virus and these appear to be ~1/40th the size of that required for
protein synthesis inhibition (Weck *et al.*, 1979; McGowan and Wagner, 1981). This would argue that VSV inhibition of host cell protein
synthesis is a separate phenomenon from that of nucleic acid synthesis inhibition by VSV.

Dunigan and Lucas-Lenard (1983) have recently published the
most definitive study to date on the effect of UV irradiation on the
capacity of VSV to shut off protein synthesis in L cells. They present
evidence for a biphasic UV inactivation curve, "suggesting that transcription of two regions of the viral genome is necessary for the virus
to become inactivated in this capacity" to shut off cellular protein
synthesis. They calculated UV target sizes of 373 nucleotides for one
transcription product that inhibits protein synthesis and 42 nucleotides for the other. UV inactivation of the larger transcription product
left the virus with 60–65% of its total capacity to shut off protein
synthesis; it required >20,000 ergs/mm² to eliminate this second
smaller protein synthesis inhibitory product, a UV dose equivalent
to that required for inactivating the leader RNA and VSV inhibition
of cellular RNA synthesis (McGowan and Wagner, 1981; Grinnell and
Wagner, 1983). Dunigan and Lucas-Lenard (1983) also tested the R1
revertant of the T1026 mutant of Stanners *et al.* (1977) and found that
its lesser capacity to inhibit cellular protein synthesis was inactivated
by low-dose UV irradiation, suggesting that the 42-nucleotide small
transcript inhibitor of cellular protein synthesis is not functional in
the VSV R1 revertant. Cell-free translation by extracts of cells infected with UV-irradiated *wt* VSV also revealed two VSV genome
targets for protein synthesis inhibition similar to that in the *in vivo*
system. These authors postulate that the larger (375 nucleotide) VSV
transcription product that inhibits protein synthesis could be the N
protein mRNA and the smaller protein synthesis inhibitor could be
the plus-strand leader RNA of *wt* VSV, possibly affecting protein
synthesis secondary to inhibition of cellular RNA synthesis (Mc-
Gowan *et al.*, 1982). These intriguing data by Dunigan and Lucas-
Lenard (1983) point the way to testing these hypotheses, principally
by assaying the protein synthesis inhibitory capacity of VSV leader

RNA and N gene mRNA in tightly controlled cell-free translation systems.

4.6. Competition between Cell and Viral Messengers?

A popular hypothesis for the mechanism by which VSV and other viruses shut off host cell protein synthesis has been that viral mRNAs might have a selective advantage over cell mRNAs for use (or abuse) of the translational machinery of the host cell. An overall decrease in initiation could explain a reduction in total protein synthesis during VSV infection. However, it does not successfully explain a switch from the synthesis of host cell specific proteins to viral specific proteins during the lytic cycle. Exposure of VSV-infected HeLa cells to hypertonic media resulted in a selective inhibition of host cell protein synthesis, indicating that the VSV mRNA could initiate protein synthesis more readily under conditions where the rate of initiation by the cellular protein synthetic machinery is reduced (Nuss *et al.*, 1975). This experiment seemed to suggest that viral mRNA has a greater intrinsic affinity for ribosomes than does cellular mRNA. Thus, a virus-induced decrease in total protein synthesis initiation following infection would result in a decrease in total protein synthesis, which, in turn, would allow the selective translation of the supposedly more favored viral mRNA by the host cell machinery. This model is attractive because it relates both inhibitory phenomena to: (1) a decrease in total protein synthesis, and (2) the selective synthesis of viral proteins.

Lodish and Porter (1980) have suggested an alternative hypothesis that inhibition of host cell protein synthesis by VSV is due to the successful competition by large amounts of VSV transcripts for a limited number of ribosomes. They concluded from *in vivo* studies that viral and cellular mRNA are about equivalent in their efficiencies of translational initiation but viral transcripts are simply in large excess and, for this reason, can successfully out-compete the cellular transcripts for available ribosomes. In a follow-up publication, Lodish and Porter (1981) described a correlation between the concentration of intracellular VSV mRNA and the extent to which cellular protein synthesis is inhibited by VSV wild-type and various mutants. Although this mechanism can successfully explain the switch from host to viral protein synthesis, it does not explain the overall reduction in total protein synthesis. In addition, a recent study by Rosen *et al.*

(1982) provided evidence that 97% of the functional VSV mRNA at 4.5 hr postinfection is sequestered in ribonucleoprotein complexes and is not able to compete with cell mRNA for ribosomes. These findings, as well as those to be presented below, are inconsistent with the model proposed by Lodish and Porter (1980, 1981).

There are a number of other experiments which fail to support the hypothesis that excess VSV mRNAs outcompete cellular messages for available ribosomes, thus inhibiting cellular protein synthesis: (1) Jaye *et al.* (1982) found that viral and cellular RNAs from infected cells are translated equally well by an *in vitro* reticulocyte extract, (2) despite a marked decline in synthesis of viral mRNA, VSV subjected to moderate doses of UV irradiation is as efficient as fully transcribable unirradiated VSV in shutting off cellular protein synthesis (Dunigan and Lucas-Lenard, 1983), and (3) the most telling argument against the Lodish–Porter hypothesis is presented by Schnitzlein *et al.* (1983) who reported quite conclusive studies that marked suppression of *wt* VSV mRNA synthesis by co-infecting DI particles has a negligible effect on the capacity of the *wt* VSV to shut off cellular protein synthesis. It seems likely, therefore, that competition between VSV and cell mRNAs does not completely, or even partially, explain VSV inhibition of cellular protein synthesis.

4.7. Viral Products as Putative Inhibitors of Cellular Protein Synthesis

From the foregoing analyses of data it seems safe to assume that a newly synthesized VSV product is the principal inhibitor of protein synthesis in infected cells. The fact that VSV transcription is required to inhibit cellular protein synthesis (McAllister and Wagner, 1976; Marvaldi *et al.*, 1977) limited the search to newly synthesized viral RNAs and proteins. Circumstantial evidence obtained from very early studies with interferon and other inhibitors appeared to rule out viral proteins as primary inhibitors of cellular protein synthesis (Yamazaki and Wagner, 1970; Wertz and Youngner, 1970; reviewed by Bablanian, 1975). Careful studies by UV inactivation of selected VSV transcripts appear to rule out the mRNAs for the M, G, and L proteins (Marvaldi *et al.*, 1977; Dunigan and Lucas-Lenard, 1983). This would seem to leave the VSV leader RNA, the N protein mRNA, and/or the NS protein mRNA as the principal candidates for inhibitors of cellular protein synthesis, or possibly one or all of these transcripts associated with template nucleocapsid. Since the likely target for protein synthesis inhibition is initiation of translation (Nuss *et al.*, 1975), it seems clear that the viral inhibitory component(s) can only be studied adequately in cell-free translation systems which permit reconstitution of mRNAs, ribosomes, and all the individual components required for initiating translation, as well as the putative inhibitors.

We are aware of only one published study (Thomas and Wagner, 1982) that focused its attention entirely on an attempt to identify a VSV gene product that interrupts initiation of *in vitro* translation, and this study is far from being completely satisfactory. In this research, we examined cell-free protein synthesis in a reticulocyte lysate to which was added polyadenylated RNA fractions from VSV-infected HeLa cells or VSV transcripts made *in vitro*. Free VSV mRNA by itself had no effect on reticulocyte lysate translation but both polyadenylated or nonpolyadenylated VSV mRNA that segregated with nucleocapsid templates did inhibit translation of endogenous mRNAs and exogenously added globin mRNA. This double-stranded VSV RNA inhibitor of translation exhibited the characteristics of other double-stranded RNA inhibitors in reticulocyte cell-free systems (Hunter *et al.*, 1975) and could be inactivated by melting and by micrococcal nuclease (Thomas and Wagner, 1982).

It is well known that synthetic or natural dsRNAs can serve as potent inhibitors of protein synthesis in various *in vitro* translation systems derived from mammalian cells (Ehrenfeld and Hunt, 1971; Kaempfer and Kaufman, 1973). A variety of other viral dsRNAs inhibit cell-free protein synthesis. The presence of dsRNA has also been detected by crosslinking experiments in HeLa cells infected with encephalomyocarditis virus or the *ts*G114(I) mutant of VSV (Nilsen *et al.*, 1981). Reovirus mRNA and polyadenylated vaccinia RNA transcribed *in vitro* have both been shown to contain dsRNAs which inhibited protein synthesis in cell-free systems (Baglioni *et al.*, 1978; McDowell *et al.*, 1972). Studies such as these with viruses other than rhabdoviruses are discussed in other chapters of this volume. The question that must be kept in mind is whether the *in vitro* inhibition of translation initiation by VSV dsRNA or any other viral dsRNA truly represents the sequence of events that takes place in the virus-infected cell.

4.8. Cellular Target for Inhibition of Protein Synthesis

Most of the circumstantial evidence indicates that inhibition of protein synthesis in VSV-infected cells takes place at the level of initiation of translation (Nuss and Koch, 1976b; Stanners *et al.*, 1977; Davis and Wertz, 1980; Centrella and Lucas-Lenard, 1982). Other investigators studying inhibition of protein synthesis by quite different positive-strand RNA viruses have used cell fractionation and *in vitro* reconstitution techniques with some success to pinpoint the target for inhibition of protein synthesis; in general, these studies reveal alterations in initiation factor(s) associated with ribosomal proteins (Kaufman *et al.*, 1976; Rose *et al.*, 1978; Helentjaris *et al.*, 1979; Steeg *et al.*, 1981; Hansen and Ehrenfeld, 1981). Thomas and Wagner (1983) applied similar techniques of fractionation and reconstitution of the cell protein-synthesizing machinery to compare mock-infected and VSV-infected mouse L cells. Their results are similar to those found for other virus–host cell systems and are as follows: (1) postmito-chondrial lysates of VSV-infected L cells show a reduction in their ability to synthesize proteins compared to mock-infected control cells, (2) this inhibition occurs at the level of initiation of translation, (3) reconstitution studies reveal that the factor(s) associated with im-paired protein synthesis appear(s) in the fraction released from in-fected cell ribosomes washed with 0.5 M KCl, (4) the responsible initiation factor(s) is present in the 0–40% ammonium sulfate precip-itate that contains predominantly eIF-3 and eIF-4B. Further studies indicated that the effect upon protein synthesis due to this altered factor(s) can be reversed by purified initiation factors eIF-3 and eIF-4B from mock-infected cells (Thomas, 1983).

5. VSV INHIBITION OF CELLULAR NUCLEIC ACID SYNTHESIS

5.1. Host Cell Responses

5.1.1. Defining the Problem and Historical Perspective

Most of the early studies of VSV effects on cellular macromo-lecular synthesis were done by measuring cell protein synthesis. We now focus our attention on the effect of VSV on cellular nucleic acids, a field which has gained momentum in recent years. There is now reasonable evidence to indicate that VSV inhibits protein and nucleic acid synthesis by two different mechanisms. Although the problems are related, we shall also examine separately (1) the viral functions required to inhibit cellular nucleic acid synthesis, and (2) the cellular targets affected during VSV inhibition of cellular nucleic acid synthesis. As indicated below, it is likely that the same VSV function inhibits RNA and DNA synthesis in the infected host cell but, for sake of clarity, these two targets (RNA and DNA) will be discussed separately.

In experiments originally designed to study the effect of VSV on interferon synthesis, Wagner and Huang (1966) found that cellular RNA synthesis was inhibited in Krebs-2 tumor cells by kinetics that mimicked the inhibitory effect of actinomycin D but not as efficiently. However, in these Krebs-2 cells, RNA synthesis was inhibited 80–85% by 3 hr postinfection and the effect was not much greater at VSV multiplicities of 50 or 0.5 PFU/cell. In earlier studies, Huang and Wagner (1965) reported that the Indiana serotype of VSV extensively UV irradiated still shuts off cellular RNA synthesis and concluded, in retrospect erroneously, that input viral protein or lipid was responsible for inhibition of nucleic acid synthesis in the infected cell. In even earlier studies, Cantell et al. (1962) described a "cytotoxic" factor in VSV based on a cytopathic effect noted in cells infected with virus which had been UV irradiated to reduce infectivity by 99%. These early studies were, of course, done without knowledge that VSV contained a transcriptase and, as even more recently described, a region of the RNA template that provides a very small UV target. Details of later experiments are presented below.

Similar experiments were performed by Yaoi et al. (1970) and Yaoi and Amano (1970) who showed that the New Jersey serotype of VSV also inhibited RNA and DNA synthesis in chick embryo fibroblasts. RNA synthesis was rapidly inhibited by VSV-New Jersey even after UV irradiation but not when the virus was heat inactivated. However, much higher multiplicities of VSV-New Jersey were required to inhibit RNA synthesis in chick embryo cells (Yaoi et al., 1970) than was required for VSV-Indiana to inhibit RNA synthesis in Krebs-2 cells (Wagner and Huang, 1966). Yamazaki and Wagner (1970) likewise reported the requirement for high multiplicities (MOI $\simeq$100) of VSV-Indiana to shut off RNA synthesis in primary rabbit kidney cells. Wertz and Youngner (1970) also reported that higher multiplicities of VSV-Indiana ($\sim$100 PFU/cell) caused only moderate inhibition of RNA synthesis in chick embryo fibroblasts, whereas a similar multiplicity of wild-type VSV resulted in rapid inhibition of RNA synthesis in L cells and their S_2 mutant was even more effective.

The importance of cell type in susceptibility to VSV inhibition of cellular RNA synthesis was also demonstrated much later by Robertson and Wagner (1981) who found a rank order in response to VSV-Indiana infection of pre-B lymphocyte 18–81 > myeloma B cells MPC-11 > HeLa cells > L cells. The importance of serotype in VSV effectiveness in inhibiting cellular RNA synthesis was demonstrated most recently by Grinnell and Wagner (1983) who reported that VSV-Indiana more rapidly and more effectively inhibited RNA synthesis in MPC-11 myeloma cell than did VSV-New Jersey. In this mouse myeloma cell type, the multiplicity of infection was found to be immaterial because 1 PFU/cell was as effective as 50 PFU/cell of VSV-Indiana (Weck and Wagner, 1978) and 10 PFU/cell was as effective as 50 PFU/cell of VSV-New Jersey (Grinnell and Wagner, 1983).

5.1.2. Relationship of Various Cellular Responses to VSV Infection

In the previous section, we dealt with the question whether VSV inhibition of cellular protein synthesis was a primary effect of the virus or was secondary to its effect on cellular RNA synthesis. We came to the conclusion, based on kinetics of the reactions and UV sensitivity, that inhibition of cellular protein synthesis was largely, if not entirely, due to a primary effect of the virus and not secondary to inhibition of cellular RNA synthesis.

We must now address the question of whether VSV inhibitions of cellular RNA and DNA synthesis are related effects and whether they are independent of viral shut-off of protein synthesis. The available evidence is not abundant, but two sets of experiments can be cited to suggest that VSV inhibitions of protein and nucleic acid synthesis are unrelated phenomena but that inhibitions of cellular RNA and DNA synthesis are probably caused by the same viral function. Again, we rely on different kinetics and extent of inhibition to indicate that inhibition of protein synthesis is different from inhibition of nucleic acid synthesis in the same cell. Figure 1, taken from a paper by McGowan and Wagner (1981), shows the lesser and slower shut-off of protein synthesis than that of RNA or DNA synthesis in MPC-11 mouse plasmacytoma cells. In sharp contrast, the same figure reveals that the rate and extent of inhibition of RNA and DNA synthesis in this cell type are almost superimposable. The second bit of evidence relates to UV inactivation of these viral functions, which will be described later in greater detail. McGowan and Wagner (1981) found that equivalent doses of UV irradiation were required to ablate the RNA- and DNA-synthesis-inhibitory activities of VSV, compared to much lower UV doses required to inhibit the effect of VSV on cellular protein synthesis (Marvaldi *et al.*, 1977; Thomas and Carroll, unpublished data cited above).

This statement still applies despite the recent elegant study of Dunigan and Lucas-Lenard (1983) which provides evidence for two sites of UV inactivation of protein synthesis inhibition, one of which may be identical to the leader sequence presumably responsible for inhibiting cellular RNA synthesis (Weck *et al.*, 1979; McGowan *et al.*, 1982). In any case, the kinetic data and UV inactivation data seem to implicate the same or closely-related viral functions for inhibiting both cellular RNA and DNA synthesis and not inhibition of cellular protein synthesis, at least not entirely.

5.2. Viral Properties as Candidates for Inhibiting Cellular Nucleic Acid Synthesis

As discussed in the previous section on VSV inhibition of cellular protein synthesis, inhibition of cellular nucleic acid synthesis could be caused by "toxic effects" of input virion components or by newly synthesized viral products. As stated above, there appears to be a multiplicity-dependent effect of VSV for inhibition of cellular RNA synthesis in different cells (Wertz and Youngner, 1970). Marked and rapid inhibition of RNA synthesis in Krebs-2 cells infected with non-replicating DI particles (Huang *et al.*, 1966) or UV-inactivated standard (B) VSV (Huang and Wagner, 1965) led to the hypothesis that structural components of input (parental) VS virions were responsible for inhibition of cellular RNA synthesis. An alternative explanation, in retrospect, is that our DI particles were heavily contaminated with infectious B virions, the transcription function of which is not compromised by DI particles (Schnitzlein *et al.*, 1983) and, as will be discussed presently, UV irradiation does not readily inactivate the capacity of standard B virions to inhibit cellular RNA synthesis (Weck *et al.*, 1979; McGowan and Wagner, 1981).

Isolated VSV glycoprotein, purified free of other virion components, was found to cause an early inhibition of nucleoside incorporation into cellular DNA and RNA (McSharry and Choppin, 1978). This inhibition of nucleic acid synthesis was not due to an effect of glycoprotein on transport or phosphorylation of DNA or RNA precursors in BHK-21 cells. The addition of VSV antiserum to whole virus or isolated glycoprotein reversed the inhibitory effect on cellular nucleic acid synthesis. Although this effect of VSV G protein can play some role in inhibiting cellular nucleic acid synthesis contributing to late cytopathology, the amount of purified VSV glycoprotein required to produce this effect was equivalent to a VSV multiplicity of 5000–10,000 PFU/cell. It is still possible that dual effects of input virion components coupled with viral products newly synthesized in infected cells could both contribute to inhibition of cellular RNA synthesis. Genty (1975) proposed such a hypothesis, similar to that suggested for VSV inhibition of cellular protein synthesis (Wertz and Youngner, 1972) by which two possible mechanisms could be responsible for the inhibition of cellular RNA synthesis, the first being multiplicity dependent and the second being due to progressive synthesis of viral RNA and/or protein (Marcus et al., 1977).

We summarize here evidence supporting the latter of these two hypotheses, based on genetic studies with temperature-sensitive mutants and DI particles of VSV and with UV inactivation of the VSV wild-type genome.

5.3. Temperature-Sensitive Mutants and DI Particles Restricted in Transcription

The purpose of these experiments was to determine whether primary transcription from the 3' end of the VSV genome was essential to express its capacity to inhibit cellular RNA synthesis. Quite obviously, such studies could not distinguish between viral RNAs and proteins as the potential inhibitors. The use of protein synthesis inhibitors, such as cycloheximide, puromycin, and amino acid analogues, always resulted in inhibition of cellular RNA synthesis as well (Weck and Wagner, unpublished data), thus, precluding this approach to the problem of identifying the viral inhibitor.

Weck and Wagner (1979a) analyzed RNA metabolism in MPC-11 cells infected with various temperature-sensitive (ts) mutants and 5'-defective-interfering particles which cannot synthesize mRNA. A group I mutant, tsG114, restricted in transcriptional activities (Hunt et al., 1976), failed to shut-off host cell RNA metabolism in MPC-11 cells incubated at the restrictive temperature of 39°C for 4 hr. At the permissive temperature (31°C), all mutants (including tsG114) were as effective as the wild-type virus in the shut-off of RNA synthesis. Because tsG114(I) did not inhibit cell RNA metabolism at the nonpermissive temperature, it was used to test for a virion structural component which could be responsible for the inhibition of host cell RNA synthesis. At the highest MOI (200 PFU/ml), *ts*G114(I) did not inhibit RNA metabolism in cells incubated at the nonpermissive temperature of 39°C. These results suggested that primary transcription of the viral genome is essential to compromise host cell RNA synthesis. Additional proof for the requirement of a critical transcriptional event was sought by temperature shift-up experiments (Weck and Wagner, 1979a). If MPC-11 cells were infected with *ts*G114(I) and kept at 31°C for 30 min or less and then shifted to 39°C, there was only a 20–25% reduction in cellular RNA synthesis which did not change during the ensuing 4 hr of incubation. However, extension of the initial incubation period with *ts*G114(I) at permissive temperature (31°C for 60 min) resulted in a rapid and profound decline in the ability of the cells to synthesize RNA following the shift up to 39°C. These experiments reveal that the cellular RNA inhibitory factor of VSV is synthesized maximally during the first hour of permissive infection.

Wu and Lucas-Lenard (1980) did similar experiments using VSV *ts* mutants from all five complementation groups. In agreement with the results of Weck and Wagner (1979a), the only mutants which failed to inhibit cellular RNA accumulation at the nonpermissive temperature were *ts*G114(I) and *ts*G22(II). Although certain group II mutants are presumed to be restricted in function of the NS protein, its presumed multiple functions in transcription and replication make interpretation of the results obtained with the *ts*G22(II) mutant somewhat speculative. Nevertheless, the results of these experiments strongly suggest that viral RNA transcription is necessary for VSV expression of cellular RNA synthesis inhibition.

Since 5'-DI particles fail to replicate without a helper and transcribe only a 46-nucleotide leader RNA (Emerson *et al.*, 1977), they provide an ideal means of examining the effects of nonreplicating truncated particles on cellular RNA metabolism. Weck and Wagner (1979a) infected MPC-11 cells with purified DI particles derived from the 5' end of the VSV genome and found no significant reduction in host cell RNA synthesis even at a MOI equivalent to 10,000 particles per MPC-11 cell. Thus, even if a very short leader sequence (Emerson *et al.*, 1977; Schubert *et al.*, 1978) is transcribed by the 5'-DI particles, it apparently plays no role in the inhibition of cellular RNA synthesis (Weck and Wagner, 1979a,b). In contrast, DI particles derived from the 3' end of genome are capable of transcribing plus-strand leader and functional mRNAs for N, NS, M, and G proteins and can inhibit cellular protein synthesis (Marcus *et al.*, 1977) and cellular DNA synthesis (McGowan and Wagner, 1981).

These experiments strongly indicate that VSV transcript(s) can inhibit cellular nucleic acid synthesis directly or by means of their translated products.

5.4. Use of UV Irradiation to Identify VSV Genetic Information Responsible for Shutting Off Cellular RNA Synthesis

Assignment of the gene order for VSV (3'-leader,N,NS,M,G,L-5') and determination of the sequential nature of transcription starting from the 3' end of the genome was made possible by the use of UV irradiation (Ball and White, 1976; Abraham and Banerjee, 1976*a,b*; Testa *et al.*, 1980). Long before this information was available, other investigators had used UV irradiation to study VSV effects on cell toxicity (Cantell *et al.*, 1962; Wagner *et al.*, 1963). Huang and Wagner (1965) found that a UV dose of 1260 ergs/mm^2 was sufficient to reduce infectivity of a viral suspension (2×10^9 PFU/ml) by 99.8%; they then irradiated VSV with a UV dose of 25,500 ergs/mm^2 and observed no loss in VSV cytotoxicity for Krebs-2 ascites cells, confirming and extending previous work by others (Cantell *et al.*, 1962; Wagner *et al.*, 1963). UV-irradiated virus at multiplicities of 50, 5, and 0.5 retained its capacity to inhibit cellular RNA synthesis (Huang and Wagner, 1965). Since the molecular mechanisms of VSV transcription were then unknown and because relatively high UV doses failed to reverse the shut-off activity of the virus, it was logically assumed that perhaps VSV structural proteins were responsible for the inhibition of cellular RNA synthesis.

Similar experiments were done by Yaoi *et al.* (1970) who irradiated VSV-New Jersey for 90 sec, which reduced infectivity levels to 10^{-5}–10^{-6}; this virus was used at MOIs of 15–500 to compare the effects on cellular RNA synthesis of UV-irradiated virus with unirradiated virus. This work confirmed previous findings (Huang and Wagner, 1965; Wagner and Huang, 1966) that UV-irradiated and unirradiated VSV exhibited an equivalent shut-off activity on chick embryo cellular RNA synthesis. Yaoi *et al.* (1970) did not observe the same inhibitory effects on the synthesis of cellular macromolecules if VSV was heated at 56°C for 20 min. Later studies showed that VSV transcription was necessary for shut-off to occur, thus, it is not surprising that heating the virus would destroy the viral transcriptase and its capacity to shut off cellular RNA synthesis (Weck and Wagner, 1978, 1979*a*). The conclusion reached from these studies (Yaoi and Amano, 1970; Yaoi *et al.*, 1970) and those done by others (Wagner and Huang, 1966; Huang and Wagner, 1965) was that a viral protein or lipid is responsible for the inhibition of RNA synthesis. Genty (1975) also noted a multiplicity-dependent UV-insensitive (10,000 ergs/mm^2) inhibition of cell RNA synthesis that she assumed to be due to a modification of chick embryo plasma membrane uptake of uridine.

The genetic evidence that transcription of VSV is required to shut off cellular RNA synthesis led two laboratories to reinvestigate the effects of UV-irradiated VS virus on cellular RNA synthesis (Weck *et al.*, 1979; Wu and Lucas-Lenard, 1980). Weck *et al.* (1979) compared the UV doses for 37% (1/e) survival levels for VSV infectivity, *in vitro* transcriptase activity, viral RNA synthesis in infected MPC-11 cells, and shut-off of host RNA synthesis. Heavily UV-irradiated virus (72,000 ergs/mm^2) retained 37% of its capacity to shut off cellular RNA synthesis and could transcribe a very limited portion of the viral genome; the transcription products were low molecular weight, nonadenylated RNA molecules. Based on the data obtained, Weck *et al.* (1979) concluded that perhaps the inhibitor of cell RNA synthesis could be the plus-strand leader RNA transcribed from the 3′ end of the *wt* genome. Others have suggested a similar mechanism for VSV inhibition of cell protein synthesis, in which at least the leader sequence, the N gene, and possibly the NS gene must be transcribed (Marvaldi *et al.*, 1978; Dunigan and Lucas-Lenard, 1983).

Wu and Lucas-Lenard (1980) did similar studies on UV irradiation of VSV to determine the target size of the VSV genome segment required for inhibition of RNA synthesis in L cells; the approximate target size in their studies appeared to be the genome length of the VSV N gene. The difference obtained in target size of the VSV genome for shut-off activity may be dependent on the type of cell or other experimental conditions used by different investigators (Robertson and Wagner, 1981; Wu and Lucas-Lenard, 1980).

Subsequent studies on UV doses required to inactivate the capacity of VSV to inhibit RNA synthesis in MPC-11 cells revealed 37% (1/e) survival rates of ~52,000 ergs/mm^2 for VSV-Indiana (McGowan and Wagner, 1981) and ~12,000 ergs/mm^2 for both VSV-Indiana and VSV-New Jersey (Grinnell and Wagner, 1983). The lower inactivating doses of UV irradiation reported by Grinnell and Wagner (1983) were obtained using exactly the same UV-light source and the same techniques as reported by Weck *et al.* (1979); these dose differences can only be attributed to the fact that Grinnell and Wagner (1983) used highly purified, serum-free preparations of VSV, essentially free of UV-absorbing nonviral proteins and other impurities. The figure of ~12,000 ergs/mm^2 of UV irradiation required to reduce the MPC-11 cell RNA synthesis activity by 37% appears to be the most reliable one. Moreover, Grinnell and Wagner (1983) showed a direct correlation between the UV dose required for reduction of VSV-Indiana and VSV-New Jersey leader RNA synthesis *in vivo* and *in vitro* and the capacity of these two viruses to shut off RNA synthesis in MPC-11 cells.

These data clearly indicate the potential relationship between synthesis of VSV leader RNA and inhibition of cellular RNA synthesis, at least in MPC-11 myeloma cells. Using a value of 104 ergs/mm^2 for the 37% (1/e) survival dose for VSV infectivity and a VSV genome size of 12,000 nucleotides, Grinnell and Wagner (1983) calculated the VSV genome UV-target size for 37% survival of cellular RNA synthesis inhibition as ~85 nucleotides, compared with the UV-target size for synthesis of the VSV leader RNA as 150 nucleotides, rather than the actual size of 47 nucleotides for the leader RNA. Such values are probably well within the range of error for such methods but are close enough to relate the two phenomena of synthesis of leader RNA and shut-off of cellular RNA synthesis by VSV. In this regard, it is interesting to compare the recent data of Dunigan and Lucas-Lenard (1983) who calculated UV-target sizes of 42 nucleotides and 373 nucleotides for the two VSV genome segments required to transcribe products that inhibit cellular protein synthesis; the smaller genome segment (42 nucleotides) is presumed to code for the leader RNA sequence and the larger (373 nucleotides) for the N protein mRNA.

5.5. Cellular Targets for VSV Inhibition of RNA Synthesis

Among the potential targets for VSV inhibition of cellular RNA synthesis are transport of nucleoside triphosphates across the cell membrane, enzymatic conversion to nucleotides, initiation/chain elongation/termination of nuclear chromatin transcription, polyadenylation, processing, transport to the cytoplasm of completed RNA transcripts, and stability. Quite obviously, RNA synthesis inhibition could be at the level of ribosomal, messenger, and/or transfer RNA catalyzed by polymerases I, II, and III, respectively.

5.5.1. VSV Effects on Uptake of Nucleoside Triphosphates

Wild-type VSV and *ts* mutants from complementation groups I (*ts*05) and III (*ts*023) were used by Genty (1975) to examine the level of uridine uptake by infected compared to uninfected chick embryo fibroblasts. In addition, Genty (1975) determined if the modification of permeability was related to MOIs ranging from 1–500. At an MOI of 1, the half-life of uridine uptake into the acid-soluble fractions was considerably greater in cells infected with *ts*023(III) (10 hr) compared to *ts*05(I) (4.5 hr) or the wild-type virus (3 hr). Genty (1975) concluded that the infection of chick embryo cells by VSV induces an alteration in the cell membranes which leads to the diminution in the specific radioactivity of the acid-soluble pool and, therefore, of the specific radioactivity of the RNA newly synthesized in the infected cells. This effect may be unique to chick embryo cells because Genty (1975) did not observe a rapid decline in uridine uptake in other cell systems (L cells or HeLa cells) even at high MOIs.

Weck and Wagner (1979*a*) used the mutant *ts*G114(I) because it synthesizes no viral mRNA or protein in L cells or MPC-11 cells infected and incubated at the nonpermissive temperature (39°C). The inability of *ts*G114(I) to inhibit cellular RNA synthesis at 39°C but not 31°C suggests that input virion proteins, at the MOIs tested (1, 20, and 200), do not significantly alter the permeability of MPC-11 cell membranes for nucleoside triphosphates sufficient to cause a reduction in RNA synthesis. MPC-11 cells also infected with nonreplicating DI particles showed no significant inhibition (10–15%) of nucleic acid synthesis and presumably of nucleoside uptake even when an extremely high multiplicity (10,000) of DI particles was used (Weck and Wagner, 1979*a*). Weck and Wagner (1978) also examined the acid-soluble and insoluble pools of [^{3}H]uridine in VSV-infected MPC-11 cells and found that the transport of [^{3}H]uridine into the VSV-infected cells was unaffected. Even though uridine transport is not altered in infected mouse myeloma cells, it still remains possible that VSV infection can distort the concentration of monovalent ions inside the cell, as suggested by Carrasco (1977).

Wu and Lucas-Lenard (1980) confirmed the findings published by Weck and Wagner (1979*a*); they used *ts* mutants G11(I), 052(II), G31(III), G33(III), G41(IV), W10(IV), 045(V), and *wt* VSV to infect L cells and determine their effects at the permissive (30°C) and nonpermissive temperature (40°C) on uridine transport. The results suggest that the decreased incorporation of [^{3}H]uridine into acid-insoluble material was not due to a decrease in the content of [^{3}H]uridine in the acid-soluble pool. In fact, they found that the amount of [^{3}H]uridine inside the cell increased at the nonpermissive temperature of 40°C.

5.5.2. RNA Processing, Transport, and Stability of RNA in Cells Infected with Vesicular Stomatitis Virus

Although transcription is the primary event in the production of cellular RNA, it has become evident that posttranscriptional events such as processing, polyadenylation, transport across the nuclear membrane, and stability of RNA also play an important role in synthesis of functional RNA (Darnell, 1976). Measurements of RNA synthesis represent the net rate of RNA transcription and degradation. Therefore, the rate of ^{3}H-RNA degradation in prelabeled infected and uninfected cells has been examined by several investigators (Weck and Wagner, 1978; Nishioka and Silverstein, 1978). Weck and Wagner (1979b) used the technique of Wertz (1975) to deplete endogenous levels of unlabeled UTP before pulsing with [^{3}H]UTP. As expected, MPC-11 cells previously infected with VSV incorporated less [^{3}H]uridine than did uninfected cells. Degradation of RNA in the uninfected cells was 52% compared to 74% in the VSV-infected cells. They considered the slightly higher amount of degradation observed in infected cells insufficient to account for the markedly reduced rate of RNA synthesis that was observed. Nishioka and Silverstein (1978) examined the size distribution of mRNA in VSV-infected Friend erythroleukemia (FL) cells to examine the possibility that the decrease in cell-specific RNA synthesis was a result of the degradation of cellular mRNA. Cytoplasmic RNA from infected cells analyzed on a 15–30% sucrose gradient indicated that, unlike infection of FL cells with herpes simplex virus type 1, VSV infection did not result in degradation of host mRNA. A more detailed analysis of globin mRNA measured by hybridization to a cDNA probe for globin sequences also showed no degradation of this cellular mRNA in VSV-infected cells (Nishioka and Silverstein, 1978).

The only report of possible degradation of cellular RNA by VSV was made by Kolakofsky and Altman (1978) who disrupted purified VSV with detergent and found endoribonucleolytic activity associated with the virus nucleocapsid when assayed by using tRNA^{tyr} as a substrate. Further characterization of this endoribonuclease activity revealed a cleavage specificity and ionic requirements similar to those for cell RNase NU.

The effect of VSV on the transport of cellular RNA from the nucleus to the cytoplasm was also examined by Weck and Wagner (1978). The RNA of infected and uninfected MPC-11 cells was labeled with [^{3}H]uridine at 2 hr postinfection and, at various intervals thereafter, the amount of acid-insoluble radioactivity in the cytoplasm was determined. Although there was a 2- to 3-fold decrease in the total amount of RNA in the cytoplasm of infected cells, the rate of transport of polyadenylated RNA from the nucleus to the cytoplasm of infected cells was no different from that in uninfected cells. Using polyacrylamide–agarose gels and linear sucrose–SDS gradients, Weck and Wagner (1978) also found no significant difference in the polydisperse nature of nuclear RNA from VSV-infected and uninfected cells.

Wu and Lucas-Lenard (1980) examined RNA precursors in L-929 cells infected with wild-type and *ts* mutants of VSV. In agreement with previous reports, they found a decrease in the synthesis of cellular RNA. A close examination of the overall distribution of cellular rRNAs by Wu and Lucas-Lenard (1980) revealed a decreased amount of synthesis but no change in the relative amount of each species of rRNA.

5.5.3. Endogeneous Levels of Cellular RNA Polymerase after VSV Infection

Weck and Wagner (1978) compared transcription of isolated nuclei from VSV-infected and uninfected MPC-11 cells to obtain a more direct measurement of nuclear polymerase activity by the method of Smith and Huang (1976). At 2 hr postinfection, they found approximately a 50% decline in the rate of RNA synthesis *in vitro* by nuclei isolated from infected cells. The toxin α-amanatin was used at a concentration of 1 μg/ml to distinguish the RNA polymerase II activity from that of the more resistant RNA polymerases I and III. During the first hour after infection, there was a rapid loss in the activities of all three RNA polymerases. Subsequently, the level of RNA polymerase II activity continued to decline until 4 hr postinfection while the level of combined RNA polymerase I and III activity remained constant.

In later studies, Weck and Wagner (1979*b*) tested solubilized RNA polymerases and chromatin templates isolated from VSV-infected and uninfected cells in order to determine whether the cellular target of VSV inhibition of cellular RNA synthesis involved an alteration in the DNA template or the RNA polymerases. In order to test these two possibilities, nuclear polymerases were solubilized and their RNA-synthesizing activities were tested by reconstitution with calf thymus DNA and with uninfected or VSV-infected cell chromatin. No difference in polymerase activity was observed with the various templates, providing suggestive evidence that viral infection does not cause irreversible damage to the polymerase molecules themselves. One must keep in mind, however, that the activities observed in this system probably did not reflect the accurate initiation of cellular transcription of the DNA templates as evidenced by the higher RNA polymerase activity observed with calf thymus DNA as template.

Weck and Wagner (1979b) also used the method of Cox (1976) to determine the actual number of active RNA polymerase units in the nuclei of uninfected compared with VSV-infected cells at intervals after infection. Quantitation of the three RNA polymerases supported earlier findings (Weck and Wagner, 1978) of a drastic and rapid loss in the quantity of RNA polymerase II units engaged in chain elongation compared to RNA polymerases I and III which were inhibited later in infection (2–4 hr). These experiments were interpreted as evidence that VSV infection of MPC-11 myeloma cells reduces the number of active cellular RNA polymerases I, II, and III that are capable of initiating transcription, but VSV infection does not affect the function of those polymerases already actively engaged in RNA chain elongation (Weck and Wagner, 1979b). Solution of this question had to await the availability of more refined methods to initiate RNA synthesis by reconstituting polymerases and cofactors with DNA templates.

5.6. Inhibition of Transcription Initiation on DNA Templates in Cell-Free Systems

The availability of procedures to isolate mammalian DNA-dependent RNA polymerases and their cofactors in order to reconstitute them with DNA templates for *in vitro* initiation of transcription provided an essential avenue for investigating VSV inhibitors of RNA synthesis. McGowan *et al.* (1982) exploited the *in vitro* transcription system developed by Manley *et al.* (1980) to test the capacity of the products of VSV transcription to inhibit the transcription of specific messengers made on SV40 DNA and of plasmids containing the adenovirus late promoter (LP) and adenovirus-associated (VA) RNA genes (Manley *et al.*, 1980; Akusjarvi *et al.*, 1980; Handa *et al.*, 1981). Relatively low concentrations (6–25 ng/μl) of the plus-strand leader RNA made *in vitro* from the 3′ end of the wild-type VSV-Indiana genome were found to inhibit the initiation of transcription catalyzed by both RNA polymerase II (LP and SV40 genes) and RNA polymerase III (VA gene). Polyadenylated VSV mRNA caused only minor inhibitory effects and only at extremely high concentrations (850–1412 ng/μl). Compared with the wild-type plus-strand RNA leader, the leader RNA synthesized *in vitro* by 5′-defective-interfering particles of VSV showed only a limited capacity to inhibit eukaryotic RNA polymerase activity and only at concentrations more than 30 times greater than that of the wild-type leader.

This system provided the first opportunity to test whether the VSV *wt* leader is the viral product responsible for the inhibition of the initiation of DNA-dependent RNA synthesis. Later studies (McGowan and Wagner, unpublished data) indicate that, when DNA templates transcribed by RNA polymerase II and III *in vitro* are used simultaneously in the same reaction, polymerase II activity is inhibited more easily than RNA polymerase III transcripts, confirming previous *in vivo* data obtained by Weck and Wagner (1978, 1979*b*).

Consistent with previous *in vivo* results (Weck and Wagner, 1978) was the finding that the leader RNA made by 5′-DI particles did not significantly inhibit DNA-dependent RNA synthesis *in vitro*. Of considerable interest was the surprising finding that synthetic poly(AU) and poly(GC), used as presumptive negative controls, both inhibited *in vitro* transcription of LP, SV40, and VA genes, although at concentrations higher than that of the *wt* leader (McGowan *et al.*, 1982). This latter finding suggested that similar AU-rich and/or GC-rich regions of the *wt* leader might be involved in transcription inhibition.

McGowan *et al.* (1982) compared the known nucleotide sequence of plus-strand *wt* leader with that of the DI leader (Colonno and Banerjee, 1978*a*; Schubert *et al.*, 1978), in an attempt to identify sequences unique to the *wt* leader that could be involved in transcription inhibition. As shown in Fig. 2, the wild-type leader RNA of VSV-Indiana has at its 5′ end one region (nucleotides 1–17) nearly identical to that of the 5′ end of the DI leader; this region is conserved both at the 3′ and 5′ ends of all strains of VSV wild-type and DI particles (Schubert *et al.*, 1978) and may serve as a nucleation signal for en-

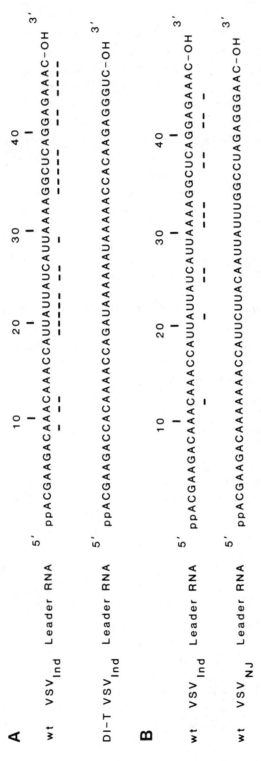

Fig. 2. Comparison of nucleotide sequences. (A) VSV-Indiana wild-type and DI-T VSV plus-strand leader RNAs. (B) Wild-type VSV-Indiana and VSV-New Jersey leader RNAs. The unique regions of the VSV-Indiana wild-type leader are highlighted by underlining. Data from VSV-Indiana wild-type leader RNA are obtained from the results of Colonno and Banerjee (1978a) and DI-T leader sequences are from the results of Schubert *et al.* (1978). The sequence data for the VSV-New Jersey (Hazelhurst) leader were kindly supplied by Jack Keene (personal communication).

capsidation with N protein (Blumberg *et al.*, 1981, 1983). Its close homology to the corresponding region of DI leader makes the first 17 nucleotides at the 5′ end of the *wt* leader an unlikely candidate for the nucleotide sequence involved in inhibition of DNA-dependent RNA synthesis.

A second region of the VSV-Indiana *wt* leader RNA (nucleotides 18–30) contains a unique sequence (AUUAUUAU), present also in the VSV-New Jersey *wt* leader (Colonno and Banerjee, 1978*b*) but not in DI leader; this nucleotide sequence struck us as being similar to the TATA or "Goldberg–Hogness box" which serves as a polII promoter (Baker and Ziff, 1981; Gruss *et al.*, 1981; Hu and Manley, 1981; Wasylyk *et al.*, 1980). Iverson and Rose (1981) noted similar AU-rich regions in the VSV genome RNA about 25 nucleotides upstream from the start of each VSV mRNA. Since the TATA box plays an important role in accurate initiation of RNA polymerase II transcription, it was tempting to speculate that this AU-rich region of *wt* leaders of VSV-Indiana and VSV-New Jersey may affect a PolII TATA recognition factor as their mechanism of transcription inhibition. More recent studies by Grinnell and Wagner (1983) show that the VSV-New Jersey leader, which has an extended AUUAUU region, is an even better inhibitor of Ad2-LP gene transcription than is the VSV-Indiana leader (Fig. 2).

McGowan *et al.* (1982) also speculated on a third region of the *wt* leader of both VSV-Indiana and VSV-New Jersey, which differs from the DI leader by unique nucleotides at positions 33–43 (Fig. 2); they noted a sequence AGGCUCAGGAG in the VSV-Indiana leader which resembled the "consensus" sequence which appears to play a role as a promoter for polymerase III in tRNA transcription (Breathnach *et al.*, 1978; Jelinek *et al.*, 1980). The degree of homology of the consensus sequence for tRNAs represents one of two sequence blocks conserved in nature as major eukaryotic promoter elements (Galli *et al.*, 1981). This and similar sequences appear at key controlling points for viral and cellular functions, not only for RNA polymerase III products such as tRNA and 7 S RNA, but also in some polymerase II transcripts (U_2, U_4, U_1) and repetitive spacer regions implicated for RNA polymerase I activity (Moss, 1983). The lack of homology in the DI leader at nucleotide positions 18–46 compared with that of the two wild-type leaders is striking and suggests a possible role for nucleotides within this region for inhibition of DNA-dependent RNA synthesis, particularly that catalyzed by PolIII.

Grinnell and Wagner (1983) have recently made an extensive study comparing the transcription-inhibiting activity of the Indiana and New Jersey serotypes of VSV and their respective leader sequences. VSV-Indiana was found to be somewhat more effective than VSV-New Jersey in shutting off RNA synthesis in MPC-11 cells and L cells. Moreover, VSV-Indiana synthesized about four times more leader RNA in infected cells than did VSV-New Jersey. However, the leader RNA of VSV-New Jersey was more effective in inhibiting transcription *in vivo* or *in vitro* than was the VSV-Indiana leader. Dose-response curves showed that polymerase III transcription of the adenovirus VA gene was more sensitive to inhibition by both New Jersey and Indiana leaders than was polymerase II transcription of the adenovirus LP gene. Most importantly, a direct correlation could be made between leader RNA synthesized in cells infected with UV-irradiated VSV and the capacity of this UV-irradiated VSV to shut off RNA synthesis in the same cells (Grinnell and Wagner, 1983).

Although the case for inhibition of DNA-dependent RNA synthesis by *wt* VSV leader RNA sequences is building, all that can really be said at this time is that *wt* VSV leader RNA contains nucleotide sequences potentially capable of interacting with promoters or with host cell protein cofactors that interact with nucleotide sequences essential for accurate transcription. Perhaps the most intriguing possibility is that VSV *wt* leader RNA can serve as a surrogate for other small RNA species found inside the cytoplasm and nucleus of cells (Lerner and Steitz, 1981; Lerner *et al.*, 1981; Zieve, 1981). Similar sequences do not appear to be present in leader RNAs transcribed from 5'-DI particles (Fig. 2), possibly explaining why DI particles do not possess the capacity to inhibit cellular RNA synthesis (Weck and Wagner, 1978).

It is of some interest that *wt* VSV transcribes leader RNA in molar amounts higher than that of any of the five mRNAs (Villareal *et al.*, 1976; Colonno and Banerjee, 1978a,b; Iverson and Rose, 1981). Kurilla *et al.* (1982) have provided further evidence that VSV leader RNA may function as a surrogate for other cellular small RNAs. Genomic RNA of VSV was labeled at its 3' end with pCp by T4 RNA ligase and then used as a hybridization probe to locate leader RNA in VSV-infected cells. In the cytoplasm, leader RNA accumulated gradually throughout the course of infection to about 200 molecules per cell at 6 hr postinfection. In the nucleus, however, there was a sharp and rapid increase in the concentration of VSV leader RNA to approximately 300 molecules per cell at 2 hr, which then decreased rapidly by 3 hr postinfection. Given the relative volume of the nucleus and cytoplasm, this represents a 30- to 50-fold concentration of VSV *wt* leader in the nucleus of infected cells. Using similar techniques, Grinnell and Wagner (1983) could detect 2900 copies/cell of VSV-Indiana leader RNA, but only 450 copies of the Indiana leader and 150 copies of the New Jersey leader could be detected in infected cells that were demonstrating 50% inhibition of RNA synthesis.

Kurilla and Keene (1983) have used anti-La specific antisera to precipitate the leader RNA transcript from VSV-infected BHK cells. The La antigen is known to bind RNA polymerase III precursors of 5 S and tRNA (Rinke and Steitz, 1982), Alu products transcribed *in vitro* by polymerase III (Shen and Maniatis, 1982), as well as adenovirus VA_I and VA_{II} RNA, and Epstein–Barr virus $EBER_1$ and $EBER_2$ RNAs (Lerner *et al.*, 1981; Rosa *et al.*, 1983). From their studies, Kurilla and Keene (1983) have concluded that VSV leader RNA appears to function in a manner similar to other La ribonucleoprotein complexes.

Of considerable importance are preliminary studies by McGowan and Wagner (unpublished data) that a Pst-1-excised cDNA copy of wild-type leader in pBR322 (obtained from Robert Lazzarini) inhibits transcription of adenovirus LP and VA genes in a Manley–Sharp reconstituted system just as well as does the *wt* leader RNA. Grinnell and Wagner (1984) have recently found that both the plus and minus strand of the VSV leader cDNA inhibit transcription by both PolII and PolIII. Moreover, a ribonuclease T_1 digest of the *wt* leader RNA, which preserves the middle UAAUAAUA region, retains its capacity to inhibit *in vitro* transcription of both LP and VA genes. The most telling preliminary experiments so far were conducted by Grinnell and Wagner (1984) who tested the transcription inhibitory activity of synthetic oligodeoxynucleotides homologous to nucleotides 18–30 (AUUAUUAU . . . region) and to nucleotides 33–44 (the GC-rich region) of the VSV-Indiana *wt* leader (kindly synthesized for us by Ernst-L. Winnacker of Munich). The ATTATTATCATTA oligodeoxynucleotide inhibited quite well transcription of *both* the adenovirus LP and VA genes, whereas, the AGGCTCAGGAGA oligodeoxynucleotide showed only minimal capacity, even at very high concentrations, to inhibit either the adenovirus LP or VA gene transcription (Grinnell and Wagner, 1984). Moreover, both strands of a Dde-1-restriction fragment of a VSV-Indiana cDNA clone, which contained all leader sequences except 12 at the 3'-purine-rich end, also inhibited transcription directed by both polymerases II and III. It now seems clear that only the middle AUUAUUAU(TATA-like) region of both VSV leaders are responsible for inhibition of DNA-dependent transcription. However, further studies indicate that flanking sequences, which provide secondary structure to the leader RNA or its cDNA, are more efficient inhibitors of transcription than the TATA sequence alone (Grinnell and Wagner, 1984).

The data presented in these reports only begin to build a case for VSV plus-strand leader RNA as the inhibitor of cellular RNA and perhaps DNA synthesis. Using the *in vitro* system and current oligonucleotide synthesis techniques, it should be possible to determine the key sequence(s) of VSV leader RNA responsible for the inhibition of cellular RNA synthesis. In addition, it is ostensibly feasible to fractionate the cytoplasmic extract from the HeLa cell system to determine what particular polymerase factor(s) or cofactors may be suppressed by interaction with VSV *wt* leader RNA. Very preliminary data (McGowan and Wagner, unpublished observations) indicate that one of four HeLa cell fractions separated on phosphocellulose columns (Segall *et al.*, 1980; kindly supplied by Andrew Fire and Phillip Sharp) could partially restore polymerase III activity that had been inactivated by VSV *wt* leader RNA. Moreover, excess amounts of partially purified polymerase II (also provided by Fire and Sharp) as well as excess DNA template were found to overcome partially the inhibition by VSV *wt* leader RNA of adenovirus LP transcription *in vitro* (McGowan and Wagner, unpublished data). The variability observed in subsequent experiments along these lines indicates the necessity for much greater refinement of techniques and the various components in the reaction mixture before conclusive experiments can be designed to identify the target of transcription inhibition by the VSV *wt* leader RNA.

5.7. Effect of VSV on Cellular and Viral DNA Synthesis

5.7.1. Parameters of Cellular DNA Synthesis Inhibition

In their studies on the effect of VSV (Indiana serotype) on interferon synthesis, Wagner and Huang (1966) noted that the incorporation of [^{3}H]thymidine into acid-precipitable material was drastically inhibited in VSV-infected compared to uninfected Krebs-2 ascites tumor cells. Yaoi and Amano (1970) infected chicken embryo cells with VSV-New Jersey at an MOI of 500 PFU/cell and observed an inhibition of DNA synthesis within 2 hr after infection. The same inhibitory effect was observed with UV-inactivated VSV but not by VSV heated at 56°C for 20 min. In much later studies, McGowan and Wagner (1981) examined the kinetics of VSV inhibition of cellular DNA synthesis at MOIs of 1, 10, and 100 in exponentially growing mouse myeloma (MPC-11) and mouse L cells. Inhibition of cellular DNA synthesis in unsynchronized MPC-11 cells was more rapid at each MOI tested than in VSV-infected L cells. For example, at 4 hr after VSV infection at an MOI of 10, DNA synthesis in MPC-11 cells was inhibited to a level of 15% that of control cells, whereas, L cell DNA synthesis decreased to a rate of not quite 35% that of uninfected cells.

Yaoi *et al.* (1970) postulated that VSV can only shut off the onset of DNA synthesis if cells are infected before they enter the S phase of the growth cycle and that cells infected in G1 did not enter the S phase; these authors had available at that time only the rather inefficient method of medium withdrawal and replacement to synchronize their chicken embryo cells and achieved only 55% cell synchrony. Yaoi *et al.* (1970) reported inhibition of the onset of DNA synthesis in these partially synchronized cells compared with the apparent inability of VSV to stop ongoing DNA synthesis. McGowan and Wagner (1981) reexamined the ability of VSV to inhibit DNA synthesis in synchronized cells just before or after the cells enter the S phase of the growth cycle; cells were synchronized by using a stringent protocol of a thymidine block for 12 hr, resuspension in fresh medium for 10 hr, followed by an additional hydroxyurea block for 12 hr before infection, resulting in >90% cell synchrony as monitored by autoradiographic analyses of [^{3}H]thymidine uptake. The kinetics of DNA synthesis inhibition in S-phase cultures of synchronized MPC-11 and L cells infected with VSV (MOI≃10) 1 hr before the release of the hydroxyurea block revealed a progressive decline in incorporation of [^{3}H]thymidine into acid-precipitable DNA of synchronized cells, reaching at 4 hr postinfection inhibitory levels of ~90% for MPC-11 cells and ~78% for L cells. Similar kinetics of inhibition of MPC-11 cell DNA synthesis was observed regardless of whether the cells were infected before or after release from the hydroxyurea block. Indeed, the patterns for the shut-off of DNA, RNA, and protein synthesis in synchronized MPC-11 cells was similar to that reported in unsynchronized cultures (see Fig. 1).

5.7.2. Viral Functions Required to Inhibit Cellular DNA Synthesis

Previous sections of this chapter discuss genetic and other evidence that VSV transcription is required to shut off cellular protein synthesis (McAllister and Wagner, 1976; Marvaldi *et al.*, 1977) and cellular RNA synthesis (Weck and Wagner, 1979*b*; Wu and Lucas-Lenard, 1980). Farmilo and Stanners (1972) reported the isolation of a VSV temperature-sensitive mutant, *ts*1026, which lacked the cytocidal properties of *wt* VSV at the nonpermissive temperature. The *wt* VSV was thought to inhibit initiation of DNA synthesis in hamster embryo fibroblasts, whereas, essentially normal DNA synthesis occurred in cells infected with the mutant despite autoradiographic evidence of viral RNA synthesis. McGowan and Wagner (1981) reported that *ts*G114(I) restricted in transcription did not significantly affect cellular DNA synthesis at 39°C but did so quite efficiently at 31°C in a manner similar to *ts*G114(I) inhibition of cellular RNA and protein synthesis. These data provide supportive evidence that VSV transcription is required to inhibit cellular DNA synthesis. It was not possible to test whether the inhibitory product is viral RNA or protein because all previous attempts to block protein synthesis by inhibitors (cycloheximide or puromycin) have also inhibited cellular nucleic acid synthesis (Weck and Wagner, 1979*a*; Weck and Wagner, unpublished data).

McGowan and Wagner (1981) also tested the DNA inhibitory action of two DI particles: VSV 5'-DI-011, which is restricted in transcription, and 3'-DI-HR-LT, which can undergo primary transcription, and the mRNA synthesized can be translated into functional N, NS, M, and G proteins but not L protein because ~6000 nucleotides are deleted from the 5' end of the genome (Johnson *et al.*, 1979). DI-011 did not inhibit cellular DNA synthesis to any significant extent, whereas, DI-HR-LT inhibited cellular DNA synthesis equally as well as *wt* VSV for the first 2 hr postinfection; however, between 2 hr and 4 hr, the observed inhibition by DI-HR-LT tapered off to ~57% compared with >90% inhibition by *wt* VSV. This latter observation is not surprising since DI-HR-LT is unable to synthesize new L protein which is required for secondary transcription and replication (Mellon and Emerson, 1978). The evidence obtained by examining the effects of DNA synthesis by the temperature-sensitive mutant and two DI particles strongly supports the hypothesis that the inhibition of cellular DNA synthesis requires at least primary transcription of VSV.

Early interesting experiments on the capacity of UV-irradiated VSV to shut off cellular DNA synthesis were reported by Yaoi *et al.* (1970) and Yaoi and Amano (1970), who found that VSV (MOI ≃500), UV irradiated to destroy all infectivity, still retains its original capacity to inhibit DNA synthesis in chick embryo cells. In sharp contrast, VSV heated to 56°C for 20 min lost all its capacity to inhibit cellular DNA synthesis. Not illogically and partly because of the high multiplicity used, these authors assumed that heat-labile components of the input virions were responsible for shutting off cellular DNA synthesis. Yaoi and Amano (1970) also examined the ability of UV-irradiated VSV to affect initiation of DNA synthesis in chick embryo cells and concluded that the onset of DNA synthesis was inhibited when cells were infected before entering the S phase but not when the cells were already in the S phase of their growth cycle. McGowan and Wagner (1981) reexamined the ability of VSV to inhibit DNA synthesis in MPC-11 and L cells infected with VSV and found no significant difference before or after synchronized cells entered the S phase.

In the same study, McGowan and Wagner (1981) determined the UV-target size for the inactivation of the VSV genome segment responsible for the inhibition of DNA synthesis by MPC-11 cells. The 37% survival rate (1/e) for the ability of VSV to inhibit cellular DNA synthesis was found to be 45,000 ergs/mm^2, and when the experiments of Weck *et al.* (1979) were repeated, McGowan and Wagner (1981) obtained a 37% activity of the UV-treated virus at 52,000 ergs/mm^2 to inhibit cellular RNA synthesis. These comparable target sizes for UV inactivation strongly suggest that the same cistron of the VSV genome codes for a single product that inhibits both RNA and DNA synthesis; as discussed above, the only logical VSV transcript that expresses these inhibitory effects is the plus-strand leader 47-nucleotide RNA made from the 3' end of the VSV genome.

5.7.3. Cellular Targets for VSV Inhibition of DNA Synthesis

The effect of VSV on cellular DNA synthesis could result from perturbation of cellular structures similar to those conceivably involved in inhibition of RNA and protein synthesis. The effect is unlikely to be at the limiting cell membrane because McGowan and Wagner (1981) found that VSV infection does not significantly alter uptake of thymidine into soluble intracellular pools of nucleosides.

McSharry and Choppin (1978) reported a similar lack of effect on thymidine uptake following exposure of cells to huge amounts of VSV glycoprotein. Yaoi *et al.* (1970) found that infection of chick embryo fibroblasts with UV-irradiated VSV did not affect the stability of preexisting DNA; this finding was confirmed by McGowan and Wagner (1981) who detected no significant shift of prelabeled ^{3}H-DNA into acid-soluble counts even 5 hr after VSV infection, at a time when DNA synthesis was reduced by more than 90%. The possibility that VSV infection might affect cellular enzymatic activity required for DNA synthesis was also examined by McGowan and Wagner (1981) who compared the thymidine kinase and DNA polymerase activity of VSV-infected and uninfected cells by established procedures (McGowan *et al.*, 1980; Allen *et al.*, 1977). The results of this study revealed no significant differences in either thymidine kinase or DNA polymerase activities at 4 hr after VSV infection. It should be kept in mind, however, that other key DNA-synthesizing enzymes could be involved and that the DNA polymerase assay may measure primarily DNA repair and chain elongation rather than initiation.

Since the effect of VSV on initiation of DNA synthesis cannot be measured directly, McGowan and Wagner (1981) were prompted to examine more closely the ability of VSV to inhibit DNA chain elongation in VSV-infected MPC-11 cells. Analyses in alkaline sucrose gradients of radioactive pulse-chased DNA from VSV-infected and uninfected cells indicated that, although [^{3}H]thymidine incorporation into DNA was inhibited by VSV infection, the process of DNA chain elongation can continue in infected cells.

We have not yet been able to identify the cellular target for VSV inhibition of DNA synthesis, but studies summarized here rule out a viral effect on nucleoside transport, DNA degradation, inactivation of DNA polymerase and thymidine kinase, or premature termination of already initiated DNA chains. By exclusion, we hypothesize that VSV inhibits DNA synthesis by blocking initiation of DNA replication, perhaps in a manner similar to its inhibition of RNA transcription (McGowan *et al.*, 1982). The ambiguities inherent in the *in vitro* DNA polymerase assay make it difficult to rule out conclusively cellular enzymes as targets for VSV inhibition of DNA replication. Further studies are necessary to test the hypothesis that a VSV product, such as the plus-strand leader RNA, inhibits initiation of both RNA transcription and DNA replication.

Currently, we are exploiting an adenovirus *in vitro* replication system (Challberg and Kelly, 1979) to elucidate the effects of VSV on eukaryotic DNA synthesis. This system was chosen because it depends, at least in part, on the host cellular DNA polymerases for replication. More importantly, it is the only system which offers an authentic representation of the *in vivo* initiation of DNA synthesis. This system may allow a clear and distinct dissection of how this negative-strand virus affects nucleic acid synthesis.

5.7.4.   Inhibition of DNA Virus Replication by VSV

5.7.4a.   Herpesviruses

Youngner *et al.* (1972) reported that VSV drastically inhibits the replication of pseudorabies (PSR) herpesvirus in RK-13 cells. In a later study, Dubovi and Youngner (1976*a*) found that the number of VSV particles capable of inhibiting the replication of PSR virus exceeds the number of plaque-forming units (PFU) by a factor of 32 to 64, a value close to the particle-to-PFU ratios reported by McCombs *et al.* (1966). This led to the conclusion that a single *wt* VSV particle is sufficient to inhibit PSR virus replication in a single cell. In contrast, 800 to 1000 UV-irradiated noninfectious VSV particles were required to inhibit PSR virus replication. Dubovi and Youngner (1976*a*) also used temperature-sensitive mutants from the five VSV genetic complementation groups and found that inhibition of PSR virus replication is correlated with their ability to synthesize VSV products. In another study, Dubovi and Youngner (1976*b*) reported that purified VSV 5'-DI-40 particles could also inhibit PSR virus replication when added at very high MOIs (340 and 3400), suggesting to them that a virion structural protein may inhibit PSR virus replication. Another possible explanation (made editorially) is that undetected contamination of DI particles with standard B virions, the transcription of which is not inhibited by DI particles (Schnitzlein *et al.*, 1983), could be responsible for inhibition of PSR virion replication.

Nishioka and Silverstein (1978) had shown that herpes simplex virus (HSV) type 1 causes rapid degradation of host cell mRNA, whereas, VSV does not. In a more recent report, Nishioka *et al.* (1983) took advantage of the difference in shut-off activities of the two viruses to examine what happens when cells are coinfected with HSV and VSV. Coinfection of cells with VSV prevents the expression of HSV-specified polypeptides, including the heat-labile HSV-associated protein responsible for HSV suppression of host cell protein synthesis. Using the HSV-specified thymidine kinase (TK) they were able to show that, if cells are infected simultaneously with VSV and HSV, very little HSV TK is made as determined by viral thymidine kinase assay. If cells are coinfected with VSV at 2 hr or 4 hr post-HSV infection, then the HSV thymidine kinase activity was 69% and 100%, respectively, of the control levels. Northern blot analysis showed that the bulk of HSV-1 TK mRNA is made by 4 hr postinfection, so it is not surprising that little inhibition by coinfecting VSV was observed at 4 hr postherpesvirus infection.

Similar experiments were done to test the ability of VSV to inhibit the transcription of the HSV TK mRNA. Nishioka et al. (1983) reported that simultaneous infection with VSV and HSV resulted in the loss of HSV-TK mRNA synthesis. Thus, the pattern of shut-off of host cellular RNA synthesis mimicked a typical VSV infection and not HSV. This was confirmed by looking at the stability of cellular mRNAs during VSV, HSV, or VSV + HSV infection of cells; as expected, they found no cellular mRNA degradation of host mRNA in VSV-infected cells, degradation of host mRNA in HSV-infected cells, and no degradation of host cell mRNA in VSV + HSV-coinfected cells. They next looked at the disassociation of the polyribosomes caused by HSV infection and found it not to be affected by VSV coinfection. The major conclusions from this study were that (1) VSV stops transcription of the HSV genome, (2) transcription of the HSV genome is required to cause the degradation of host cell mRNA, (3) disruption of the host cell polyribosome does not require HSV transcription, and (4) in cells coinfected with VSV and HSV, the shut-off activity of VSV predominates.

5.7.4b. Simian Virus 40

Kranz and Reichmann (1983) undertook to study the capacity of VSV to inhibit cellular DNA synthesis by studying, as a model system, in vivo inhibition of SV40; the genome of this virus is considerably smaller than that of cellular chromatin, has a single origin of replication, and replicates as a minichromosome in association with cellular DNA polymerases and histones. They found that VSV inhibited SV40 DNA synthesis to about the same extent as cellular DNA synthesis in synchronized and unsynchronized Vero cells, suggesting that cellular and SV40 DNA synthesis were inhibited by similar molecular mechanisms. The principal effect of VSV was to block su-

percoiling of SV40 DNA, thus, resulting in accumulation of monomeric forms. These data led to the conclusion that VSV either inhibits
the topoisomerases responsible for nicking and resealing SV40 DNA
or by inhibition of protein synthesis leading to inhibition of nucleo-
some formation in the SV40 minichromosome. Kranz and Reichmann
(1983) came to the conclusion that VSV inhibits DNA synthesis indirectly through inhibition of cellular protein synthesis in a manner
similar to that of cycloheximide inhibition of SV40 replication, terminating in covalently closed DNA circles lacking superhelical turns
(Borgaux and Borgaux-Ramoisy, 1972; White and Eason, 1973). The
incomplete supercoiled SV40 DNA molecules could not be chased
into SV40 chromatin, suggesting that inhibition of protein synthesis
resulted in abortion of DNA replication.

6. SUMMARY

Vesicular stomatitis virus provides an interesting model system
for exploring the mechanisms by which certain viruses interfere with
metabolic processes of host cells culminating in death. VSV is highly
cytopathic for many species of animals and inhibits cellular macromolecular synthesis to varying degrees depending on species and type
of cell. Although there is clear evidence for a temporal relationship
between shutting off cellular RNA, DNA, and protein synthesis, it
is important to keep in mind that these effects cannot be causally
related to cell death. A variety of cellular functions is compromised
by VSV infection and it is difficult to be certain which are primary
and which are secondary. The earliest host response is decline in the
rate of RNA, DNA, and protein synthesis which seems to suggest
that these events are of paramount significance. Certain differences
in the kinetics of inhibition and other viral properties strongly indicate
that different viral products cause inhibition of cellular protein synthesis and cellular nucleic acid synthesis.

There is some evidence that structural components of the invading parental virion at high concentrations can subvert cellular
functions; but, by far, the greatest influence on inhibiting cellular
macromolecular synthesis can be attributed to endogenous VSV metabolic functions, which are limited to synthesis of viral RNA and
proteins. The weight of recent evidence appears to rule out viral proteins as inhibitors of cell functions and, thus, leaves newly synthesized viral RNA(s) as the culprit(s). Genetic data and UV inactivation

studies clearly indicate that VSV transcription is required to inhibit cellular protein and nucleic acid synthesis. Several laboratories have shown that unimpaired capacity to transcribe the VSV N gene and possibly the leader gene is required for inhibition of host protein synthesis by the invading virus. Preliminary data also suggest that the translational inhibitory component is double-stranded mRNA complexed with viral nucleocapsid. The viral inhibitor of cellular protein synthesis appears to act by blocking initiation of translation presumably by inactivating eukaryotic initiation factor(s).

VSV appears to inhibit cellular RNA and DNA synthesis by the same mechanism as judged by similar kinetics of the reaction and a similar dose response to UV inactivation of the viral inhibitory activity. The target size for UV inactivation of this viral inhibitory factor corresponds to the gene coding for the plus-strand leader sequences of both VSV-Indiana and VSV-New Jersey. Greater inhibition of cellular RNA synthesis by infection with VSV-Indiana appears to be a function of increased synthesis of leader that migrates to the nucleus. The leader RNAs of both VSV-Indiana and VSV-New Jersey inhibit DNA-dependent RNA synthesis in a reconstituted *in vitro* transcription system containing templates for either polymerase II or III. A greater transcription inhibitory activity of VSV-New Jersey leader may be related to more secondary structure than that of VSV-Indiana leader. Studies with a T_1 ribonuclease-resistant segment of the leader as well as with leader cDNA clones and synthetic oligodeoxynucleotides indicate that the nucleotide sequence responsible for inhibition of DNA-dependent transcription directed by both polymerases II and III is nucleotides 18–30, 5′-AUUAUUAUCAUUA-3′, resembling the TATA promoter.

Much more research is required for definitive identification of the leader RNA sequences which inhibit RNA synthesis and particularly those that inhibit cellular DNA synthesis. Little is known about the cellular targets for virus inhibition of RNA and DNA synthesis.

ACKNOWLEDGMENTS

This manuscript was prepared during the tenure of R. R. Wagner as a U.S. Senior Scientist Awardee of the Alexander von Hunboldt Foundation at the Universities of Giessen and Würzburg. The support and encouragement of the Hunboldt Foundation and Professors Rudolf Rott and Volker ter Meulen are gratefully acknowledged. We are grateful to Paige Hackney for excellent preparation of the manuscript. The research reported herein was supported by grants from the National Institute of Allergy and Infectious Diseases (AI-11112), the American Cancer Society (MV-9E), and the National Science Foundation (PCM-88-00494).

7. REFERENCES

Abraham, G., and Banerjee, A. K., 1976a, Sequential translation of the genes of vesicular stomatitis virus, *Proc. Natl. Acad. Sci. USA* **73**:1504.

Abraham, G., and Banerjee, A. K., 1976b, The nature of the RNA products synthesized *in vitro* by subviral components of vesicular stomatitis virus, *Virology* **71**:230.

Abraham, G., Rhodes, D. P., and Banerjee, A. K., 1975, The 5' terminal structure of the methylated mRNA synthesized *in vitro* by vesicular stomatitis virus, *Cell* **5**:51.

Akusjarvi, G., Mathews, M. B., Andersson, P., Vennstrom, B., and Pettersson, U., 1980, Structure of genes for virus-associated RNA_I and RNA_{II} of adenovirus type 2, *Proc. Natl. Acad. Sci. USA* **77**:2424.

Allen, G. P., O'Callaghan, D. J., and Randall, C. C., 1977, Purification and characterization of equine herpes virus-induced DNA polymerase, *Virology* **76**:395.

Bablanian, R., 1975, Structural and functional alterations in cultured cells infected with cytocidal viruses, *Prog. Med. Virol.* **19**:40.

Baglioni, C., Lenz, J. R., Maroney, P. A., and Weber, L. A., 1978, Effect of double-stranded RNA associated with viral mRNA on *in vitro* protein synthesis, *Biochemistry* **17**:3257.

Baker, C. C., and Ziff, E. B., 1981, Promoters and heterogeneous 5' termini of the messenger RNAs of adenovirus serotype 2, *J. Mol. Biol.* **149**:189.

Ball, L. A., and Wertz, G. W., 1981, VSV RNA synthesis: How can you be positive?, *Cell* **26**:143.

Ball, L. A., and White, C. N., 1976, Order of transcription of genes of vesicular stomatitis virus, *Proc. Natl. Acad. Sci. USA* **73**:442.

Baltimore, D., Huang, A. S., and Stampfer, M., 1970, Ribonucleic acid synthesis of vesicular stomatitis virus. II. An RNA polymerase in the virion, *Proc. Natl. Acad. Sci. USA* **66**:572.

Banerjee, A. K., and Rhodes, D. P., 1976, *In vitro* synthesis of RNA that contains polyadenylate by virion-associated RNA polymerase of vesicular stomatitis virus, *Proc. Natl. Acad. Sci. USA* **70**:3566.

Banerjee, A. K., Abraham, G., and Colonno, R. J., 1977, Vesicular stomatitis virus: Model of transcription, *J. Gen. Virol.* **34**:1.

Baxt, B., and Bablanian, R., 1976a, Mechanisms of vesicular stomatitis virus-induced cytopathic effects. I. Early morphologic changes induced by infectious and defective-interfering particles, *Virology* **72**:370.

Baxt, B., and Bablanian, R., 1976b, Mechanism of vesicular stomatitis virus-induced cytopathic effects. II. Inhibition of macromolecular synthesis induced by infectious and defective-interfering particles, *Virology* **72**:383.

Bishop, D. H. L., 1979, *Rhabdoviruses,* 3 Volumes, CRC Press, West Palm Beach.

Bishop, D. H. L., Emerson, S. U., and Flamand, A., 1974, Reconstitution of infectivity and transcriptase activity of homologous and heterologous viruses, vesicular stomatitis (Indiana serotype), Chandipura, vesicular stomatitis (New Jersey serotype), and Cocal virus, *J. Virol.* **14:**139.

Bishop, D. H. L., Repik, P., Obijeski, J. F., Moore, N. F., and Wagner, R. R., 1975, Restitution of infectivity to spikeless vesicular stomatitis virus by solubilized virus components, *J. Virol.* **16:**75.

Blumberg, B. M., and Kolakofsky, D., 1981, Intracellular vesicular stomatitis leader RNAs are found in nucleocapsid structures, *J. Virol.* **40:**568.

Blumberg, B. M., Leppert, M., and Kolakofsky, D., 1981, Interaction of VSV leader RNA and nucleocapsid protein may control VSV genome replication, *Cell* **23:**837.

Blumberg, B. M., Giorgi, M. C. and Kolakofsky, D., 1983, VSV N protein selectively encapsidates VSV leader RNA *in vitro, Cell* **32:**559.

Borgaux, P., and Borgaux-Ramoisy, D., 1972, Is a specific protein responsible for the supercoiling of polyoma DNA? *Nature* **235:**105.

Both, G. W., Banerjee, A. K., and Shatkin, A. J., 1975, Methylation-dependent translation of viral messenger RNAs *in vitro, Proc. Natl. Acad. Sci. USA* **72:**1189.

Breathnach, R., Benoist, C., O'Hare, K., Cannon, F., and Chambon, P., 1978, Ovalbumin gene: Evidence for a leader sequence in mRNA and DNA sequences at the exon–intron boundaries, *Proc. Natl. Acad. Sci. USA* **75:**4853.

Breindl, M., and Holland, J. J., 1975, Coupled *in vitro* transcription and translation of vesicular stomatitis virus messenger RNA, *Proc. Natl. Acad. Sci. USA* **72:**2545.

Cantell, K., Skurska, Z., Paucker, K., and Henle, W., 1962, Quantitative studies on viral interference in suspended L-cells. II. Factors affecting interference by UV-irradiated Newcastle disease virus, *Virology* **17:**312.

Carrasco, L., 1977, The inhibition of cell function after viral infection, *FEBS Lett.* **76:**11.

Carroll, A. R., and Wagner, R. R., 1978, Reversal by certain polyanions of an endogenous inhibitor of the vesicular stomatitis virus-associated transcriptase, *J. Biol. Chem.* **253:**3361.

Carroll, A. R., and Wagner, R. R., 1979, Role of membrane (M) protein in endogenous inhibition of *in vitro* transcription of vesicular stomatitis virus, *J. Virol.* **29:**134.

Cartwright, B., and Brown, F., 1977, Role of sialic acid in infection with vesicular stomatitis virus, *J. Gen. Virol.* **35:**197.

Centrella, M., and Lucas-Lenard, J., 1982, Regulation of protein synthesis in vesicular stomatitis virus-infected mouse L-929 cells by decreased protein synthesis initiation factor 2 activity, *J. Virol.* **41:**781.

Challberg, M. D., and Kelly, T. J., Jr., 1979, Adenovirus DNA replication *in vitro*: Origin and direction of daughter strand synthesis, *J. Mol. Biol.* **135:**999.

Clinton, G. M., Little, S. P., Hagen, F. S., and Huang, A. S., 1979, The matrix (M) protein of vesicular stomatitis virus regulates transcription, *Cell* **15:**1455.

Colonno, R. J., and Banerjee, A. K., 1976, A unique RNA species involved in initiation of vesicular stomatitis virus RNA transcription *in vitro, Cell* **8:**197.

Colonno, R. J., and Banerjee, A. K., 1977, Mapping and initiation studies on the leader RNA of vesicular stomatitis virus, *Virology* **77:**260.

Colonno, R. J., and Banerjee, A. K., 1978*a*, A complete nucleotide sequence of the leader RNA synthesized *in vitro* by vesicular stomatitis virus, *Cell* **15:**93.

Colonno, R. J., and Banerjee, A. K., 1978*b*, Nucleotide sequence of the leader RNA of the New Jersey serotype of vesicular stomatitis virus, *Nucl. Acids Res.* **51**:4165.

Colonno, R. J., Lazzarini, R. A., Keene, J. D., and Banerjee, A. K., 1977, *In vitro* synthesis of messenger RNA by a defective-interfering particle of vesicular stomatitis virus, *Proc. Natl. Acad. Sci. USA* **74**:1888.

Cooper, P. D., and Bellett, A. J. D., 1959, A transmissible interfering component of vesicular stomatitis virus preparations, *J. Gen. Microbiol.* **21**:485.

Cox, R. F., 1976, Quantitation of elongating form A and B RNA polymerases in chick oviduct nuclei and the effects of estradiol, *Cell* **7**:455.

Darnell, J. E., 1976, mRNA structure and function, *Prog. Nucl. Acid Res. Mol. Biol.* **19**:493.

Davis, N. L., and Wertz, G. W., 1980, A VSV mutant synthesizes a large excess of functional mRNA but produces less viral protein than its wild-type parent, *Virology* **103**:21.

Davis, N. L., and Wertz, G. W., 1982, Synthesis of vesicular stomatitis virus negative-strand RNA *in vitro:* Dependence on viral protein synthesis, *J. Virol.* **41**:821.

Dille, B. J., Hughes, J. V., Johnson, T. C., Rabinowitz, S. G., and Dal Canto, M. C., 1981, Cytopathic effects in mouse neuroblastoma cells during a nonpermissive infection with a mutant of vesicular stomatitis virus, *J. Gen. Virol.* **55**:343.

Dubovi, E. J., and Youngner, J. S., 1976*a*, Inhibition of pseudorabies virus replication by vesicular stomatitis virus. I. Activity of infectious and inactivated B particles, *J. Virol.* **18**:526.

Dubovi, E. J., and Youngner, J. S., 1976*b*, Inhibition of pseudorabies virus replication by vesicular stomatitis virus. II. Activity of defective interfering particles, *J. Virol.* **18**:534.

Dunigan, D. D., and Lucas-Lenard, J. M., 1983, Two transcription products of the vesicular stomatitis virus genome may control L-cell protein synthesis, *J. Virol.* **45**:618.

Ehrenfeld, E., and Hunt, T., 1971, Double-stranded poliovirus RNA inhibits initiation of protein synthesis by reticulocyte lysates, *Proc. Natl. Acad. Sci. USA* **68**:1075.

Ehrenfeld, E., and Lund, H., 1977, Untranslated vesicular stomatitis virus mRNA after poliovirus infection, *Virology* **80**:297.

Emerson, S. U., 1982, Reconstitution studies detect a single polymerase entry site on the vesicular stomatitis virus genome, *Cell* **31**:635.

Emerson, S. U., and Wagner, R. R., 1972, Dissociation and reconstitution of the transcriptase and template activities of vesicular stomatitis B and T virion, *J. Virol.* **10**:297.

Emerson, S. U., and Wagner, R. R., 1973, L protein requirement for *in vitro* RNA synthesis by vesicular stomatitis virus, *J. Vriol.* **12**:1325.

Emerson, S. U., and Yu, Y. H., 1975, Both NS and L proteins are required for *in vitro* RNA synthesis by vesicular stomatitis virus, *J. Virol.* **15**:1348.

Emerson, S. U., Dierks, P. M., and Parsons, J. T., 1977, *In vitro* synthesis of a unique RNA species by a T particle of vesicular stomatitis virus, *J. Virol.* **23**:708.

Epstein, D. A., Herman, R. C., Chien, I., and Lazzarini, R. A., 1980, Defective interfering particles generated by internal deletions of the vesicular stomatitis virus genome, *J. Virol.* **33**:818.

Farmilo, A. J., and Stanners, C. P., 1972, Mutant of vesicular stomatitis virus which allows DNA synthesis and division in cells synthesizing viral RNA, *J. Virol.* **10:**605.

Flamand, A., and Bishop, D. H. L., 1973, Primary *in vivo* transcription of vesicular stomatitis virus and temperature-sensitive mutants of five vesicular stomatitis virus complementation groups, *J. Virol.* **12:**1238.

Follett, E. A. C., Pringle, C. R., Wunner, W. H., and Skehel, J. J., 1974, Virus replication in enucleated cells: Vesicular stomatitis virus and influenza virus, *J. Virol.* **13:**394.

Francoeur, A. M., and Stanners, C. P., 1978, Evidence against the role of K$^+$ in the shut-off of protein synthesis by vesicular stomatitis virus, *J. Gen. Virol.* **39:**551.

Galli, G., Hofstetter, H., and Birnstiel, M. L., 1981, Two conserved sequence blocks within eukaryotic tRNA genes are major promoter elements, *Nature* **294:**626.

Gallione, C. J., and Rose, J. K., 1983, Nucleotide sequence of a cDNA clone encoding the entire glycoprotein from the New Jersey serotype of vesicular stomatitis virus, *J. Virol.* **46:**162.

Gallione, C. J., Greene, J. R., Iverson, L. E., and Rose, J. K., 1981, Nucleotide sequences of the mRNA's encoding the vesicular stomatitis virus N and NS proteins, *J. Virol.* **39:**529.

Genty, N., 1975, Analysis of uridine incorporation in chicken embryo cells infected with vesicular stomatitis virus and its temperature-sensitive mutants: Uridine transport, *J. Virol.* **15:**8.

Genty, N., and Berreur, P., 1975, Metabolisme des acide ribonucleique et des proteines de cellules de l'embryon de poulet infectées par le virus de la stomatite vesiculaire: Études des effects de mutants thermosensible, *Ann. Microbiol. (Inst. Pasteur)* **124A:**135.

Giorgi, C., Blumberg, B., and Kolakofsky, D., 1983, Sequence determinations of the (+) leader RNA regions of the vesicular stomatitis virus Chandipura, Cocal and Piry serotype genomes, *J. Virol.* **46:**125.

Grinnell, B. W., and Wagner, R. R., 1983, Comparative inhibition of cellular transcription by vesicular stomatitis virus serotypes New Jersey and Indiana: Role of each viral leader RNA, *J. Virol.* **48:**88.

Grinnell, B. W., and Wagner, R. R., 1984, Nucleotide sequence and secondary structure of VSV leader RNA and homologous DNA involved in inhibition of DNA-dependent transcription, *Cell* **36:**533.

Gruss, P., Dhar, R., and Khoury, G., 1981, Simian virus 40 tandem repeated sequences as an element of the early promoter, *Proc. Natl. Acad. Sci. USA* **78:**943.

Handa, H., Kaufman, F. J., Manley, J., Gefter, M., and Sharp, P. A., 1981, Transcription of simian virus 40 DNA in a HeLa cell extract, *J. Biol. Chem.* **256:**478.

Hansen, J., and Ehrenfeld, E., 1981, Presence of the cap-binding protein in initiation factor preparations from poliovirus-infected HeLa cells, *J. Virol.* **38:**438.

Heine, J. W., and Schnaitman, C. A., 1971, Entry of vesicular stomatitis virus into L cells, *J. Virol.* **8:**786.

Helenius, A., and Marsh, M., 1982, Endocytosis of enveloped animal viruses, in: *Membrane Recycling,* Ciba Foundation Symposium 92, pp. 59–76, Pitman Books, London.

Helentjaris, T., Ehrenfeld, E., Brown-Leudi, M. L., and Hershey, J. W., 1979, Alterations in initiation factors activity from poliovirus-infected HeLa cells, *J. Biol. Chem.* **254:**10973.

Herman, R., 1983, Conditional synthesis of an aberrant glycoprotein mRNA by the internal deletion mutant of vesicular stomatitis virus, *J. Virol.* **46**:709.

Herman, R. C., Schubert, M., Keene, J. D., and Lazzarini, R. A., 1980, Polycistronic vesicular stomatitis virus RNA transcripts, *Proc. Natl. Acad. Sci. USA* **77**:4662.

Hill, V. M., Marnell, L., and Summers, D. F., 1981, *In vitro* replication and assembly of vesicular stomatitis virus nucleocapsids, *Virology* **113**:109.

Holland, J. J., Kennedy, S. I. T., Semler, B., Jones, J. L., Roux, L., and Grabau, E. A., 1980, Defective interfering RNA viruses and the host cell response, in: *Comprehensive Virology,* Vol. 16 (H. Fraenkel-Conrat and R. R. Wagner, eds.), pp. 137–192, Plenum Press, New York.

Hsu, C.-H., Kingsbury, D. W., and Murti, K. G., 1979, Assembly of vesicular stomatitis nucleocapsids *in vivo:* A kinetic analysis, *J. Virol.* **32**:304.

Hu, S. L., and Manley, J. L., 1981, DNA sequence required for initiation of transcription *in vitro* from the major late promoter of adenovirus type 2, *Proc. Natl. Acad. Sci. USA* **78**:820.

Huang, A. S., and Baltimore, D., 1977, Defective interfering animal viruses, in: *Comprehensive Virology* Vol. 10 (H. Fraenkel-Conrat and R. R. Wagner, eds.), pp. 73–116, Plenum Press, New York.

Huang, A. S., and Manders, E., 1972, Ribonucleic acid synthesis of vesicular stomatitis virus. IV. Transcription by standard virus in the presence of defective-interfering particles, *J. Virol.* **9**:909.

Huang, A. S., and Wagner, R. R., 1965, Inhibition of cellular RNA synthesis by nonreplicating vesicular stomatitis virus, *Proc. Natl. Acad. Sci. USA* **54**:1579.

Huang, A. S., and Wagner, R. R., 1966a, Defective T particles of vesicular stomatitis virus. II. Biologic role in homologous interference, *Virology* **30**:173.

Huang, A. S., and Wagner, R. R., 1966b, Comparative sedimentation coefficients of RNA extracted from plaque-forming and defective particles of vesicular stomatitis virus, *J. Mol. Biol.* **22**:381.

Huang, A. S., Greenawalt, J. W., and Wagner, R. R., 1966, Defective T particles of vesicular stomatitis virus. I. Preparation, morphology, and some biological properties, *Virology* **30**:161.

Hunt, D. M., Emerson, S. U., and Wagner, R. R., 1976, RNA⁻ temperature-sensitive mutants of vesicular stomatitis virus: L protein thermosensitivity accounts for transcriptase restriction of group I mutants, *J. Virol.* **18**:596.

Hunter, T., Hunt, T., and Jackson, R. J., 1975, The characteristics of inhibition of protein synthesis by double-stranded ribonucleic acid in reticulocyte lysates, *J. Biol. Chem.* **250**:409.

Imblum, R. L., and Wagner, R. R., 1974, Protein kinase and phosphoproteins of vesicular stomatitis virus, *J. Virol.* **13**:113.

Iverson, L. E., and Rose, J. K., 1981, Localized attenuation and discontinuous synthesis during vesicular stomatitis virus transcriptions, *Cell* **22**:477.

Iverson, L. E., and Rose, J. K., 1982, Sequential synthesis of 5′-proximal vesicular stomatitis virus mRNA sequences, *J. Virol.* **44**:356.

Jacobs, B. L., and Penhoet, E. E., 1982, Assembly of vesicular stomatitis virus: Distribution of the glycoprotein on the surface of infected cells, *J. Virol.* **44**:1047.

Jaye, M. C., Wu, F.-S., and Lucas-Lenard, J. M., 1980, Inhibition of synthesis of ribosomal protein and ribosome assembly after infection of L cells with vesicular stomatitis virus, *Biochim. Biophys. Acta.* **606**:1.

Jaye, M. C., Godchaux, W., and Lucas-Lenard, J., 1982, Further studies on the inhibition of cellular protein synthesis by vesicular stomatitis virus, *Virology* **116**:148.

Jelinek, W. R., Toomey, T. P., Leinwand, L., Duncan, C. H., Biro, P. A., Choudary, P. V., Weissman, S. M., Rubin, C. M., Houck, C. M., Deininger, P. L., and Schmid, C. W., 1980, Ubiquitous interspersed repeated sequences in mammalian genomes, *Proc. Natl. Acad. Sci. USA* **77**:1398.

Johnson, L. D., Binder, M., and Lazzarini, R. A., 1979, A defective interfering vesicular stomatitis virus particle that directs synthesis of functional proteins in the absence of helper virus, *Virology* **99**:203.

Kaempfer, R., and Kaufman, J., 1973, Inhibition of cellular protein synthesis by double-stranded RNA: Inactivation of an initiation factor, *Proc. Natl. Acad. Sci. USA* **70**:222.

Kaufman, T., Goldstein, E., and Penman, S., 1976, Poliovirus-induced inhibition of polypeptide initiation *in vitro* on native polyribosomes, *Proc. Natl. Acad. Sci. USA* **73**:1834.

Keene, J. D., Chien, I. M., and Lazzarini, R. A., 1981, Vesicular stomatitis virus defective particle contains a muted internal leader RNA gene, *Proc. Natl. Acad. Sci. USA* **78**:2090.

Kelley, J. M., Emerson, S. U., and Wagner, R. R., 1972, The glycoprotein of vesicular stomatitis virus is the antigen that gives rise to and reacts with neutralizing antibody, *J. Virol.* **10**:1231.

Kiley, M. P., Bowen, E. T. W., Eddy, G. A., Isaacson, M., Johnson, K. M., McCormick, J. B., Murphy, F. A., Pattyn, S. R., Peters, D., Prozesky, O. W., Regsery, R. L., Sampson, D. I. H., Slencska, W., Sureau, P., van der Gruen, G., Webb, P. A., and Wulff, H., 1982, Filoviridae: A taxonomic home for Marburg and Ebola viruses, *Intervirology* **18**:24.

Kingsford, L., and Emerson, S. U., 1980, Transcriptional activity of different phosphorylated species of NS protein purified from vesicular stomatitis virus and cytoplasm of infected cells, *J. Virol.* **33**:1097.

Knipe, D. M., Baltimore, D., and Lodish, H. F., 1977, Separate pathways of maturation of the major structural proteins of vesicular stomatitis virus, *J. Virol.* **21**:1128.

Kolakofsky, D., and Altman, S., 1978, Endonuclease activity associated with animal RNA viruses, *J. Virol.* **25**:274.

Kranz, D., and Reichmann, M. E., 1983, Inhibition of SV40 DNA synthesis by vesicular stomatitis virus in doubly infected monkey kidney cells, *Virology* **128**:418.

Kurilla, M. G., and Keene, J. D., 1984, The leader RNA of vesicular stomatitis virus is bound by a cellular protein reactive with anti-La lupus antibodies, *Cell* **34**:837.

Kurilla, M. G., Piwnica-Worms, H., and Keene, J. D., 1982, Rapid and transient localization of the leader RNA of vesicular stomatitis virus in the nuclei of infected cells, *Proc. Natl. Acad. Sci. USA* **79**:5240.

Lazzarini, R. A., Keene, J. D., and Schubert, M., 1981, The origin of defective interfering particles of the negative strand RNA viruses, *Cell* **26**:145.

Leamnson, R. N., and Reichmann, M. E., 1974, The RNA of defective vesicular stomatitis virus particles in relation to viral cistrons, *J. Mol. Biol.* **85**:551.

Lerner, M. R., and Steitz, J. A., 1981, Snurps and scyrps, *Cell* **25**:298.

Lerner, M. R., Andrews, N. C., Miller, G., and Steitz, J. A., 1981, Two small RNAs encoded by Epstein–Barr virus and complexed with proteins are precipitated by antibodies from patients with systemic lupus erythematosus, *Proc. Natl. Acad. Sci. USA* **78**:805.

Lodish, H. F., and Froshauer, S., 1977, Relative rates of initiation of translation of different vesicular stomatitis messenger RNAs, *J. Biol. Chem.* **252**:8804.

Lodish, H. F., and Porter, M., 1980, Translational control of protein synthesis after infection by vesicular stomatitis virus, *J. Virol.* **36**:719.

Lodish, H. F., and Porter, M., 1981, Vesicular stomatitis virus mRNA and inhibition of translation of cellular mRNA—Is there a P function in vesicular stomatitis virus?, *J. Virol.* **38**:504.

Madore, H. P., and England, J. M., 1975, Selective suppression of cellular protein synthesis in baby hamster kidney (BHK-21) cells infected with rabies virus, *J. Virol.* **16**:1351.

Manders, E. K., Tilles, J. G., and Huang, A. S., 1972, Interferon-mediated inhibition of virion transcription, *Virology* **49**:573.

Manley, J. L., Fire, A., Cano, A., Sharp, P. A., and Gefter, M. L., 1980, DNA-dependent transcription of adenovirus genes in a soluble whole cell extract, *Proc. Natl. Acad. Sci. USA* **77**:3855.

Marsh, M., Bolzau, E., and Helenius, A., 1983, Penetration of Semliki Forest virus from acidic prelysosomal vacuoles, *Cell* **32**:931.

Marcus, P. I., and Sekellick, M. J., 1974, Cell Killing by viruses. I. comparison of cell-killing, plaque-forming, and defective-interfering particles of vesicular stomatitis virus, *Virology* **57**:321.

Marcus, P. I., and Sekellick, M. J., 1975, Cell killing by viruses. II. Cell killing by vesicular stomatitis virus: A requirement for virion-derived transcription, *Virology* **63**:176.

Marcus, P. I., and Sekellick, M. J., 1976, Cell killing by viruses. III. The interferon system and inhibition of cell-killing by vesicular stomatitis virus, *Virology* **63**:378.

Marcus, P. I., Engelhardt, D. L., Hunt, J. M., and Sekellick, M. J., 1971, Interferon action. Inhibition of vesicular stomatitis virus RNA synthesis induced by virion-bound polymerases, *Science* **174**:593.

Marcus, P. I., Sekellick, M. J., Johnson, L. D., and Lazzarini, R. A., 1977, Cell killing by viruses. V. Transcribing defective-interfering particles of vesicular stomatitis virus function as cell-killing particles, *Virology* **82**:242.

Marvaldi, J., and Lucas-Lenard, J., 1977, Differences in the ribosomal protein gel profile after infection of L cells with wild-type or temperature-sensitive mutants of vesicular stomatitis virus, *Biochemistry* **16**:4320.

Marvaldi, J., Lucas-Lenard, J., Sekellick, M., and Marcus, P. I., 1977, Cell killing by viruses. IV. Cell killing and the inhibition of cell protein synthesis requires the same gene function of vesicular stomatitis virus, *Virology* **79**:267.

Marvaldi, J., Sekellick, M. J., Marcus, P. I., and Lucas-Lenard, J., 1978, Inhibition of mouse cell protein synthesis by ultraviolet-irradiated vesicular stomatitis virus requires viral transcription, *Virology* **84**:127.

Matlin, K., Reggio, H., Simons, K., and Helenius, A., 1982, The pathway of vesicular stomatitis virus entry leading to infection, *J. Mol. Biol.* **95**:676.

Matsumoto, S., 1974, Morphology of rabies virion and cytopathology of virus infected cells, *Symp. Ser. Immunobiol. Strand.* **21**:25.

Matthews, R. E. F., 1982, Classification and nomenclature of viruses: Fourth Report of the International Committee on Taxonomy of Viruses, *Intervirology* **17**:1.

McAllister, P. E., and Wagner, R. R., 1975, Structural proteins of two salmonid rhabdoviruses, *J. Virol.* **15**:733.

McAllister, P. E., and Wagner, R. R., 1976, Differential inhibition of host protein synthesis in cells infected with RNA⁻ temperature-sensitive mutants of vesicular stomatitis virus, *J. Virol.* **18**:550.

McCombs, R. M., Benyesh-Melnick, M., and Brunschwig, J. P., 1966, Biophysical studies of vesicular stomatitis virus, *J. Bacteriol.* **91**:803.

McCoombs, K., Mann, E., Edwards, J., and Brown, D. T., 1981, Effects of chloroquine and cytochalasin B on the infection of cells by Sindbis virus and vesicular stomatitis virus, *J. Virol.* **37**:1060.

McDowell, M. J., Joklik, W. K., Villa-Konisoff, L., and Lodish, H. F., 1972, Translation of reovirus mRNAs synthesized *in vitro* into reovirus polypeptides in several mammalian cell-free extracts, *Proc. Natl. Acad. Sci. USA* **69**:2649.

McGeoch, D. J., and Dolan, A., 1979, Sequence of 200 nucleotides at the 3' terminus of the genome RNA of vesicular stomatitis virus, *Nucl. Acids Res.* **6**:3199.

McGeoch, D. J., Dolan, A., and Pringle, C. R., 1980, Comparison of nucleotide sequences in the genome of the New Jersey and Indiana serotypes of vesicular stomatitis virus, *J. Virol.* **33**:69.

McGowan, J. J., and Wagner, R. R., 1981, Inhibition of cellular DNA synthesis by vesicular stomatitis virus, *J. Virol.* **38**:356.

McGowan, J. J., Allen, G. P., Barnett, J. M., and Gentry, G. A., 1980, Biochemical characterization of equine herpes virus type 3-induced deoxythymiidine kinase purified from lytically infected horse embryo dermal fibroblasts, *J. Virol.* **34**:474.

McGowan, J. J., Emerson, S. U., and Wagner, R. R., 1982, The plus-strand leader RNA of VSV inhibits DNA-dependent transcription of adenovirus and SV40 genes in a soluble whole cell extract, *Cell* **28**:325.

McSharry, J. J., and Choppin, P. W., 1978, Biological properties of the VSV glycoprotein. I. Effects of the isolated glycoprotein on host macromolecular synthesis, *Virology* **84**:172.

McSharry, J. J., and Wagner, R. R., 1971a, Lipid composition of purified vesicular stomatitis virus, *J. Virol.* **7**:59.

McSharry, J. J., and Wagner, R. R., 1971b, Carbohydrate composition of vesicular stomatitis virus, *J. Virol.* **7**:412.

Mellon, M. G., and Emerson, S. U., 1978, Rebinding of transcriptase components (L and NS proteins) to the nucleocapsid template of vesicular stomatitis virus, *J. Virol.* **27**:560.

Morrison, T. G., and Lodish, H. F., 1975, Sites of synthesis of membrane and non-membrane proteins of vesicular stomatitis virus, *J. Biol. Chem.* **250**:6955.

Moss, T., 1983, A transcription function for the repetitive ribosomal spacer in *Xenopus laevis, Nature* **302**:223.

Mudd, J. A., and Summers, D. F., 1970, Protein synthesis in vesicular stomatitis virus-infected HeLa cells, *Virology* **42**:928.

Murphy, F. A., 1977, Rabies pathogenesis, *Arch. Virol.* **54**:279.

Nilsen, T. W., Wood, D. L., and Baglioni, C., 1981, Cross-linking of viral RNA by 4'-aminoethyl-4,5,8-trimethyl-psorate in HeLa cells infected with encephalomyocarditis virus and the *ts*G114 mutant of vesicular stomatitis virus, *Virology* **109**:82.

Nishioka, Y., and Silverstein, S., 1978, Alteration in the protein synthetic apparatus of Friend erythroleukemia cells infected with vesicular stomatitis virus or herpes simplex virus, *J. Virol.* **25**:422.

Nishioka, Y., Jones, G., and Silverstein, S., 1983, Inhibition by vesicular stomatitis virus of herpes simplex virus directed protein synthesis, *Virology* **124**:238.

Nowakowski, M., Bloom, B. R., Ehrenfeld, E., and Summers, D. F., 1973, Restricted replication of vesicular stomatitis virus in human lymphoblastoid cells, *J. Virol.* **12**:1272.

Nuss, D. L., and Koch, G., 1976*a*, Differential inhibition of vesicular stomatitis polypeptide synthesis by hypertonic initiation block, *J. Virol.* **17**:283.

Nuss, D. L., and Koch, G., 1976*b*, Translation of individual host mRNA's in MPC-11 cells is differentially suppressed after infection by vesicular stomatitis virus, *J. Virol.* **19**:572.

Nuss, D. L., and Koch, G., 1976*c*. Variations in the relative synthesis of immunoglobulin G and non-immunoglobulin G proteins in cultured MPC-11 cells with changes in the overall rate of polypeptide chain initiation and elongation, *J. Mol. Biol.* **102**:601.

Nuss, D. L., Opperman, H., and Koch, G., 1975, Selective blockage of initiation of host protein synthesis in RNA virus-infected cells, *Proc. Natl. Acad. Sci. USA* **72**:1258.

Obijeski, J. F., and Simpson, R. W., 1974, Conditional lethal mutants of vesicular stomatitis virus. II. Synthesis of virus-specific polypeptides in nonpermissive cells infected with RNA⁻ host-restricted mutants, *Virology* **57**:369.

Oldstone, M. B. A., Tishon, A., Dutko, F. J., Kennedy, S. I. T., Holland, J. J., and Lampert, P. W., 1980, Does the major histocompatibility complex serve as a specific receptor for Semliki Forest virus?, *J. Virol.* **34**:256.

Otto, M. J., and Lucas-Lenard, J., 1980, The influence of the host cell on the inhibition of virus protein synthesis in cells infected with vesicular stomatitis virus and mengovirus, *J. Gen. Virol.* **50**:29.

Patton, J. T., Davis, N. L., and Wertz, G. W., 1983, Cell-free synthesis and assembly of vesicular stomatitis virus nucleocapsids, *J. Virol.* **45**:155.

Patzer, E. J., Wagner, R. R., and Dubovi, E. J., 1979, Viral membranes: Model systems for studying biological membranes, *Crit. Rev. Biochem.* **6**:165.

Perrault, J., and Holland, J. J., 1972, Absence of transcriptase activity or transcription inhibiting ability in defective-interfering particles of vesicular stomatitis virus, *Virology* **50**:159.

Pinney, D. F., and Emerson, S. U., 1982, Identification and characterization of a group of discrete oligonucleotides transcribed *in vitro* from the terminus of the N-gene of vesicular stomatitis virus, *J. Virol.* **42**:889.

Pringle, C. R., 1977, Genetics of rhabdoviruses, in: *Comprehensive Virology,* Vol. 9 (H. Fraenkel-Conrat and R. R. Wagner, eds.), pp. 239–289, Plenum Press, New York.

Pringle, C. R., 1978, The *td*CE and *hr*CE phenotype: Host range mutants of vesicular stomatitis virus in which polymerase function is affected, *Cell* **15**:597.

Rabinowitz, S. G., Huprika, J., and Dal Canto, M. C., 1981, Comparative neurovirulence of selected vesicular stomatitis virus temperature-sensitive mutants of complementation groups II and III, *Infect. Immunol.* **33**:120.

Repik, P., Flamand, A., and Bishop, D. H. L., 1974, Effect of interferon upon the primary and secondary transcription of vesicular stomatitis and influenza viruses, *J. Virol.* **14**:1169.

Rhodes, D. P., Moyer, S. A., and Banerjee, A. K., 1974, *In vitro* synthesis of methylated messenger RNA by the virion-associated RNA polymerase of vesicular stomatitis virus, *Cell* **3**:327.

Rinke, F., and Steitz, J. A., 1982, Precursor molecules of both human ribosomal RNA and tRNA are bound by a cellular protein reactive with anti-La lupus antibodies, *Cell* **29**:149.

Robertson, B. H., and Wagner, R. R., 1981, Host cell variation in response to vesicular stomatitis virus inhibition of RNA synthesis, in: *The Replication of Negative Strand Viruses* (D. H. L. Bishop and R. W. Compans, eds.), pp. 955–963, Elsevier North Holland, Amsterdam.

Rosa, M. D., Hendrick, J. P., Lerner, M. R. and Steitz, J. A., 1983, A mammalian tRNA containing antigen is recognized by the polymyositis-specific antibody anti-Jo-1, *Nucl. Acids Res.* **11**:853.

Rose, J. K., 1980, Complete intergenic and flanking gene sequences from the genome of vesicular stomatitis virus, *Cell* **19**:415.

Rose, J. K., 1982, Expression from cloned cDNA of cell-surface secreted forms of the glycoprotein of vesicular stomatitis virus in eukaryotic cells, *Cell* **30**:753.

Rose, J. K., and Gallione, C. J., 1981, Nucleotide sequence of the mRNA's encoding the vesicular stomatitis virus G and M proteins determined from cDNA clones containing the complete coding regions, *J. Virol.* **39**:519.

Rose, J. K., and Lodish, H. F., 1976, Translation *in vitro* of vesicular stomatitis virus mRNA lacking 5'-terminal 7-methylguanosine, *Nature* **262**:32.

Rose, J. K., Trachsel, H., Leong, K., and Baltimore, D., 1978, Inhibition of translation by poliovirus: Inactivation of a specific initiation factor, *Proc. Natl. Acad. Sci. USA* **75**:2732.

Rose, J. K., Doolittle, R. F., Anilonis, A., Curtis, P. J., and Wunner, W. H., 1982, Homology between the glycoproteins of vesicular stomatitis and rabies virus, *J. Virol.* **43**:361.

Rosen, C. A., Ennis, H. L., and Cohen, P. S., 1982, Translational control of vesicular stomatitis virus protein synthesis: Isolation of an mRNA-sequestering particle, *J. Virol.* **44**:932.

Rothman, J. E., Bursztyn-Pettegrew, H., and Fine, R. E., 1980, Transport of the membrane glycoprotein of VSV to the cell surface in two stages, *J. Cell Biol.* **86**:162.

Rubio, C., Kolakofsky, D., Hill, V. M., and Summers, D. F., 1980, Replication and assembly of VSV nucleocapsids: Protein association with RNPs and effects of cycloheximide on replication, *Virology* **105**:123.

Schloemer, R. H., and Wagner, R. R., 1975, Sialoglycoprotein of vesicular stomatitis virus: Role of the neuraminic acid in infection, *J. Virol.* **14**:270.

Schnitzlein, W. M., O'Banion, M. K., Poirot, M. K., and Reichmann, M. E., 1983, Effect of intracellular vesicular stomatitis virus mRNA concentration on the inhibition of host cell protein synthesis, *J. Virol.* **45**:206.

Schubert, M., and Lazzarini, R. A., 1981a, Structure and origin of a snapback defective interfering particle RNA of vesicular stomatitis virus, *J. Virol.* **37**:661.

Schubert, M., and Lazzarini, R. A., 1981b, *In vitro* transcription of the 5' terminal extracistronic region of vesicular stomatitis virus RNA, *J. Virol.* **38**:256.

Schubert, M., Keene, J. D., Lazzarini, R. A., and Emerson, S. U., 1978, The complete sequence of a unique RNA species synthesized by a DI particle of VSV, *Cell* **15**:103.

Schubert, M., Harmison, G. G., Sprague, J., Condra, C., and Lazzarini, R. A., 1982, *In vitro* transcription of vesicular stomatitis virus initiation with GTP at a specific site within the N cistron, *J. Virol.* **43**:166.

Segall, J., Matsuk, T., and Roeder, R. G., 1980, Multiple factors are required for the accurate transcription of purified genes by RNA polymerase III, *J. Biol. Chem.* **255**:11906.

Sekellick, M. J., and Marcus, P. I., 1979, Persistant infection. II. Interferon-inducing temperature-sensitive mutants as mediators of cell spacing. Possible role of persistent infection by vesicular stomatitis virus, *Virology* **95**:36.

Sen, G. C., 1982, Mechanisms of interferon action: Progress toward its understanding, *Prog. Nucl. Acid. Res. Mol. Biol.* **17**:106.

Shen, C.-K., and Maniatis, T., 1982, The organization, structure, and *in vitro* transcription of the ALU family RNA polymerase III transcription units in the human α-like globin gene cluster: Precipitation of *in vitro* transcripts by lupus and La antibodies, *J. Mol. Appl. Gen.* **1**:343.

Simpson, R. W., and Obijeski, J. F., 1974, Conditional lethal mutants of vesicular stomatitis virus. I. Phenotypic characterization of single and double mutants exhibiting host restriction and temperature sensitivity, *Virology* **57**:357.

Simpson, R. W., Hausen, R. E., and Dales, S., 1969, Viropexis of vesicular stomatitis virus by L cells, *Virology* **17**:285.

Smith, M. M., and Huang, R. C. C., 1976, Transcription *in vitro* of immunoglobulin kappa light chain genes in isolated mouse myeloma nuclei and chromatin, *Proc. Natl. Acad. Sci. USA* **73**:775.

Sprague, J., Condra, J. H., Arnheiter, H., and Lazzarini, R. A., 1982, Expression of a recombinant DNA gene coding for the vesicular stomatitis virus nucleocapsid protein, *J. Virol.* **45**:773.

Stanners, C. P., Francoeur, A. M., and Lam, T., 1977, Analysis of VSV mutant with attenuated cytopathogenicity in viral function, P, for inhibition of protein synthesis, *Cell* **11**:273.

Steeg, H. V., Thomas, A., Verbeck, S., Kasparaitis, M., Vooma, H. O., and Benne, R., 1981, Shut-off of neuroblastoma cell protein synthesis by Semliki Forest virus: Loss of ability of crude initiation factors to recognize early Semliki Forest virus and host mRNAs, *J. Virol.* **38**:728.

Testa, D., Chanda, P. K., and Banerjee, A. K., 1980, Unique model of transcription *in vitro* by vesicular stomatitis virus, *Cell* **21**:267.

Thomas, J. R., 1983, Inhibition of translation by vesicular stomatitis virus, Ph.D. dissertation, University of Virginia, Charlottesville, Virginia.

Thomas, J. R., and Wagner, R. R., 1982, Evidence that vesicular stomatitis virus produces double-stranded RNA that inhibits protein synthesis in a reticulocyte lysate, *J. Virol.* **44**:189.

Thomas, J. R., and Wagner, R. R., 1983, Inhibition of translation in lysates of mouse L cells infected with vesicular stomatitis virus: Presence of a defective ribosome-associated factor, *Biochemistry* **22**:1540.

Trachsel, H., Sonenberg, N., Shatkin, A. J., Rose, J. K., Leong, K., Bergman, J. E., Gordon, J., and Baltimore, D., 1980, Purification of a factor that restores translation of vesicular stomatitis virus mRNA in extracts from poliovirus-infected HeLa cells, *Proc. Natl. Acad. Sci. USA* **77**:770.

Villareal, L. P., Breindl, M., and Holland, J. J., 1976, Determination of molar ratios of vesicular stomatitis virus-induced RNA species in BHK-21 cells, *Biochemistry* **15**:1633.

Volk, W. A., Snyder, R. M., Benjamin, D. C., and Wagner, R. R., 1982, Monoclonal antibody to the glycoprotein of vesicular stomatitis virus: Comparative neutralizing activity, *J. Virol.* **42**:220.

Wagner, R. R., 1974, Pathogenicity and immunogenicity for mice of temperature-sensitive mutants of vesicular stomatitis virus, *Infect. Immunol.* **10**:309.

Wagner, R. R., 1975, Reproduction of rhabdoviruses, in: *Comprehensive Virology,* Vol. 4 (H. Fraenkel-Conrat and R. R. Wagner, eds.), pp. 1–93, Plenum Press, New York.

Wagner, R. R., 1984, Cytopathic effects of viruses: A general survey, in: *Comprehensive Virology,* Vol. 19 (H. Fraenkel-Conrat and R. R. Wagner, eds.), pp. 1–63, Plenum Press, New York.

Wagner, R. R., and Huang, A. S., 1966, Inhibition of RNA and interferon synthesis in Krebs-2 cells infected with vesicular stomatitis virus, *Virology* **28**:1.

Wagner, R. R., Levy, A. H., Snyder, R. M., Ratcliff, G. A., Jr., and Hyatt, D. F., 1963, Biologic properties of two plaque variants of vesicular stomatitis virus (Indiana serotype), *J. Immunol.* **91**:112.

Wagner, R. R., Snyder, R. M., and Yamazaki, S., 1970, Proteins of vesicular stomatitis virus: Kinetics and cellular sites of synthesis, *J. Virol.* **5**:548.

Wagner, R. R., Heine, J. W., Goldstein, G., and Schnaitman, C. A., 1971, Use of antiviral-antiferritin hybrid antibody for localization of viral antigen in plasma membrane, *J. Virol.* **7**:274.

Wasylyk, B., Derbyshire, R., Guy, A., Molko, D., Roget, A., Teaule, R., and Chambon, P., 1980, Specific *in vitro* transcription of a conalbumin gene is drastically decreased by single-point mutation in TATA box homology sequence, *Proc. Natl. Acad. Sci. USA* **77**:7024.

Weck, P. K., and Wagner, R. R., 1978, Inhibition of RNA synthesis in mouse myeloma cells infected with vesicular stomatitis virus, *J. Virol.* **25**:770.

Weck, P. K., and Wagner, R. R., 1979*a*. Transcription of vesicular stomatitis virus is required to shut off cellular RNA synthesis, *J. Virol.* **30**:410.

Weck, P. K., and Wagner, R. R., 1979*b*, Vesicular stomatitis virus infection reduces the number of active DNA-dependent RNA polymerases in myeloma cells, *J. Biol. Chem.* **254**:5430.

Weck, P. K., Carroll, A. R., Shattuck, D. M., and Wagner, R. R., 1979, Use of UV irradiation to identify the genetic information of vesicular stomatitis virus responsible for shutting off cellular RNA synthesis, *J. Virol.* **30**:746.

Wertz, G. W., 1975, Method of examining viral RNA metabolism in cells in culture: Metabolism of vesicular stomatitis virus RNA, *J. Virol.* **16**:1340.

Wertz, G. W., 1978, Isolation of possible replicative intermediate structures from vesicular stomatitis virus infected cells, *Virology* **85**:271.

Wertz, G. W., 1980, RNA replication, in: *Rhabdoviruses,* Vol. II (D. H. L. Bishop, ed.), pp. 75–94, CRC Press, Cleveland.

Wertz, G. W., 1983, Replication of vesicular stomatitis virus defective interfering particle RNA *in vitro:* Transition from synthesis of defective interfering leader RNA to synthesis of full-length defective-interfering RNA, *J. Virol.* **46**:513.

Wertz, G. W., and Levine, M., 1973, RNA synthesis by vesicular stomatitis virus and a small plaque mutant: Effects of cycloheximide, *J. Virol.* **12:**253.

Wertz, G. W., and Youngner, J. S., 1970, Interferon production and inhibition of host synthesis in cells infected with vesicular stomatitis virus, *J. Virol.* **6:**476.

Wertz, G. W., and Youngner, J. S., 1972, Inhibition of protein synthesis in L cells infected with vesicular stomatitis virus, *J. Virol.* **9:**85.

White, M., and Eason, R., 1973, Supercoiling of SV40 DNA can occur independently of replication. *Nature New Biol.* **241:**46.

Wiener, J. R., Pal, R., Barenholz, Y., and Wagner, R. R., 1983, Influence of the peripheral matrix protein of vesicular stomatitis virus on the membrane dynamics of mixed phospholipid vesicles: Fluorescence studies, *Biochemistry* **22:**2162.

Wiktor, T. J., and Koprowski, H., 1974, Rhabdovirus replication in enucleated host cells, *J. Virol.* **14:**300.

Wilson, T., and Lenard, J., 1981, Interaction of wild-type and mutant M protein of VSV with nucleocapsids *in vitro, Biochemistry* **20:**1349.

Wilusz, J., Kurilla, M. G., and Keene, J. D., 1984, La protein binds to a unique species of minus sense leader RNA during the replication of vesicular stomatitis virus, *Proc. Natl. Acad. Sci. USA* **80:**5827.

Wu, F. S., and Lucas-Lenard, J. M., 1980, Inhibition of ribonucleic acid accumulation in mouse L cells infected with vesicular stomatitis virus requires viral ribonucleic acid transcription, *Biochemistry* **19:**804.

Yamazaki, S., and Wagner, R. R., 1970, Action of interferon: Kinetics and differential effects on viral functions, *J. Virol.* **6:**421.

Yang, F., and Lazzarini, R. A., 1983, Analysis of the recombination event generating a vesicular stomatitis virus defective interfering particle, *J. Virol.* **45:**766.

Yaoi, Y., and Amano, M., 1970, Inhibitory effect of ultraviolet inactivated vesicular stomatitis virus on inhibition of DNA synthesis in cultured chick embryo cells, *J. Gen. Virol.* **9:**69.

Yaoi, Y., Mitsui, H., and Amano, M., 1970, Effect of UV-irradiated vesicular stomatitis virus on nucleic acid synthesis in chick embryo cells, *J. Gen. Virol.* **8:**165.

Youngner, J. S., and Preble, O. T., 1980, Viral persistence: Evolution of viral populations, in: *Comprehensive Virology,* Vol. 16, (H. Fraenkel-Conrat and R. R. Wagner, eds.), pp. 73–135, Plenum Press, New York.

Youngner, J. S., and Wertz, G. W., 1968, Interferon production in mice by vesicular stomatitis virus, *J. Virol.* **2:**1360.

Youngner, J. S., Thacore, H. R., and Kelly, M. E., 1972, Sensitivity of RNA and DNA viruses to different species of interferon in cell cultures, *J. Virol.* **10:**171.

Zakowski, J. J., and Wagner, R. R., 1980, Localization of membrane-associated proteins in vesicular stomatitis virus by the use of hydrophobic probes and cross-linking reagents, *J. Virol.* **36:**93.

Zakowski, J. J., Petri, W. A., Jr., and Wagner, R. R., 1981, Role of matrix protein in assembling the membrane of vesicular stomatitis virus: Reconstitution of matrix protein with negatively charged phospholipid vesicles, *Biochemistry* **20:**3902.

Zieve, G. W., 1981, Two groups of small stable RNAs, *Cell* **25:**296.

CHAPTER 6

Adenovirus Cytopathology

S. J. Flint

Molecular Biology
Princeton University
Princeton, New Jersey 08544

1. INTRODUCTION

Human adenoviruses were originally recognized by virtue of their cytopathic effect, the ability to induce degenerative changes in human epithelial cells (Rowe *et al*, 1953; Hilleman and Werner, 1954). The study of adenovirus productive infections, and transformation of rodent cells, has continually accelerated during the 30 years that have passed since these first descriptions. The application of ever more sophisticated molecular techniques has led to the construction of an extremely detailed picture of the viral genome, including the complete nucleotide sequence of more than 36,000 base pairs in the case of adenovirus type 2 (Ad 2), and of its expression and replication in permissive cells (see Tooze, 1980; Ziff, 1980; Flint, 1982, for recent reviews). The popularity and value of the adenovirus system can largely be attributed to the virus' dependence on its host cell for the molecular machinery that permits progression through the replication cycle. It is not surprising, then, that much information of general relevance to the understanding of gene expression in eukaryotic cells has emerged from investigations of adenovirus infected cells, perhaps the most striking example being the discovery of spliced mRNA species (Berget *et al.*, 1977; Chow *et al.*, 1977; Klessig and Grodzicker, 1977).

The history of investigation of the consequences of an adenovirus infection for the host cell must stand in stark contrast to this happy situation. After the initial burst of interest, adenovirus cytopathology has inspired relatively little enthusiasm, compared to other topics within the field. Consequently, it has often appeared as an arcane corner, rarely approached with the fervor applied to elucidation of, say, viral gene expression in productively-infected cells. The past few years have witnessed a revival of interest in certain aspects of this subject, spurred in part by the reagents and methods generated by recombinant DNA technology. It is, nevertheless, important to emphasize that even the cellular responses to adenoviruses that have been most thoroughly investigated (the disruptions of cellular RNA metabolism and protein synthesis in productively-infected cells and the induction of cellular DNA synthesis in quiescent cells) are far from understood; at this juncture, it is possible to provide little more than a description of the ways in which cells respond to adenovirus infection and sift the few clues that might hint at the underlying mechanisms.

2. ADENOVIRUS–MAMMALIAN CELL INTERACTIONS

The interaction between a human adenovirus and a mammalian cell is a complicated business; the response of the cell as well as the outcome of the infection, that is, whether it is completely or partially productive or abortive, are influenced by such factors as the growth state of the host cell, the species of the cell and the serotype of the infecting virus. Thus, although adenoviruses are generally considered to be efficient pathogens, they can, under appropriate circumstances, induce infected cells to replicate their DNA and progress through the cell cycle or to undertake the increased synthesis of certain mRNA species. The ability of human adenoviruses to inhibit or to induce production of cellular macromolecules forms the subject of this review.

2.1. The Adenovirus Productive Cycle

The adenovirus productive cycle has most usually been studied following infection of established lines of human cells, such as HeLa, with sufficiently high multiplicities of a human serotype to achieve a

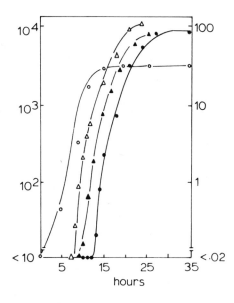

Fig. 1. Time course of synthesis of viral DNA, RNA, and proteins and virion assembly in Ad2-infected cells. The measurements of intracellular virus and viral DNA (pfu/cell) are from Green *et al.* (1971), and those of viral mRNA, measured by hybridization of [^{3}H]-RNA to Ad2 DNA, and late protein, and hexon antigen measured by complement fixation from Philipson and Lindberg (1974). Modified from Tooze (1980).

more or less synchronous infection. Serotypes 2 and 5, belonging to the nononcogenic subgroup C (see Tooze, 1980) have been most widely used. Unless otherwise stated, it is such conditions that will be designated by the term "productive infection" in this and subsequent sections. During such an infection, viral DNA synthesis, the event that delineates the early and late phases of infection, begins 8–10 hr after infection and is followed successively, as illustrated in Fig. 1, by synthesis of viral late mRNA, the viral structural proteins, and assembly of infectious particles (see Tooze, 1980 for details). Thus, by 12–15 hr after infection, all the components required to synthesize new virions are produced in increasing quantities as the infected cell devotes its biosynthetic machinery to the replication and expression of viral genetic information. Synthesis and assembly of components of adenovirions continue for many hours but eventually cease, probably because the inhibition of host cell metabolism is sufficiently drastic to kill the cell. The kinetics of the infectious cycle are influenced by both the multiplicity of infection and the growth state of the infected cell; not surprisingly, the cycle unfolds more rapidly in cells infected at high multiplicities. It has also been routinely observed that a typical productive infection proceeds more slowly in HeLa cells maintained in monolayer culture than those maintained in suspension (see, for example, Chow *et al.*, 1979; Yoder *et al.*, 1983; Flint *et al.*, 1983; Kao and Nevins, 1983). This phenomenon does not appear to

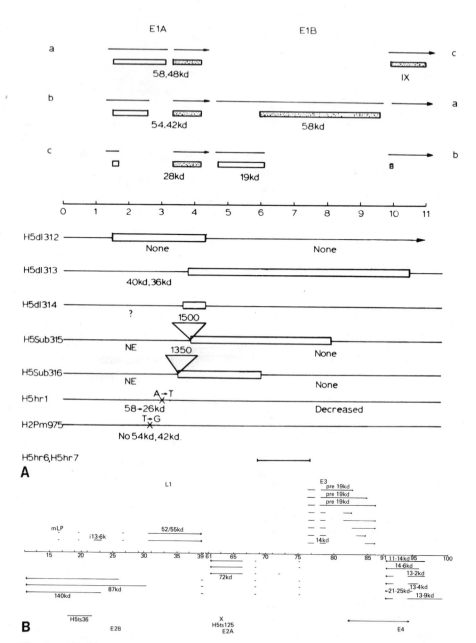

Fig. 2. (A) Structure of the adenovirus E1A and E1B early genes. The region 0–11 units (1 unit = 366 bp) of the subgroup C adenovirus genome is represented by the solid horizontal line in the center of the figure. The mRNA species complementary to each region are depicted by horizontal arrows above the genome drawn in the direction of transcription, in which the solid arrows and gaps represent sites of polyadenylation and sequences removed during splicing, respectively. The structures

reflect the specific delay of any one step, but rather an expansion of all phases in the cycle. This difference may reflect the less active growth of cells in monolayer culture, which are generally infected when close to confluence, or any one of a number of other differences such as in cell shape or attachment.

The program of viral gene expression that defines a productive infection has been subject to intense scrutiny, but will be described only briefly here; the interested reader is referred to Tooze (1980), Ziff (1980), or Flint (1982) for more detailed information. Early genes, defined as those expressed prior to the onset of viral DNA replication, are expressed as six transcriptional units E1A, E1B, E2, E3, E4,

---

of the mRNA species are from Berk and Sharp (1978), Chow *et al.* (1979), Perricaudet *et al.* (1979, 1980), and Aleström *et al.* (1980) and they are designated according to the nomenclature of Chow *et al.* (1979). The polypeptides specified by each species and their coding regions are shown below each mRNA; the stippled boxes show reading frames shifted by two nucleotides relative to frames represented by the open boxes (Perricaudet *et al.*, 1979, 1980; Halbert *et al.*, 1979; Aleström *et al.*, 1980; Esche *et al.*, 1980; Bos *et al.*, 1981). The molecular weight shown for E1A proteins are those of Esche *et al.* (1980). The sites of several mutations discussed in the text and their consequences for production of E1A or E1B proteins are shown below the genome. Only those polypeptides that differ from wild type are indicated. Deletions, substitutions, and point mutants are represented by □, ▽, and X, respectively. The limits within which the H5hr6 and H5hr7 mutations have been mapped are indicated by the bar. This summary is based on data from Frost and Williams (1978), Jones and Shenk (1979*a*), Ross *et al.* (1980), Esche *et al.* (1980), Ricciardi *et al.* (1981), and Montell *et al.* (1982). NE = not examined. (B) Organization of adenovirus early genes in the region 11–100 map units. The adenovirus genome is represented by the horizontal line, 11–100 map unit (1 mu = 0.366 kb) in which the regions 39–61 units and 86–90 units are omitted. The mRNA species encoded by regions E2A, E2B, E3, E4, and L1 are represented by the horizontal lines drawn above or below the genome in the direction of transcription. The sites at which poly(A) is added are represented by the solid arrows and sequences removed during splicing as gaps. The structures of the mRNA species are based on the data of Berk and Sharp (1978), Chow *et al.* (1979), and Stillman *et al.* (1980). For E2B (Stillman *et al.*, 1980; Binger *et al.*, 1982; Gingeras *et al.*, 1982; Aleström *et al.*, 1982; Friefeld *et al.*, 1983), L1 (Lewis and Mathews, 1980; Miller *et al.*, 1980; Lewis and Anderson, 1983), and E2A (Lewis *et al.*, 1976), the protein products are shown above or below the appropriate mRNA species. The proteins that have been assigned to the multiple mRNA species encoded by E3 (Persson *et al.*, 1980) or E4 (Downey *et al.*, 1983; Tigges and Raskas, 1982) are also listed.

The positions of several mutations discussed in the text are also indicated; the limits, ⊢——⊣, within which the H5ts36 mutation has been mapped by marker rescue (Galos *et al.*, 1979), the site X of the H5ts125 mutation in E2A (Kruijer *et al.*, 1981), and the deletion, ↔, carried by H2d1807 (Challberg and Ketner, 1981).

and L1. With the possible exception of L1, transcripts of each of the early regions are fashioned into multiple mRNA species that differ in the ways in which they are spliced and also, in certain cases (E2, E3, and E4), the sites at which the primary transcript is polyadenylated. It is not, therefore, surprising to learn that each early region encodes more than one polypeptide, although the array of products that has been assigned to each of E3 and E4 is startling. As illustrated in Fig. 2, considerable progress has been made in the laborious task of assignment of individual, early polypeptides to their mRNA species. Much less is known about the functions of early proteins. Those encoded by E3 are nonessential for growth of the virus in tissue culture and have not received much attention. Several that participate in viral DNA replication have been identified; these include the 72K DNA binding protein (DBP) of region E2A and the E2B products, the 87K precursor to the terminal protein which serves as the primer for DNA synthesis (Challberg et al., 1980, 1982; Lichy et al., 1981), and a DNA polymerase of 140K (Stillman et al., 1980, 1982; Binger et al., 1982; Friefeld et al., 1983; van Bergen and van der Vliet, 1983). Products of both the E1A and E1B regions function in transformation, but also play important roles in viral gene expression in permissive cells. The products(s) of the largest E1A mRNA potentiate transcription of all other genes expressed during the early phase (Berk et al., 1979; Jones and Shenk, 1979a; Ricciardi et al., 1981; Nevins, 1981), whereas, an E1B product permits the production of the large quantities of viral, late proteins that characterize the late phase of infection (B. Karger, Y-S. Ho, C. L. Castiglia, J. Williams, and S. J. Flint, in preparation). Despite the great amount of work that remains to establish the role of every early polypeptide, it is apparent that the adenovirus early functions form a more complex set than might have been supposed on the basis of the complexity of the genome and the fraction devoted to early genes.

Once the accumulation of the relevant early gene products permits adenoviral DNA replication to begin, a striking alteration in the pattern of viral gene expression also takes place. The major change is complete transcription of the major late transcriptional unit from its promoter site near 16.45 units (mLP; Ziff and Evans, 1978) to the right-hand end of the genome (see Fig. 3); early in infection, transcription initiated at the mLP halts somewhere in the middle of the genome (Shaw and Ziff, 1980; Akusjärvi and Persson, 1981; Nevins and Wilson, 1981). Only viral DNA molecules that have replicated can support complete transcription of the late, r-strand transcriptional

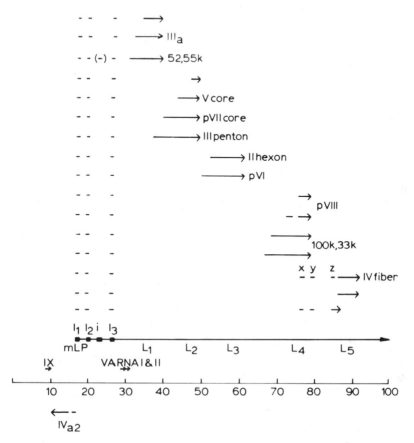

Fig. 3. Organization of adenvirus type 2 late genes. The adenovirus type 2 genome is represented by the horizontal line divided into 100 map units at the bottom of the figure. Transcripts of the r and l strands are shown as arrows, above and below the genome, respectively, drawn in the direction of transcription. The mRNA species fashioned from transcripts initiated at the major late promoter site (mLP; Ziff and Evans, 1978) by addition of poly(A) (→) and splicing are illustrated above the primary transcript. The sites of polyadenylation define the five families of late mRNA L1 to L5 as shown (Fraser and Ziff, 1978; Nevins and Darnell, 1978; McGrogan and Raskas, 1978). Sequences removed during splicing are shown as gaps (Berget et al., 1977; Chow et al., 1977; Klessig, 1977). The segments of the major late transcript that comprise 5'-terminal leader sequences of the mRNA species, l1, l2, l3, and i (Chow et al., 1979) are greatly exaggerated in size for clarity. The assignments of late polypeptides to individual mRNA species shown are based on hybridization selection and in vitro translation (Lewis et al., 1975, 1977) or hybrid-arrested translation (Paterson et al., 1977; Miller et al., 1980).

unit (Thomas and Mathews, 1980). As illustrated in Fig. 3, such late transcripts can be polyadenylated at five sites. Each poly(A)-containing product is then subjected to a series of splicing reactions that fashion the 5'-terminal leader segment, most commonly a tripartite leader comprising the segments 11, 12, and 13 shown in Fig. 3, and ligate it to one of several sites within each polyadenylated transcript. The net result is the synthesis of 16 or 17 late mRNA species that fall into five families of 3' coterminal species (see Fig. 3). The great majority, as summarized in Fig. 3, encode polypeptides that form the capsomers and internal core of adenovirions. Two virion polypeptides, IX and IVa$_2$, are, however, encoded by independent transcriptional units (see Fig. 3).

In addition to the early and late genes that are expressed as mRNA and are transcribed by the host cells' RNA polymerase form II, the adenoviral genome encodes two small RNA species, termed VA-RNA$_I$ and VA-RNA$_{II}$. These viral genes are transcribed by RNA polymerase III. Transcription of both occurs during the early phase of infection, but that of the VA-RNA$_I$ gene accelerates once an infection enters the late phase (Söderlund et al., 1976). Indeed, VA-RNA$_I$ is produced in much larger quantities than VA-RNA$_{II}$ and accumulates in the cytoplasm (Mathews and Pettersson, 1978), where it is essential for the efficient translation of viral mRNA species (Thimmappaya et al., 1982).

2.2. Abortive and Partially-Productive Infections

Many mammalian cells of nonhuman species permit entry of human adenoviruses. The success of the subsequent replication cycle varies with both the nature of the host cell and the serotype of the infecting virus. Rodent cells, mouse, rat, or hamster embryo fibroblasts for example, are semipermissive for the growth of Ad2 or Ad5 (members of subgroup C), permitting reproduction of these viruses to yields that are reduced up to 10^4-fold, compared to those attained in fully-permissive human cells (Takahashi, 1972; Williams, 1973; Gallimore, 1974; Younghusband et al., 1979; M. Zucker and S. J. F., unpbulished observations). By contrast, replication of Ad12, a member of subgroup A, is completely blocked at an early stage in rodent cells; no replication of viral DNA takes place, although early genes are expressed (Doerfler 1968, 1969; zur Hausen and Sokoh, 1969; Doerfler and Lundholm, 1970; Ortin et al., 1976). It is interesting that, even in permissive human cells, subgroup A serotypes such as Ad12 exhibit decreased cytopathenogenicity, a longer growth cycle and a more restricted range of cells in which they grow efficiently compared to Ad2 or Ad5 (Ledinko, 1967; Yamashita and Shimojo, 1969; Green *et al.*, 1971; J. Williams, personal communication). Such striking differences in the biological properties of subgroup A and subgroup C human adenoviruses have never been explained.

Indeed, the whole question of the variable responses of mammalian cells to different human adenovirus serotypes is poorly understood, despite the fascinating range of interactions displayed. The variable abilities of cells derived from different species, or different tissues, to support adenovirus replication is generally interpreted in terms of the presence, or lack of, cellular factors that the virus needs to complete various steps in the replication cycle. More precisely, putative factors that permit human adenovirus replication in human cells may be present in, say, murine cells, but sufficiently divergent that they fail to interact optimally with the relevant viral components. Thus, an understanding in molecular terms of the steps in the viral replication cycle that are blocked in non- or semipermissive cells should provide important information about the host cell components utilized by the virus and, thus, the molecular interactions among viral and cellular products.

An interesting example is provided by the replication of type C adenovirus 2 and 5 in simian cells. The latter do permit production of progeny virus, but the yield is reduced approximately 100- to 1000-fold compared to that obtained in human cells (Rabson *et al.*, 1964; Friedman *et al.*, 1970). The defect in certain Ad2-infected monkey cells, such as CV-1, appears to lie in the processing and translation of viral, late mRNA species, especially that encoding fiber polypeptide IV (Klessig and Anderson, 1975; Klessig and Chow, 1980). Interestingly, host-range mutants of Ad2 that have acquired the ability to grow as well in simian cells as in human cells (Klessig, 1977) carry mutations within the E2A gene (Klessig and Grodzicker, 1977) encoding the 72K DNA-binding protein (see Fig. 2B). These observations, therefore, implicate an interaction between the DBP and one or more components of the cellular processing or translational machinery in the efficient expression of viral, late mRNA species such as that specifying polypeptide IV. Alternatively, (but less likely) the altered DBP that permits efficient growth of Ad2 in monkey cells could provide a function supplied by a cellular factor in human cells, but that is lacking (at least in a functional sense) in simian cells. Be that as it may, these studies of the replication of human adenovirus in monkey cells revealed a previously unknown function of the DBP. In a similar vein, the properties of another class of host-range mutants of Ad5 suggest that some viral functions may be redundant in certain cell types. The host-range mutants in question were originally selected for loss of the ability to grow efficiently in typical host cells, such as HeLa, being complemented in transformed human cells that retain and express E1A and E1B sequences (Harrison *et al.*, 1977). However, host-range mutants that carry lesions within E1B can replicate in human embryonic kidney cells and certain human tumor cells almost as well as in adenovirus-transformed human cells (Harrison *et al.*, 1977; J. Williams, personal communication) suggesting that a cellular product of the latter cell types (which is not present in HeLa cells) can complement the E1B lesions.

2.3. General Features of the Response of Permissive Cells to Adenovirus Infection

Some of the pioneering studies of adenovirus infection performed during the decade following the virus' discovery elucidated the general properties of the response of permissive cells to adenovirus infection as descriptions in morphological terms were rapidly succeeded by investigations of alterations in macromolecular synthesis.

A day or so after an adenovirus has been adsorbed to human cells (such as HeLa) in monolayer culture and the virus innoculum removed, characteristic changes in cell morphology can be discerned; infected cells appear more rounded and refractile than their uninfected counterparts and tend to clump and detach from the substratum (Hilleman and Werner, 1954; Rowe *et al.*, 1958; Periera, 1958). Characteristic cytopathic changes induced by adenoviruses are illustrated in Fig. 4. If the infected culture is left for a sufficient period, usually two to several days, all cells in the culture will display the characteristics of infected cells and detach from the substratum in clumps. The rate and extent of the initial appearance of the cytopathic response are determined by the multiplicity with which the culture was infected; it is obvious that this parameter will determine the number of cells initially infected and, thus, the number that show cytopathic changes as a result of the first cycle of virus replication. However, as discussed previously, cells that have been infected at higher multiplicities also progress through the replication cycle more rapidly than those infected at lower multiplicities.

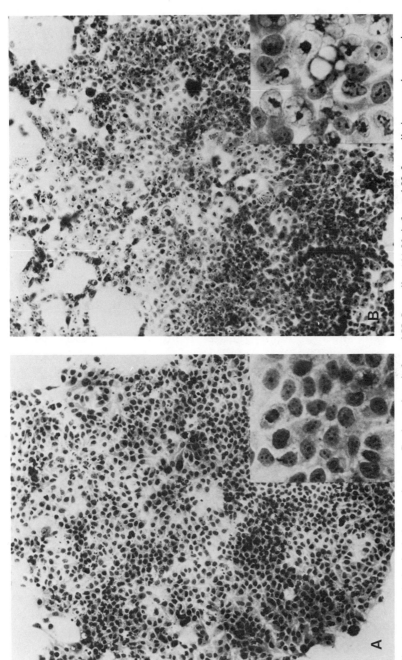

Fig. 4.   Morphological changes in type C adenovirus-infected HeLa cells. (A) Uninfected HeLa cells in monolayer culture. Magnification, × 1.2; insert, × 52. (B) HeLa cell culture 3 days after Ad5 infection. Magnification, × 1.2; insert, × 52. From Periera (1958).

Distinct morphological features are difficult to discern in cells maintained in suspension culture. Nevertheless, it is possible with some experience to recognize suspended cells that have been infected by an adenovirus; after about 24 hr, infected cells display enlarged nuclei and surface alterations. The failure of such a culture to continue growth (Green and Daesch, 1961) provides another marker for adenovirus infection.

A number of early studies suggested that a structural protein of the virion contributed to the cytopathic effect (cpe); a factor that induced cells to display typical cpe could be separated from infectious virus by ultracentrifugation (see for example, Periera, 1958). Moreover, when virion preparations were treated with trypsin or removed from infected culture after an appropriate period of adsorption, appearance of the cytopathic changes, visible in a few hours when untreated preparations were left in contact with the host cells, was substantially delayed (Periera, 1958; Rowe *et al.*, 1958; Everett and Ginsberg, 1958). These observations suggested that a protein present in virions and made relatively late in the infectious cycle was responsible for the gross morphological changes that characterize adenovirus-infected cells. Subsequently, Pettersson and Höglund (1969) demonstrated that the highly purified penton of Ad2 virions (native molecular weight 400–515K and comprising polypeptide III, 85K; see Tooze, 1980), was sufficient to induce human cells in monolayer culture to become refractile and swollen and to detach from the substratum.

The alterations in nuclear morphology that characterize adenovirus-infected cells, including a substantial enlargement of that organelle, have been compared in some detail among a number of human serotypes (see Brandon and McClean, 1962; Norrby, 1971 for reviews). The characteristic changes in the infected cell nucleus, including its enlargement, can largely be ascribed to the presence of crystalline arrays of newly-assembled virus, which are readily demonstrated by appropriate staining techniques (Kjellen *et al.*, 1955; Harford *et al.*, 1956; Morgan *et al.*, 1957), but may also reflect alterations in cellular RNA metabolism.

Adenovirus infection of rapidly growing permissive cells, for example, HeLa cells in suspension culture, leads to the cessation of cell division in 10–12 hr (Green and Daesch, 1961). Nevertheless, such infected cells do continue macromolecular synthesis and have been reported to contain twice the mass, compared to uninfected cells, of DNA and protein by 32 hr after infection (Green and Daesch, 1961;

Ginsberg *et al.*, 1967; Piña and Green, 1969). Similarly, the mass of RNA per infected cell as well as the rate of total RNA synthesis appear to increase by 30–80% during the first 24 hr of an adenovirus infection. Both parameters subsequently decline to levels that are substantially lower than those measured in uninfected control cells (Ginsberg *et al.*, 1967; Piña and Green, 1969). The first investigators of these phenomena were limited to the distinction of viral from cellular macromolecules, relying, for example, on hybridization methods, or the distinct properties of adenoviral DNA (see Tooze, 1980). Nevertheless, the dramatic consequences of an adenovirus infection for the host cell were unequivocally demonstrated. The most striking of these must be the inhibition of host cell macromolecule production that characterizes productive infection of fully permissive cells, the complete usurpation of the cells' biosynthetic machinery by the virus. But, under other circumstances, adenoviruses can induce infected cells to synthesize their DNA (or certain gene products) and enter an aberrant cell cycle.

3. EFFECTS OF ADENOVIRUS INFECTION ON CELLULAR DNA METABOLISM

3.1. Inhibition of Cellular DNA Synthesis in Productively-Infected Cells

The best known response of cellular DNA synthesis to adenovirus infection must be the inhibition typically seen when permissive human cells are infected. In growing cells, cellular DNA synthesis is inhibited by more than 50% by 6–8 hr and completely by 10–13 hr after an adenovirus infection (Ginsberg *et al.*, 1967), concomitant with the initiation and acceleration of viral DNA synthesis. Unfortunately, very few studies have concentrated on elucidation of this inhibitory process. Hodge and Scharff (1969) presented some evidence to suggest that initiation of new rounds of cellular DNA synthesis was prevented by adenovirus infection. Thus, when synchronized human cells were infected at a time in their cell cycle that placed viral DNA replication in the G1 phase, neither premature, nor normal S phase, initiation of cellular DNA synthesis were observed. Alternatively, infection to place initiation of viral DNA synthesis after the onset of the host cells' S phase permitted completion of those rounds of cellular DNA synthesis begun before the inhibitory mechanism exerted its effect, but no initiation of new cycles of cellular DNA replication: in other words, a rapid decrease in cellular DNA synthesis was observed as viral DNA replication began. These observations implicate competition between cellular and viral templates for components of the host replication machinery. Conversely, the ability of cells in the G1 phase of the cell cycle to support viral DNA synthesis, also reported by Hodge and Scharff (1969), might seem to indicate that initiation of viral DNA replication is relatively independent of those host cell replication factors that participate in initiation of cellular DNA synthesis. The recent elucidation of the mechanism of initiation of adenoviral DNA synthesis emphasizes, at least at first sight, the independence of the virus during this step; priming of viral DNA synthesis is the function of the 87K precursor to the terminal protein (Fig. 5) (Challberg et al., 1980) which covalently binds the first nucleotide of the new DNA chain, dCMP in the case of Ad2 or Ad5 (Lichy et al., 1981; Challberg et al., 1982; Ikeda et al., 1982). This reaction also requires the adenoviral DNA polymerase, a 140K protein encoded by early region E2B (Lichy et al., 1982; Enomoto et al., 1981; Stillman et al., 1982; Friefeld et al., 1983; van Bergen and van der Vliet, 1983). Indeed, initiation of DNA synthesis, at least in vitro, can take place in the presence of only a complex of the DNA polymerase with the terminal protein precursor and, as template, the viral genome (or terminal fragments) bearing the mature, 55K form (Robinson et al., 1973; Rekosh et al., 1977) of the terminal protein. Nevertheless, a host nuclear protein termed nuclear factor I stimulates the covalent linkage of dCMP to the 87K preterminal protein and is especially necessary to this reaction when the 72K DBP is also present (Nagata et al., 1982). Neither the requirement for nuclear factor I during adenovirus DNA replication in vivo nor its role in cellular DNA synthesis have been established. It is, consequently, difficult to judge the significance of the observations made with in vitro replication systems in relation to inhibition of cellular DNA synthesis in vivo.

It seems quite clear that cellular factors participate in later steps in adenoviral DNA synthesis, for the in vitro systems that permit complete and faithful replication of a viral DNA-terminal protein template are dependent on them (Nagata et al., 1983). Thus, competition for a cellular protein that functions beyond initiation also remains a real possibility. Unfortunately, the detailed examinations of the properties of cellular DNA made as the inhibitory effects of an adenovirus infection take hold, which are necessary to assess such an hypothesis, have not been performed. In the one study of this kind reported, Pater

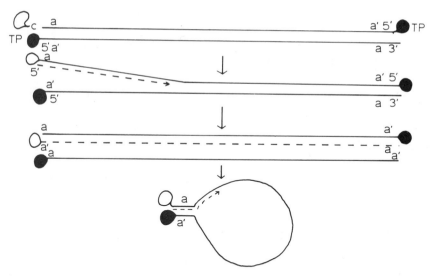

Fig. 5. Schematic representation of adenovirus replication. The linear viral genome with the 5′-linked, terminal protein, ●, 55K is shown associated with the 87K precursor to the terminal protein, 0, to which CMP has become covalently linked in the first step of the initiation reaction, priming. A new DNA chain is then elongated from this primer, displacing its parental homologue (steps 1 and 2 in the figure). This displaced strand is believed to circularize by virtue of the inverted terminal repetition, represented as a-a′, to form a double-stranded, terminal segment that is identical to that of the parental DNA. Such a scheme was originally proposed by Rekosh et al. (1977) and has been modified to include the priming role of the precursor to the terminal protein discussed in the text.

and Mak (1975) observed that ligation of Okazaki fragments of cellular DNA took place more slowly in adenovirus-infected cells. Adenovirus DNA synthesis occurs by a strand displacement mechanism, illustrated in Fig. 5, in which each daughter strand is synthesized in a continous fashion; no viral DNA segments corresponding to Okazaki fragments are made. As the mechanisms of cellular and viral DNA replication are, in this respect, so different, it is difficult to see how the observation of Pater and Mak (1975) can provide an explanation, based on competition for common elements, for the effect of the virus on host cell DNA synthesis. Indeed, Pater and Mak (1975) suggested that the slower rate of ligation of Okazaki fragments was probably a general consequence, rather than a cause, of inhibition of cellular DNA synthesis, for the same phenomenon was induced by inhibitors other than adenovirus.

Our understanding of the mechanism(s) whereby adenovirus infection of growing, permissive cells inhibits replication of cellular

DNA clearly leaves much to be desired. Nor has a great deal been learned about the viral products that might mediate the profound inhibition typically observed in these circumstances, for little concerted effort has been applied to this question. Ginsberg and colleagues (1967) originally suggested that the fiber antigen might cause the inhibition of cellular DNA synthesis. Although it has not been established absolutely that an adenoviral, early product is responsible, the kinetics of the inhibitory process are most consistent with this notion and difficult to reconcile with a role for the fiber, which does not accumulate in substantial quantities until relatively late in the infectious cycle; as mentioned previously, inhibition of cellular DNA synthesis is complete some 10–12 hr after an infection has been initiated.

Inhibition of cellular DNA synthesis has generally been examined during the phenotypic screening of collections of adenovirus mutants. Temperature-sensitive mutants of Ad5 bearing lesions in the E2A gene product, the 72K DNA-binding protein (Levine *et al.*, 1975; Grodzicker *et al.*, 1977), or the 140K DNA polymerase encoded by region E2B (see Fig. 2B) fail to replicate their DNA under nonpermissive conditions of infection. Nevertheless, they do induce inhibition of cellular DNA synthesis (Ensinger and Ginsberg, 1972; Wilkie *et al.*, 1973). Adenovirus 12 DNA-negative mutants appear to exhibit similar properties (Ledinko, 1974; Shiroki and Shimojo, 1975). These phenotypes establish that viral DNA replication *per se* is not necessary to the inhibition of cellular DNA synthesis. Group II host-range mutants of Harrison *et al.* (1977) that carry lesions within early region E1B (Frost and Williams, 1978) also inhibit cellular DNA synthesis in nonpermissive cells (Lassam *et al.*, 1978; Karger *et al.*, in preparation). Indeed, the only mutants that have been found to be defective in inhibition of cellular DNA synthesis are host-range and host-range, cold-sensitive mutants of Ad5 whose lesions lie within early region E1A (Harrison *et al.*, 1977; Frost and Williams, 1978; Ho *et al.*, 1982; Karger *et al.*, in preparation). Of the several E1A mutants that fail to inhibit cellular DNA synthesis in their host cells, all are also negative for viral DNA synthesis (Lassam *et al.*, 1978; Karger *et al.*, in preparation). Only the mutation carried by H5hrl has been mapped precisely, to a site within the region that is uniquely expressed in the largest of the three E1A mRNA species (Ricciardi *et al.*, 1981; see Fig. 2A for a summary of the organization and expression of adenovirus 5 early region E1A).

As the H5hrl mutation, like deletions within E1A, also prevents the expression of other early genes (Berk *et al.*, 1979; Jones and

Shenk, 1979*b*), it might be argued that the apparent requirement of the 58K and 48K polypeptides encoded by the largest E1A mRNA (Halbert *et al.*, 1979; Esche *et al.*, 1980; Ricciardi *et al.*, 1981; see Fig. 2A), in fact, reflects the role of some other early gene product that cannot be made in cells infected by such E1A mutants. As we have seen, this argument cannot apply to regions E2A or E2B. Similarly, cells infected by a viable deletion mutant of Ad2, H2dl807, in which substantial portions of the C-terminal regions of early regions E3 and E4 are deleted, are not only wild type with respect to viral DNA synthesis, but also inhibit cellular DNA synthesis normally (Challberg and Ketner, 1981). On the other hand, each of these two early regions encodes a surprisingly large number of polypeptides (and, corresponding, mRNA species; see Fig. 2B) not all of which would be affected by the H2dl807 deletion. Thus, it seems premature to conclude that the products of the largest E1A mRNA species are directly required to inhibit cellular DNA synthesis. Clearly, almost everything remains to be learned about the roles of adenoviral gene products in inhibition of cellular DNA synthesis; significant progress will, however, require an acceleration in the snails' pace that has characterized research into this topic ever since the adenoviruses were discovered in 1953.

Over the years, reports linking adenoviruses with DNase activities have appeared with some regularity. Unfortunately, the significance of many of these observations with respect to inhibition of cellular DNA synthesis remains difficult to evaluate. The purified penton base or penton base-containing structures of several adenovirus serotypes have been found to possess DNase activity (Burlingham *et al.*, 1971; Burlingham and Doerfler, 1972; Marusyk *et al.*, 1975). Burlingham and Doerfler (1972) described an activity both in such structures and extracts of Ad2 or Ad12-infected cells that cleaved double-stranded viral DNA into segments of approximately 5×10^6 daltons. By contrast, Marusyk *et al.* (1975) detected only a single-stranded specific endonuclease in purified pentons. In view of these discrepancies, the status of the reported endonuclease activities must be considered open to question.

More recent observations have suggested that adenovirus infection specifically inhibits one or more cellular DNases. In cells infected with H2*ts*111, a DNA-negative early mutant, at a nonpermissive temperature, both viral and cellular DNA become considerably more degraded than in corresponding wild-type infections (D'Halliun *et al.*, 1981). Somewhat enhanced degradation of viral DNA was also observed under permissive conditions. This observation implies the continued activity of a nonspecific (with respect to viral and cellular DNA sequences) DNase in H2ts111 compared to wild-type infected cells. Similarly, Nass and Frenkel (1978) have observed the inhibition of cellular, alkaline DNases III and IV, which recognize single-stranded DNA substrates, during the course of Ad5 infections. No similar inhibition of the DNases was observed in cells infected by H5ts125 under nonpermissive conditions, a phenotype that Nass and Frenkel (1980) attributed to the failure of the mutated 72K DBP specified by H5ts125 to bind to, and thus protect, single-stranded DNA substrates of the cellular enzymes. In other words, cellular DNases do not appear to be inhibited directly, rather their substrates are simply sequestered by the large quantities of viral DNA-binding protein made in infected cells. The results of D'Halliun *et al.* (1981) also implicate the product of the gene mutated in H2ts111, as yet unidentified, in protection of both viral and cellular DNA sequenced from DNase degradation.

No evidence for the selective protection of parental and newly-replicated viral DNA, a property that might partially account for the reduced levels of cellular DNA made in infected cells, has emerged from the studies on H5ts125 or H2ts111. Nevertheless, a clear link between inhibition of DNA degradation and the cytopathogenicity of adenoviruses has been established, the result of characterization of the *cyt* mutants of Ad12. These interesting mutants, the first of which were isolated by Takemori *et al.* (1968), exert a far stronger cytopathic effect than wild-type Ad12; they induce the formation of large, clear plaques that differ markedly from the small, fuzzy variety typical of wild-type Ad12. As might be expected, cells infected by *cyt* mutants are also more rapidly destroyed than those infected by wild-type Ad12. An interesting property of the *cyt* mutants is their low oncogenicity *in vivo* (Takemori *et al.*, 1968; Yamamoto *et al.*, 1972). Some, but not all, of the *cyt* isolates that have been examined also transform rat cells *in vitro* with considerably reduced efficiency (Takemori *et al.*, 1968; Takemori, 1972; Mak and Mak, 1983). It was established, soon after this class of mutants was described, that revertants of *cyt* mutants that had recovered wild-type plaque morphology and cyto-pathenogenicity had also regained the high oncogenicity characteristic of wild-type Ad12 (Takemori, 1972). Thus, it is difficult to escape the conclusion that a single genetic change is responsible for the apparently unrelated phenotypes displayed by *cyt* mutants.

Human cells infected with Ad12 *cyt* mutants produce far fewer virions than those infected by wild-type Ad12 (Ezoe and Mak, 1974), a phenotype that results, at least in part, from the extensive degradation of newly-synthesized DNA that takes place in *cyt* mutant infected cells; cells infected by the mutant H12*cyt*70, for example, accumulate only 50% of the wild-type level of viral DNA, and most of that is much shorter than intact Ad12 DNA as judged by sedimentation in alkaline or neutral sucrose gradients (Ezoe *et al.*, 1981). Pulse-chase experiments established that viral DNA was synthesized normally in Ad12 *cyt* mutant-infected cells, but degraded within a few hours of its synthesis (Ezoe *et al.*, 1981). The observation that newly-synthesized cellular DNA shares the same fate in *cyt* mutant, but not wild-type Ad12-infected cells (Ezoe *et al.*, 1981), provides one explanation for the rapid cell-killing characteristic of this class of mutants and implies that the viral product(s) mutated in the *cyt* viruses normally prevents the degradation of both adenoviral and cellular DNA.

The sites of the lesions carried by certain *cyt* mutants have been located by complementation tests. When human cells were coinfected with H12*cyt*70 and either wild-type Ad12, wild-type Ad5, or Ad5 mutants whose lesions lie within region E1A, such as H5hr1 and H5d1312 (see Fig. 2A), no DNA Degradation was observed (Ezoe *et al.*, 1981; Lai Fatt and Mak, 1982), indicating that the *cyt* defect can be complemented in trans and does not affect region E1A. By contrast, H5d1313, from which the C-terminal portion of E1A and most of region E1B are deleted (Jones and Shenk, 1979*a*), not only failed to complement H12*cyt*70 in mixed infections, but itself induced degradation of cellular DNA in single infections (Lai Fatt and Mak, 1982). Thus, the viral product involved in inhibition of DNA degradation and defective in Ad12 *cyt* mutants can confidently be assigned to early region E1B. As illustrated in Fig. 2A, this region encodes at least three polypeptides, the 58K and 19K early proteins, and the virion component polypeptide IX. Human 293 cells carry integrated viral DNA sequences of regions E1A and E1B but do not express the gene for polypeptide IX (Aiello *et al.*, 1979; Spector *et al.*, 1980; Colby and Shenk, 1981). The ability of 293 cells to complement H12*cyt*70 eliminates a role for protein IX. Similarly, the properties of the host-range mutant H5hr6, whose lesion prevents the expression of the E1B 58K protein (Lassam *et al.*, 1978; Ross *et al.*, 1980; Frost and Williams, 1978) indicate that the 58K protein is not involved; H5hr6 can supply the function that inhibits DNA degradation and that is lacking in H5d1313 or H12cyt70. Nor does H5hr6 itself induce DNA degradation in infected cells. These observations would certainly seem to point to the 19K protein, the remaining well-characterized product of region E1B, as the most likely candidate for a role in inhibition of DNase activity in adenovirus-infected cells. Indeed, some evidence that cyt-mutant-infected cells fail to synthesize the 19K protein has been collected (cited in Lai Fatt and Mak, 1982). It is, however, important to bear in mind that it is not known whether a truncated version of the 58K protein is produced in H5hr6-infected cells. The examination of the phenotypes of additional E1B mutants and of the products whose synthesis they direct should establish unequivocally which of the region E1B products protects viral and cellular DNA from degradation in adenovirus-infected cells.

Despite the uncertainty with regard to this question, analysis of the Ad12 cyt mutants has established that degradation of cellular DNA in infected cells can make a marked contribution to the cytopathogenicity displayed by the virus. This property, combined with the inhibition of cellular DNA synthesis, which appears to be mediated by distinct viral products from those altered in the cyt mutants, surely accelerates cell death and release of progeny virions. What is perhaps the central aspect of this phenomenon, the underlying cause of the susceptibility of cellular DNA to degradation in adenovirus-infected cells when functional E1B or E2A products are not made, is, however, still a complete mystery.

3.2. The Induction of Cellular DNA Synthesis in Quiescent Cells

By contrast to the complete inhibition of cellular DNA synthesis that follows adenovirus infection of growing, permissive cells, stationary, confluent human cells infected by Ad2 or Ad12 display an induction of cellular DNA synthesis (see for example, Ledinko, 1967; Takahashi et al., 1969; Yamashita and Shimojo, 1969). In this respect, adenoviruses resemble papovaviruses such as SV40 or polyomavirus (Minowada and Moore, 1963; Winocour et al., 1965; Dulbecco et al., 1965; Weil et al., 1965). Interestingly, a similar response is invoked in rodent cells, regardless of whether they are partially permissive or completely nonpermissive for the growth of the particular serotype with which they have been infected. As rodent cells are those that become transformed by human adenoviruses (see Tooze, 1980), the phenomenon of induction of cellular DNA synthesis has been viewed as an early, important step in the transformation process and consequently examined in some detail.

Induction of cellular DNA synthesis has been reported to occur in confluent or serum-starved (and therefore growth-arrested) hamster (BHK or their derivatives) cells infected by Ad12 (Strohl, 1969a,b; Zimmerman et al., 1970; Laughlin and Strohl, 1976a) or Ad2 (Laughlin and Strohl, 1976b; Rossini et al., 1979), in serum-starved rat or mouse fibroblasts infected by Ad5 (Younghusband et al., 1979; Braithewaite et al., 1981), and, as mentioned previously, in Ad2 or 12-infected stationary human cells. Thus, the response appears to be quite general among cells of different mammalian and rodent species and independent of the serotype of the infecting adenovirus. When the vast majority of rodent cells in a culture have been arrested and are not engaged in DNA synthesis, Ad12 or Ad2 infection induces DNA synthesis in at least 70% of the cells in 20–30 hr, as judged by [^{3}H]thymidine labeling followed by autoradiography (Takahashi et al., 1969; Strohl, 1969a,b; Rossini et al., 1979). In these experiments as well as others in which the amount of labeled DNA made after infection was measured (Rossini et al., 1979, for example), little difference was observed in the response of arrested cells to Ad2 (group C) or Ad12 (group A). However, for reasons that have never been clarified, the actual magnitude of the induction reported in different experiments has varied from only a 1.6- to threefold increase in incorporation of [^{3}H]thymidine when Ad2 infects arrested mouse cells to some tenfold in Ad2-infected hamster or Ad12-infected human cells (Shimojo and Yamashita, 1968; Yamashita and Shimojo, 1976; Rossini et al., 1979). The variable efficiences of induction described do not simply reflect the inadequacy of assays that count the number of cells induced to synthesize DNA or measure total DNA synthesis, important because many rodent cell lines permit some degree of group C adenovirus replication, but no replication of group A adenovirus DNA (see Section 2.2), for in many experiments viral and cellular DNA were distinguished by equilibrium centrifugation in CsCl gradients or by other methods (Shimojo and Yamashita, 1968; Rossini et al., 1979; Younghusband et al., 1979; Braithewaite et al., 1981). Of course, it remains possible that the differences reflect the use of different cell lines from different rodent species or subtle differences in the degree of arrest of the cells at the time of infection.

The induction of cellular DNA synthesis that occurs in adenovirus-infected rodent cells, and their subsequent progression through the cell cycle, can be distinguished from the response of the same arrested cells to serum stimulation by many criteria, including the 5–10 greater magnitude of the response to serum (Braithewaite *et al.*, 1981). Adenovirus type 5 can, for example, induce cellular DNA synthesis in rat cells in the presence of concentrations of dibutyrl cAMP that inhibit serum stimulation (Braithewaite *et al.*, 1981). Similarly, arrested BHK cells are induced by Ad2 to synthesize cellular DNA in the presence of 0.03 µg/ml actinomycin D, conditions that completely prevent stimulation by serum (Laughlin and Strohl, 1976*b*) or, surprisingly, Ad12 (Laughlin and Strohl, 1976*a,b*). Adenovirus type 2 infection of a temperature-sensitive derivative of BHK cells, *ts*AF8, that become arrested in the G1 phase of the cell cycle at a nonpermissive temperature of 40.5°C, can, by contrast to treatment with 10% serum, infection with the papovavirus polyoma, or infection with Ad12, overcome the block and induce cellular DNA synthesis (Rossini *et al.*, 1979). No Ad2 DNA is made under these conditions. These results clearly distinguish the response of arrested cells to adenovirus infection from that following serum stimulation. It, therefore, appears that the common response, initiation of cellular DNA synthesis, is achieved by quite different mechanisms, at least when considering subgroup C adenovirus infection and stimulation by serum. Whether the apparent similarity in the response induced by Ad12 and serum observed in experiments described in this paragraph extends to the mechanism of induction has not been established, but seems unlikely for reasons discussed subsequently.

The induction of arrested mammalian cells into the cell cycle has also been well characterized with respect to events that precede DNA synthesis itself. These include synthesis of RNA and the induction of synthesis of such enzymes as thymidine kinase and ornithine decarboxylase (see Baserga, 1976; Pardee *et al.*, 1978, for reviews). Quiescent cells that have been infected by an adenovirus do not display the accumulation of RNA that is typical of serum-stimulated cells (Pochron *et al.*, 1980), nor is synthesis of ornithine decarboxylase induced (Cheetham and Bellett, 1982). Both serum stimulation and adenovirus infection do induce synthesis of thymidine kinase (Kit *et al.*, 1965; Takahashi *et al.*, 1966; Ledinko, 1967; Zimmerman *et al.*, 1970), but the effects of simultaneous treatment with α-methyl-ornithine indicate that the pathways of induction are not the same; this agent has little effect on induction of thymidine kinase by Ad5, but inhibits the response to serum (Cheetham and Bellett, 1982). It, therefore, seems fair to conclude from this brief survey of the properties of the process whereby adenoviruses induce cellular DNA synthesis that it is not mediated in the same way as the "normal" induction of quiescent cells into the cell cycle, for example, following addition of serum to their medium. Indeed, it is clear that adenovirus-infected cells can bypass several events that define the normal G1 to S phase progression. Moreover, later events in the cell cycle are also abnormal in adenovirus-infected cells.

Adenovirus infection of G1-arrested cells does induce complete progression through the cell cycle, but also leads to the accumulation of aneuploid and polyploid cells, with greater than the G2 content of DNA (Braithewaite *et al.*, 1981; Murray *et al.*, 1982*a*). Flow cytometry, performed 1, 2.5, or 5 days after Ad5 infection of growing, unsynchronized rat cells, has established that an increased number of cells exhibit a G2 diploid or greater content of DNA; by 5 days, one-third of the infected cells displayed aneuploid or polyploid quantities of DNA and one-half had G2 or greater contents. Corresponding changes were not observed in serum-stimulated, resting cultures or uninfected growing cultures (Braithewaite *et al.*, 1981; Murray *et al.*, 1982*a*). Thus, the induction into the cell cycle mediated by adenovirus infection can be considered to be grossly abnormal. Examination of the infected cells suggests that cell fusion is not the origin of polyploid or aneuploid cells. Rather, mitosis appears to begin normally, but initiation of a new round of DNA synthesis takes place prematurely, before chromosomal segregation during the first round is complete (Murray *et al.*, 1982*a*). Both this type of abnormality, as well as those of the S phase discussed previously, presumably contribute to the accelerated cell cycle characteristic of adenovirus-infected cells (Braithewaite *et al.*, 1981; Murray *et al.*, 1982*a*). The molecular mechanisms that cause the earlier and later aberrations and thus the question of whether the S phase and later abnormalities bear any mechanistic relationship to one another are not yet understood.

The aberrant mitoses that occur in adenovirus-infected rodent cells suggest that chromosomal damage might be a common feature of adenovirus infections. Indeed, such consequences have been seen consistently in both infected and transformed cells. The subgroup A member Ad12 appears to be particularly efficient in this respect; Ad12-infected hamster BHK21 cells, which are completely nonpermissive for replication of this serotype (see Section 2.2), begin to display a very disorganized mitosis, characterized by highly fragmented, over-condensed chromsomes as well as chromatid exchanges, by 30 hr after infection (zur Hausen, 1967; Stich and John, 1967; Cooper *et al.*, 1968; Strohl, 1969*a*,*b*). Such severe chromosome damage unquestionably contributes to the killing of these nonpermissive cells by Ad12. Indeed, cell survival, measured as the cloning efficiency of the infected cells, is inversely related to the incidence of chromosomal damage (zur Hausen, 1967; Strohl, 1969a). Similar chromosomal aberrations, including the formation of chromosomal bridges and fragments and coiling deficiences during anaphase have been observed in Ad5-infected, growing rat embryo cells (Stich and John, 1967; Stich, 1973; Murray *et al.*, 1982a). Murray *et al.*, (1982a) described successive waves of chromosomal damage with a cyclic period corresponding to the shortened cell cycle typical of adenovirus-infected cells. A comparison of the time at which damage was first detected to the synthesis of DNA first revealed damage at the second mitosis. Murray and colleagues (1982a), therefore, suggested that cells in the asynchronous population that they were studying were most susceptible to virus-induced damage when in S phase at the time of infection, such that a peak of damage was observed every time that cohort underwent mitosis.

Such chromosomal damage as the induction of gaps, breaks, or fragmentation is not restricted to cells that cannot support efficient adenovirus replication, for it has also been observed in permissive cells infected by Ad2 or Ad12 (zur Hausen, 1967; McDougall, 1971, 1975). In addition, more specific effects have been described. McDougall (1971), for example, reported that the three group A members Ad12, 18, and 31 damage the same site on human chromsome 17. In order to identify more precisely the site of this apparently specific chromosomal lesion, a mouse–human hybrid cell line carrying a translocation of part of human chromosome 17 to a mouse chromosome was infected with Ad12. The infected, hybrid cells exhibited both random chromosomal damage and nonrandom gaps within the 17q region of the translocated portion of human chromosome 17 (McDougall *et al.*, 1973). The latter abberation was observed in 32% of the metaphase spreads examined. The human thymidine kinase (tk) gene was retained by the mouse–human hybrid cell line used, which had no human genetic information except for the translocated portion of chromosome 17. It was, therefore, possible to correlate the retention of tk activity with specific break-points introduced after virus infection (McDougall *et al.*, 1973); when the region 17q22 was lost, the cells reverted to a thymidine kinase negative phenotype. In other clones established after Ad12 infection, tk activity was retained. Thus, although a specific chromosomal region was particularly susceptible to adenovirus-induced damage, that specificity did not extend to the level of the DNA sequence. The mechanism whereby this and other more consistently observed chromosomal abberations, for example, on human chromosome 1 (McDougall, 1971), are induced is not understood. The sites affected do not, however, appear to correspond to sites of preferential integration of viral DNA (McDougall et al., 1972). Indeed, evidence for site-specific integration of adenoviral DNA has never been produced (see Tooze, 1980).

The mutagenic capacity of adenovirus has also been examined at the hypoxanthine phosphoribosyl transferase (HPRT) locus of rodent cells (Marengo et al., 1981; Paraskeva et al., 1983). Adenovirus 2 infection of HPRT(−) cells causes an increase in the frequency of reversion to the HPRT(+) phenotype of up to nine-fold measured in fluctuation tests. A similar effect was induced by complete or incomplete particles. The resulting revertants were stable for many generations when cultured in nonselective medium (Marengo et al., 1981) implying that the virus had caused a stable change in the genotype of the cell. The nature of this change has not been investigated.

Efforts to identify viral functions that are responsible for induction of cellular DNA synthesis have made use of mutants of Ad5. The mutants H5ts36 and H5ts125 that bear lesions in early regions E2A and E2B, respectively (see Fig. 2B, Galos et al., 1979; Grodizker et al., 1977; Gingeras et al., 1982; Aleström et al., 1982), are indistinguishable from wild-type (wt) Ad5 in their ability to induce cellular DNA synthesis in quiescent rodent cells under nonpermissive conditions of infection (Braithewaite et al., 1981; Cheetham and Bellett, 1982). Like wt Ad5, H5ts36- and H5ts125-infected cells also enter an aberrant cell cycle (Murray et al., 1982b). However, the population of cells infected by H5ts125 at the nonpermissive temperature of 39.5°C and induced to enter the cell cycle displayed peaks of cells with 8n, 16n, and 32n DNA contents, compared to peaks of between 4n and 8n observed at the permissive temperature of 33°C or in wild-type infected cells (Murray et al., 1982b). Despite this unexplained difference, these experiments establish that neither the E2A nor E2B gene products affected by the mutations tested (see Fig. 2B) are required for the induction of cellular DNA synthesis or entry into the cell cycle. By contrast, several mutants carrying lesions that completely or partially abolish E1A function fail to induce synthesis of cellular DNA or thymidine kinase in infected rat cells or the altered cell cycle progression characteristic of wild-type infection (Rossini et al., 1981; Braithewaite et al., 1983); this phenotype has been observed for cells infected with H5dl312 and H5dl314, whose deletions are confined to the E1A region (Jones and Shenk, 1979*a*; see Fig. 2A) and with H5hr1 and H5hr3, mutants that fail to grow efficiently in permissive HeLa cells (Harrison *et al*, 1977). The DNA of the H5hr1 variant carries a point mutation in that portion of the E1A gene that is uniquely expressed in the largest E1A mRNA and its protein products (Ricciardi *et al*., 1981; see Fig. 2A). Thus, the polypeptides of 58K and 48K encoded by this mRNA might be ascribed a role in the induction process. These same 58K and 48K products have also been shown to be required for production of the normal levels of other early mRNA species in infected cells (Jones and Shenk, 1979*b*; Berk *et al*., 1979; Esche *et al*., 1980; Nevins, 1981; Ricciardi *et al*., 1981). It could, therefore, be argued as in Section 2.2 that the apparent requirement for these E1A proteins is, in fact, an indirect one and that a product of early region E1B, E3, or E4, expressed in significant amounts only in the presence of functional E1A-a mRNA products (see Fig. 2A), is the direct inducer. None of the viruses carrying mutated E1B genes that have been examined, however, differed significantly from wild-type Ad5 in their ability to induce cellular DNA synthesis and cell cycle progression (Rossini *et al*., 1981; Braithewaite *et al*., 1983). Moreover, two substitution mutants of Ad5, H5sub315 and H5sub316, each of which carries a deletion of viral and an insertion of cellular sequences across the E1A/E1B junction (Jones and Shenk, 1979*a*; see Fig. 2A), exhibit phenotypes that argue for a direct role of E1A products in induction; these mutants appear to be relatively normal with respect to potentiated expression of early genes such as E3 and E4 (Jones and Shenk, 1979*a,b*; Shenk *et al*., 1979; Ross *et al*., 1980), but fail to cause aberrant cell cycle progression when they infect rat cells (Braithewaite *et al*., 1983). The nature of the mutant E1A proteins made in substitution mutant-infected cells has not been investigated nor can it be predicted, for the deletions and substitutions have not been sequenced. The deletions carried by these two substitution mutants do not, however, appear to extend into the region uniquely expressed in the largest E1A mRNA in which the point mutation of H5hr1 lies (see Fig. 2A). Thus, the precise relationship of the observations made on H5sub315 and H5sub316-infected cells to those on H5hr1-infected cells remains obscure. The data presently available do not, for example, preclude a role for products of E1A-b mRNA (see Fig. 2A) which would be altered by the deletions present in H5sub315 or H5sub316 DNA, acting in concert with the products of E1A-a mRNA. An E1A mutation that specifically abolishes synthesis of the E1A-b mRNA and its products has been constructed (Montell *et al.*, 1982), but has not been examined for its ability to induce cellular DNA synthesis and cell-cycle progression in quiescent or nonpermissive cells (A. Berk, personal communication). Similarly, it is not yet known whether E1A products interact with components of the cellular replication machinery itself or whether their effect is exerted by stimulation of expression of cellular genes (see Section 4.2) whose products then permit initiation of cellular DNA synthesis.

As we have seen (Section 3.1), it is possible that an E1A product may also mediate, or participate in, inhibition of cellular DNA synthesis in actively growing, permissive cells. In this case, evidence for a direct role, rather than indirect activation of expression of a second early gene product, has not been collected. Nevertheless, it must be pertinent to consider the activities of the E1A products discussed in this section in relation to their well-known participation in transformation; E1A DNA of adenovirus subgroup C or subgroup A serotypes is sufficient to cause an incomplete or partial transformation of rodent cells (van der Eb *et al.*, 1979; Houweling *et al.*, 1980; Shiroki *et al.*, 1981), implicating E1A gene products directly in the acquisition of the new set of phenotypes displayed by the resulting transformants. Examination of the properties of certain E1A mutants has led to the same conclusion (see, for example, Shiroki *et al.*, 1981; Solnick, 1981; Solnick and Anderson, 1982; Ho *et al.*, 1982). It, therefore, comes as little surprise that many lines of adenovirus-transformed rodent cells are aneuploid and display chromsome aberrations (see Tooze, 1980), abnormalities that are induced shortly after infection of quiescent or growing rodent cells. In the case of papovaviruses, it is firmly established that a minimum of one cycle of cellular DNA synthesis, induced by the A-gene product large T antigen, is necessary to "fix" the transformed state (see Tooze, 1980). The same can be said of transformation of non- or semipermissive cells by adenoviruses, although the phenomenon has been less well studied in the adenovirus system; when replication of hamster cells was inhibited by the addition of excess thymidine to the medium for two days following infection by simian adenovirus 7, transformation was completely prevented (Castro, 1973). Moreover, certain mutants that alter expression of the E1A region, such as H5hr1, are deficient in both induction of cellular DNA synthesis and transformation (Harrison *et al.*, 1977; Graham *et al.*, 1978; Ho *et al.*, 1982; Braithewaite *et al.*, 1983). As must be apparent, quite a large number of mutants carrying lesions within the E1A gene have now been isolated and at least partially characterized. The transformation phenotypes of some of these are cold sensitive or host dependent (Graham *et al.*, 1978; Jones and Shenk, 1979*a*; Shenk *et al.*, 1979; Shiroki *et al.*, 1981; Ruben *et al.*, 1982; Ho *et al.*, 1982). It should, therefore, be possible, despite the extreme inefficiency of the transformation process (see Tooze, 1980), to establish whether a direct relationship exists between the ability of an adenovirus to induce cellular DNA synthesis and cell cycle progression and the ultimate appearance of transformed descendants of the infected cells. Although E1B gene products also clearly play an important role in acquisition of the fully transformed phenotype, particularly tumorigenicity (Bernards *et al.*, 1982, 1983; Jochemsen *et al.*, 1982; Shiroki *et al.*, 1981), it is difficult to discern any relationship between this role and that of protection of viral and cellular DNA from degradation discussed in Section 3.1.

It seems only reasonable to suppose that the ability of adenoviruses to induce cellular DNA synthesis and entry into the cell cycle in mammalian cells is of advantage to the virus' replication cycle; such a property could hardly have evolved for any other purpose. Bellett and colleagues (see, for example, Murray *et al.*, 1982*a*) have pointed out the particular relevance of this ability to the natural conditions of infection. By contrast to the artificial laboratory situation in which the virus is usually provided with cells that are partially transformed (they are immortal) and undergoing rapid growth and division, most cells encountered by the virus in the natural host are likely to be arrested at the G1/G0 boundary. Thus, it would seem to be of considerable advantage to the virus to possess a mechanism that, soon after an adenovirus enters such a relatively inactive host cell, induces that cell to enter its most active biosynthetic state and, thus, provide maximal quantities of those cellular proteins upon which viral DNA synthesis and gene expression depend.

4. CELLULAR RNA METABOLISM IN ADENOVIRUS-INFECTED CELLS

As mentioned in Section 2.3, some of the very first examinations of adenovirus-infected, permissive cells revealed large quantitative changes in total RNA synthesis. Subsequently, it was established that the metabolism of both cellular mRNA and rRNA are severely disrupted as a full-fledged productive infection enters the late phase. These changes, coupled with the selective translation of viral mRNA species to be considered in the next section, ensure that, beyond 16–18 hr after a typical infection, only adenovirus-specific proteins are made. These responses are limited to permissive cells as nonpermissive cells, for one reason or another (see Tooze, 1980), fail to mount an effective late phase.

4.1. Inhibition of Production of Cellular mRNA

It has been known for some time that, once an adenovirus infection reaches the late phase, the great majority of newly-synthesized, cytoplasmic RNA sequences are viral in origin. Lindberg *et al.* (1972), for example, estimated that 85% of the labeled, poly(A)-containing RNA isolated from polysomes was virus-specific by 14–16 hr after infection, an observation consistent with the exclusive synthesis of adenoviral proteins in infected cells (see Section 5). Similar values have been reported by others (Bhaduri *et al.*, 1972; Tal *et al.*, 1975). More recent experiments have established that 98% of the poly(A)-containing, cytoplasmic RNA labeled in a 30 min pulse and reaching the cytoplasm during a subsequent chase hybridized to adenoviral DNA (Beltz and Flint, 1979; Castiglia and Flint, 1983). The greater extent of hybridization observed in the latter experiments probably reflects the more efficient method of hybridization employed, as well as the complete infection of the culture. These results imply that newly-made cellular RNA sequences do not enter the cytoplasm during the late phase of adenovirus infection (Beltz and Flint, 1979). Analysis of the expression of several cellular genes in group C adenovirus-infected human cells has amply confirmed this inference. The genes examined include the SV40 early gene integrated into human DNA in the SV40-transformed human cell line SV80 (Flint *et al.*, 1983), amplified dihydrofolate reductase (DHFR) genes present in methotrexate-resistant lines derived from HeLa S3 cells (Yoder *et al.*, 1983), eight HeLa cell genes that encode abundant, poly(A)-containing, cytoplasmic mRNA species, identified by their cDNA clones (Babich *et al.*, 1983), and human β-tubulin genes (J. J. Plunkett and S. J. F, unpublished observations). In all cases, it has been observed that cytoplasmic, poly(A)-containing RNA populations labeled during the late phase of adenovirus infection do not contain sequences that hybridize to cloned DNA probes for the gene(s) under study. The implications of such selective appearance in the cytoplasm of viral mRNA species for synthesis of cellular proteins will be discussed in Section 5.

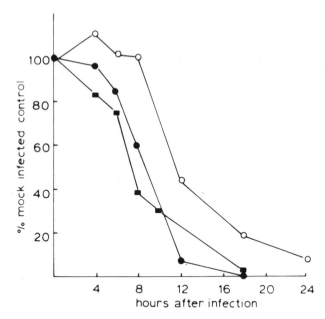

Fig. 6.   Inhibition of appearance in the cytoplasm of cellular mRNA and rRNA during adenovirus infection. The levels of labeled cytoplasmic mRNA, ■——■, 28 S rRNA, ●——●, and 18 S rRNA, ○——○, entering the cytoplasm in a 2-hr labeling period at various times after Ad5-infection of HeLa cells are expressed as percentages of the mock-infected control values. Reprinted from Castiglia and Flint (1983).

The time course of this dramatic change in the normal pattern of cellular mRNA production has been examined in a few cases. One example is shown in Fig. 6 in which cellular mRNA sequences were identified as those that fail to hybridize to adenoviral DNA; few cellular sequences enter the cytoplasm by 12 hr after infection, a point in the infectious cycle that corresponded to the beginning of the late phase as judged by the onset of synthesis of viral, late proteins. By 18 hr after infection, 96% inhibition was achieved (Castiglia and Flint, 1983). Similarly, Yoder *et al.* (1983) described a severe inhibition of the appearance in the cytoplasm of newly-synthesized DHFR RNA sequences, concomitant with the onset, and increasing production, of viral, late mRNA and structural proteins. One interpretation of such a temporal pattern of inhibition might be that the vast amounts of viral, late RNA that are produced as an infection progresses through the late phase compete effectively with the cellular RNA population for some component essential to the transcription, processing, or transport from the nucleus of both classes of RNA. Attempts to elucidate the mechanism underlying the failure of cellular mRNA sequences to enter the cytoplasm have, therefore, been directed toward identification of that step in the normal pathway of cellular mRNA production that is disrupted by adenovirus infection.

The simplest way in which adenovirus infection could be imagined to inhibit production of cellular mRNA would be by inhibition of transcription of cellular genes. Such a mechanism has several precedents among bacteriophage infections (reviewed by Losick and Pero, 1976; Bautz, 1976). The vast quantities of viral DNA that accumulate in infected cells also give this notion some plausibility. The first examinations of the nature of nuclear RNA sequences pulse labeled during the late phase of adenovirus infection produced contradictory findings; when using actinomycin D to inhibit rRNA synthesis, Price and Penman (1972) estimated that 14% of the pulse-labeled hnRNA population was adenovirus specific. Conversely, Phillipson *et al.*, (1975) concluded that such RNA was exclusively viral, based on their observation that 50–60% of nuclear RNA labeled in the absence of any drug was rRNA. The former value is likely to be an underestimate for even low doses of actinomycin D have been shown to inhibit the transcription of G + C-rich adenoviral DNA (Lindberg *et al.*, 1972). Nevertheless, more recent experiments have established that the lower estimate of Price and Penman (1972) is the more correct one, that is, that transcription of cellular pre-mRNA sequences indeed continues during the late phase of adenovirus infection. Hybridization of hnRNA, isolated after the removal of nucleoli containing rRNA, to viral DNA or to cDNA made from HeLa cytoplasmic, poly(A)-containing RNA templates indicated that 45–50% of the pulse-labeled hnRNA was viral in origin (Beltz and Flint, 1979). Similarly, pulse-labeled transcripts of the prototypic cellular gene represented by integrated SV40 DNA in the SV80 cell line (Flint *et al.*, 1983), of DHFR genes (S. S. Yoder and S. M. Berget, personal communication), and of five HeLa genes identified by cloned cDNA probes (Babich *et al.*, 1983) are synthesized during the late phase in amounts that cannot be distinguished from those of uninfected control cells. It, therefore, seems clear that adenovirus infection does not inhibit transcription of cellular pre-mRNA during the period in which inhibition of appearance in the cytoplasm of the corresponding mRNA sequences is complete.

The processing of cellular pre-mRNA sequences has not yet been as thoroughly examined as their transcription. Such processing reactions as capping of the 5′ end of newly-synthesized pre-mRNA and

addition of poly(A) to create the 3' terminus of the mature mRNA
have been examined in total hnRNA populations isolated from adenovirus-infected HeLa cells. Neither appear to be defective; no enrichment of viral RNA sequences could be discerned in the poly(A)-
containing compared to the poly(A)-lacking, or total, fractions of infected cell RNA sequences labeled in a 30-min period (Beltz and Flint,
1979). Similarly, Ad2 or Ad5-infected SV80 cells produce the same
amount of nuclear, poly(A)-containing, SV40-specific RNA in a 30-
min period as do their mock-infected counterparts (Flint *et al.*, 1983).
The population of capped T1-oligonucleotides newly labeled in Ad2-
infected HeLa cells during the late phase comprises the population
made in uninfected HeLa cells plus the undecanucleotide (Gelinas
and Roberts, 1977; Lockard *et al.*, 1979) diagnostic of viral, late
mRNA species (G. A. Beltz and S. J. Flint, manuscript in preparation). Investigation of the splicing of a specific, newly-synthesized,
cellular transcript presents technical difficulties. However, the unlabeled population of nuclear, poly(A)-containing RNA molecules
purified from cells harvested during the late phase of adenovirus infection has been shown to contain species that exhibit the same length
as the corresponding cytoplasmic, poly(A)-containing RNA from uninfected cells. This is true of five HeLa genes examined by virtue of
cDNA clones (Babich *et al.*, 1983) and SV40 transcripts in the SV80
cell line (Flint *et al.*, 1983). The latter cell line produces large quantities of the SV40 large T antigen (Henderson and Livingston, 1974)
and SV40 mRNA species; some 1.8% and 0.4% of the newly-synthesized, cytoplasmic, and nuclear, poly(A)-containing fractions, respectively, hybridize to SV40 DNA (Flint *et al.*, 1983). Both the small
and large T-antigen early mRNA species, as well as a species of 3.7K
that has no counterpart in SV40-infected monkey cells, are abundant
in SV80 cells (Flint and Beltz, 1979; Flint *et al.*, 1983). These same
three species, as well as an additional, larger species were readily
detected in nuclear, poly(A)-containing RNA prepared from Ad2 or
Ad5-infected SV80 cells (Flint *et al.*, 1983). Furthermore, SV40-spe-
cific RNA species of the same size as fully processed mRNA species
were present in the newly-synthesized, i.e., pulse-labeled, nuclear
RNA population obtained from infected cells. Clearly, more extensive
studies of the processing of specific cellular transcripts in adenovirus-
infected cells need to be performed, not only to generalize the observations made using SV80 cells, but also to follow at least one
cellular transcript through all steps in its biogenesis. The evidence
presently available does, however, seem to be sufficient to warrant

the conclusion that splicing of cellular pre-mRNA transcripts, like other enzymatic processing reactions, is not deranged during the late phase of an adenovirus productive infection (Babich *et al.*, 1983; Flint *et al.*, 1983). Nor has any evidence of selective degradation of cellular RNA sequences within the nucleus been forthcoming. Indeed, such an explanation for the failure for cellular mRNA sequences to reach the cytoplasm in infected cells would invoke the existence of an RNase activity, recognizing only cellular RNA sequences, for which no precedent whatsoever can be invoked.

The arguments given in the preceding paragraph assume that adenovirus infection disrupts one specific step in the normal pathway of cellular mRNA production. An important caveat should, therefore, be mentioned. Whereas there is agreement that the rate of transcription of cellular genes is not reduced during the relevant period of the late phase of infection (Beltz and Flint, 1979; Babich *et al.*, 1983; Flint *et al.*, 1983; S. S. Yoder and S. M. Berget, personal communication), it is less easy to eliminate the possibility that the large quantities of viral RNA compete to some degree with cellular RNA species at all subsequent processing steps. In other words, it is possible that the cumulative consequences of such competition are seen in the failure of cellular mRNA species to leave the nucleus. The limited information collected to date, and discussed in previous paragraphs, does not tend to support this view, but a complete and quantitative description of the synthesis and processing of a specific cellular transcript during the late phase of adenovirus infection will be required to counter it unequivocally.

If none of the normal processing reactions that modify a pre-mRNA chain are inhibited by adenovirus infection, then it would seem that the virus must induce either selective transport of viral mRNA or a mechanism that discriminates among viral and cellular species at some prior step, upon which transport itself depends. Examples of the latter class of mechanism might include incomplete or incorrect packaging of cellular transcripts with nuclear proteins for form hnRNP or the displacement of cellular RNA molecules from their normal site(s) of synthesis or processing within the nucleus (for example, attached to a structure such as the nuclear matrix; Herman *et al.*, 1978; Miller *et al.*, 1978; van Eekelen and van Venrooij, 1981; Jackson *et al.*, 1981, 1982; Maniman *et al.*, 1982.). Experimental evidence that addresses either of these possibilities directly is not available. It does, however, seem clear that hnRNP populations isolated from adenovirus-infected cells do not comprise solely viral RNA sequences (Gattoni *et al.*, 1980; Munroe, 1982; S. Kiode and S. J. Flint, unpublished observations) as the simplest formulation of the first mechanism would imply. Nor can such a mechanism be reconciled with the continued processing of cellular transcripts; it is well established that nascent RNA becomes packaged into RNP before transcription is completed (for example, McKnight and Miller, 1976; Laird and Chooi, 1976; Foe *et al.*, 1976). Thus, RNP must be the substrate of posttranscriptional processing reactions, some of which have indeed been observed to take place upon nascent RNP structures (Beyer *et al.*, 1980, 1981). We are, therefore, forced to the surprising conclusion that adenovirus infection inhibits the appearance in the cytoplasm of newly-synthesized cellular mRNA species by a mechanism that alters nuclear to cytoplasmic transport or induces some relatively subtle alteration in the structural or spatial requirements of that process. Little precedent for such a regulatory mechanism comes to mind, although there does seem to be some analogy to changes in macromolecular synthesis seen when anchorage-dependent fibroblasts are placed in suspension culture; under such conditions, the production of new mRNA sequences declines by a factor of approximately five in the absence of any equivalent reduction in synthesis of hnRNA (Benecke *et al.*, 1978, 1980).

The question of whether a specific viral gene product or products mediates the selective appearance in the cytoplasm of viral mRNA species is also unanswered at present. The kinetics of the inhibitory process, discussed previously, suggest that any such viral product should be encoded by an early gene, or at least one whose product is made in maximal amounts towards the beginning of the late phase of infection. However, none of the mutants that affect early gene function have been screened for alterations in cellular mRNA metabolism in infected cells. Nor has it been possible to test whether inhibition of synthesis of viral mRNA restores the synthesis of cytoplasmic, cellular mRNA, as a competition model would predict, for no means to halt viral, late mRNA production once the late phase is initiated have been discovered.

4.2. Induction of Cellular Gene Expression in Adenovirus-Infected Cells

We have seen that adenovirus infection of quiescent human or rodent cells induces the synthesis of such enzymes as thymidine kinase (Section 3.2). Whether the increased enzyme activity measured under such circumstances is the result of transcriptional activation of the thymidine kinase gene or regulation of some other level has never been investigated. The former mechanism, however, seems likely in view of several recently documented examples of transcriptional activation of cellular genes during the early phase of a typical productive infection.

When the proteins labeled 5 hr after an Ad5 infection of HeLa cells in suspension culture were examined carefully, one exhibiting an apparent molecular weight of 70K was found to be especially prominent in infected compared to mock-infected cells (Nevins, 1982). This polypeptide comigrated with a 70K protein whose synthesis was also induced by heat shock. The identity of the virus-induced and heat-shock-induced 70K polypeptides was established by comparison of the products of partial proteolysis (Nevins, 1982). Using a cloned cDNA copy of the 70K heat-shock protein (hsp) mRNA, Nevins and colleagues have measured a 100-fold increase in the concentration of 70K hsp mRNA in Ad5-infected cells and a substantial increase in the transcription of the 70K hsp gene (Kao and Nevins, 1983). No equivalent accumulations of the 70K hsp mRNA or protein were observed in cells infected by H5d1312, which cannot express early region E1A (Jones and Shenk, 1979a). In this experiment, high multiplicities of H5d1312 infection were employed to circumvent (Nevins, 1981) the low levels of expression of other early genes typical of H5d1312 infections (see Sections 3.1 and 3.3). The failure of the mutant virus to activate transcription of the 70K hsp gene can, therefore, be attributed directly to its lack of E1A coding sequences and E1A proteins (Nevins, 1982; Kao and Nevins, 1983). One additional HeLa gene, also examined by Kao and Nevins (1983), showed no corresponding activation following Ad5 infection.

The expression of several human genes during the early phase of infection has also been examined in cells infected with H5d1312 at low multiplicity and superinfected with H5d1327, a mutant from which early region E3 has been deleted (Jones and Shenk, 1978). Under these circumstances, no expression of viral genes is seen in the absence of the superinfecting helper H5d1327 virus, which provides the E1A products that cannot be supplied by H5d1312. When transcription of viral and cellular genes was examined by pulse labeling and hybridization, the expression of human β-tubulin and histone genes was found to be substantially enhanced by 4–6 hr after the superinfection (R. Stein and E. Ziff, personal communication).

The activation of expression of viral E3, which must be from the original H5dl312 genomes, exhibited a similar time course, implying that the E1A product(s) supplied by superinfecting H5dl327 activated transcription of both viral (E3 of H5dl312) and cellular genes (R. Stein and E. Ziff, personal communication). However, the role of E1A products has yet to be directly demonstrated. The transcription of globin genes occurs at a very low rate in HeLa cells and was not enhanced under these circumstances. Conversely, actin transcripts are synthesized in respectable amounts in uninfected cells but their expression too was indifferent to adenovirus infection (R. Stein and E. Ziff, personal communication). Although the number of these examples is small, these observations and those of Kao and Nevins suggest that adenovirus infection does not gratuitously activate the expression of all cellular genes.

Activation of the human DHFR gene has also been observed in Ad2-infected human cells in monolayer culture. For about the first 15 hr of infection, the steady-state level of cytoplasmic DHFR RNA, species of 3.8, 1.0, and 0.8kb, increased with a corresponding increase in the amount of DHFR protein made (Yoder et al., 1983). It should be noted that these changes occurred during the early phase of infection, for no late proteins were detected until 18 hr in these experiments. As discussed previously, it has been quite generally observed that the productive cycle proceeds more slowly in cells maintained in monolayer compared to suspension culture (see Section 2.1). No equivalent accumulation took place when the cells were infected by H5dl312 (S. S. Yoder and S. M. Berget, personal communication). Surprisingly, hybridization of nuclear RNA labeled in 5-min pulses, administered at intervals following wild-type Ad2 infection, to DHFR DNA indicated that activation of expression of DHFR genes was not mediated at a transcriptional level, that is, no increase in the concentrations of pulse-labeled DHFR RNA sequences was observed during the period in which the concentration of cytoplasmic DHFR mRNA increased so substantially (S. S. Yoder and S. M. Berget, personal communication). The apparent contradiction between this result and the transcriptional activation of the 70K hsp, and β-tubulin and histone genes is unresolved at present. It is, however, worth noting that evidence for both transcriptional and posttranscriptional effects of E1A on the expression of the other viral early genes has been presented.

We have seen that the properties and location of the H5hr1 mutation implicate the products of the E1A-a mRNA (see Fig. 2A) in activation of viral gene expression. The mechanism of action of these products is, however, a subject shrouded in confusion. Nevins (1981) reported that the low levels of early mRNA seen in H5d1312-infected cells were reflected in substantially lower rates of transcription of such early genes as E3 and E4. Moreover, he observed that the addition of cycloheximide 45 min before infection restored transcription in H5d1312-infected cells, at least in the case of E4 and L1. It was, therefore, concluded that the E1A products acted indirectly at the level of transcription, perhaps by counteracting a short-lived, cellular inhibitor of transcription of viral early genes (Nevins, 1981). In independent experiments, Philipson and colleagues reported that addition of anisomycin 30 min before H5hr1 infection restored the cytoplasmic levels of E3 and E4 mRNA (Katze *et al.*, 1981) and, thus, like Nevins (1981), concluded that the E1A product(s) must act indirectly. However, under these conditions, no significant changes compared to the wild-type control in the rate of transcription of E3 or E4 were observed in H5hr1-infected cells in the absence or presence of the drug. These authors concluded that the E1A products mutated in H5hr1 act posttranscriptionally (Katze *et al.*, 1981, 1983). Most recently, Gaynor and Berk (1983) found *no* amelioration of the H5d1312 defect in the presence of cycloheximide. As these authors point out, the effect of inhibitors of protein synthesis on adenovirus early gene expression is itself a complicated subject; the nature of the response appears to vary with such paramets as the time at which the inhibitor is added and the stringency of the inhibition (see Flint, 1982 for a review). It is not, therefore, unexpected that the addition of inhibitors of protein synthesis to E1A mutant-infected cells has failed to provide clear information. In light of such uncertainty about the mechanism(s) whereby products of the E1A-a mRNA influence viral gene expression, it is not surprising that the more recently reported effects on cellular gene expression are not fully understood.

The relevance of the examples of activated cellular gene expression that have been reported so far to the life cycle of the virus has also to be established. Adenovirus depends on its hose cell for a large number of products, whose synthesis might be expected to be enhanced soon after infection to prime the cell for the period when massive quantities of viral DNA, late mRNA, and structural proteins are made. At this juncture, it is, however, difficult to appreciate the significance of increased expression of such genes as those encoding β-tubulin and the 70K heat-shock protein. This may reflect our ignorance about many aspects of the virus–cell interaction. Alternatively, it may well be that these responses are in a sense gratuitous, the secondary consequences of properties of the viral E1A products that were evolved to overcome restraints on viral gene expression imposed by the intracellular milieu. Undoubtedly, the activation of cellular gene expression in adenovirus-infected cells will continue to be studied with enthusiasm, leading to the resolution of these issues.

4.3. Posttranscriptional and Transcriptional Regulation of Ribosomal RNA Synthesis in Adenovirus-Infected Cells

Adenovirus infection has long been known to disrupt production of ribosomal RNA (rRNA) species; more than 10 years ago, it was reported that the appearance in the cytoplasm of newly-synthesized rRNA is strongly inhibited during the late phase of Ad2 or Ad12 infection (Raskas *et al.*, 1970; Ledinko, 1972). More recent experiments have attempted to elucidate this inhibitory mechanism and have provided evidence for both transcriptional and posttranscriptional regulation of rRNA gene expression in adenovirus-infected, permissive cells. When the cytoplasmic, poly(A)-lacking RNA species labeled in 2-hr periods at intervals following Ad2 or Ad5 infection of HeLa cells in suspension were examined by electrophoresis, the complete inhibition of appearance of both 28 S and 18 S rRNA species by 24 hr after infection was confirmed (Castiglia and Flint, 1983). However, the inhibition of appearance in the cytoplasm of newly-synthesized 28 S rRNA was, as illustrated in Fig. 6, complete at an earlier time than that of 18 S rRNA; the amount of cytoplasmic 28 S rRNA labeled in a 2-hr period is reduced to 35% of the control, mock-infected value by 8 hr after infection, a time at which no change in the quantity of labeled, cytoplasmic 18 S rRNA could be detected (Castiglia and Flint, 1983; see Fig. 6). This is a most unexpected observation, for which no precedent can be found among the results of the numerous studies of inhibition of rRNA synthesis and maturation that have been performed (see Hadjiolev and Nikolaev, 1976 for review). By the time inhibition of appearance in the cytoplasm of 28 S rRNA was complete, the levels of nucleolar 45 S, 32 S, and 28 S rRNA species were barely altered (45 S pre-rRNA) or reduced by less than 30% (32 S and 28 S rRNA; Castiglia and Flint, 1983). Moreoever, the 28 S rRNA made in the nucleolus of infected cells was packaged into 80 S and 55 S ribonucleoprotein particles that are indistinguishable, in sedimentation analyses, from their counterparts produced in mock-infected cells

(C. L. Castiglia and S. J. Flint, unpublished observations). Infected cell 60 S particles prepared from the nucleoplasm do, however, differ in their polypeptide composition in one striking respect, the presence of a polypeptide of apparent molecular weight 25K that is exclusive to adenovirus-infected cells, a product of early region E4 (C. L. Castiglia and S. J. Flint, unpublished observations).

As 28 S rRNA and the ribonucleoprotein in which it resides are produced at 70% the level of mock-infected cells when a 90% inhibition of appearance in the cytoplasm of 28 S rRNA has been attained, it is difficult to escape the conclusion that adenovirus infection does not, at least within this period, disrupt to any great extent the transcription or processing of 45 S pre-rRNA. In this respect, the effects of infection upon pre-rRNA and rRNA metabolism are directly analogous to those exerted upon cellular pre-mRNA and mRNA production (see Section 4.1). However, rRNA and mRNA are the products of completely independent transcription and processing pathways which are not known to share any components and indeed are segregated within the nucleus. These properties and the similar kinetics of inhibition of appearance in the cytoplasm of cellular mRNA and rRNA (Castiglia and Flint, 1983) suggest that it might be the normal transport mechanism that becomes deranged in adenovirus-infected cells. This hypothesis has the virtue of an efficient explanation of the disparate changes in cellular RNA metabolism. Nevertheless, it has yet to be confirmed by direct experiment.

The production of cytoplasmic 18 S rRNA is also disrupted in adenovirus-infected cells but, as illustrated in Fig. 6, no change is observed until later than 8 hr after infection, and inhibition is not complete until 24 hr after infection. The exact timing of these events is variable and influenced by the multiplicity of infection, but the pattern of inhibition of 28 S relative to that of 18 S rRNA is perfectly reproducible. The inhibition of appearance in the cytoplasmic of 18 S rRNA takes place during the period of inhibition of cellular protein synthesis (see Section 5). The production of such nucleolar pre-rRNA species as 41 S and 32 S rRNA also declines at this time, and by 16 hr after infection a reduction in the synthesis of 45 S pre-rRNA can be detected (Castiglia and Flint, 1983). These later changes in rRNA metabolism in adenovirus-infected cells resemble those reported when cellular protein synthesis is inhibited. The addition of cycloheximide, for example, induces a more rapid inhibition of processing of 45 S pre-rRNA than of its transcription. Consequently, the amounts of newly-synthesized rRNA entering the cytoplasm are reduced to very low levels, compared to untreated cells, significantly before transcription is inhibited to the same degree (Higashi *et al.*, 1968; Willems *et al.*, 1969; Craig and Perry, 1970). Processing of rRNA precursors and their assembly into ribonucleoprotein particles are coupled processes (Phillips and McConkey, 1976; Hadjiolev, 1980; Lastick, 1980), so it is not difficult to see how reduction in the quantities of ribosomal proteins available during the late phase of adenovirus infection could disrupt processing. Similarly, Tijan and colleagues have recently presented evidence that Ad2-infected HeLa cells are deficient in a factor required for transcription of rRNA genes.

Cloned human rDNA segments containing the 5′ end of 45 S pre-rRNA can be efficiently transcribed in appropriate extracts of mammalian cells (Miller and Sollner-Webb, 1981; Mishima *et al.*, 1981; Grummt, 1981; Miesfeld and Arnheim, 1982). Extracts prepared from adenovirus-infected cells at 19 hr after infection were, however, found to be 1/20th as active in run-off transcription of human rDNA as uninfected HeLa cell extracts, under conditions in which both extracts contained approximately the same amounts of RNA polymerase I (Learned *et al.*, 1983). The effect was specific for RNA polymerase I; transcription of an SV40 DNA template by RNA polymerase II was, if anything, more efficient in infected compared to uninfected extracts (Learned *et al.*, 1983). In principle, the inhibition of rDNA transcription observed in the Ad2-infected cell extracts could reflect the synthesis of a specific repressor of rRNA transcription, encoded or induced by the virus, or the deficiency of some cellular factor crucial to initiation. The latter mechanism seems most plausible in view of the well-known inhibition of cellular protein synthesis during the late phase of adenovirus infection and the previous conclusion that short-lived cellular factors are required for RNA polymerase I activity (Muramatsu *et al.*, 1970; Mishima *et al.*, 1979). It is also consistent with the finding that a 1:1 mixture of infected and uninfected HeLa cell extracts transcribed rDNA as efficiently as the one unit of uninfected cell extract alone (Learned *et al.*, 1983). The ability of uninfected HeLa cell extracts to complement infected cell extracts will certainly permit the purification and characterization of the factor, or factors, deficient (or induced) in adenovirus-infected cells and, thus, a definitive test of the mechanism whereby transcription of rDNA is inactivated in infected cell.

5. SELECTIVE TRANSLATION OF VIRAL mRNA IN PRODUCTIVELY-INFECTED CELLS

By the time the late phase of an adenovirus infection is well underway, later than 16–18 hr after infection under typical conditions, infected cells translate exclusively viral mRNA species (Bello and Ginsberg, 1967; Ginsberg et al., 1967; White et al., 1969; Russell and Skehel, 1973; Anderson et al., 1973). As illustrated in Fig. 7, the major protein products of the late phase are the structural proteins of the virion. Synthesis of cellular proteins is dramatically, if not completely, inhibited. This inhibition can first be detected when translation of viral, late mRNA begins and is complete a few hours thereafter, 6–8 hr in the experiment shown in Fig. 7 (for example, Anderson et al., 1973; Beltz and Flint, 1979). Provided that every cell in the culture becomes infected, the degree of inhibition is independent of the multiplicity of infection, although the kinetics of the inhibition of cellular protein synthesis, as those of the entire replication cycle, are multiplicity dependent (see, for example, Castiglia and Flint, 1983). No examples of cellular proteins whose synthesis is resistant to the inhibitory effects of an adenovirus infection have been documented, although the one-dimensional, SDS–polyacrylamide gels used in all experiments of this kind do not permit a particularly sensitive, or exhaustive, search for exceptions.

Such a profound inhibition of cellular protein synthesis is not an uncommon feature of virus infections of mammalian cells. Several lines of research have recently converged to demonstrate that adenoviruses, like certain other viruses that inhibit cellular protein synthesis, such as the picornaviruses poliomyelitis virus and encephalomyocarditis virus, have evolved a mechanism to permit the selective translation of viral mRNA species. Although both adenoviruses and picornaviruses can efficiently redirect the translational machinery of their host cells, the actual molecular mechanisms employed appear to be quite distinct.

The first evidence for control of translation in adenovirus-infected cells came when mRNA isolated from infected cells harvested during the late phase was translated in heterologous, cell-free systems, usually derived from reticulocyte lysates; such mRNA preparations direct the synthesis of quite substantial quantities of cellular proteins in addition to large amounts of viral structural proteins (see, for example, Anderson et al., 1974; Lewis et al., 1975; Paterson et al., 1977). The mRNA was prepared from cells collected at a time

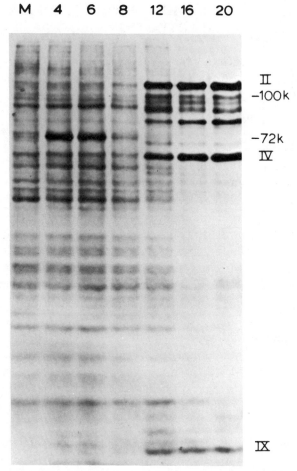

Fig. 7.   Protein synthesis in adenovirus-infected cells. Ad5-infected HeLa cells were labeled with [³H]leucine for 2-hr periods starting at the times after infection shown. M = mock infected. Cytoplasmic proteins are shown in this autoradiogram. Modified from Castiglia and Flint (1983).

when inhibition of cellular protein was complete, or nearly complete, so these observations imply that cellular mRNA species are present, but not translatable, within the infected cell. More recent studies of the expression of individual cellular genes have provided direct demonstrations that at least certain cellular mRNA species indeed persist in the cytoplasm of adenovirus-infected cells until well into the late phase.

Adenovirus infection of human cells disrupts the entry into the cytoplasm of newly-synthesized cellular mRNA sequences (see Section 4.1). Thus, in the absence of any mechanism to degrade preexisting cellular mRNA specifically, the steady-state level of each cellular mRNA species during the late phase of an adenovirus infection would be expected to be determined by the intrinsic half-life of that mRNA species. The available data, although limited, support such a view and, therefore, imply that the virus does not induce the selective degradation of preexisting cellular mRNA sequences (Philipson *et al.*, 1975). In uninfected SV80 cells, for example, the SV40 early, cytoplasmic, poly(A)-containing RNA sequences exhibit a half-life of 40–60 min (Flint *et al.*, 1983). It is, therefore, not surprising that by the time inhibition of cellular protein synthesis is complete after Ad2 or Ad5 infection, SV40-specific mRNA species can barely be detected in the cytoplasm (Flint *et al.*, 1983). Similarly, under appropriate conditions of infection, the steady-state level of histone mRNA species is reduced on the order of 100-fold by 18 hr after infection (G. Stein, M. Plumb, J. Stein, and S. J. Flint, unpublished observations); human histone mRNA sequences have been estimated to possess half-lives of 40 min in exponentially growing cells (Heintz *et al.*, 1983). These observations might suggest that the normally rapid turnover of such mRNA sequences coupled with the failure of infected cells to replenish their cytoplasmic pools is sufficient to account for the inhibition of translation of cellular mRNA species. However, a substantial fraction of cellular RNA sequences are not short-lived (see for example, Greenberg, 1972) Perry and Kelly, 1968; Singer and Penman, 1973). In adenovirus-infected cells specifically, it has been shown that mRNA species complementary to a human DHFR probe, while reduced in amount, can be easily detected during the late phase of infection (Yoder *et al.*, 1983). Similarly, the steady-state levels of two human β-tubulin mRNA species are reduced by less than a factor of two by 18.5 hr after infection, when inhibition of cellular protein synthesis was more than 90% complete (S. J. Flint, unpublished observations). These results, like those of the cell-free translations cited previously, imply a specific inhibition of translation of cellular mRNA species in adenovirus-infected cells, but are subject to the criticism that both the steady-state levels and translatability of individual mRNA species were not examined in the same experiments. Fortunately, such careful experiments have been performed by Babich and colleagues (1983). Four HeLa mRNA species that are abundant in uninfected HeLa cell, cytoplasmic, poly(A)-containing fractions were found to be present at 15 hr after Ad2 infection at concentrations indistinguishable from those of uninfected cells (Babich *et al.*, 1983).

Each of these four mRNA species is stable in uninfected HeLa cells, with a half-life of at least several hours (Babich *et al.*, 1983). Their high concentration in the cytoplasm, despite the inhibition of appearance in the cytoplasm of newly-synthesized RNA complementary to the four clones (Babich *et al.*, 1983) is, therefore, not surprising and is in line with arguments given in the previous paragraph. The same result was obtained when the steady-state levels of actin mRNA in uninfected and infected cells were examined (Babich *et al.*, 1983) Moreover, actin mRNA extracted 15 hr after infection was translated efficiently in a reticulocyte lysate. These results establish beyond any reasonable doubt that the host cells' translational machinery becomes modified by the late phase of adenovirus infection such that only viral mRNA species can be recognized.

Independent demonstration of selective translation, as well as the first clues about the mechanism responsible, have come from a completely different quarter, the detailed analysis of the phenotypes of certain mutants of Ad5, those that cannot express their VA-RNA$_I$ gene is transcriptionally inactivated, produce normal amounts of cytoplasmic, viral, late mRNA species. Nevertheless, no virus-specific, late proteins are made (Thimmappaya *et al.*, 1982). All structural features of the viral, late mRNA tested proved to be indistinguishable from those of wild-type, late mRNA. Moreover, H5d1330-infected cell, cytoplasmic, poly(A)-containing RNA was efficiently translated in reticulocyte lysates into both cellular and viral late proteins (Thimmappaya *et al.*, 1982; Babich *et al.*, 1983). It is, therefore, clear that VA-RNA$_I$ is one adenovirus product that mediates selective translation of viral mRNA; in the absence of this small RNA species, normal, intrinsically translatable, viral mRNA species are produced by the infected cell, but cannot be utilized by its translational machinery. Whether VA-RNA$_{II}$ also plays some role in translation is not yet known. The H5d1330 and H5d1331 mutants direct the synthesis of normal quantities of VA-RNA$_{II}$ in infected cells, so it can be concluded that VA-RNA$_{II}$ cannot substitute for VA-RNA$_I$ in the efficient translation of viral mRNA species (Thimmappaya *et al.*, 1982), despite the quite extensive sequence and structural homology of one to the other (Mathews and Pettersson, 1978; Akusjärvi *et al.*, 1980). Mutants that cannot express their VA-RNA$_{II}$ gene are viable, exhibiting growth properties quite similar to those of the wild-type parent (Thimmappaya *et al.*, 1982), implying that VA-RNA$_{II}$ does not fulfill an essential role. However, the construction of double mutants expressing neither VA-RNA$_I$ nor VA-RNA$_{II}$ will be required to exclude the possibility that VA-RNA$_{II}$ does make some contribution to the translation of viral mRNA species.

The mechanism of action of VA-RNA$_I$ is not yet completely understood. However, the observations that polysomes contain fewer ribosomes in H5d1330- or H5d1331-infected compared to wild-type infected cells (Thimmappaya *et al.*, 1982; Babich *et al.*, 1983) and that the rate of elongation of nascent polypeptides is the same in H5d1331 and wild-type-infected cells suggest that VA-RNA$_I$ is required for initiation of translation of viral mRNA. The recent finding that H5d1330-infected cells fail to accumulate 48 S, mRNA-containing initiation complexes (R. J. Schneider, C. Weinberg, and T. E. Shenk, unpublished observations) provides strong support for this view. The obvious question that follows the demonstration that VA-RNA$_I$ is necessary for the efficient translation of viral mRNA species must be whether it is also sufficient. Cells infected by H5d1330 or H5d1331 do not continue to translate cellular mRNA species during the late phase of infection; when mutant-infected cells are exposed to [^{35}S]methionine during the late phase, neither viral nor cellular proteins are labeled (Thimmappaya *et al.*, 1982; Babich *et al.*, 1983). It can, therefore, be concluded that some additional viral product, or products, participate in the selective and efficient translation of viral mRNA species.

Several lines of evidence point to the involvement of additional viral RNA sequences, a subset of those that form the tripartite leader. This leader comprises the 5'-terminal 202 nucleotides (Akusjärvi and Persson, 1979; Zain *et al.*, 1979) of Ad2 late mRNA species processed from transcripts initiated at the major late promoter site near 16.45 units (see Fig. 3). The same general approach has been employed in several independent investigations of tripartite leader function, the construction of adenoviruses in which the major late promoter site (mLP) and various elements of the tripartite leader are placed upstream of a viral or cellular gene whose expression as mRNA and polypeptide can be readily assayed. Logan and Shenk (personal communication), for example, constructed a matched set of viruses, all containing the mLP in place of E1A transcriptional control signals, but with either no leader elements or the variations 11 alone, 11 + 12 + 13, 11 + 12, or 11 + 13. All viruses directed eight- to ten-fold more efficient expression of E1A mRNA than their unmodified counterparts. However, those that possessed no leader elements, or only 11, displayed a ten-fold reduced level of E1A protein synthesis, assayed by immunoprecipitation and two-dimensional electrophoresis, compared to the virus in which all three leader elements were inserted upstream of E1A coding sequences (J. Logan and T. Shenk, personal communication). Moreover, the E1A mRNA species are found on faster-sedimenting polysomes in cells infected by the virus that has all three leader segments upstream of E1A, directly implicating leader sequences in enhanced translation. It appears that the presence of all three leaders at the 5′ side of the E1A coding sequences is required to obtain the ten-fold increase in synthesis of E1A proteins; variants containing 11 and 12, 12 and 13, or 1 and 13 lead to the production of only three-fold higher levels of these proteins (J. Logan and T. Shenk, personal communication). Protein and mRNA synthesis induced by hybrid viruses containing mouse DHFR cDNA under the control of the mLP with 11 alone or all three leader segments or by Ad-SV40 hybrid viruses in which SV40 early sequences are expressed in mRNA species that include different combinations of the 11, 12, i (Chow et al., 1979), and 13 segments have also been examined (K. Berkner and P. A. Sharp, unpublished observations; Thummel et al., 1983). The results obtained were similar to those of Logan and Shenk in that incomplete leader segments, for example, 11 alone (K. Berkner and P. A. Sharp, unpublished observations; Thummel et al., 1983) or 11 plus 12 plus i (Thummel et al., 1983) are not sufficient for efficient translation in vivo. The various combinations of leader segments tested by Logan and Shenk, combined with their quantitative analysis, emphasize that each of the three leaders makes a significant contribution to the efficient, intracellular translation of mRNA species on which they are present.

This conclusion is satisfying for several reasons. Adenoviruses have evolved a baroque transcriptional unit under the control of the mLP, to ensure that each of the 16 or 17 mRNA species processed from its transcripts contains the tripartite leader. The mechanism of synthesis of one mRNA species from transcripts initiated at the mLP appears to be one of the most wasteful encountered; because each mature mRNA contains the tripartite leader fashioned during splicing (Berget et al., 1977; Chow et al., 1977, Klessig, 1977), only one of the 16 or 17 potential mRNA species can be processed from each approximately 25kb transcript. Such apparent inefficiency has always been taken as evidence of the importance of the leader sequences, so it is pleasing to have that inference confirmed. Furthermore, VA-RNA$_1$ can base pair with cDNA copies of the tripartite leader but not with genomic DNA fragments in which leader elements are widely separated (Mathews, 1980) suggesting a simple model for the preferential translation of viral, late mRNA species in which the VA-RNA$_I$–leader sequence interaction mediates the recognition of this specific class of mRNA species by the modified translational machinery. This mechanism is consistent with the failure of H5d1330-infected cells to form mRNA-containing, 48 S preinitiation complexes and predicts a similar failure for mutated mRNA species that no longer possess the relevant leader sequences. It also predicts that there must be an interaction between VA-RNA$_I$ and 40 S ribosomal subunits in infected cells to direct the latter to viral mRNA species, a prediction that has yet to be tested. Such a model cannot, however, be quite complete, for viral mRNA species that are not encoded as part of the major, late transcriptional unit are translated efficiently during the late phase of infection, for example, those specifying polypeptides IX and IVa$_2$ (see Fig. 3). It seems likely, if not a certainty, that such viral mRNA species will be found to contain sequences that confer recognition by the modified translational apparatus of infected cells. Clearly, a great deal more work is required to establish both the identity of the sequences that comprise the recognition signal of the tripartite leader and the precise mechanism whereby it is recognized in adenovirus-infected cells. But, even at this early stage, these are exciting observations that lay the groundwork for an excellent system in which to elucidate mechanisms that regulate protein synthesis in mammalian cells.

The interaction of VA-RNA$_I$ with specific mRNA signals such as those within the tripartite leader element would, in principle, appear to be sufficient to guarantee selective and, therefore, efficient translation of viral mRNA species. Nevertheless, this interaction (and, therefore, the requirement for the two sets of viral RNA sequences) is not sufficient to explain the previously cited observation that the H5d1330 mutation fails to restore translation of cellular mRNA species; in the absence of participation of any other viral products, the lack of VA-RNA$_I$ to direct initiation of translation of viral mRNA would be expected to permit continued synthesis of celullar proteins. The phenotype of H5d1330-infected cells, therefore, indicates that the translational apparatus of infected cells must be modified by additional viral products, but these have not yet been identified.

6. SUMMARY AND CONCLUSIONS

The adenovirus–host cell interactions discussed in the preceding sections can be placed into one of two obvious categories, those that lead to repression and those that permit induction of cellular macromolecular synthesis. The former class of responses, inhibition of cellular DNA synthesis, disruption of cellular RNA metabolism, and reduced translation of cellular mRNA species, are observed only in permissive cells. Thus, such changes can be readily rationalized in terms of the advantage to the virus of such redirection of the host cells' resources and biosynthetic machinery. The current state of knowledge of the molecular mechanisms that mediate inhibition of cellular gene expression and DNA synthesis in adenovirus-infected cells ranges from fair, and improving rapidly (selective translation of viral mRNA species during the late phase of infection) to abysmal (inhibition of cellular DNA synthesis). Much further work is needed to elucidate completely any one of these mechanisms and, thus, to describe the precise functions of relevant viral products, that is, to establish how specific viral proteins or RNA sequences interact with components of the cellular replication, translation, and RNA production systems to achieve their efficient redirection toward production of viral macromolecules. Nevertheless, the work reviewed here does illustrate that the means by which adenovirus usurps the biosynthetic machinery of its host cells are not only fascinating in their own right, but also likely to be of general relevance to the ways in which normal cells conduct their affairs. A case in point is the role of RNA–RNA interactions, those between VA-RNA-I and sequences of the tripartite leader of viral, late mRNAs, in the selective translation of viral mRNA characteristic of the late phase of infection (see Section 5). It seems likely that this interaction, whose molecular details should be elucidated in the near future, will prove to be representative of a more general phenomenon; small, cellular RNA species of the *Alu* family (7 S RNA) are, for example, now known to be components of "signal recognition particles" that mediate the specific attachment of polysomes encoding polypeptides bearing certain signal peptides to microsomal membranes (Walter and Blobel, 1982).

Less well-defined at present are the processes that lead to the failure of cellular mRNA and 28 S rRNA to leave the nucleus of adenovirus-infected cells during the late phase. Indeed, the changes in cellular RNA metabolism observed in adenovirus-infected cells not only seem to be without precedent, but are also difficult to explain when an efficient mechanism has been evolved by the virus to induce preferential translation of viral mRNA species. On the other hand, the alterations in 28 S rRNA metabolism that characterize infected cells (Section 4.3) may prove to be a part, or consequence, of the mechanism that permits selective and efficient translation of viral mRNA; we have, for example, seen that additional modifications of the translational mechinery must be invoked to account for the failure of cells infected with adenovirus mutants that cannot express the VA-RNA$_I$ gene to continue the translation of cellular mRNA species during the late phase (Section 5). Clearly, further work is needed to pinpoint the block to production of cellular mRNA and rRNA in infected cells more precisely and to determine whether specific viral proteins (or RNA sequences) mediate these inhibitory processes. One of the most interesting conclusions to emerge from studies of the positive responses of cells to adenovirus infection is that products of the E1A region mediate both activation of cellular DNA synthesis in both permissive and non- or semipermissive cells (Section 3.2) and enhanced expression of certain cellular genes, at least in human cells (Section 4.1). These observations are of obvious importance in relation to the mechanisms, not yet understood, of adenovirus transformation. It is, therefore, to be hoped that studies of induction of cellular gene expression will be extended, for example, to rodent cells. A description of the functions of the individual products of the E1A region as well as elucidation of the mechanisms whereby they act also remain important goals; despite the great attention this region has received, it is not yet clear which E1A product(s) mediates induction of cellular DNA synthesis (Section 3.2), nor do the results of studies of the mechanism whereby the larger E1A products potentiate gene expression tell a coherent story (Section 4.2).

In summary, the studies of the ways in which adenovirus infection can alter the program of the host cell that have been performed in the 30 years since the discovery of this virus group have established that a fascinating variety of interactions can take place. Nor are these of merely esoteric interest, for the few molecular details now available suggest direct relationships to either normal mechanisms of control employed by eukaryotic cells or to the acquisition of the transformed cell phenotype. Thus, the task of obtaining a molecular picture of adenovirus–mammalian cell interactions, likely to occupy a good few years, should prove to be well worth the effort.

ACKNOWLEDGMENTS

I am grateful to J. Nevins, S. Berget, R. Stein, E. Ziff, J. Hogan, and T. Shenk for the communication of results prior to publication.

I also thank Cathy Castiglia and Mike Vayda for helpful discussions and Noël Mann for preparation of the manuscript. The unpublished work performed in the author's lab was supported by a Public Health Service Grant Number A117265.

7. REFERENCES

Aiello, L., Guilfoyle, R., Huebner, K., and Weinmann, R., 1979, Adenovirus 5 DNA sequences present and RNA sequences transcribed in transformed human embryo kidney cells (HEK - Ad5 or 293), *Virology* **94**:450.

Akusjärvi, G., and Persson, H., 1981, Controls of RNA splicing and termination in the major late adenovirus transcription unit, *Nature (London)* **292**:420.

Akusjärvi, G. and Pettersson, U., 1979, Sequence analysis of adenovirus DNA: complete nucleotide sequence of the spliced 5' non-coding region of adenovirus 2 hexon messenger RNA, *Cell* **16**:841.

Akusjärvi, G., Mathews, M. B., Andersson, P., Vennströmm, B., and Pettersson, U., 1980, Structure of genes for virus-associated RNA_I and RNA_{II} of adenovirus type 2, *Proc. Natl. Acad. Sci. USA* **77**:2424.

Aleström, P., Akusjärvi, G., Perricaudet, M., Mathews, M. B., Klessig, D., and Pettersson, U., 1980, The gene for polypeptide IX of adenovirus type 2 and its unspliced messenger RNA, *Cell* **19**:671.

Aleström, P., Akusjärvi, G., Pettersson, M., and Pettersson, U., 1982, DNA sequence analysis of the region encoding the terminal protein and the hypothetical N-gene product of adenovirus type 2, *J. Biol. Chem.* **257**:13492.

Anderson, C. W., Baum, P. R., and Gesteland, R. F., 1973, Processing of adenovirus 2 induced proteins, *J. Virol.* **12**:241.

Anderson, C. W., Lewis, J. B., Atkins, J. F., and Gesteland, R. F., 1974, Cell-free synthesis of adenovirus 2 proteins programmed by fractionated mRNA: A comparison of polypeptide products and mRNA lengths, *Proc. Natl. Acad. Sci. USA* **71**:2756.

Babich, A., Feldman, L. T., Nevins, J. R., Darnell, J. E., and Weinberger, C., 1983, Effects of adenovirus on metabolism of specific host mRNAs: Transport control and specific translational discrimination, *Mol. Cell. Biol.* **3**:1212.

Baserga, R., 1976, *Multiplication and Division in Mammalian Cells*, p. 239, Marcel Dekker, New York.

Bautz, E. K. F., 1976, Bacteriophage-induced DNA-dependent RNA polymerase, in: *RNA Polymerase* (R. Losick, and M. Chamberlin, eds.), p. 273, Cold Spring Harbor Laboratory, New York.

Bello, L. J., and Ginsberg, H. S., 1967, Inhibition of host protein synthesis in type 5 adenovirus-infected cells, *J. Virol.* **1**:843.

Beltz, G., and Flint, S. J., 1979, Inhibition of HeLa cell protein synthesis during adenovirus infection: Restriction of cellular messenger RNA sequences to the nucleus, *J. Mol. Biol.* **131**:353.

Benecke, B.-J., BenZe'ev, A., and Penman, S., 1978, The control of mRNA production, translation and turnover in suspended and reattached anchorage-dependent fibroblasts, *Cell* **14**:931.

Benecke, B.-J., BenZe've, A., and Penman, S., 1980, The regulation of RNA metabolism in suspended and reattached anchorage-dependent 3T6 fibroblasts, *J. Cell Physiol.* **103**:247.

Berget, S. M., Moore, C., and Sharp, P. A., 1977, Spliced segments at the 5′-terminus of adenovirus 2 late mRNA, *Proc. Natl. Acad. Sci. USA* **74**:3171.

Berk, A. J., and Sharp, P. A., 1978, Structure of the adenovirus 2 early mRNAs, *Cell* **14**:695.

Berk, A. J., Lee, F., Harrison, T., Williams, J. F., and Sharp, P. A., 1979, Pre-early adenovirus 5 gene product regulate synthesis of viral early messenger RNAs, *Cell* **17**:935.

Bernards, R., Houweling, A., Schrier, P. J., Bos, J. L., and van der Eb, A. J., 1982, Characterization of cells transformed by Ad5/Ad12 hybrid early region 1 plasmids, *Virology* **120**:422.

Bernards, R., Schrier, P. I., Bos, J. L., and van der Eb, A. J., 1983, Roles of types 5 and 12 early 1b tumor antigens in oncogenic transformation, *Virology* **127**:45.

Beyer, A. L., Miller, O. L., and McKnight, S. J., 1980, Ribonucleoprotein structure in nascent RNA is non-random and sequence dependent, *Cell* **20**:75.

Beyer, A. L., Bouton, A. H., and Miller, O. L., 1981, Correlation of hnRNP structure and nascent transcript cleavage, *Cell* **26**:155.

Bhaduri, S., Raskas, H. J., and Green, M., 1972, Procedure for the preparation of milligram quantities of adenovirus messenger RNA, *J. Virol.* **10**:1126.

Binger, M. H., Flint, S. J., and Rekosh, D., 1982, Expression of the gene encoding the adenovirus DNA terminal protein precurosor in productively-infected and transformed cells, *J. Virol.* **42**:488.

Bos, J. L., Polder, L. J., Bernards, R., Schrier, P. I., van den Elsen, P. J., van der Eb, A. J., and van Ormondt, H., 1981, The 2.2kb. E1B mRNA of human Ad12 and Ad5 codes for two tumor antigens starting at different AUG triplets, *Cell* **27**:121.

Braithewaite, A. W., Murray, J. D., and Bellett, A. J. D., 1981, Alterations to controls of cellular DNA synthesis by adenovirus infection, *J. Virol.* **39**:331.

Braithewaite, A. W., Cheetham, B. F., Li, P., Parish, C. R., Waldron-Stevens, L. K., and Bellett, A. J. D., 1983, Adenovirus-induced alterations of the cell growth cycle: A requirement for expression of E1A but not E1B, *J. Virol.* **45**:192.

Brandon, F. B. and McLean, I. W., 1962, Adenovirus, *Adv. Virus Res.* **9**:157.

Burlingham, B. T., and Doerfler, W., 1972, An endonuclease in cells infected with adenoviruses and associated with adenovirions, *Virology* **48**:1.

Burlingham, B. T., Doerfler, W., Pettersson, U., and Phillipson, L., 1977, Adenovirus endonuclease: Association with penton of adenovirus type 2, *J. Mol. Biol.* **60**:45.

Castiglia, C. L., and Flint, S. J., 1983, Effects of adenovirus infection on rRNA synthesis and maturation in HeLa cells, *Mol. Cell Biol.* **3**:662.

Castro, B. C., 1973, Biologic parameters of adenovirus transformation, *Prog. Exp. Tumor Res.* **18**:166.

Challberg, M. D., Desiderio, S. V., and Kelly, T. J., 1980, Adenovirus DNA replication *in vitro*: Characterization of a protein covalently linked to nascent DNA strands, *Proc. Natl., Acad. Sci. USA* **77**:5105.

Challberg, M. D., Castrove, J. M., and Kelly, T. J., 1982, Initiation of adenovirus DNA replication: Detection of covalent complexes between nucleotide and the 80 kilodalton terminal proteins, *J. Virol.* **41**:265.

Challberg, S. S., and Ketner, G., 1981, Deletion mutants of adenovirus 2: Isolation and initial characterization of virus carrying mutations near the right end of the viral genome, *Virology* **114**:196.

Cheetham, B. F., and Bellett, A. J. D., 1982, A biochemical investigation of the adenovirus-induced G1 to S phase progression: Thymidine kinase, ornithine decarbozylase and inhibitors of polyamine biosynthesis, *J. Cell. Physiol.* **110**:114.

Chow, L. T., Gelinas, R. E., Broker, T. R., and Roberts, R. J., 1977, An amazing sequence arrangement at the 5' end of adenovirus 2 messenger RNA, *Cell* **12**:1.

Chow, L. T., Broker, T. R., and Lewis, J. B., 1979, Complex splicing patterns of RNAs from the early regions of adenovirus-2, *J. Mol. Biol.* **134**:265.

Colby, W., and Shenk, T., 1981, Adenovirus type 5 virions can be assembled *in vivo* in the absence of polypeptide IX, *J. Virol.* **39**:977.

Cooper, J. E. K., Yohn, D. S., and Stich, H. F., 1968, Viruses and mammalian chromosomes: A comparative study of the chromosome damage induced by human and simian adenoviruses, *Exp. Cell Res.* **53**:225.

Craig, N., and Perry, R. P., 1970, Aberrant intranucleolar maturation of ribosomal precursors in the absence of protein synthesis, *J. Cell. Biol.* **45**:554.

D'Halliun, J-C., Allart, C., Cousin, C., Boulanger, P. S., and Martin, G. R., 1981, Adenovirus early function required for protection of viral and cellular DNA, *J. Virol.* **32**:61.

Doerfler, W., 1968, The fate of the DNA of adenovirus 12 in baby hamster kidney cells, *Proc. Natl. Acad. Sci. USA* **60**:636.

Doerfler, W., 1969, Non-productive infection of baby hamster kidney cells with adenovirus type 12, *Virology* **38**:387.

Doerfler, W., and Lundholm, U., 1970, Absence of the replication of the DNA of adenovirus type 12 in BHK 21 cells, *Virology* **40**:754.

Downey, J. F., Rowe, D. T., Bacchetti, S., Graham, F. L., and Bayley, S. J., 1983, Mapping of a 14,000 dalton antigen to early region 4 of the human adenovirus 5 genome, *J. Virol.* **45**:514.

Dulbecco, R., Hartwell, L. H., and Vogt, M., 1965, Induction of cellular DNA synthesis by polyoma virus, *Proc. Natl. Acad. Sci. USA* **53**:403.

Enomoto, T., Lichy, J. H., Ikeda, J. E., and Herwitz, J., 1981, Adenovirus DNA replication *in vitro*: Purification of the terminal protein in a functional form, *Proc. Natl. Acad. Sci. USA* **78**:6779.

Ensinger, M., and Ginsberg, H. S., 1972, Selection and preliminary characterization of temperature-sensitive mutants of type 5 adenovirus, *J. Virol.* **10**:328.

Esche, H., Mathews, M. B., and Lewis, J. B., 1980, Proteins and messenger RNAs of the transforming region of wild-type and mutant adenovirus, *J. Mol. Biol.* **142**:399.

Everett, S. J., and Ginsberg, H. S., 1958, A toxin-like material separable from type 5 adenovirus particles, *Virology* **6**:770.

Ezoe, H., and Mak, S., 1974, Comparative studies on functions of human adenovirus type 12 and its low oncogenic mutant virions, *J. Virol.* **14**:713.

Ezoe, H., Lai Fatt, R. B., and Mak, S., 1981, Degradation of intracellular DNA in KB cells infected with *cyt* mutants of human adenovirus type 12, *J. Virol.* **40**:20.

Flint, S. J. 1982, Expression of adenoviral genetic information in productively-infected cells, *Biochem. Biophys. Acta* **651**:175.

Flint, S. J., and Beltz, G. A., 1979, Expression of transforming viral genes in semipermissive cells transformed by SV40 or adenovirus type 2 or type 5, *Cold Spring Harbor Symp. Quant. Biol.* **44**:89.

Flint, S. J., Beltz, G. A., and Linzer, D. I. H., 1983, Synthesis and processing of simian virus 40-specific RNA in adenovirus-infected, simian virus 40-transformed human cells, *J. Mol. Biol.* **167**:335.

Foe, V. E., Wikinson, L. A., and Laird, C. D., 1976, Comparative organization of active transcription units from oncopeltus fasciatus, *Cell* **9**:131.

Fraser, N., and Ziff, E., 1978, RNA structures near poly(A) of adenovirus 2 late messenger RNA, *J. Mol. Biol.* **124**:27.

Friedman, M. P., Lyons, M. J., and Ginsberg, H. S., 1970, Biochemical consequences of type 2 adenovirus and SV40 double infection of African green monkey cells, *J. Virol.* **5**:586.

Friefeld, B. R., Lichy, J. H., Hurwitz, J., and Horwitz, M. S., 1983, Evidence for an altered adenovirus DNA polymerase in cells infected with the mutant H5ts149, *Proc. Natl. Acad. Sci. USA* **80**:1589.

Frost, E., and Williams, J. F., 1978, Mapping temperature-sensitive and host-range mutants of adenovirus type 5 by marker rescue, *Virology* **91**:39.

Gallimore, P. H., 1974, Interactions of adenovirus type 2 with rat embryo cells: Permissiveness, transformation and *in vitro* characterization of adenovirus 2 transformed rat embryo cells, *J. Gen. Virol.* **25**:263.

Galos, R., Williams, J. F., Binger, M-H., and Flint, S. J., 1979, Location of additional early gene sequences in the adenoviral chromosome, *Cell* **17**:945.

Gattoni, R., Stévinin, J., and Jacob, M., 1980, Comparison of the nuclear ribonucleoproteins containing transcripts of adenovirus 2 and HeLa cell DNA, *Eur. J. Biochem.* **108**:203.

Gaynor, R. B., and Berk, A. J., 1983, Cis-acting induction of adenovirus transcription, *Cell* **33**:683.

Gelinas, R. E., and Roberts, R. J., 1977, One predominant 5'-undecanucleotide in adenovirus 2 late messenger RNA, *Cell* **11**:533.

Gingeras, T. R., Sciaky, D., Gelinas, R. E., Bing-Dong, J., Yen, C. E., Kelly, M. E., Bullock, P. A., Parsons, B. L., O'Niell, K. E., and Roberts, R. J., 1982, Sequences from the adenovirus-2 genome, *J. Biol. Chem.* **257**:13475.

Ginsberg, H. S., Bello, L. J., and Levine, A. J., 1967, Control of biosynthesis of host macromolecules in cells infected with adenovirus, in: *The Molecular Biology of Viruses* (J. S. Colter, and W. Paranchych, eds.), p. 547, Academic Press, New York.

Graham, F. G., Harrison, T. J., and Williams, J. F., 1978, Defective transforming capacity of adenovirus type 5 host-range mutants, *Virology* **86**:10.

Green, M., and Daesch, G. E., 1961, Biochemical studies on adenovirus multiplication II kinetics of nucleic acid and protein synthesis in suspension cultures, *Virology* **13**:169.

Green, M., Parsons, J. T., Piña, M., Fujinaga, K., Caffier, H., and Langraf-Leurs, I., 1971, Transcription of adenovirus genes in productively-infected and in transformed cells, *Cold Spring Harbor Symp. Quant. Biol.* **35**:803.

Greenberg, J. R., 1972, High stability of mRNA in growing cultured cells, *Nature (London)* **242**:102.

Grodzicker, T., Anderson, C. W., Sambrook, J., and Mathews, M. B., 1977, The physical location of structural genes in adenovirus DNA, *Virology* **80**:111.

Grummt, I., 1981, Mapping of a mouse ribosomal DNA promoter by *in vitro* transcription, *Nucl. Acids Res.* **9**:6093.

Hadjiolev, A. A., 1980, Biogenesis of ribosomes in eukaryotes, *Subcell. Biochem.* **7**:1.

Hadjiolev, A. A., and Nikolaev, N., 1976, Maturation of ribosomal RNAs and the biogenesis of ribosomes, *Prog. Biophys. Mol. Biol.* **31**:95.

Halbert, D. N., Spector, D. J., and Raskas, H. J., 1979, *In vitro* translation products specified by the transforming region of adenovirus type 2, *J. Virol.* **31**:621.

Harford, C. G., Hamlin, A., Parker, E., and van Rauensway, T., 1956, Electron microscopy of HeLa cells infected with adenoviruses, *J. Exp. Med.* **104**:443.

Harrison, T. J., Graham, F., and Williams, J. F., 1977, Host range mutants of adenovirus type 5 defective for growth on HeLa cells, *Virology* **77**:319.

Heintz, N., Sive, H. L., and Roeder, R. G., 1983, Regulation of human histone gene expression: Kinetics of accumulation and changes in the rate of synthesis and in the half-lives of individual histone mRNAs during the HeLa cell cycle, *Mol. Cell Biol.* **3**:539.

Henderson, I. C., and Livingston, D. M., 1974, Partial purification and characterization of SV40 T-antigen, *Cell* **3**:65.

Herman, R., Weymouth, L., and Penman, S., 1978, Heterogeneous nuclear RNA-protein fibers in chromatin-depleted nuclei, *J. Cell. Biol.* **78**:663.

Higashi, K., Matsukisa, T., Kitao, A., and Sakamoto, Y., 1968, Selective supression of nucleolar RNA metabolism in the absence of protein synthesis, *Biochim. Biophys. Acta* **166**:388.

Hilleman, M. R., and Werner, J. H., 1954, Recovery of a new agent from patients with acute respiratory illness, *Proc. Soc. Exp. Biol. Med.* **85**:183.

Ho, Y-S., Galos, R., and Williams, J. F., 1982, Isolation of type 5 adenovirus mutants with a cold-sensitive phenotype: Genetic evidence of an adenovirus transformation maintainence function, *Virology* **122**:109.

Hodge, L. D., and Scharff, M. D., 1969, Effect of adenovirus on host cell DNA synthesis in synchronized cells, *Virology* **37**:554.

Houweling, A., van den Elsen, P. J., and van der Eb, A. J., 1980, Partial transformation of primary rat cells by the left most 4.5% fragment of adenovirus 5 DNA, *Virology* **105**:537.

Ikeda, J.-E., Enomoto, T., and Hurwitz, J., 1982, Adenoviral protein-primed initiation of DNA chains *in vitro*, *Proc. Natl. Acad. Sci. USA* **79**:2442.

Jackson, D. A., McCready, S. J., and Wok, I. D., 1981, RNA is synthesized at the nuclear cage, *Nature (London)* **292**:352.

Jackson, D. A., Carton, A. J., McCready, S. J., and Cook, P., 1982, Influenza RNA is synthesized at fixed sites in the nucleus, *Nature (London)* **296**:366.

Jochemsen, H., Daniels, G. S. G., Hertoghs, J. J. L., Schrier, P. I., van den Elsen, P. M., and van der Eb, A. J., 1982, Identification of adenovirus 12 gene products involved in transformation and oncogenesis, *Virology* **122**:15.

Jones, N., and Shenk, T., 1978, Isolation of deletion and substitution mutants of adenovirus type 5, *Cell* **13**:181.

Jones, N., and Shenk, T., 1979*a*, Isolation of adenovirus type 5' host-range delection mutants defective for transformation of rat embryo cells, *Cell* **17**:683.

Jones, N., and Shenk, T., 1979*b*, An adenovirus type 5 early gene function regulates expression of other early genes, *Proc. Natl. Acad. Sci. USA* **76**:3665.

Kao, H-T., and Nevins, J. R., 1983, Transcriptional activation and subsequent control of the human heat shock gene during adenovirus infection, *Mol. Cell. Biol.* **3**:2058.

Katze, M. G., Persson, H., and Phillipson, L., 1981, Control of adenovirus gene expression: Post-transcriptional control mediated by both viral and cellular gene products, *Mol. Cell. Biol.* **1**:807.

Katze, M. G., Persson, H., Johansson, B-M., and Philipson, L., 1983, Control of adenovirus gene expression: Cellular gene products restrict expression of adenovirus host range mutants in non-permissive cells, *J. Virol.* **46**:50.

Kit, S., Dubbs, D. R., deTorres, R. A., and Melnick, J. L., 1965, Enhanced thymidine kinase activity following infection of green monkey kidney cells by simian adenoviruses, simian papovaviruses and an adenovirus-SV40 "hybrid" virus, *Virology* **27**:453.

Kjellen, L., Lagermalm, G., Svedmyr, A., and Thorsson, K. G., 1955, Crystalline-like patterns in the nuclei of cells infected with an animal virus, *Nature (London)* **175**:505.

Klessig, D. F., 1977, Two adenovirus mRNAs have a common 5' terminal leader sequence encoded at least 10kb upstream from their main coding regions, *Cell* **12**:9.

Klessig, D. F., and Anderson, C. W., 1975, Block to multiplication of adenovirus serotype 2 in monkey cells, *J. Virol.* **16**:1650.

Klessig, D. F., and Chow, L. T., 1980, Incomplete splicing and deficient accumulation of the fiber messenger RNA in monkey cells infected by human adenovirus type 2, *J. Mol. Biol.* **139**:221.

Klessig, D. F., and Grodzicker, T., 1977, Mutations that allow human Ad2 and Ad5 to express late genes in monkey cells map in the viral gene encoding the 72kd DNA binding protein, *Cell* **17**:957.

Kruijer, W., van Schaik, F. M. A., and Sussenbach, J. S., 1981, Structure and organization of the gene coding for the DNA binding protein of adenovirus type 5, *Nucl. Acids Res.* **9**:4439.

Lai Fatt, R. B., and Mak, S., 1982, Mapping of an adenovirus function involved in the inhibition of DNA degradation, *J. Virol* **42**:969.

Laird, C. D., and Chooi, W. Y., 1976, Morphology of transcription units in *D. Melanogaster, Chromosoma* **58**:193.

Lassam, N. J., Bayley, S. T., and Graham, F. L., 1978, Synthesis of DNA, late polypeptides and infectious virus by host-range mutants of adenovirus 5 in nonpermissive cells, *Virology* **87**:463.

Lastick, S. M., 1980, The assembly of ribosomes in HeLa cell nucleoli, *Eur. J. Biochem.* **111**:175.

Laughlin, C., and Strohl, W. A., 1976a, Factors regulating cellular DNA synthesis induced by adenovirus infection I. The effects of actinomycin D or G-arrested BHK21 cells abortively-infected by type 12 adenovirus or stimulated by serum, *Virology* **74**:30.

Laughlin, C., and Strohl, W. A., 1976b, Factors regulating cellular DNA synthesis induced by adenovirus infection II. The effects of actinomycin D on productive virus-cell systems, *Virology* **74**:44.

Learned, R. M., Smale, S. T., Haltiner, M. M., and Tjian, R., 1983, Regulation of human ribosomal RNA transcription, *Proc. Natl. Acad. Sci. USA* **80**:3558.

Ledinko, N., 1967, Stimulation of DNA synthesis and thymidine kinase activity in human embryonic kidney cells infected by adenovirus 2 or 12, *Cancer Res.* **27**:1459.

Ledinko, N., 1972, Nucleolar ribosomal precursor RNA and protein synthesis in HEK cultures infected with Ad12, *Virology* **49**:79.

Ledinko, N., 1974, Temperature-sensitive mutants of adenovirus type 12 defective in viral DNA synthesis, *J. Virol.* **14**:457.

Levine, A. J., van der Vliet, P. C., Rosenwirth, B., Rabek, J., Frenkel, G., and Ensinger, M., 1975, Adenovirus-infected cell-specific DNA binding proteins, *Cold Spring Harbor Symp. Quant. Biol.* **39**:559.

Lewis, J. B., and Anderson, C. W., 1983, Proteins encoded near the adenovirus late messenger RNA leader segments, *Virology* **127**:112.

Lewis, J. B., and Mathews, M. B., 1980, Control of adenovirus early gene expression: A class of immediate early products, *Cell* **21**:303.

Lewis, J. B., Atkins, J. F., Anderson, C. W., Baum, P. R., and Gesteland, R. F., 1975, Mapping of late adenovirus genes by cell-free translation of RNA selected by hybridization to specific DNA fragments, *Proc. Natl. Acad. Sci. USA* **72**:1344.

Lewis, J. B., Atkins, J. F., Baum, P. R., Solem, R., Gesteland, R. F., and Anderson, C. W., 1976, Location and identification of the genes for adenovirus type 2 early polypeptides, *Cell* **7**:141.

Lewis, J. B., Anderson, C. W., and Atkins, J. F., 1977, Further mapping of late adenovirus genes by cell-free translation of RNA selected by hybridization to specific DNA fragments, *Cell* **12**:37.

Lichy, J. H., Horwitz, M. S., and Hurwitz, J., 1981, Formation of a covalent complex between the 80,000 dalton adenovirus terminal protein and 5′ dCMP *in vitro, Proc. Natl. Acad. Sci. USA* **78**:3678.

Lichy, J. H., Field, J., Horwitz, M. S., and Hurwitz, J., 1982, Separation of the adenoviral terminal protein precursor from its associated DNA polymerase: Role of both proteins in the initiation of adenovirus DNA replication, *Proc. Natl. Acad. Sci. USA* **79**:5225.

Lindberg, U., Persson, T., and Philipson, L., 1972, Isolation and characterization of adenovirus messenger RNA in productive infection, *J. Virol.* **10**:909.

Lockard, R. E., Berget, S. M., RajBhandary, U. L., and Sharp, P. A., 1979, Nucleotide sequence of the 5′-terminus of adenovirus 2 late messenger RNA, *J. Biol. Chem.* **254**:587.

Losick, R., and Pero, J., 1976, Regulatory subunits of RNA polymerase, in *RNA polymerase* (R. Losick, and M. Chamberlin, eds.), p. 227, Cold Spring Harbor Laboratory, New York.

Mak, I., and Mak, S., 1983, Transformation of rodent cells by *cyt* mutants of adenovirus type 12 and mutant of adenovirus type 5, *J. Virol.* **45**:1107.

Mariman, E. C. M., van Eekelen, C. A. G., Reinders, R. J., Berns, A. J. M., and van Venrooij, W. J., 1982, Adenoviral heterogeneous nuclear RNA is associated with the host nuclear matrix during splicing, *J. Mol. Biol.* **154**:103.

Marengo, C., Mbikay, M., Weber, J., and Thirion, J-P., 1981, Adenovirus-induced mutations at the hpoxanthine phosphoribosyl transferase locus in Chinese hamster cell, *J. Virol.* **38**:184.

Marusyk, R. G., Morgan, A. R., and Wadell, G., 1975, Association of endonuclease activity with serotypes belong to the three subgroups of human adenoviruses, *J. Virol.* **16**:456.

Mathews, M. B., 1980, Binding of adenovirus YARNA to mRNA: A possible role in splicing? *Nature (London)* **285:**575.

Mathews, M. B., and Pettersson, U., 1978, The low molecular weight RNAs of adenovirus 2 infected cells, *J. Mol. Biol.* **119:**293.

McDougall, J. K., 1971, Adenovirus-induced chromosome aberrations in human cells, *J. Gen. Virol.* **12:**43.

McDougall, J. K., 1975, Adenoviruses-interaction with the host cell genome, *Prog. Med. Virol.* **21:**118.

McDougall, J. K., Dunn, A. R., and Jones, K. W., 1972, *In situ* hybridization of adenovirus RNA and DNA, *Nature (London)* **236:**346.

McDougall, J. K., Kucherlapati, R., and Ruddle, F. H., 1973, Localization and induction of the human thymidine kinase gene by adenovirus type 12, *Nature New Biol.* **245:**172.

McGrogan, M., and Raskas, H. J., 1978, Two regions of the adenovirus 2 genome specify families of late polysomal RNA containing common sequences, *Proc. Natl. Acad Sci. USA* **75:**625.

McKnight, S. L., and Miller, O. L., 1976, Ultrastructural patterns of RNA synthesis during early embryogenesis of *Drosophila melanogaster, Cell* **8:**305.

Miesfeld, R., and Arnheim, N., 1982, Identification of the *in vivo* and *in vitro* origin of transcription in human rDNA, *Nucl. Acids Res.* **10:**3933.

Miller, J. S., Ricciardi, R. P., Roberts, B. E., Paterson, B. M., and Mathews, M. B., 1980, Arrangement of messenger RNAs and protein coding sequences in the major late transcription unit of adenovirus, *J. Mol. Biol.* **142:**455.

Miller, K., and Sollner-Webb, B., 1981, Transcription of mouse rRNA genes by RNA polymerase I: *in vitro* and *in vivo* initiation and processing sites, *Cell* **27:**165.

Miller, T. E., Huang, C., and Pogo, A. O., 1978, Rat liver nuclear skeleton and ribonucleo-protein complexes containing hnRNA, *J. Cell. Biol.* **76:**675.

Minowada, J., and Moore, G. E., 1963, DNA synthesis in X-irradiated cultures infected with polyoma virus, *Exp. Cell Res.* **29:**31.

Mishima, Y., Mastui, T., and Muramatsu, M., 1979, The mechanism of decrease in nucleolar RNA synthesis by protein synthesis inhibition, *J. Biochem.* **85:**807.

Mishima, Y., Yamamoto, O., Komihami, R., and Muramatsu, M., 1981, *In vitro* transcription of a cloned mouse ribosomial RNA gene, *Nucl. Acids Res.* **9:**6773.

Montell, C., Fisher, E. F., Caruthers, M. H., and Berk, A. J., 1982, Resolving the functions of overlapping viral genes by site-specific mutagenesis at a mRNA splice site, *Nature (London)* **295:**380.

Morgan, C., Goodman, G. C., Rose, H. M., Howe, C., and Huang, J. S., 1957, Electron microscopic and histochemical studies of unusual crystalline protein occurring in cells infected by type 5 adenovirus, *J. Biophys. Biochem. Cytol.* **3:**505.

Munroe, S., 1982, Ribonucleoprotein structure of adenovirus nuclear RNA probed by nuclease digestion, *J. Mol. Biol.* **162:**535.

Muramatsu, M., Shimada, N., and Higashiankagawa, T., 1970, Effect of cycloheximide on nucleolar RNA synthesis in rat liver, *J. Mol. Biol.* **53:**91.

Murray, J. D., Bellett, A. J. D., Braithewaite, A. W., Waldron, L. K., and Taylor, I. W., 1982*a*, Altered cell cycle progression and aberrant mitosis in adenovirus-infected rodent cells, *J. Cell. Physiol.* **11:**89.

Murray, J. D., Braithewaite, A. W., Taylor, I. W., and Bellett, A. J. D., 1982*b*, Adenovirus-induced alterations of the cell growth cycle: Effects of mutations in early regions E2A and E2B, *J. Virol.* **44:**1072.

Nagata, K., Guggenheimer, R. A., Enomoto, T., Lichy, J. H. and Hurwitz, J., 1982, Adenovirus DNA in replication: Identification of a host factor that stimulates synthesis of the preterminal protein-dCMP complex, *Proc. Nat. Acad. Sci. USA* **79**:6438.

Nagata, K., Guggenheimer, R. A. and Hurwitz, J., 1983, Adenovirus DNA replication *in vitro*: Synthesis of full length DNA with purified proteins, *Proc. Nat. Acad. Sci. USA* **80**:4266.

Nass, K., and Frenkel, G. D., 1978, Adenovirus-induced inhibition of cellular DNAase, *J. Virol.* **26**:540.

Nass, K., and Frenkel, G. D., 1980, Adenovirus-specific DNA binding protein inhibits the hydrolysis of DNA by DNase *in vitro*, *J. Virol.* **35**:314.

Nevins, J. R., 1981, Mechamism of activation of early viral transcription by the adenovirus E1A gene product, *Cell* **26**:213.

Nevins, J. R., 1982, Induction of the synthesis of a 70,000 dalton mammalian heat shock protein by the adenovirus E1A gene products, *Cell* **29**:913.

Nevins, J. R., and Darnell, J. E., 1978, Groups of adenovirus type 2 mRNAs derived from a large primary transcript: Probable nuclear origin and possible common 3' ends, *J. Virol.* **25**:811.

Nevins, J. R., and Wilson, M. C., 1981, Regulation of adenovirus-2 gene expression at the level of transcriptional termination and RNA processing, *Nature (London)* **290**:113.

Norrby, E., 1971, Adenoviruses, in: *Comprehensive Virology* (Maramorosch, E. and Kurstak, E., ed.), Academic Press, New York, p. 105.

Ortin, J., Scheidtmann, K. H., Greenberg, R., Westphal, M., and Doerfler, W., 1976, Transcription of the genome of adenovirus type 12. III. Transcription maps in productively-infected human cells and abortively-infected and transformed hamster cells, *J. Virol.* **20**:355.

Paraskeva, C., Roberts, C., Briggs, P. and Gallimore, P. H., 1983, Human adenovirus type 2 but not adenovirus type 12 is mutagenic at the hypoxanthine phosphoribosyl transferase locus of cloned rat liver epithelial cells, *J. Virol.* **46**:131.

Pardee, A. B., Dubrow, R., Hamlin, J. F., and Kleitzien, R. F., 1978, Animal cell cycle, *Annu. Rev. Biochem.* **47**:715.

Pater, M. M., and Mak, S., 1975, Replication of human KB cell DNA after infection by adenovirus type 12, *Nature (London)* **258**:636.

Paterson, B. M., Roberts, B., and Kuff, E. L., 1977, Structural gene identification and mapping by DNA mRNA hybrid-arrested cell-free translation, *Proc. Natl. Acad. Sci. USA* **74**:4370.

Periera, H. G., 1958, A protein factor responsible for the early cytopathic effect of adenovirus, *Virology* **6**:601.

Perricaudet, M., Akusjärvi, G., Virtanen, A., and Pettersson, U., 1979, Structure of two spliced mRNAs from the transforming region of human subgroup C adenoviruses, *Nature (London)* **281**:694.

Perricaudet, M., LeMoullec, J. M., and Pettersson, U., 1980, Predicted structures of two adenovirus tumor antigens, *Proc. Natl. Acad. Sci. USA* **77**:3778.

Perry, R. D., and Kelley, D. E., 1968, Messenger RNA-protein complexes and newly-synthesized ribosomal subunits: Analysis of free particles and componets of polyribosomes, *J. Mol. Biol.* **35**:37.

Persson, H., Jörnvall, J., and Zabielski, J., 1980, Multiple mRNA species for the precursor to an adenovirus-encoded glycoprotein: Identification and structure of the signal sequence, *Proc. Natl. Acad. Sci. USA* **77**:6349.

Pettersson, U., and Höglund, L., 1969, Structural proteins of adenoviruses III. Purification and characterization of the adenovirus type 2 penton antigen, *Virology* **39**:90.

Philipson, L., and Lindberg, U., 1974, Reproduction of adenoviruses, in: *Comprehensive Virology*, Vol. 3 (H. Fraenkel-Conrat, and R. R. Wagner, eds.), pp. 143, Plenum Press, New York.

Philipson, L., Pettersson, U., Lindberg, U., Tibbetts, C., Vennström, B, and Persson, T., 1975, RNA synthesis and processing in adenovirus infected cells, *Cold Spring Harbor Symp. Quant. Biol.* **39**:447.

Phillips, W. F., and McConkey, E. H., 1976, Relative stoichiometry of ribosomal proteins in HeLa nucleoli, *J. Biol. Chem.* **251**:2876.

Piña, M., and Green, M., 1969, Biochemical studies on adenovirus multiplication: XIV. Macromolecule and enzyme synthesis in cells replicating oncogenic and non-oncogenic human adenovirus, *Virology* **38**:573.

Pochron, S., Rossini, M., Darzynkiewicz, Z., Traganos, F., and Baserga, R., 1980, Failure of accumulation of cellular RNA in hamster cells stimulated to synthesize DNA by infection with adenovirus 2, *J. Biol. Chem.* **255**:4411.

Price, R., and Penman, S., 1972, Transcription of the adenovirus genome by an α-amanitin-sensitive RNA polymerase in HeLa cells, *J. Virol.* **9**:621.

Rabson, A. S., O'Connor, G. T., Berezesky, I. K., and Paul, F. J., 1964, Enhancement of adenovirus growth in green monkey cells by SV40, *Proc. Soc. Exp. Biol. Med.* **116**:187.

Raskas, H. J., Thomas, D. C., and Green, M., 1970, Biochemical studies on adenovirus multiplication. XVII. Ribosome synthesis in uninfected and infected KB cells, *Virology* **40**:893.

Rekosh, D. M. K., Russell, W. C., Bellett, A. J. D., and Robinson, A. J., 1977, Identification of a protein linked to the ends of adenovirus DNA, *Cell* **11**:283.

Ricciardi, R. P., Jones, R. L., Cepko, C. T., Sharp, P. A., and Roberts, B. E., 1981, Expression of early adenovirus genes requires a viral encoded acidic polypeptide, *Proc. Natl. Acad. Sci. USA* **78**:6121.

Robinson, A. J., Younghusband, H. B., and Bellett, A. J. D., 1973, A circular DNA-protein complex from adenoviruses, *Virology* **56**:54.

Ross, S. R., Levine, A. J., Galos, R. S., Williams, J., and Shenk, T., 1980, Early viral proteins in HeLa cells infected with adenovirus type 5 host range mutants, *Virology* **163**:475.

Rossini, M., Weinmann, R., and Baserga, R., 1979, DNA synthesis in temperature-sensitive mutants of the cell cycle infected by polyoma virus and adenovirus, *Proc. Natl. Acad. Sci. USA* **76**:4441.

Rossini, M., Jonak, G. J., and Baserga, R., 1981, Identification of adenovirus 2 early genes required for induction of cellular DNA synthesis in resting hamster cells, *J. Virol.* **38**:982.

Rowe, W. F., Huebner, R. J., Gilmore, L. K., Parrott, R. H., and Ward, T. G., 1953, Isolation of a cytogenic agent from human adenoids undergoing spontaneous degradation in tissue culture, *Proc. Soc. Exp. Biol. Med.* **84**:570.

Rowe, W. P., Hartley, J. W., Roizmann, B., and Levy, H. B., 1958, Characterization of a factor formed in the course of adenovirus infection of tissue cultures causing detachment of cells from glass, *J. Exp. Med.* **108**:713.

Ruben, M., Bacchetti, S., and Graham, F. L., 1982, Integration and expression of viral DNA in cells transformed by host-range mutant of adenovirus type 5, *J. Virol.* **41**:674.

Russell, W. C., and Skehel, J. J., 1973, The polypeptides of adenovirus-infected cells, *J. Gen. Virol.* **15**:45.

Shaw, A. R., and Ziff, E., 1980, Transcripts from the adenovirus 2 major late promoter yield a single early family of 3' coterminal mRNAs and five late families, *Cell* **22**:905.

Shenk. T., Jones, N., Colby, W., and Fowlkes, D., 1979, Functional analysis of adenovirus 5 host-range deletion mutants defective for transformation of rat embryo cells, *Cold Spring Harbor Symp. Quant. Biol.* **44**:567.

Shimojo, H., and Yamashita, T., 1968, Induction of DNA synthesis by adenoviruses in contact-inhibited hamster cells, *Virology* **36**:422.

Shiroki, K., and Shimojo, H., 1975, Analysis of adenovirus 12 temperature-sensitive mutants defective in viral DNA replication, *Virology* **61**:674.

Shiroki, K., Maruyama, K., Saito, I., Fukui, Y., and Shimojo, H., 1981, Incomplete transformation of rat cells by a deletion mutant of adenovirus type 5, *J. Virol.* **38**:1048.

Singer, R. H., and Penman, S., 1973, Messenger RNA in HeLa cells: Kinetics of formation and decay, *J. Mol. Biol.* **78**:321.

Söderlund, H., Pettersson, U., Vennströmm, B., Philipson, L., and Mathews, M. B., 1976, A new species of virus-coded, low molecular weight RNA from cells infected with adenovirus type 2, *Cell* **7**:585.

Solnick, D., 1981, An adenovirus mutant defective in splicing RNA from early region E1A, *Nature (London)* **291**:508.

Solnick, D., and Anderson, M. A., 1982, Transformation-deficient adenovirus mutant defective in expression of region 1A but not region 1B, *J. Virol.* **42**:106.

Spector, D. J., Halbert, D. N., and Raskas, H. J., 1980, Regulation of integrated adenovirus sequences during adenovirus infection of transformed cells, *J. Virol.* **36**:860.

Stich, H. F., 1973. Oncogenic and non-oncogenic mutants of adenovirus 12: Induction of chromosome aberrations and cell division, *Prog. Exp. Tumor Res.* **18**:260.

Stich, H. F., and John, D. S., 1967, The mutagenic capacity of adenoviruses for mammalian cells, *Nature (London)* **216**:1292.

Stillman, B. W., Lewis, J. B., Chow, L. T., Mathews, M. B., and Smart, J. E., 1980, Identification of the gene and mRNA for adenovirus terminal protein precursor, *Cell* **23**:497.

Stillman, B. W., Tamanoi, F., and Mathews, M. B., 1982, Purification of an adenvorus-encoded DNA polymerase that is required for initiation of DNA replication, *Cell* **31**:613.

Strohl, W. A., 1969a, The response of BHK21 cells to infection with adenovirus 12.1, Cell-killing and T-antigen synthesis as correlated viral genome functions, *Virology* **39**:642.

Strohl, W. A., 1969b, The response of BHK21 cells to infection with adenovirus 12 II. The relationship of virus-stimulated DNA synthesis to other viral functions, *Virology* **39**:653.

Takahashi, M., 1972, Isolation and characterization of conditional lethal (temperature-sensitive and host-range) mutants of adenovirus type 5, *Virology* **49**:815.

Takashashi, M., van Hoosier, G. L., and Trentin, J. J., 1966, Stimulation of DNA synthesis in human and hamster cells by human adenovirus types 12 and 5, *Proc. Soc. Exp. Biol. Med.* **132**:740.

Takahashi, M., Ogino, T., Baba, K., and Onaka, M., 1969, Synthesis of deoxyribonucleic acid in human and hamster kidney cells infected with human adenovirus types 5 and 12, *Virology* **37**:513.

Takemori, N., 1972, Genetic studies with tumorigenic adenoviruses III. Recombination in adenovirus type 12, *Virology* **47**:157.

Takemori, N., Riggs, J. L., and Aldrich, C., 1968, Genetic studies with tumorigenic adenoviruses 1. Isolation of cytocidal (*cyt*) mutants of adenovirus type 12, *Virology* **36**:575.

Tal, J., Craig, E. A., and Raskas, H. J., 1975, Sequence relationships between adenovirus 2 early RNA and viral RNA size classes synthesized at 18 hours after infection, *J. Virol.* **15**:137.

Thimmappaya, B., Weinberger, C., Schneider, R. J., and Shenk, T., 1982, Adenovirus VA$_1$RNA is required for efficient translation of viral mRNA at late times after infection, *Cell* **31**:543.

Thomas, G. P., and Mathews, M. B., 1980, DNA replication and the early to late transition in adenovirus infection, *Cell* **22**:523.

Thummel, C., Tjian, R., Hu, S.-L., and Grodzicker, T., 1983, Translational control of SV40 T antigen expressed from the adenovirus late promoter, *Cell* **33**:455.

Tigges, M. A., and Raskas, H. J., 1982, Expression of adenovirus 2 of the early region 4 polypeptides to their respective mRNA using *in vitro* translations, *J. Virol.* **44**:907.

Tooze, J. ed., 1980, *The Molecular Biology of Tumor Viruses 2. DNA Tumor Viruses*, 2nd Ed., Cold Spring Harbor Laboratory, New York.

van Bergen, B. G. M., and van der Vliet, P. C., 1983, Temperature-sensitive initiation and elongation of adenovirus DNA replication *in vitro* with nuclear extracts from H5ts36, H5ts149 and H5ts125-infected HeLa cells, *J. Virol.* **46**:642.

van der Eb, A. J., van Ormondt, H., Schrier, P. I., Lupker, J. H., Jochemsen, H., van den Elsen, P. J., DeLeys, R. J., Maat, J., van Beveran, C. P., Dijkema, R., and de Waard, A., 1979, Structure and function of the transforming genes of human adenoviruses and SV40, *Cold Spring Harbor Symp. Quant. Biol.* **44**:383.

van Eekelen, C. A. G., and van Venrooij, W. J., 1981, HnRNA and its attachement to a nuclear matrix protein, *J. Cell. Biol.* **88**:554.

Walter, P., and Blobel, G., 1982, Signal recognition particle contains a 7S RNA essential for protein translocation across the endoplasmic reticulum, *Nature (London)* **299**:691.

Weil, R., Michel, M. R., and Ruschman, G. K., 1965, Induction of cellular DNA synthesis by polyoma virus, *Proc. Natl. Acad. Sci. USA* **53**:1468.

White, D. I., Scharff, M. D., and Maizel, J. V., 1969, The polypeptides of adenovirus III. Synthesis in infected cells, *Virology* **38**:395.

Wilkie, N. M., Ustacelabi, S., and Williams, J. F., 1973, Characterization of temperature-sensitive mutants of adenovirus type 5: Nucleic acid synthesis, *Virology* **51**:499.

Willems, M., Penman, M., and Penman, S., 1969, The regulation of RNA synthesis and processing in the nucleolus during inhibition of protein synthesis, *J. Cell. Biol.* **4**:177.

Williams, J. F., 1973, Oncogenic transformation of hamster embryo cells *in vitro* by adenovirus type 5, *Nature (London)* **243**:162.

Winocour, E., Kaye, A. M., and Stollar, V., 1965, Synthesis and transmethylation of DNA in polyoma-infected cultures, *Virology* **27**:156.

Yamamoto, H., Shimojo, H., and Hamada, C., 1972, Less tumorigenic (*cyt*) mutants of adenovirus 12 defective in induction of cell surface change, *Virology* **50**:743.

Yamashita, T., and Shimojo, H., 1969, Induction of cellular DNA synthesis by adenovirus 12 in human embryo kidney cells, *Virology* **38**:351.

Yoder, S. S., Robberson, B. L., Leys, E. J., Hook, A. G., Al-Ubaidi, M., Yeung, K-Y., Kellems, R. E., and Berget, S. M., 1983, Control of cellular gene expression during adenovirus infection: Induction and shut-off of dihydrofolate reductase gene expression by adenovirus type 2, *Mol. Cell Biol.* **3**:819.

Younghusband, H. B., Tyndall, C., and Bellett, A. J. D., 1979, Replication and interaction of virus DNA and cellular DNA in mouse cells infected by a human adenovirus, *J. Gen. Virol.* **45**:455.

Zain, S., Sambrook, J., Roberts, R. J., Keller, W., Fried, M. and Dunn, A. R., 1979, Nucleotide sequence analysis of the leader segment in a cloned copy of adenovirus 2 fiber mRNA, *Cell* **16**:851.

Ziff, E., 1980, Transcription and RNA processing by the DNA tumor viruses, *Nature (London)* **287**491.

Ziff, E., and Evans, R., 1978, Coincidence of the promoter and capped 5' terminus of RNA from the adenovirus 2 major late transcription unit, *Cell* **15**:1463.

Zimmerman, J. E., Raška, K., and Strohl, W. A., 1970, The response of BHK21 cells to infection with type 12 adenovirus IV. Activation of DNA-synthesizing apparatus, *Virology* **42**:1147.

zur Hausen, H., 1967. Induction of specific chromosomal aberrations by adenovirus type 12 in human embryonic kidney cells, *J. Virol.* **1**:1174.

zur Hausen, H., and Sokol, F., 1969, Fate of adenovirus 12 genomes in nonpermissive cells, *J. Virol.* **4**:255.

The Effects of Herpesviruses on Cellular Macromolecular Synthesis

Michael L. Fenwick

The Sir William Dunn School of Pathology
University of Oxford
United Kingdom

1. INTRODUCTION

Infection of cultured permissive animal cells with herpes simplex virus typically results in the initiation of the synthesis of the first ("α" or "immediate-early") viral mRNA within 1–2 hr. This is soon followed by production of virus-specific proteins in a regulated sequence of stages (Honess and Roizman, 1974, 1975; Spear and Roizman, 1981). The α-mRNA is transcribed from the infecting genome by the cellular RNA polymerase II, which is normally concerned with the synthesis of cellular mRNA. In the presence of cycloheximide, an inhibitor of protein synthesis, α-mRNA accumulates and is translated as soon as the inhibitor is removed. Functional α-polypeptides are required to initiate and maintain the synthesis of β ("early") polypeptides and, thereafter, the production of α-polypeptides decreases. The γ or "late" polypeptides are largely, though not exclusively, made after the replication of viral DNA has begun, from about 4 hr after infection. Mature progeny may accumulate from about 6 hr, reaching a maximum 10–12 hr after infection. These events usually occur against a declining background of host macromolecular synthesis. The sharpness of the decline varies with different virus strains and possibly with different host cells but substantial effects on host synthesis within 2–4 hr of infection are common.

The state of knowledge of the effects of infection with herpes-virus on the synthesis of macromolecules in the host cell was reviewed and discussed by Kaplan (1973a,b) and, in a wider context, by Roizman and Furlong (1974) in an early volume of this series. This chapter starts at the beginning again but concentrates more on the findings that have been published in the last 10 years.

2. EARLY STUDIES

A rapid interference with cellular growth processes was evident when it was shown that herpes simplex virus (HSV*) prevented cell division in synchronized HeLa cells if they were infected as little as 1 hr before mitosis was expected to begin (Stoker and Newton, 1959).

The earliest studies using radioactive precursors to measure the overall rates of synthesis of nucleic acids showed that in growing rabbit kidney cells infected with pseudorabies virus (PRV) the rate of DNA synthesis declined steadily over the first 5 hr and then increased but the total quantity of DNA in the cells did not change significantly (Kaplan and Ben-Porat, 1960). Newton et al. (1962) observed a similar decline in incorporation of radioactive thymidine during the first 6 hr of infection with HSV followed by a gradual rise to 10 hr after infection. The decline was accompanied by release of radioactive material from cells prelabeled before infection with [³H]thymidine (Newton, 1964).

The rate of RNA synthesis also declined, although less sharply, to about 50% of normal within 6 hr after infection with PRV (Kaplan and Ben-Porat, 1960), with HSV (Aurelian and Roizman, 1965), or equine abortion virus (EAV; O'Callaghan et al., 1968b).

Similar early experiments with radioactive precursors of proteins showed that in cells infected with HSV-1 an initial decrease in the overall rate of protein synthesis to 70% of normal at 3 hr after infection was followed by an increase (as viral proteins were made) and a final decline (Roizman et al., 1965). Although there was no change in the rate of incorporation of [³H]leucine in cells infected with PRV, the

---

* Abbreviations: HSV-1, herpes simplex virus type 1; HSV-2, herpes simplex virus type 2; PRV, pseudorabies virus; EAV, equine abortion virus; EBV, Epstein–Barr virus; AdV, adenovirus; VSV, vesicular stomatitis virus.

proportion of the labeled protein that could be precipitated by cell-specific antibodies declined, while the proportion of virus-specific protein increased with time (Hamada and Kaplan, 1965). Subsequently, it was found that lysine uptake was inhibited soon after infection while arginine uptake was stimulated (Kaplan *et al.*, 1970), reflecting the switch from cellular to viral protein synthesis.

More detailed descriptions of these phenomena became possible as improved techniques for distinguishing viral from cellular macromolecules were introduced. However, the mechanisms involved in the specific suppression of normal cellular processes (while very similar reactions resulting in the production of viral macromolecules are allowed to proceed) are still poorly understood although they are clearly of great interest.

3. EFFECTS ON DNA SYNTHESIS

Herpesvirus DNA has a higher content of G-C pairs than cellular DNA and, consequently, a higher buoyant density in the presence of CsCl. This property enabled Kaplan and Ben-Porat (1963) and Roizman and Roane (1964) to separate the two by CsCl gradient centrifugation and to show that, as infection progressed, the rate of accumulation of radioactive thymidine in cellular DNA declined while that in viral DNA increased. O'Callaghan *et al.* (1968a,b) reported a similar transition from cellular to viral DNA synthesis in cells infected with EAV. They achieved a separation of the two types of DNA by chromatography on MAK columns.

The decline, in PRV-infected cells, was not due to degradation of cellular DNA but may have been related to the margination of chromatin that was observed at about the same time (Kaplan and Ben-Porat, 1963). It was not accompanied by the appearance of viral DNA polymerase with a higher affinity for viral DNA than cellular DNA *in vitro*. Nor was it due to increased competition of newly formed viral DNA with cellular DNA as a template, since it occurred even if viral DNA synthesis was blocked with FUdR (Ben-Porat and Kaplan, 1965). The decline was, however, prevented by adding puromycin at the time of infection, suggesting (after taking into account the inhibitory effect of the drug itself on cellular DNA synthesis) that a newly made viral protein was responsible for suppressing host synthesis in normal infection (Ben-Porat and Kaplan, 1965). When cells were infected and incubated for 4 hr in the presence of cycloheximide,

DNA synthesis was strongly inhibited within an hour after removal of the drug, suggesting that an early (possibly "immediate-early") viral protein was involved (Ben-Porat et al., 1974). On the other hand, experiments with UV-inactivated HSV-1 and HSV-2 have shown that host DNA synthesis can be suppressed without detectable synthesis of viral proteins (Newton, 1969; Fenwick and Walker, 1978).

4. EFFECTS ON RNA SYNTHESIS

4.1. General

Sucrose gradient analysis of pulse-labeled RNA extracted from infected cells (Hay et al., 1966; Flanagan, 1967) revealed a decline in the rate of synthesis of rapidly labeled high molecular weight RNA (precursors of rRNA and probably mRNA).

Wagner and Roizman (1969) examined nuclear and cytoplasmic RNA separately using gel electrophoresis. The processing of nuclear 45 S ribosomal precursor RNA was studied after a chase period in the presence of actinomycin. The radioactivity incorporated into 45 S RNA during a 20 min pulse at 5 hr after infection was 40% of that in uninfected cells. Eighty percent of these counts were lost during a 45 min chase but none appeared in ribosomal RNA (in contrast to uninfected cells in which about 70% of the 45 S radioactivity was chased into ribosomal sizes). Incorporation of [^{3}H]uridine into cytoplasmic 4 S RNA was inhibited somewhat less than into ribosomal RNA. It was concluded that infection with HSV-1 resulted in inhibition of the synthesis of both small (4 S) and large (45 S) cellular RNA and also the processing of large precursor molecules to form ribosomal RNA.

Rakusanova et al. (1972) examined the pulse-labeled RNA that annealed with cell DNA (cRNA). Both production and turnover of nuclear cRNA were inhibited in cells infected with PRV. Only 10% of the cRNA detected in the cytoplasm of infected cells was found associated with a polysome fraction as compared to 40% in normal cells. The remaining 90% of cytoplasmic cRNA that was not associated with polysomes nevertheless contained poly(A). The implication is that both the synthesis and processing of cellular mRNA were interrupted in infected cells and that those molecules that entered the cytoplasm, although carrying poly(A), were not able to associate with polysomes. However, the techniques available did not permit a clear identification of mRNA.

4.2. Messenger RNA

Pizer and Beard (1976) sought a less complex probe than total cell DNA with which to identify cellular mRNAs by hybridization. They used as a model system BHK cells transformed by polyoma virus, in which polyoma mRNA could be specifically detected by hybridization with polyoma DNA. The rate of synthesis of this RNA was estimated by pulse labeling with [^{3}H]uridine and measuring the radioactivity in DNA–RNA hybrid form after annealing whole cell RNA with excess polyoma DNA. After infection with HSV-1, the rate fell to 25% of normal by 2.5 hr and to 5% within 5.5 hr. Relative levels of polyoma mRNA in the cytoplasm were estimated using labeled single strands of polyoma DNA as a probe. The level fell to 20% of normal within 5 hr after infection with HSV-1. It was not determined whether this loss was promoted directly by the virus or resulted from natural decay following the inhibition of transcription.

That this effect may not be characteristic of all classes of mRNA was suggested by later work with the same virus–cell system (Nakai *et al.*, 1982). An attempt was made to study the fate of cellular mRNA using cDNA complementary to polyadenylated cytoplasmic RNA as a probe. They found that 5 hr after infection with HSV-1, the level of poly(A)-containing cellular RNA in the cytoplasm had declined significantly but that of nonadenylated RNA had risen four-fold, suggesting some degradation or improper processing of cellular mRNA.

In rat cells transformed by adenovirus (AdV), stable AdV-specific nuclear RNA was detected by labeling and hybridization with AdV DNA and a less stable polyadenylated species appeared in the cytoplasm. These cells were semipermissive for HSV-1 (final yield 5 PFU/cell). Infection had little effect on the synthesis of the nuclear AdV RNA but inhibited the accumulation of poly(A)-containing AdV RNA in the cytoplasm by 60% by 6.5–8.0 hr after infection (Spector and Pizer, 1978). In similar experiments with fully permissive human cells transformed by adenovirus (Stenberg and Pizer, 1982), both synthesis and migration to the cytoplasm were inhibited after infection with HSV-1. By 3 hr after infection, the rate of incorporation of [^{3}H]uridine into AdV-specific nuclear RNA had fallen to 28% of normal while the extent of accumulation of labeled AdV RNA in the cytoplasm was only 10% of normal at the same stage of infection.

4.3. Functional mRNA

Messenger RNA activity, as opposed to hybridizable mRNA sequences, has been measured by extracting RNA, translating it into proteins with the aid of a rabbit reticulocyte lysate, and characterizing the proteins by electrophoresis and autoradiography. Cellular mRNA activity was lost after infection with PRV (McGrath and Stevely, 1980). During the first 6 hr of infection with HSV-1, the amount of functional cellular mRNA fell steadily to a low level as viral mRNA activity increased, in synchrony with the changing pattern of protein synthesis in intact infected cells (Inglis and Newton, 1981).

5. EFFECTS ON PROTEIN SYNTHESIS

5.1. Polyribosomes

Sydiskis and Roizman (1966, 1967, 1968) studied the effects of infection with HSV-1 on the state of aggregation of ribosomes by centrifuging cell lysates through sucrose gradients. The early decline in the overall rate of protein synthesis during the first 3 hr could be accounted for by a progressive breakdown of polysomes to single ribosomes. The specific activity of the surviving polysomes, as judged by pulse labeling with [^{14}C]amino acids, remained approximately constant. The breakdown of polysomes was not prevented by actinomycin, azauridine, or fluorophenylalanine but did not occur if the virus was inactivated by irradiation with UV light (9600 ergs/mm^2) before infecting the cells (Sydiskis and Roizman, 1967). Later, as the rate of viral protein synthesis increased, larger polysomes were formed.

Ben-Porat et al. (1971) showed that the larger polysomes that are characteristic of cells infected with PRV contained more RNA that would hybridize to viral DNA and a lower proportion of lysine in nascent polypeptides than the smaller polysomes of infected or uninfected cells. Using the declining ratio of lysine to leucine as a measure of the "shut-off" of host protein synthesis, they found that the decline was prevented by actinomycin added at the time of infection and arrested by actinomycin added at 3 hr after infection. The effect of cycloheximide on shut-off was deduced from the distribution of newly made RNA in polysomes. These experiments showed that virus-specific and cell-specific RNA that was made between 5 and 6.5 hr (in the presence of cycloheximide added at the beginning of infection) became associated with polysomes within 3 min after removing cycloheximide. By 15 min, the amount of cell-specific RNA associated with the polysomes had declined substantially. Actinomycin, added at the same time as removing the cycloheximide, did not prevent this decline. No such fall was observed in uninfected controls. It was proposed that "in cycloheximide-treated infected cells, mRNA molecules which code for a species of [immediate-early] protein responsible for the inhibition of cell-specific protein synthesis, are formed and that they are translated during the first few minutes after removal of cycloheximide." The observation that actinomycin added at 3 hr after normal infection arrested the decline of cellular protein synthesis suggested that the mRNA for this supposed inhibitor was unstable and that a continuing supply of inhibitor was needed to sustain the progress of the decline.

The early breakdown of polysomes that followed infection with HSV-2 was not prevented by cycloheximide (Fenwick and Walker, 1978) indicating that it did not depend on newly made protein, nor on continuing translation with its associated termination of polypeptide chains and detachment of ribosomes from the messenger.

Silverstein and Engelhardt (1979) measured the protein-synthesizing activity of polysomes and rates of chain elongation of polypeptides, and found that the translation rate did not alter during the time that the polysomes were declining. They also concluded that a substantial proportion of the larger polysomes that formed after the early breakdown were inactive in protein synthesis and suggested a second block which resulted in inhibition of translation but not breakdown of polysomes and affected specifically cellular and early viral protein synthesis.

5.2. Polypeptides

The earliest reports of attempts to analyze the proteins made in infected cells by gel electrophoresis (Spear and Roizman, 1968; Shimono *et al.*, 1969) provided further evidence of the declining synthesis of cellular proteins and a concomitant rise in the production of new and different proteins, some of which were recognizable as virus structural proteins. These changing patterns of protein synthesis were seen in clearer detail using autoradiography of dried polyacrylamide gels (Honess and Roizman, 1973). Honess and Roizman (1975) also used cycloheximide to confine viral mRNA synthesis to the production of immediate-early (α) mRNA. They found that, after reversal of cycloheximide, host protein synthesis declined but that the decline was less rapid if actinomycin was added at the same time as removing the cycloheximide, suggesting that β-polypeptides were at least partly responsible for the inhibition of cellular protein synthesis.

Host protein synthesis was also suppressed in cells infected in the presence of canavanine, an analogue of arginine (Honess and Roizman, 1975). In these circumstances, α-proteins were made but were apparently defective because β-protein synthesis was severely restricted. This implied either that the host-suppressing activity of α- or β-proteins was not eliminated by substituting canavanine for arginine or that the observed suppression, like the breakdown of polysomes, did not depend on proteins synthesized after infection, as if some component of the virus inoculum itself was also involved.

5.3. Enzymes

The inhibition of cellular protein synthesis is inevitably followed by the decline of the functions of unstable proteins with possible widespread effects on the cell. It has been observed that the activity of enzymes controlling polyamine synthesis (McCormick and Newton, 1975) and of RNA polymerases I and II, measured in isolated nuclei, declined after infection (Preston and Newton, 1976). However, the rate of decline was faster than occurred after adding cycloheximide to uninfected cells, suggesting that the polymerases may have been inactivated by interaction with some virus-specific protein. Lowe (1978) could detect no changes in the activities of purified RNA polymerases in infected cells but extraction of the enzymes may possibly restore their activity by dissociation of an inhibitor. It would be interesting to have a direct comparison between the polymerase activities measured in intact nuclei and after purification.

5.4. A Virion-Associated Inhibitor

Experiments in which expression of viral genes was prevented have supported the proposition that a virion component can, in some cases, initiate the suppression of cellular protein synthesis. Infection with HSV-1 that had been inactivated by irradiation with UV light caused disaggregation of polysomes and a decline in protein synthesis (Nishioka and Silverstein, 1978). Similarly, infection of cells with UV-inactivated HSV-2 or with intact virus in the presence of actinomycin, or infection of enucleated cytoplasts with intact HSV-2, caused inhibition of cellular protein synthesis without detectable viral protein synthesis (Fenwick and Walker, 1978). The experiments with cytoplasts showed that the inhibition was a cytoplasmic phenomenon and did not involve the nucleus. The inhibitory activity of the inoculum could not be distinguished from infectious virions by differential centrifugation or by dilution or antibody neutralization.

The action of a herpes virion-associated inhibitor probably also explains the result of mixed infection with vesicular stomatitis virus (VSV; Nishioka *et al.*, 1983). VSV alone shuts off cellular protein synthesis but less rapidly than HSV and without causing the disaggregation of polysomes. In mixed infection, VSV was dominant in that it prevented the production of HSV-specific proteins, apparently by inhibiting transcription, but the polysomes were dissociated and host protein synthesis ceased.

5.5. Two Distinct Stages

The effect of infection on cellular mRNA has been studied using murine erythroid cells transformed by Friend leukemia virus. When grown in the presence of dimethylsulphoxide these cells make unusually large amounts of globin mRNA which can readily be identified by hybridization to globin cDNA. Such experiments showed that 85% of the globin mRNA had been degraded within 4 hr after infection with HSV-1 (Nishioka and Silverstein, 1977). It appeared, however, that this attack on mRNA was not the primary cause of the disaggregation of polysomes which follows infection. The rate of dissociation of globin mRNA from polysomes was much faster than the rate of degradation of the total globin mRNA. Furthermore, infection with UV-inactivated virus caused breakdown of polysomes, with a concomitant decline in protein synthesis but this was not accompanied by loss of hybridizable mRNA (Nishioka and Silverstein, 1978). Thus, there were two distinct stages in the suppression of host protein synthesis: dissociation of mRNA from polysomes and subsequent degradation of the mRNA. Only the second stage depended on expression of viral genes.

The suggestion that there are two distinct mechanisms of attack on the cellular protein-synthesizing system was supported by work with a temperature-sensitive mutant of HSV-1, *ts*B7, which does not grow or inhibit cellular protein synthesis at its nonpermissive temperature (39°C). A revertant form derived from this mutant was able to grow at 39°C and to suppress host synthesis slowly. After UV inactivation, the revertant virus could shut off the host at 34°C but not at 39°C, suggesting that a temperature-sensitive virion component was responsible for "early shut-off," while "delayed shut-off" was caused by a protein made after infection (Fenwick and Clark, 1982). Early shut-off was apparently not an essential function for the growth of the virus in tissue culture. The two stages, early and delayed, may correspond to the disaggregation of polysomes and the degradation of cellular mRNA, respectively (Nishioka and Silverstein, 1978). A strain of HSV-1, HFEM, which is deficient in virion-mediated shut-off (unpublished observation) caused a progressive loss of functional cellular mRNA (Inglis and Newton, 1981). However, our recent experiments with extracted cytoplasmic RNA have revealed a substantial loss of functional cellular mRNA within 2 hr after infection with HSV-2 in the presence of actinomycin or with UV-inactivated virus (Fenwick and McMenamin, in preparation), indicating that early shut off also involves some alteration of host mRNA to the extent that it is no longer active in an *in vitro* translation system.

The distinction between early and delayed shut-off has been further clarified by Read and Frenkel (1983) who isolated a series of temperature-sensitive mutants, designated *vhs* for virion-associated host shut-off. One of these, *vhs*1, was defective in early shut-off, i.e., suppression of protein synthesis in the presence of actinomycin, at both permissive (34°C) and nonpermissive (39°C) temperatures, again demonstrating that early shut-off is not essential for growth. *vhs*1 Growing at 34°C nevertheless caused a delayed or "secondary" shut-off which was prevented by actinomycin. Another mutant, *vhs*4, exhibited temperature-sensitive early shut-off (as well as growth). The infectivity of *vhs*4 was relatively unstable at 45°C, confirming the presence of an abnormal virion component, which may be the same as the early shut-off protein.

Pseudorabies virus apparently causes delayed, but not early, virion-associated shut-off. The inhibition of cellular protein synthesis by PRV is prevented by actinomycin (Ben-Porat *et al.*, 1971; Ihara *et al.*, 1984). By labeling proteins during the hour immediately following reversal of a cycloheximide block it was shown that host protein synthesis was quickly suppressed (Ihara *et al.*, 1984), presumably as a result of the production of immediate-early or early polypeptides. Whether or not this involved the degradation of cellular mRNA is not clear, although it was largely intact at the time of removal of cyclo-heximide, as demonstrated by translation *in vitro* of extracted RNA. McGrath and Stevely (1980) observed the loss of functional cellular mRNA after normal infection with PRV.

5.6. Specificity of Shut-off

Autoradiography of electrophoretically separated polypeptides after pulse labeling in infected cells indicates that a wide range of cellular proteins are similarly affected. However, Isom *et al.* (1983) reported that, in rat hepatoma cells, the synthesis of several plasma proteins (identified by antibody precipitation) was specifically inhib-ited early after infection with HSV-2 at a time when general cellular protein synthesis was still unaffected and before any new viral pro-teins were detected. It did not occur, however, if the virus was in-activated by irradiation with UV light and was, therefore, thought to require expression of early viral genes. The decline was accompanied by a decrease in the level of functional mRNA for the plasma proteins, as measured by translation *in vitro* of poly(A)-containing mRNA, and a decrease in the amount of RNA that hybridized to cDNA specific for the plasma proteins. This effect has the characteristics of "delayed shut-off," described above, i.e., it depends on the expression of viral genes and is accompanied by degradation of cellular mRNA. On the other hand, it occurred early, before the general decline in cellular protein synthesis and so may possibly represent yet another distinct specific effect.

5.7. Cellular Stress Proteins

Cells respond to a variety of stimuli by actively transcribing genes that are normally expressed at a very low level or not at all (Kelley and Schlesinger, 1978). One of the stimulants to the production of "heat shock" or "stress" proteins is azetidine. It is not surprising that infection with HSV-1 prevented the synthesis of stress proteins, as well as suppressing cellular protein synthesis in general, in cells treated with azetidine (Fenwick *et al.*, 1980). However, in cells infected for 4 hr with the mutant tsK of HSV-1(17) at its nonpermissive temperature, cellular protein synthesis was not shut off and production of stress proteins was stimulated (Notarianni and Preston, 1982). This mutant makes a defective α-polypeptide, ICP4, and is severely restricted in its ability to initiate β-protein synthesis. UV-inactivated virus did not produce this effect and it was concluded that the stimulation was caused by one of the small range of viral polypeptides (almost exclusively immediate-early) that are made at the nonpermissive temperature. Examination of a series of mutants showed that induction of stress proteins was correlated with overproduction of immediate-early proteins. Presumably, it also depends on absence of early shut-off. A similar phenomenon has been observed in cells infected with adenovirus (Nevins, 1982). Other experiments with the HSV-1 mutant tsK have suggested that the shut-off of host transcription is also mediated by a viral α-polypeptide (Stenberg and Pizer, 1982; see Section 7.6).

5.8. Differences between HSV-1 and HSV-2

Several strains of HSV-2 have been reported to cause a more rapid decline of host cell protein synthesis than certain HSV-1 strains (Powell and Courtney, 1975; Pereira *et al.*, 1977; Hill *et al.*, 1983) suggesting the generalization that HSV-2 causes early, virion-associated shut-off and HSV-1 only delayed, expression-dependent shut-off. However, the work described above (Nishioka and Silverstein, 1978; Fenwick and Clark, 1982; Read and Frenkel, 1983) has demonstrated that some strains of HSV-1 can cause early shut-off. Sometimes contrasting results have been obtained. HSV-1(KOS) failed to cause early shut-off in Friend leukemia cells (Hill *et al.*, 1983) but did so in HEp-2 cells (Read and Frenkel, 1983). One strain of HSV-2, HG52, does not cause early shut-off, at least in Vero cells (unpublished observation). The possible role of the host in succumbing to or resisting early shut-off has not been studied systematically and temperature may also be critical (Hill *et al.*, 1983). HSV-1(F) caused early shut-off at 37°C but not at 39°C, whereas, HSV-2(G) was fully effective at 39°C (Fenwick and Clark, 1982).

In FL cells mixedly infected with HSV-1(KOS) and HSV-2(186), the early suppression of globin synthesis was less marked than in cells infected with HSV-2 alone. UV-inactivated HSV-1 was also effective in reducing the early shut-off by HSV-2 (T. M. Hill and J. R.

Sadler, personal communication). It is possible, therefore, that HSV-1 carried a nonfunctional virion protein analogous to the HSV-2 early shut-off protein, that nevertheless competed with the latter for a critical site in the host cell.

5.9. Genetic Mapping

Differences between the rates of suppression of cellular protein synthesis following infection with HSV-1 or HSV-2 have been exploited in experiments with recombinant viruses designed to locate the gene controlling the type 2 function responsible for "rapid shut-off" on the genetic map (Morse *et al.*, 1978). The composition of HSV-1 × HSV-2 recombinant DNA was deduced from the restriction enzyme sites inherited from the type 1 and type 2 parental viruses. The ability of a series of recombinants to suppress host protein synthesis rapidly was correlated with the possession of type 2 DNA in the region between 0.52 and 0.59 units on the genetic map. Subsequently, the same region was shown to be correlated with rapid inhibition of (1) cellular protein synthesis in the presence of actinomycin, and (2) cellular DNA synthesis after inactivation of the virus with UV light (Fenwick *et al.*, 1979), suggesting that the same gene, or a pair of closely linked genes, was involved in controlling the two aspects of early shut-off. This section of the viral genome (about 100 kbp of DNA) does not contain any α-genes although several others of unknown functions have been assigned map positions within it.

6. OTHER HERPESVIRUSES

6.1. Epstein–Barr Virus

Primary infection of human B lymphocytes with Epstein–Barr virus (EBV) results in transformation to established cell lines, such as the Raji line, that carry EBV DNA but do not produce virus. Treatment of these cells with various chemical inducing agents provokes the synthesis of a group of early EBV antigens, but if the cells are superinfected with EBV, a lytic virus growth cycle ensues. Gergely *et al.* (1971) detected the small proportion of antigen-producing cells with fluorescent antibody after infecting Raji cells with EBV and showed by autoradiography of pulse-labeled cells that the overall rate of DNA, RNA, and protein synthesis in these cells was inhibited between 24 and 48 hr after infection. By using higher multiplicities of infection, Nonoyama and Pagano (1972) induced early antigen production in 70% of Raji cells and observed an overall inhibition of DNA synthesis from 15 hr onward and degradation of cellular DNA.

Using gel electrophoresis and autoradiography to distinguish cellular from viral proteins, it was found that cellular protein synthesis was substantially suppressed within the first 4 hr of superinfection and this was not prevented by phosphonoacetic acid (Bayliss and Nonoyama, 1978) and so did not depend on replication of viral DNA. Others observed a slight initial stimulation of cellular protein synthesis, followed by a progressive decline between 6 and 24 hr after infection (Feighny *et al.*, 1980; Bayliss and Wolf, 1981).

The timing of this effect suggests delayed (expression-dependent) shut-off since virus-specific proteins were being made from 2 hr after infection. This is supported by the observation that shut-off did not occur if viral protein synthesis was confined to a group of early proteins by incubating the cells in medium in which arginine was replaced by its analogue canavanine (Bayliss and Wolf, 1982). It is interesting that in medium lacking both arginine and canavanine, in which viral protein synthesis was also restricted, host synthesis was suppressed (Bayliss and Wolf, 1982). Chemical induction of Raji cells also resulted in production of a limited range of viral proteins and did not result in shut-off (H. Wolf, personal communication).

6.2. Herpesvirus Saimiri

Herpesvirus saimiri grows slowly in owl monkey kidney cells without affecting the protein synthesis of its host until late in infection. Cytopathic changes are only evident after several days. The growth cycle was accelerated by adding a phorbol ester (Modrow and Wolf, 1983), a tumorigenic agent which had previously been found to enhance the production of viral antigens in cell lines transformed by EBV. Under these conditions, the suppression of host protein synthesis was also more marked and was complete within 24 hr.

6.3. Cytomegalovirus

Human cytomegalovirus (CMV) is another slow-growing herpesvirus which has very little inhibitory effect on host macromolecular synthesis until several days after infection. In resting cells, cellular DNA synthesis (St. Jeor *et al.*, 1974), RNA synthesis, and polysome formation (Tanaka *et al.*, 1975) were stimulated during the first 2–3 days after infection. UV-inactivated virus did not induce these changes. In growing cells, cellular protein synthesis declined to 50% of normal by 8 hr after infection but then increased to uninfected control levels by 30 hr and continued to constitute a substantial proportion of total protein synthesis for several days, while viral protein synthesis increased (Stinsky, 1977). The initial decline occurred after infection with UV-inactivated virus but the subsequent stimulation did not. However, the formation of cellular glycoproteins, detected by labeling with [^{3}H]glucosamine, was suppressed as early as 14 hr after infection (Stinsky, 1977). Cellular protein synthesis was suppressed in cells infected and incubated for 8 hr in the presence of cycloheximide and then labeled in hypertonic medium after removing cycloheximide (Stinsky, 1978), but the relative influences of the virus, the cycloheximide, and the high ionic strength in this situation are not clear.

7. MECHANISMS

7.1. Multiplicity of Infection

If early shut-off is caused by a component of the virion, the response might be expected to be related to the dose of virus received, even beyond that required to infect all the cells. However, the evidence suggests that a maximum rate of early inhibition of protein synthesis is obtained with an added multiplicity of about 4 PFU/cell (Fenwick and Walker, 1978), i.e., enough to infect 98% of the cells, and that increasing the proportion to 80 or 110 PFU/cell had no further effect (Fenwick *et al.*, 1979; Read and Frenkel, 1983). If the particle-to-PFU ratio is, say, 10, this implies either that one infectious particle is enough to cause maximum shut-off and the nine extra particles are not involved or that, by a coincidence, ten particles per cell are needed. That the virus must actually penetrate the cell to cause early shut-off is not certain, though often assumed. All that can be said is that heating it to 56°C (Nishioka and Silverstein, 1978) or treating it with γ-globulin (Fenwick and Walker, 1978), unlike UV inactivation, eliminated its ability to suppress host protein synthesis. Heated virus adsorbed normally to the cells. Antibody-treated virus may have adsorbed but probably did not penetrate.

7.2. Breakdown of Polysomes

As an early consequence of the interaction of some component of the virus inoculum with the cell, the polysomes become progressively disaggregated during incubation for 2 hr at 37°C, but not at 4°C (Fenwick and Walker, 1978). The breakdown of polysomes might be caused by inhibiting the initiation step of polypeptide chain formation or by ribonuclease attack on the mRNA. The single ribosomes that accumulate as a result of normal polypeptide chain termination, after blocking initiation with sodium fluoride, dissociate into subunits if exposed to a high concentration of sodium ions. In contrast, those obtained by RNase treatment of polysomes, bearing fragments of mRNA, are stable in the presence of a high salt concentration. The ribosomes that accumulated after infection with HSV-2 were susceptible to high salt (Fenwick and Walker, 1978), favoring an initiation block rather than RNase attack, and the work of Nishioka and Silverstein (1978) showed that early shut-off does not necessarily involve degradation of cellular mRNA. However, it was also observed (Fenwick and Walker, 1978) that cycloheximide did not prevent the decline, whereas such an inhibitor of chain elongation should protect polysomes from breakdown following inhibition of initiation, by preventing the ribosomes from running off the mRNA.

The available evidence thus suggests that the ribosomes are detached from cellular mRNA without finishing the polypeptide chain that they were engaged in synthesizing. It is possible that the virion-associated inhibitor produces, either directly or indirectly, a change in the ribosomes, causing them to lose their affinity for cellular mRNA while retaining or acquiring an affinity for viral mRNA. A protein associated with ribosomes became phosphorylated soon after infection with HSV-1 or HSV-2 (Fenwick and Walker, 1979) but this required expression of viral genes and, therefore, could not be related to early shut-off. Kennedy et al. (1981) found an increase in the phosphorylation of a protein of the small ribosomal subunit, S6, after infection with HSV-1 or PRV. Since PRV does not cause early shut-off, this effect was probably also dependent on prior viral protein synthesis although this was not established. It is known, however, that protein kinase activity is associated with herpes simplex virions (Rubenstein et al., 1972; Lemaster and Roizman, 1980).

7.3. Reversibility of Early Shut-off

Read and Frenkel (1983) examined the reversibility of early shut-off using the mutant vhs4 which is able to suppress cellular protein synthesis in the presence of actinomycin at 34°C but not at 39°C. At various times after infection with *vhs*4 in the presence of actinomycin at 34°C the temperature was raised to 39°C. The rate of host protein synthesis thereafter remained approximately constant, indicating that the continuous functioning of the *ts* virion component was required for early shut-off to proceed but inactivation of the inhibitory agent did not reverse the shut-off already achieved at the time the temperature was raised.

It was reported (Fenwick and Clark, 1982) that early shut-off by the UV-inactivated mutant *ts*B7 could be reversed by raising the temperature from 34 to 39°C. Subsequent experiments (unpublished) have shown that, in agreement with Read and Frenkel (1983), recovery of host protein synthesis, which was nearly complete within 90 min at 39°C, was prevented by actinomycin. This suggests that the ribosomes could be re-used after early shut-off but that new mRNA was needed to initiate cellular protein synthesis. The work of Nishioka and Silverstein (1978) has shown that globin mRNA is not extensively degraded in FL cells as a result of early shut-off. However, a small change in the structure of the mRNA might result in loss of its function without appreciable loss of sequences hybridizing to a cDNA probe. Indeed, as mentioned above (section 5.5), our recent experiments have indicated that early shut off involves an effect on host mRNA to the extent that it is no longer functional in an *in vitro* translation system.

7.4.   Ionic Imbalance

When cells are incubated in hypertonic medium, initiation of new polypeptide chains stops and polysomes are quickly disaggregated. The synthesis of viral proteins in cells infected with poliovirus, however, is appreciably less sensitive to such treatment than is host protein synthesis (Nuss *et al.*, 1975).

After infection with a variety of viruses, cell membranes become more permeable (Kohn, 1979). It has been suggested that the consequent disturbance of the intracellular balance of ions may account for the interference with the synthesis of cellular macromolecules (Carrasco and Smith, 1976; Carrasco, 1977; Durham, 1977). It was reported, in support of this mechanism, that the optimum concentration of monovalent ions for the translation of picornaviral mRNA *in vitro* was higher than that required for translation of cellular mRNA, and that the shut-off of host protein synthesis by EMC virus could be reversed by resuspending the infected cells in medium lacking NaCl (Alonso and Carrasco, 1981).

Among the molecules that are normally excluded from healthy cells but can pass through membranes rendered permeable by virus infection is Hygromycin B, an inhibitor of protein synthesis. This suggests the interesting possibility of preventing the spread of viral infection by specifically inhibiting protein synthesis in infected cells (Carrasco, 1978). Hygromycin B has been shown to penetrate monkey kidney cells infected with HSV-1 (Benedetto *et al.*, 1980). In this system, the decline of total protein synthesis started about 6 hr after infection and fell to 50% of normal by 10 hr and, therefore, probably represents delayed, expression-dependent shut-off. Between 5 and 30 hr the remaining protein synthesis became increasingly susceptible to 5×10^{-4} M Hygromycin B. Mature progeny appeared from 10 hr, reaching a maximum level about 25 hr after infection. Thus, the delayed shut-off of host protein synthesis was accompanied by a gradually increasing permeability of the plasma membrane. Whether this increased permeability is the cause of suppression is not clear.

Other experiments with herpesviruses have lent only limited support to the theory that ionic imbalance causes selective shut-off of cellular protein synthesis after infection. A differential sensitivity to hypertonic medium has been reported in cells infected with human or murine cytomegaloviruses which cause only a very slow decline late in infection (Gupta and Rapp, 1978; Chantler, 1978), but in HSV-2-infected cells (Fenwick and Walker, 1978) and PRV-infected cells (Stevely and McGrath, 1978) viral protein synthesis was as susceptible as protein synthesis in uninfected cells to increased concentrations of NaCl. No differences have been found between the concentrations of K^+ ions needed for optimal translation *in vitro* of viral (HSV-1 or PRV) and cellular mRNAs, respectively (Inglis and Newton, 1981; McGrath and Stevely, 1980). Attempts to prevent the early shut-off of protein synthesis by HSV-2 by reducing the Na^+ or Ca^{2+} concentration in the medium were not successful (Fenwick and Walker, 1978), but these experiments were done at times which probably preceded the permeability changes described by Benedetto *et al.* (1980).

There is, thus, at present, no reason to suppose that early shut-off by HSV, involving displacement of ribosomes from mRNA, is caused by changes in intracellular ion concentrations. Delayed shut-off by those viruses that do not cause early virion-associated shut-off may possibly be related to a late increase in permeability of the cell membrane. It should be borne in mind, however, that superinfection with poliovirus 5 hr after infection with HSV-1 has been reported to cause the disaggregation of herpes-specific polysomes (Saxton and Stevens, 1972), indicating that the two viruses do not employ the same mechanism to break down polysomes.

7.5. Delayed Shut-off

Delayed shut-off of protein synthesis appears to involve the degradation of cellular mRNA and is conditional upon the synthesis of viral RNA and protein. Which viral protein is concerned is not known; all three classes, α, β, and γ, have been suspected of being involved.

Honess and Roizman (1974) found that, in cells infected with HSV-1 and incubated in the presence of cycloheximide, after removing the inhibitor, the rate of synthesis of cellular proteins declined as viral proteins began to be made. If actinomycin was added when cycloheximide was removed, allowing the translation of α-mRNA but preventing the synthesis of β- and γ-mRNA, host protein synthesis declined more slowly, suggesting that β- or γ-proteins were at least partly responsible for the shut-off. Host protein synthesis also declined after infection in the presence of canavanine, an arginine analogue, which restricted viral protein synthesis to α- and a few β-polypeptides (Honess and Roizman, 1975). They considered the possibility that two mechanisms were involved, only one of which required newly synthesized viral protein, and it now seems likely that they were indeed observing a combination of early, virion-associated, and delayed shut-off.

Nishioka and Silverstein (1978) avoided this problem by specifically observing delayed shut-off, which they distinguished from early shut-off by measuring the loss of hybridizable globin mRNA in Friend leukemia cells infected with HSV-1. The level of this cellular mRNA began to decline soon after reversal of a cycloheximide block and the rate of decline was reduced if cycloheximide was added again 2 hr later. They concluded that an early viral protein was responsible for the degradation but did not test the effect of adding actinomycin at the time of cycloheximide reversal to prevent the synthesis of any β- or γ-mRNAs.

Read and Frenkel (1983) addressed the same question using a mutant strain of HSV-1, $vhs1$, which was defective in early shut-off, i.e., it failed to suppress host synthesis in the presence of actinomycin but did so in the course of normal infection. After reversal of a cy-cloheximide block with this virus, host synthesis declined, but the decline was completely prevented by adding actinomycin shortly before reversal, showing that β- or γ-proteins were required and that α-proteins alone were not effective.

Other experiments revealed that delayed shut-off by *vhs*1 was reduced considerably, but not entirely prevented by treating the cells with either phosphonoacetic acid or canavanine. The former inhibits viral DNA replication, resulting in reduced production of most of the γ-polypeptides. One of these was therefore implicated. Canavanine may also act by preventing the synthesis of the polypeptide concerned or by being incorporated into it and rendering it nonfunctional.

Ihara *et al.* (1983), working with PRV, found that it differs from HSV-1 in that only one immediate-early polypeptide, with an electrophoretic mobility similar to that of ICP4 of HSV, was made after reversing a cycloheximide block. Cellular protein synthesis was inhibited soon after reversal. *ts*G$_1$ is a mutant of PRV with a defect in the immediate-early protein which (like *ts*K of HSV-1; see Section 7.6) is unable to progress from immediate-early to early protein synthesis at the nonpermissive temperature (41°C). This mutant caused significant shut-off of cellular protein and DNA synthesis at 41°C but less than wild-type virus. It was concluded that the immediate-early protein is involved in the shut-off, but either the mutant form is partially defective in this function or some later viral proteins also contribute to the shut-off by wild-type virus. The case for the host-suppressing function of the immediate-early protein would be strengthened if it were confirmed that shut-off occurred after reversal of cycloheximide in the presence of actinomycin, as was reported for polysome breakdown (Ben-Porat *et al.*, 1971).

7.6. Inhibition of Transcription

Stenberg and Pizer (1982), studying the transcription and accumulation of AdV-specific mRNA in AdV-transformed cells infected with HSV-1 and a mutant, *ts*K, with a defective α-ICP4, have concluded that an α-polypeptide is involved in the suppression of transcription. The mutant virus was able to prevent accumulation of AdV mRNA (measured by labeling for 90 min, followed by hybridization to AdV DNA) at the nonpermissive temperature (39°C). This effect was blocked by cycloheximide and occurred after removal of the inhibitor, showing that a protein made after infection at 39°C was responsible. Measurement of transcription rate after short pulses of labeled RNA precursor showed that at 39°C synthesis of AdV mRNA was inhibited to the same extent (73%) by *ts*K as by wild-type virus. The interpretation that an α-polypeptide is responsible for inhibiting host transcription is based on the fact that the transition from α- to β-polypeptide synthesis in cells infected with *ts*K is impaired at 39°C. It is subject to the reservation that β-ICP6 and a small amount of γ-ICP5, at least, may also have been produced (Preston, 1979*a*). Alpha-ICP4 itself is an unlikely candidate for the role for inhibitor of transcription because at 39°C it does not accumulate in the nucleus as does the ICP4 of wild-type virus (Preston, 1979*b*).

The mechanism of the inhibition of transcription is a matter for speculation. It may involve the declining activity of an unstable RNA polymerase I (which mediates the synthesis of ribosomal RNA) following the suppression of protein synthesis (Preston and Newton, 1976), although Lowe (1978) could detect no changes in the activities of purified polymerases after infection. Perhaps a more likely mechanism would involve the interaction of an early viral protein with cellular polymerases (Preston and Newton, 1976; Spector and Pizer, 1978). It is possible that such interactions may also confer new specificities on cellular polymerases enabling them to recognize regulatory sequences of viral genes. Transcription of viral α-genes is probably mediated by cellular RNA polymerase II (Costanzo *et al.*, 1977) and is enhanced by a virion-associated factor (Post *et al.*, 1981). Herz and Roizman (1983) have shown that expression of a viral gene is determined by specific base sequences upstream from the coding region of the gene and not by its location or structural organization. They attached the regulatory sequences of the α-ICP4 gene to a chick ovalbumin gene and inserted it into the genome of a mammalian cell. When the cells were infected with HSV-1, the ovalbumin gene was expressed transiently, synchronously with authentic viral α-genes.

7.7. Inhibition of DNA Synthesis

The cause of the decline in host DNA synthesis is still not understood in any detail. An early effect of infection with HSV-1 is displacement of some of the cellular DNA from its attachment to the nuclear membrane (Newton, 1972), a possible site of replication (McCready *et al.*, 1980). Another relevant observation, whose significance is not at present clear, is that the suppression was accelerated by prostaglandin E (1 μg/ml) and prevented by cyclic GMP (300 μg/ml; Newton, 1978).

Daksis *et al.* (1982) isolated a phosphonoacetic acid (PAA)-resistant mutant of HSV-1(KOS) and found that it also had a temperature-sensitive mutation in a second gene. This PAA^r/*ts* virus was defective both in viral DNA synthesis and in the suppression of cellular DNA synthesis (measured 6–24 hr after infection) at the nonpermissive temperature (39°C). Both functions were normal at the permissive temperature (34°C). PAA^r/*ts*$^+$ and PAA^s/*ts* recombinant viruses both made viral DNA at 39°C but were both still defective in the ability to shut off host DNA synthesis, i.e., a mutation in either gene prevented shut-off at 39°C, but mutations in both were required to stop viral DNA synthesis. The authors proposed that both viral DNA replication and the inhibition of cellular DNA synthesis by HSV-1 require an interaction between the viral DNA polymerase and a second virus-specified polypeptide. Viral DNA synthesis at 39°C was significantly defective only when both gene products were mutated, perhaps due to the instability of the interaction between the two altered polypeptides at the restrictive temperature. However, modification of the structure of either polypeptide may give rise to an abnormal interaction in the complex, leading to a defect in the shut-off of cellular DNA synthesis. It may be relevant that PAA has been reported to reduce delayed shut-off of host protein synthesis (Read and Frenkel, 1983; see Section 7.5).

7.8. Relationship between DNA and Protein Shut-off

In deciding at what stage the pathways from the adsorption of the virus to the various end results of shut-off diverge, possible interdependencies between the three main synthetic processes in healthy cells must be considered. Of these, the relationship between RNA and protein synthesis and that between protein and DNA synthesis are possibly relevant. Inhibition of mRNA synthesis will cause a decline in protein synthesis as the existing mRNA decays. This can be ruled out as a main cause of the suppression of protein synthesis by herpesviruses since the decline is often more rapid after infection than after administration of actinomycin, a much more efficient inhibitor of mRNA synthesis, although the effect must be taken into account in cases of slow decline of protein synthesis. Conversely, the inhibition of protein synthesis may cause a slow decline in the rate of synthesis of RNA mediated by an unstable RNA polymerase.

The synthesis of cellular DNA, however, requires concomitant protein synthesis. The rate of DNA chain growth begins to fall within minutes after adding cycloheximide (Planck and Mueller, 1977; Stimac et al., 1977), suggesting the existence of a small pool of essential protein with a high rate of turnover. It is possible, therefore, that the decline of cellular DNA synthesis after infection with herpesvirus is a consequence of the inhibition of protein synthesis (Kaplan, 1973a). The fact that viral DNA may be synthesized after the early suppression of protein synthesis implies that, unlike cellular DNA synthesis, the replication of viral DNA does not depend on ongoing protein synthesis. This was indeed observed in early work with HSV (Newton, 1964; Roizman and Roane, 1964), PRV (Kaplan, 1973b) and adenovirus (Horwitz et al., 1973).

Consistent with a single cause for the early shut-off of both protein and DNA synthesis are the observations that both can be caused by UV-inactivated virus (Fenwick and Walker, 1978) and that rapid virion-mediated shut-off of DNA synthesis was correlated with rapid shut-off of protein synthesis in a series of recombinant viruses (Fenwick et al., 1979). This suggested that the two effects were controlled by the same gene or by closely linked genes.

We have found (S. J. Foote and M. L. Fenwick, unpublished experiments) that after exposing cells to UV-inactivated HSV-2(G) the rate of DNA synthesis declines in parallel with that of protein synthesis just as it does after adding a low dose of cycloheximide. However, after infection with HSV-2(HG52), although no shut-off of cellular protein synthesis occurred within 10 hr, the rate of incorporation of [^{3}H]thymidine into cellular DNA fell to 75% of normal by 7 hr and to 25% of normal by 10 hr.

In no case of early shut-off by several strains of HSV have we seen a more rapid effect on host DNA synthesis than would be expected from the observed inhibition of protein synthesis. We, therefore, conclude that early shut-off probably operates on cellular protein synthesis and not primarily on DNA synthesis. A more direct delayed attack on DNA synthesis may occur later if it has not been forestalled by the early virion-mediated effect on protein synthesis.

7.9. Inhibition of α-Protein Synthesis

The transient expression of the α-genes of herpesvirus during the early stages of infection, which was mentioned in the introduction, is evidence of another specific shut-off phenomenon. When cells were enucleated 2 hr after infection with HSV-1, by centrifuging them in the presence of cytochalasin B, the cytoplasts continued to make host and viral proteins. The rate of synthesis of the viral α-polypeptide ICP4 declined as in intact cells, while synthesis of β- and γ-polypeptides was maintained (Fenwick and Roizman, 1977), suggesting that the agent responsible for the decline was present and active in the cytoplasm 2 hr after infection. Infection of anucleate cytoplasts has shown that early shut-off of host protein synthesis is accomplished by a virion component in the cytoplasm (see Section 5.4). Delayed shut-off is probably also a cytoplasmic process, involving degradation of cytoplasmic mRNA (Nishioka and Silverstein, 1977, 1978). It is of interest to compare these shut-off phenomena with the regulatory process by which viral α-polypeptide synthesis is suppressed after an early period of activity.

Honess and Roizman (1974) incubated cells infected with HSV-1 in the presence of cycloheximide to allow accumulation of α-mRNA. On removing the inhibitor, viral α-polypeptides as well as cellular polypeptides were made and the rates of synthesis of both declined as β- and γ-polypeptides were produced. The rates of decline were less if actinomycin was added at the time of removal of cycloheximide, suggesting that β- or γ-polypeptides were concerned with the inhibition of both host and α-protein synthesis.

The amino acid analogues canavanine and azetidine had the effect of restricting the synthesis of viral β- and γ-polypeptides and prolonging α-polypeptide synthesis (Honess and Roizman, 1975), consistent with the idea that functional β- or γ-polypeptides were responsible for the inhibition of α-polypeptide synthesis, but not excluding the possibility than an α-polypeptide might inhibit its own synthesis. Experiments with HSV-1 mutants that produce a temperature-sensitive β-ICP8 have suggested that this polypeptide (which is a DNA-binding protein) is concerned in restricting the accumulation of cytoplasmic viral mRNA, including α-mRNA, probably by inhibiting transcription (Godowski and Knipe, 1983). On the other hand, an autoregulatory control of transcription is supported by work with mutants that make a temperature-sensitive α-ICP4. Watson and Clements (1980) found that in infected cells incubated at the nonpermissive temperature an immediate-early pattern of transcription of the viral DNA was maintained, in contrast to the more complex "early" pattern seen in cells kept at permissive temperature. If the temperature was raised to the nonpermissive level 3 hr after infection transcription reverted to the immediate-early pattern, implying that a temperature-sensitive protein was suppressing the transcription of α-genes.

At the nonpermissive temperature such mutants overproduce α-polypeptides and if the temperature is raised from permissive to nonpermissive after α-polypeptide synthesis has been turned off, synthesis of ICP4 starts again, while synthesis of β- and γ-polypeptides continues (Dixon and Schaffer, 1980; Herz and Roizman, 1983). A similar observation has been made with a *ts* mutant of PRV. Actinomycin prevented the reexpression of the immediate-early gene after raising the temperature, again implying that the immediate-early protein (analogous to α-ICP4 of HSV) suppressed the synthesis of its own mRNA (Ihara *et al.*, 1983).

Thus, it appears that functional ICP4 is required not only to initiate and maintain transcription of β-genes but also to inhibit the transcription of α-genes. On raising the temperature, existing ICP4 was inactivated and synthesis of β-mRNA ceased. Synthesis of α-mRNA was resumed and it was translated despite the continuing production of β- and γ-polypeptides (presumably using preexisting mRNA). This casts some doubt on the proposed role of β-polypeptides in shutting off α-polypeptide synthesis.

The work of Read and Frenkel (1983) suggests that virion proteins are responsible for both early shut-off of host protein synthesis and shut-off of α-protein synthesis. The mutant *vhs*1 (which was defective in the virion-associated host shut-off factor) made α-mRNA that was functionally more stable than that of the parental wild-type HSV-1. If cells infected in the presence of cycloheximide, in which *vhs*1 α-mRNA had accumulated, were then treated with actinomycin and superinfected with wild-type HSV-1, the α-polypeptide synthesis was prevented and at the same time host protein synthesis was shut off. It was suggested that the wild-type virus carried in its virion an agent that was responsible for inactivating preformed α-mRNA, or restricting its translation, whereas, *vhs*1 was defective in this respect and, hence, that the same virion-associated protein may possibly cause both early host shut-off and the decline of α-polypeptide synthesis. If so, α-mRNA presumably has some structural feature in common with cellular mRNA and different from later viral mRNA and it would not be surprising if the β- or γ-protein that causes delayed shut-off may also (or in some cases, alternatively) be involved in the inactivation of α-mRNA.

8. CONCLUSION

Infection of animal cells with herpesviruses is commonly, though not invariably, followed by inhibition of the synthesis of cellular DNA, RNA, and protein, phenomena collectively and colloquially known as "host shut-off." Host shut-off precedes most of the more obvious cytopathic effects which are characteristic of the late stages of a lytic infection and often precedes the replication of viral DNA. In the 20 years or so that have passed since the phenomena were first observed, little positive information has accumulated concerning the means by which the virus suppresses the synthesis of cellular macromolecules while directing the synthesis of its own.

Two distinct phases can be distinguished: "early," "primary," or "virion-associated" shut-off and "delayed," "secondary," or "expression-dependent" shut-off. Some viruses carry in their virions an agent or agents capable of suppressing host DNA and protein synthesis within 2 hr after infection and RNA synthesis to a lesser extent. Experiments with recombinant viruses and measurements of rates of suppression suggest that DNA synthesis declines as a result of a primary effect on protein synthesis. Whether or not inhibition of RNA synthesis is a distinctly separate process is not clear. The possibility of an effect on RNA polymerases by a virus-induced protein during delayed shut-off has not been fully investigated. The virion-mediated early shut-off of RNA synthesis is perhaps more difficult to envisage in these terms and may be a secondary effect.

The early shut-off of protein synthesis apparently involves detachment of ribosomes from mRNA. The mRNA is apparently not massively degraded but is rendered non-functional (i.e., not translatable in an *in vitro* system). The detachment probably does not occur in the course of normal polypeptide chain termination since it is not prevented by cycloheximide. Either the detached ribosomes or others previously free are able to associate productively with viral mRNA. A specific, limited enzymic attack on cellular mRNA is a possibility as is an alteration in a recognition site on the ribosomes. Perhaps slightly more conceivable, but no better supported by evidence, is the suggestion that a change in the intracellular ionic environment might promote the detachment of cellular mRNA and the association of viral mRNA, as well as a change in the affinity of the DNA for RNA polymerase.

The secondary or delayed phase of shut-off occurs only if synthesis of viral proteins is allowed to occur. It results in the degradation of cellular mRNA to nonhybridizable fragments and the decline of DNA synthesis even in the case of a virus that does not shut off protein synthesis. Delayed shut-off of protein synthesis is probably mediated by viral β- or early γ-proteins although there is some evidence from experiments with *ts* mutants that α-proteins are responsible in PRV infection and for inhibition of transcription in HSV infection. The shut-off of host protein synthesis and the regulation of viral α-protein synthesis may possibly be related phenomena.

The recognition of two phases in the suppression of host synthetic processes (neither of which is an essential function for the growth of the virus in tissue culture) should facilitate investigations of the mechanisms involved. It is important in future research to study the processes separately, i.e., early shut-off with UV-inactivated virus or in the presence of inhibitors of viral gene expression and delayed shut-off with viruses that do not cause virion-mediated shut-off. The third decade of research on host shut-off will surely lead to a better understanding of both virus and cell.

9. REFERENCES

Alonso, M. A., and Carrasco, L., 1981, Reversion by hypotonic medium of the shutoff of protein synthesis induced by EMC virus, *J. Virol.* **37**:535.

Aurelian, L., and Roizman, B., 1965, Abortive infection of canine cells by HSV. II. The alternative suppression of synthesis of interferon and viral constituents, *J. Mol. Biol.* **11**:539.

Bayliss, G. J., and Nonoyama, M., 1978, Mechanisms of infection with EBV. III. The synthesis of proteins in superinfected Raji cells, *Virology* **87**:204.

Bayliss, G. J., and Wolf, H., 1981, The regulated expression of EBV. III. Proteins specified by EBV during the lytic cycle, *J. Gen. Virol.* **56**:105.

Bayliss, G. J., and Wolf, H., 1982, Effect of the arginine analogue canavanine on the synthesis of EBV-induced proteins in superinfected Raji cells, *J. Virol.* **41**:1109.

Benedetto, A., Rossi, G. B., Amici, C., Bellardelli, F., Cioè, L., Carruba, G., and Carrasco, L., 1980, Inhibition of animal virus production by means of translation inhibitors unable to penetrate normal cells, *Virology* **106**:123.

Ben-Porat, T., and Kaplan, A. S., 1965, Mechanism of inhibition of cellular DNA synthesis by pseudorabies virus, *Virology* **25**:22.

Ben-Porat, T., Rakusanova, T., and Kaplan, A. S., 1971, Early functions of the genome of herpesvirus. II. Inhibition of the formation of cell-specific polysomes, *Virology* **46**:890.

Ben-Porat, T., Jean, J-H., and Kaplan, A. S., 1974, Early functions of the genome of herpesvirus. IV. Fate and translation of immediate-early RNA, *Virology* **59**:524.

Carrasco, L., 1977, The inhibition of cell functions after viral infection. A proposed general mechanism, *FEBS Lett.* **76**:11.

Carrasco, L., 1978, Membrane leakiness after viral infection and a new appraoch to the development of antiviral agents, *Nature (London)* **272**:694.

Carrasco, L., and Smith, A. E., 1976, Sodium ions and the shut-off of host cell protein synthesis, *Nature (London)* **264**:807.

Chantler, J. K., 1978, The use of hypertonicity to selectively inhibit host translation in murine cytomegalovirus-infected cells, *Virology* **90**:166.

Costanzo, F., Campadelli-Fiume, G., Foa-Tomasi, L., and Cassai, E., 1977, Evidence that HSV DNA is transcribed by cellular RNA polymerase B, *J. Virol.* **21**:996.

Daksis, J. I., Priemer, M. M., and Chan, V.-L., 1982. Isolation and preliminary characterization of a phosphonoacetic acid-resistant and temperature-sensitive mutant of HSV-1, *J. Virol.* **42**:20.

Dixon, R. A. F., and Schaffer, P. A., 1980, Fine structure mapping and functional analysis of temperature-sensitive mutants in the gene encoding the HSV-1 immediate-early protein VP175, *J. Virol.* **36**:189.

Durham, A. C. H., 1977, Do viruses use calcium ions to shut off host cell functions? *Nature (London)* **267**:375.

Feighny, R. J., Farrel, M. P., and Pagano, J. S., 1980, Polypeptide synthesis and phosphorylation in EBV-infected cells, *J. Virol.* **34**:455.

Fenwick, M. L., and Clark, J., 1982, Early and delayed shut-off of host protein synthesis in cells infected with HSV, *J. Gen. Virol.* **61**:121.

Fenwick, M. L., and Roizman, B., 1977, Regulation of herpesvirus macromolecular synthesis. VI. Synthesis and modification of viral polypeptides in enucleated cells, *J. Virol.* **22**:720.

Fenwick, M. L., and Walker, M. J., 1978, Suppression of the synthesis of cellular macromolecules by HSV, *J. Gen. Virol.* **41**:37.

Fenwick, M. L., and Walker, M. J., 1979, Phosphorylation of a ribosomal protein and of virus-specific proteins in cells infected with HSV, *J. Gen. Virol.* **45**:397.

Fenwick, M. L., Morse, L. S., and Roizman, B., 1979, Anatomy of HSV DNA. XI. Apparent clustering of functions effecting rapid inhibition of host DNA and protein synthesis, *J. Virol.* **29**:825.

Fenwick, M. L., Walker, M. J., and Marshall, L., 1980, Some characteristics of an early protein (ICP 22) synthesized in cells infected with HSV, *J. Gen. Virol.* **47**:333.

Flanagan, J. F., 1967, Virus-specific ribonucleic acid synthesis in KB cells infected with HSV, *J. Virol.* **1**:583.

Gergely, L., Klein, G., and Ernberg, I., 1971, Host cell macromolecular synthesis in cells containing EBV-induced early antigens studied by combined immunofluorescence and autoradiography, *Virology* **45**:22.

Godowski, P. J., and Knipe, D. M., 1983, Mutations in the major DNA-binding protein gene of HSV-1 result in increased levels of viral gene expression, *J. Virol.* **47**:478.

Gupta, P., and Rapp, F., 1978, Cyclic synthesis of human cytomegalovirus-induced proteins in infected cells, *Virology* **84**:199.

Hamada, C., and Kaplan, A. S., 1965, Kinetics of synthesis of various types of antigenic proteins in cells infected with pseudorabies virus, *J. Bacteriol.* **89**:1328.

Hay, J., Koteles, G. J., Keir, H. M., and Subak-Sharpe, H., 1966, Herpesvirus-specified ribonucleic acids, *Nature (London)* **210**:387.

Herz, C., and Roizman, B., 1983, The promoter-regulator-ovalbumin chimeric gene resident in human cells is regulated like the authentic 4 gene after infection with HSV-1 mutants in 4 gene, *Cell* **33**:145.

Hill, T. M., Sinden, R. R., and Sadler, J. R., 1983, Herpes simplex virus types 1 and 2 induce shutoff of host protein synthesis by different mechanisms in Friend erythroleukemia cells, *J. Virol.* **45**:241.

Honess, R. W., and Roizman, B., 1973, Proteins specified by HSV. XI. Identification and relative molar rates of synthesis of structural and non-structural herpesvirus polypeptides in the infected cell, *J. Virol.* **12**:1347.

Honess, R. W., and Roizman, B., 1974, Regulation of herpesvirus macromolecular synthesis. I. Cascade regulation of the synthesis of three groups of viral proteins, *J. Virol.* **14**:8.

Honess, R. W., and Roizman, B., 1975, Regulation of herpesvirus macromolecular synthesis: Sequential transition of polypeptide synthesis requires functional viral polypeptides, *Proc. Natl. Acad. Sci. USA* **72**:1276.

Horwitz, M. S., Brayton, C., and Baum, S. G., 1973, Synthesis of type 2 adenovirus DNA in the presence of cycloheximide, *J. Virol.* **11**:544.

Ihara, S., Feldman, L., Watanabe, S., and Ben-Porat, T., 1984, Characterization of the immediate-early functions of pseudorabies virus, *Virology* **131**:437.

Inglis, M. M., and Newton, A. A., 1981, Comparison of the activities of HSV-1 and cellular mRNAs as templates for *in vitro* translation, *Virology* **110**:1.

Isom, H. C., Liao, W. S. L., Taylor, J. M., Willworth, G. E., and Eadline, T. S., 1983, Rapid and selective shutoff of plasma protein production in HSV-2-infected hepatoma cells, *Virology* **126**:548.

Kaplan, A. S., 1973a, A brief review of the biochemistry of herpesvirus–host cell interaction, *Cancer Res.* **33**:1393.

Kaplan, A. S., 1973b, *The Herpesviruses* (A. S. Kaplan, ed.), p. 177, Academic Press, New York.

Kaplan, A. S., and Ben-Porat, T., 1960, The incorporation of C^{14}-labelled nucleosides into rabbit kidney cells infected with pseudorabies virus, *Virology* **11**:12.

Kaplan, A. S., and Ben-Porat, T., 1963, The pattern of viral and cellular DNA synthesis in pseudorabies virus-infected cells in the logarithmic phase of growth, *Virology* **19**:205.

Kaplan, A. S., Shimono, H., and Ben-Porat, T., 1970, Synthesis of proteins in cells infected with herpesvirus. III. Relative amino acid content of various proteins formed after infection, *Virology* **40**:90.

Kelley, P. M., and Schlesinger, M. J., 1978, The effect of amino acid analogues and heat shock on gene expression in chick embryo fibroblasts, *Cell* **15**:1277.

Kennedy, I. M., Stevely, W. S., and Leader, D. P., 1981, Phosphorylation of ribosomal proteins in hamster fibroblasts infected with Pseudorabies virus or Herpes Simplex virus, *J. Virol.* **39**:359.

Kohn, A., 1979, Early interactions of viruses with cellular membranes, *Adv. Virol. Res.* **24**:223.

Lemaster, S., and Roizman, B., 1980, Herpesvirus phosphoproteins, II. Characterization of the virion protein kinase and the polypeptides phosphorylated in the virion, *J. Virol.* **35**:798.

Lowe, P. A., 1978, Levels of DNA-dependent RNA polymerases in HSV-infected BHK cells, *Virology* **86**:577.

McCormick, F. P., and Newton, A. A., 1975, Polyamine metabolism in cells infected with HSV, *J. Gen. Virol.* **27:**25.

McCready, S. J., Godwin, J., Mason, D. W., Brazell, I. A., and Cook, P. R., 1980, DNA is replicated at the nuclear cage, *J. Cell. Sci.* **46:**365.

McGrath, B. M., and Stevely, W. S., 1980, The characteristics of the cell-free translation of mRNA from cells infected with the herpes virus Pseudorabies virus, *J. Gen. Virol.* **49:**323.

Modrow, S., and Wolf, H., 1983, Herpesvirus Saimiri-induced proteins in lytically infected cells. I. Time-ordered synthesis, *J. Gen. Virol.* **64:**37.

Morse, L. S., Pereira, L., Roizman, B., and Schaffer, P. A., 1978, Anatomy of HSV DNA. X. Mapping of viral genes by analysis of polypeptides and functions specified by HSV-1 × HSV-2 recombinants, *J. Virol.* **26:**389.

Nakai, H., Maxwell, I. H., and Pizer, L. I., 1982, Herpesvirus infection alters the steady state levels of cellular polyadenylated RNA in polyoma virus-transformed BHK cells, *J. Virol.* **42:**1131.

Nevins, J. R., 1982, Induction of the synthesis of a 70,000 dalton mammalian heat shock protein by the adenovirus E1A gene product, *Cell* **29:**913.

Newton, A. A., 1964, Synthesis of DNA in cells infected by virulent DNA viruses, in: *Acidi Nucleici e Loro Funzione Biologica,* p. 109, Istituto Lombardo; Fondazione Baselli, Milano.

Newton, A. A., 1969, Internal virology I, in: *Proceedings of the 1st International Congress of Virology* (J. L. Melnick, ed.), p. 253, S. Karger, Basel.

Newton, A. A., 1972, The involvement of nuclear membrane in the synthesis of herpes-type viruses, in: *Oncogenesis and Herpes Viruses I* (P. M. Biggs, G. de Thé, and L. N. Payne, eds.), p. 489, International Agency for Research on Cancer Scientific Publications, No. 2, Lyon.

Newton, A. A., 1978, Effect of cyclic nucleotides on the response of cells to infection by various herpes viruses, in: *Oncogenesis and Herpesviruses III* (G. de Thé, W. Henle, and F. Rapp, eds.), p. 381, International Agency for Research on Cancer Scientific Publications, No. 24, Lyon.

Newton, A. A., Dendy, P. P., Smith, C. L., and Wildy, P., 1962, A pool size problem associated with the use of tritiated thymidine, *Nature (London)* **194:**886.

Nishioka, Y., and Silverstein, S., 1977, Degradation of cellular mRNA during infection by HSV, *Proc. Natl. Acad. Sci. USA* **74:**2370.

Nishioka, Y., and Silverstein, S., 1978, Requirement of protein synthesis for the degradation of host mRNA in Friend erythroleukemia cells infected with HSV-1, *J. Virol.* **27:**619.

Nishioka, Y., Jones, G., and Silverstein, S., 1983, Inhibition by Vesicular Stomatitis virus of HSV-directed protein synthesis, *Virology* **124:**238.

Nonoyama, M., and Pagano, J. S., 1972, Replication of viral DNA and breakdown of cellular DNA in Epstein–Barr virus infection, *J. Virol.* **9:**714.

Notarianni, E. L., and Preston, C. M., 1982, Activation of cellular stress protein genes by HSV *ts* mutants which overproduce immediate early polypeptides, *Virology* **123:**113.

Nuss, D. L., Oppermann, H., and Koch, G., 1975, Selective blockage of initiation of host protein synthesis in RNA virus-infected cells, *Proc. Natl. Acad. Sci. USA* **72:**1258.

O'Callaghan, D. J., Cheevers, W. P., Gentry, G. A., and Randall, C. C., 1968*a*, Kinetics of cellular and viral DNA synthesis in equine abortion (herpes) virus infection of L-M cells, *Virology* **36**:104.

O'Callaghan, D. J., Hyde, J. M., Gentry, G. A., and Randall, C. C., 1968*b*, Kinetics of viral deoxyribonucleic acid, protein and infectious particle production and alterations in host macromolecular syntheses in equine abortion (herpes) virus-infected cells, *J. Virol.* **2**:793.

Pereira, L., Wolff, M. H., Fenwick, M., and Roizman, B., 1977, Regulation of herpesvirus macromolecular synthesis. V. Properties of α polypeptides made in HSV-1 and HSV-2 infected cells, *Virology* **77**:733.

Pizer, L. I., and Beard, P., 1976, The effect of herpesvirus infection on mRNA in polyoma virus-transformed cells, *Virology* **75**:477.

Planck, S. R., and Mueller, G. C., 1977, DNA chain growth and organization of replicating units in HeLa cells, *Biochemistry* **16**:1808.

Post, L. E., Mackem, S., and Roizman, B., 1981, Regulation of α genes of HSV; expression of chimeric genes produced by fusion of thymidine kinase with α gene promoters, *Cell* **24**:555.

Powell, K. L., and Courtney, R. J., 1975, Polypeptides synthesized in HSV-2-infected HEp-2 cells, *Virology* **66**:217.

Preston, C. M., 1979*a*, Control of HSV-1 mRNA synthesis in cells infected with wild-type virus or the temperature-sensitive mutant *ts*K, *J. Virol.* **29**:275.

Preston, C. M., 1979*b*, Abnormal properties of an immediate-early polypeptide in cells infected with the HSV-1 mutant *ts*K, *J. Virol.* **32**:357.

Preston, C. M., and Newton, A. A., 1976, The effects of HSV-1 on cellular DNA-dependent RNA polymerase activities, *J. Gen. Virol.* **33**:471.

Rakusanova, T., Ben-Porat, T., and Kaplan, A., 1972, Effects of herpesvirus infection on the synthesis of cell-specific RNA, *Virology* **49**:537.

Read, S., and Frenkel, N., 1983, HSV mutants defective in the virion-associated shutoff of host polypeptide synthesis and exhibiting abnormal synthesis of α (immediate early) viral polypeptides, *J. Virol.* **46**:498.

Roizman, B., and Furlong, D., 1974, The replication of herpesviruses, in: *Comprehensive Virology,* Vol. 3 (H. Fraenkel-Conrat and R. R. Wagner, eds.), p. 357, Plenum Press, New York.

Roizman, B., and Roane, P. H. J., 1964, The multiplication of HSV. II. The relation between protein synthesis and the duplication of viral DNA in infected HEp-2 cells, *Virology* **22**:262.

Roizman, B., Borman, G. S., and Rousta, M., 1965, Macromolecular synthesis in cells infected with HSV, *Nature (London)* **206**:1374.

Rubenstein, A. S., Gravell, M., and Darlington, R., 1972, Protein kinase in enveloped Herpes Simplex virions, *Virology* **50**:287.

Saxton, R. E., and Stevens, J. G., 1972, Restriction of HSV replication by poliovirus: A selective inhibition of viral translation, *Virology* **48**:207.

Shimono, H., Ben-Porat, T., and Kaplan, A. S., 1969, Synthesis of proteins in cells infected with herpesvirus. I. Structural viral proteins, *Virology* **37**:49.

Silverstein, S., and Engelhardt, D. L., 1979, Alterations in the protein synthetic apparatus of cells infected with HSV, *Virology* **95**:334.

Spear, P. G., and Roizman, B., 1968, The proteins specified by HSV. I. Time of synthesis, transfer into nuclei and properties of proteins made in productively infected cells, *Virology* **36**:545.

Spear, P. G., and Roizman, B., 1981, The Herpesviruses, in: *Molecular Biology of Tumor Viruses: DNA tumor viruses* (J. Tooze, ed.), p. 215, Cold Spring Harbor Laboratory, New York.

Spector, D., and Pizer, L. I., 1978, Herpesvirus infection modified adenovirus RNA metabolism in adenovirus type 5-transformed cells, *J. Virol.* **27**:1.

St. Jeor, S. C., Albrecht, T. B., Funk, F. D., and Rapp, F., 1974, Stimulation of cellular DNA synthesis by human cytomegalovirus, *J. Virol.* **13**:353.

Stenberg, R. M., and Pizer, L. I., 1982, HSV induced changes in cellular and adenovirus RNA metabolism in an adenovirus type 5-transformed human cell line, *J. Virol.* **42**:474.

Stevely, S., and McGrath, B. M., 1978, The effect of hypertonic conditions on protein synthesis in cells infected with herpesvirus, *FEBS Lett.* **87**:308.

Stimac, E., Housman, D., and Huberman, J. A., 1977, Effects of inhibition of protein synthesis on DNA replication in cultured mammalian cells, *J. Mol. Biol.* **115**:485.

Stinsky, M. F., 1977, Synthesis of proteins and glycoproteins in cells infected with human cytomegalovirus, *J. Virol.* **23**:751.

Stinsky, M. F., 1978, Sequence of protein synthesis in cells infected by human CMV: Early and late virus-induced polypeptides, *J. Virol.* **26**:686.

Stoker, M. G. P., and Newton, A. A., 1959, Mitotic inhibition of HeLa cells caused by herpesvirus, *Ann. N.Y. Acad. Sci.* **81**:129.

Sydiskis, R. J., and Roizman, B., 1966, Polysomes and protein synthesis in cells infected with a DNA virus, *Science* **153**:76.

Sydiskis, R. J., and Roizman, B., 1967, The disaggregation of host polyribosomes in productive and abortive infection with HSV, *Virology* **32**:678.

Sydiskis, R. J., and Roizman, B., 1968, The sedimentation profiles of cytoplasmic polyribosomes in mammalian cells productively and abortively infected with HSV, *Virology* **34**:562.

Tanaka, S., Furukawa, T., and Plotkin, S. A., 1975, Human cytomegalovirus-stimulated host cell DNA synthesis, *J. Virol.* **15**:297.

Wagner, E. K., and Roizman, B., 1969, Ribonucleic acid synthesis in cells infected with HSV, *J. Virol.* **4**:36.

Watson, R. J., and Clements, J. B., 1980, A herpes simplex type 1 function continuously required for early and late virus RNA synthesis, *Nature* (*London*) **285**:329.

CHAPTER 8

Poxvirus Cytopathogenicity: Effects on Cellular Macromolecular Synthesis

Rostom Bablanian

SUNY, Downstate Medical Center
Department of Microbiology and Immunology
Brooklyn, New York 11203

1. INTRODUCTION

1.1. Virus Cytopathogenicity

The initial observations by Enders (1954) and his collaborators that virus–cell interaction culminates in the destruction of the infected cells in culture made it possible to study these alterations at both the morphological and biochemical levels. Highly cytocidal viruses, during the process of either productive or abortive virus replication, ultimately kill the cell. Notable examples of this type of viruses are the picornaviruses, the rhabdoviruses, the herpesviruses, and the poxviruses. Early studies of virus cytopathology were mainly descriptive in nature, documenting the various types of virus-induced morphological alterations to cells. With the advent of new biochemical techniques, it became possible to scrutinize the mechanisms by which these viruses altered the infected cells. The main objective of these studies has been to define the initial step of virus-induced alteration and to characterize the nature of the viral product(s) which initiates the steps leading to the ultimate death of the cell.

The most prominent feature of cytocidal virus–cell interaction is the inhibition of host macromolecular synthesis. The understanding of the molecular basis of this inhibition where the host DNA, RNA, and protein synthesis is selectively blocked while virus replication goes on is the central theme of this chapter. The mechanisms by which particular virus–cell combinations accomplish this inhibition appear to be quite variable, and a number of hypotheses have been proposed to explain this phenomenon. It has been suggested that inhibition of host macromolecular synthesis by cytocidal viruses may be initiated by (1) structural components of the input virion, (2) virus-induced products synthesized in the course of infection, (3) viral nucleic acids as either replicative intermediates (double-stranded RNA) or as multiple copies of the viral genome (single-stranded RNA), or (4) the results of perturbations of the cellular membrane which allow for changes in the normal ion concentrations within the cell.

Of the various marcromolecular changes induced by cytocidal viruses, the mechanism of inhibition of host cell protein synthesis (shut-off) has been most thoroughly studied. While several virus–cell models have been used in these investigations, poliovirus-infected cells have been the most extensively studied model. Early work suggested that during poliovirus infection, inhibition of cellular protein synthesis is brought about by a viral protein (Steiner-Pryor and Cooper, 1973) although this polypeptide has not yet been identified. It has long been known that this inhibition occurs at the level of initiation of protein synthesis (Willems and Penman, 1966) and is not caused by degradation (Colby *et al.*, 1974) or modification of host mRNA (Fernandez-Munoz and Darnell, 1976). More recent work has indicated that the inactivation of initiation factors is related to shut-off (Rose *et al.*, 1978; Helentjaris *et al.*, 1979). Sonenberg *et al.* (1978) discovered a 24,000 dalton host polypeptide associated with eukaryotic initiation factors which specifically binds to the 5'-cap structure typical of eukaryotic mRNA molecules and potentiates the capacity of mRNAs to form initiation complexes. It has been shown that this cap-binding polypeptide (Sonenberg *et al.*, 1978; Trachsel *et al.*, 1980) is inactivated during poliovirus infection (Hansen *et al.*, 1982). The implication from these studies is that poliovirus infection hinders the process of initiation of capped mRNAs and since poliovirus mRNA does not possess a cap (Nomoto *et al.*, 1976; Hewlett *et al.*, 1976), it can selectively initiate and translate. Not all picornaviruses studied thus far, however, cause shut-off by this mechanism since a comparative study between poliovirus and EMC virus has shown that unlike lysates from poliovirus-infected cells, which initiate poorly the capped mRNAs, lysates from EMC virus-infected cells initiate capped mRNAs better than uninfected cell lysates (Jen *et al.*, 1980; Jen and Thach, 1982; for recent reviews, see Shatkin, 1983; Ehrenfeld, this volume.).

A block at the level of initiation of protein synthesis has also been suggested as the mechanism of shut-off by vesicular stomatitis virus (VSV; Stanners *et al.*, 1977; Jaye *et al.*, 1982; Gillies and Stollar, 1982). As in poliovirus-induced shut-off, degradation of host mRNAs does not seem to play a role in VSV-induced shut-off since host mRNA can be extracted from infected cells and translated *in vitro* (Ehrenfeld and Lund, 1977). However, recent work has demonstrated that the synthesis of VSV leader RNA is directly related to inhibition of host RNA synthesis (Grinnell and Wagner, 1983). Unlike poliovirus mRNAs, VSV mRNAs are capped and require cap-binding protein for translation (Banerjee, 1980; Rose *et al.*, 1978). The mechanism of VSV-induced shut-off is presently under active investigation to determine if competition between mRNAs (Lodish and Porter, 1980, 1981), alteration of initiation factors (Centrella and Lucas-Lenard, 1982), and/or inhibition by viral transcripts (Schnitzlein et al., 1983) is the cause of this inhibition.

Herpesvirus-induced inhibition of host protein synthesis has been primarily associated to a viral component which is either synthesized during the replication of herpesvirus type 1 (Stenberg and Pizer, 1982; Hill *et al.*, 1983) or is a component of the parental herpesvirus type 2 (Fenwick and Walker, 1978). In addition, herpesvirus type 1 causes the degradation of host cell mRNA (Nishioka and Silverstein, 1978; Inglis, 1982; Stenberg and Pizer, 1982). All the three virus–cell systems mentioned above also cause inhibition of host RNA and DNA synthesis during virus replication which may contribute to shut-off.

From this brief summary, it is evident that a multiplicity of strategies are used by cytocidal viruses to take over the host cell for their own replication. In this review, the inhibition of host cell protein synthesis caused by vaccinia virus (a member of the poxvirus group) will be discussed in greater detail.

1.2. Poxvirus Structure

Several excellent reviews on the reproduction of poxviruses are available (Moss, 1974, 1978). A brief synopsis of the morphology of the virus and its mode of replication is included in this chapter to point out virion components and events in virus replication likely to be associated with virus cytopathology. Poxviruses are the largest in size and most complex of all animal viruses. Vaccinia virus, the prototype of these viruses, has been widely used to study the molecular biology of this major group.

Electron microscopy shows that vaccinia virus resembles a brick-shaped ellipsoid approximately 200 × 300 nm. The virion is complex in symmetry and consists of three major architectural elements: a biconcave protein-bounded core, two lateral bodies juxtaposed to the concavities of the core, and a surrounding lipoprotein membrane. The core contains a linear double-stranded DNA genome, crosslinked at both ends (Berns and Silverman, 1970; Geshelin and Berns, 1974; Esteban and Holowczak, 1977; Esteban *et al.*, 1977). It has a molecular weight of approximately $1.2–1.3 \times 10^8$ daltons (Geshelin and Berns, 1974; Cabrera and Esteban, 1978; McCarron *et al.*, 1978). Although by definition viruses contain either DNA or RNA, in the case of vaccinia virus, in addition to its DNA genome, some RNA has also been detected within the virus particles (Planterose *et al.*, 1962; Roening and Holowczak, 1974). It is not known whether this RNA is an accidental packaging of viral transcripts present in the cytoplasm during assembly of the virion or if it is a product formed within the virus particle by the endogenous RNA polymerase. Up to now, no function has been assigned to this virion-associated RNA, but as we shall see later in this chapter, it may be involved in the selective inhibition of host protein synthesis. No function has been attributed to the lateral bodies, again reflecting the complexity of the virion. The projections seen by electron microscopy on the surface of the virus particle give it the so-called mulberry appearance. These surface tubules are the major protein component of the membrane with a molecular weight of 58,000 daltons (Stern and Dales, 1976a; Moss, 1978). Vaccinia virus has a coding capacity for at least 150–300 polypeptides as predicted from its genome size. Analysis of the virion particles by polyacrylamide–gel electrophoresis (PAGE) reveals only 56 bands ranging in molecular weight from 10,000–200,000 (Holowczak and Joklik, 1967; Moss and Salzman, 1968; Sarov and Joklik, 1972; Moss *et al.*, 1973; Stern and Dales, 1976b). More recent work using two-dimensional gel electrophoresis demonstrates at least 111 proteins with a molecular weight range of 8000–96,000 (Essani and Dales, 1979). Some of these structural proteins are glycosylated, with the carbohydrate portion being glucosamine (Sarov and Joklik, 1972), or phosphorylated (Garon and Moss, 1971). The purified virions contain a multitude of enzyme activities including DNA-dependent RNA polymerase (RNA nucleotidyl transferase; Munyon *et al.*, 1967), poly-A polymerase (terminal riboadenylate transferase; Moss *et al.*, 1973), guanylyl and methyl transferases (Ensinger *et al.*, 1975), nucleoside triphosphate phosphohydrolases (Munyon *et al.*, 1968, exo and endodeoxyribonucleases (Pogo and Dales, 1969), a protein kinase (Paoletti and Moss, 1972), and a nicking and closing enzyme (Bauer *et al.*, 1977).

1.3. Poxvirus Replication

There are two schools of thought concerning the adsorption and penetration of poxviruses, neither of which relies on the use of specific receptor sites. The virion may enter the cell by (1) fusion of its lipid membrane with that of the host cell, thereby releasing the viral core into the cytoplasm (Armstrong *et al.*, 1973; Payne and Norrby, 1978), or by (2) viropexis (engulfment by the cellular membrane) resulting in the appearance of the virion in phagocytic vacuoles (Dales, 1965). Most likely, virus penetration is accomplished by both mechanisms. Regardless of its mode of penetration, the progressive degradation of the viral particle through its subviral particle or core stages and the subsequent replication of the virus take place almost entirely within the cytoplasm, although some active nuclear participation has been implicated for the maturation of particles (Walen, 1971; La Colla and Weisbach, 1975; Bolden *et al.*, 1979; Hruby *et al.*, 1979). Once the outer viral membrane is removed, probably by host enzymes (Dales, 1965), the cores become exposed and transcription proceeds in a biphasic pattern corresponding to early and late classes of mRNA. The early mRNAs, characterized by a sedimentation value of 10–14 S are synthesized by the virion polymerase and can be amplified in the presence of inhibitors of protein synthesis (Kates and McAuslan, 1967*a*), or produced by cores *in vitro* (Kates and Beeson, 1970). Late mRNAs can be distinguished from early mRNAs by their sedimentation rate (16–23 S) and by hybridization competition (Oda and Joklik, 1967). Three necessary events precede the transcription of late mRNAs: (1) early protein synthesis (Joklik and Becker, 1964), (2) exposure of the genome by secondary uncoating (Joklik and Becker, 1964), and (3) replication of the DNA which provides additional templates (Holowczak, 1982). Hybridization studies indicate that early mRNA sequences are a subset of late mRNA competing with roughly half of these sequences, and that portions of both strands are transcribed (Oda and Joklik, 1967; Kaverin *et al.*, 1975). Overlapping sequences used for transcription may account for the small percentage of double-stranded RNA (dsRNA) detected in infected cells (Colby and Duesberg, 1969). Both classes of mRNAs can be modified at the 5′ end by methylated or guanylylated cap structures (Moss *et al.*, 1976) and at the 3′ end by polyadenylation (Kates and Beeson, 1970; Nevins and Joklik, 1975).

As can be expected from the temporally distinct classes of transcripts, viral proteins synthesized in the infected cells also fall into early and late classes; thus, 30 polypeptides are seen prior to and 50 after the onset of DNA replication (switchover; Pennington, 1974). Among the structural and nonstructural proteins, there appear to be three subsets of early proteins which show the following temporal characteristics: (1) immediate synthesis and abrupt cessation, (2) progressively increased synthesis, and (3) prolonged synthesis (Esteban and Metz, 1973). Late proteins are synthesized (1) immediately following DNA replication, and (2) in gradually increasing amounts for prolonged periods (Moss and Salzman, 1968; Pennington, 1974).

Replication of the viral DNA occurs only after a second stage uncoating and this step requires protein synthesis (Joklik, 1964) because the addition of cycloheximide at this stage of replication strongly inhibits DNA replication (Kates and McAuslan, 1967*b*). The double-stranded, crosslinked genome is replicated from each end semiconservatively (Esteban and Holowczak, 1977; Esteban *et al.*, 1977) and is localized within membrane-associated cytoplasmic viral factories (Cairns, 1960). DNA synthesis continues in these factories for a period up to 5 hr postinfection at which time limiting membranes begin to form around areas of condensed filaments of DNA. In the following period and up to 10 hr, this assembly results in the formation of spherical immature particles which then proceed to the slightly smaller, brick-shaped mature particles by means of continued eccentric condensation of the enclosed components (Sarov and Joklik, 1973). The mature particles diffuse away from the factories and are released from the cytoplasm, without lysis, as complex structures containing the undefined lateral bodies and a double outer membrane.

All of the host cell macromolecular synthetic processes are inhibited by infection with members of the poxvirus group. The rapidity with which these events occur is dependent on the parameters of the infection, most importantly, the multiplicity of infection, the particular strain of virus, and the cell line used.

2. POXVIRUS CYTOPATHOGENICITY

2.1. Mechanisms of Morphological Lesions

The types of morphological changes seen during infection with cytocidal viruses is quite variable. Usually these changes are observed in productive viral infections; however, it is also quite common to see identical changes during abortive infections either in nonproductive cell lines or when virus replication is restricted with chemical inhibitors. (For reviews, see Bablanian, 1975; Schrom and Bablanian, 1981).

Poxviruses induce a variety of morphological changes in infected cells. Very little is presently known of the biochemical and biophysical virus-induced lesions which bring about these morphological changes in infected cells. Three types of changes have been delineated which may be associated with a certain phase of virus–cell interaction: (1) early cell rounding, (2) cell fusion, and (3) late cellular changes.

2.1.1. Early Cell Rounding

Soon after infection with high multiplicities of poxviruses, infected cells become rounded. This so-called "toxic effect" occurs so rapidly that it was thought to be due to a component of the infecting virus (Brown *et al.*, 1959; Hanafusa, 1960a; Appleyard *et al.*, 1962). Early cell rounding also occurs under restrictive conditions where some viral antigens are made without the synthesis of progeny virus (Appleyard *et al.*, 1962; Shatkin, 1963; Loh and Payne, 1965). Early cell rounding, however, is prevented when inhibitors of protein and RNA synthesis are present during infection with high multiplicities (Bablanian, 1968, 1970). These results clearly demonstrate that a component of the virion is not the cause of early cell rounding but that *de novo* synthesis of viral products is necessary for this phenomenon to occur. This product does not seem to be viral mRNA since it is

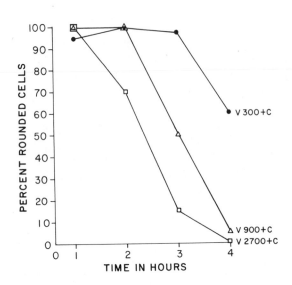

Fig. 1. Rate of early cell rounding after the removal of cycloheximide from LLC-MK$_2$ cells infected with vaccinia virus. Cells were infected at various multiplicities in the presence of cycloheximide (300 μg/ml) as indicated. At hourly intervals, for a period of 4 hr, the compound was removed by washing the cultures three times with warm PBS from three plates infected at 300, 900, or 2700 particles/cell. Cell rounding was quantified 1 hr after the removal of the compound. From Bablanian *et al.* (1978*b*).

well known that, in the presence of inhibitors of protein synthesis, early mRNAs are transcribed in infected cells but not translated (Munyon and Kit, 1966; Woodson, 1967; Kates and McAuslan, 1967*a*; Metz and Esteban, 1972). Therefore, it seemed more likely that the cause of early cell rounding is a translational product. A systematic study with three cell types and various multiplicities of infection indicated that early cell rounding is associated with vaccinia virus polypeptide synthesis (Bablanian *et al.*, 1978*a*), and the evidence is as follows. In a monkey kidney cell line (LLC-MK$_2$) early cell rounding is observed within 30–60 min after infection, and the time of appearance is directly dependent on the multiplicity of infection. In the presence of cycloheximide (300 μg/ml), however, cell rounding is prevented regardless of the multiplicity of infection used (300, 900 or 2700 particles/cell). Cell rounding in these infected cells does occur soon after cycloheximide is washed out. The time of appearance and extent of cell rounding is directly related to the time of removal of the inhibitor and the multiplicity of infection used. If the drug is removed 1 or 2 hr after infection, viral polypeptide synthesis is evident and early cell rounding occurs in cells infected with all three virus multiplicities. However, if cycloheximide is washed out from cells 4 hr after infection, cell rounding and viral polypeptide synthesis are seen only in cultures infected with the lowest multiplicity used (300 particles/cell; Figs. 1 and 2). This virus-induced early cell rounding takes place faster in HeLa cells than in LLC-MK$_2$ cells and even faster

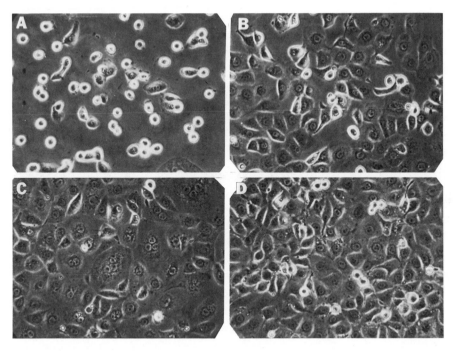

Fig. 2. Early cell rounding in vaccinia virus-infected LLC-MK$_2$ cells after the removal of cycloheximide. Cells were infected at (A) 300, (B) 900, and (C) 2700 particles/cell in the presence of cycloheximide (300 μg/ml). Four hours after infection the compound was removed by washing the cultures three times with warm PBS and cell rounding was recorded 1 hr later. (D) Uninfected cells from which cycloheximide was removed. From Bablanian *et al.* (1978*b*).

in L cells than in HeLa cells. Thus, if HeLa and L cells are infected at a multiplicity of 300 particles/cell in the presence of cycloheximide, early cell rounding will occur in HeLa cells if the compound is removed at 2 hr after infection but not later, and in L cells when the drug is washed out at 40 min after infection and not later. Cell rounding is not seen if cycloheximide is removed at later times. In HeLa and L cells, similar to LLC-MK$_2$ cells, this early virus-induced rounding is seen only when viral polypeptide synthesis takes place. Thus, these results establish a strong correlation between viral polypeptide synthesis and the virus-induced early cell rounding. It remains to be determined which viral polypeptide(s) is directly responsible for this effect. Additional support for this point of view comes from the work of McFadden *et al.* (1979) who describe an abortive infection of cells by molluscum contagiosum virus, a poxvirus which produces benign skin tumors in humans. Primate cells infected with the virus show an early cell rounding similar to that seen with vaccinia virus. As in the case of vaccinia virus-infected cells, early cell rounding is prevented by inhibitors of protein and RNA synthesis. Unlike vaccinia virus-infected cells, however, in molluscum contagiousum virus-infected cells, this early cell rounding is reversible. If these cells are reinfected, they undergo the same cycle of early cell rounding and recovery. It is tempting to speculate that this difference between vaccinia virus and molluscum contagiosum virus infection may be due to the abortive nature of the infection of molluscum contagiousum where the virus does not undergo second stage uncoating and the early mRNAs responsible for synthesizing this cell rounding protein is degraded. Reinfection renews the transcription of the early mRNAs responsible for producing the cell rounding protein, and a second cycle of cell rounding and recovery occurs. We still do not understand how viruses derange normal cellular structure. Intuitively, it seems that the interaction of viruses or their components with cytoskeletal structures must be associated with the early morphological alterations.

2.1.2. Cell Fusion

Another effect of vaccinia virus infection is the formation of polykaryocytes. This phenomenon of cell fusion usually occurs late in infection and is primarily associated with the production of nonglycosylated forms of the virus hemagglutinin produced by some strains of virus such as the IHD-W strain (Dales *et al.*, 1976; Ichihashi and Dales, 1971; Stern and Dales, 1976a; Weintraub *et al.*, 1977). Although this glycoprotein is present on the surface of the virions (Ichihashi and Dales, 1971), cell fusion "from without" like that seen with paramyxoviruses (Bratt and Gallaher, 1969) is not observed early in vaccinia virus infections. Early fusion in vaccinia virus-infected cells was noticed when we used very high multiplicity of infection (3000 particles/cell). In these experiments, we were trying to prevent early cell rounding by the addition of protein and RNA inhibitors. Under these conditions where early cell rounding was prevented, cell fusion became apparent (Bablanian *et al.*, 1978b) and since it occurred in the presence of both inhibitors of RNA and protein synthesis, it was concluded that it probably was due to a virion component. To test this conclusion further, we used two strains of vaccinia virus kindly provided by Dr. S. Dales, the IHD-W, which produces the nongly-

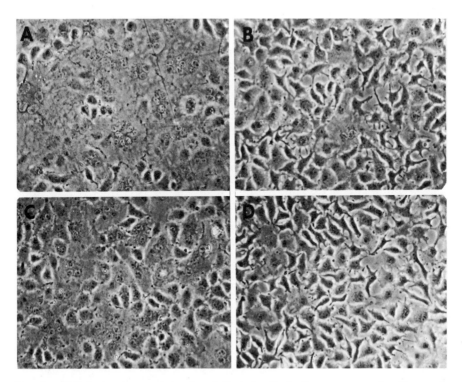

Fig. 3. Cell fusion in vaccinia virus-infected cells in the presence of actinomycin D. L cells were infected with the IHD-J or the IHD-W strains of vaccinia virus (3000 particles/cell) in the presence of actinomycin D (20 μg/ml). The phase contrast photomicrographs were taken at 3 hr after infection and treatment. (A) IHD-W with actinomycin D, (B) actinomycin D, (C) IHD-J with actinomycin D, and (D) cell control. From unpublished data of R. Bablanian.

cosylated viral hemagglutinin and causes extensive late fusion, and the IHD-J which produces the glycosylated hemagglutinin and causes substantially less fusion. When we infected L cells (3000 particles/cell) with the purified virus of these two strains in the presence of either cycloheximide (300 μ/ml) or actinomycin D (20 μg/ml) early cell fusion (2–3 hr after infection) was seen in cells infected with both strains, but cell fusion in cultures infected with the IHD-W strain was strikingly greater than in cultures infected with the IHD-J strain (Fig. 3; Bablanian, unpublished observations). It is important to emphasize at this point that the cell-fusing capacity of vaccinia viruses "from without" is lower than that found with paramyxoviruses (Bratt and Gallaher, 1969).

2.1.3. Late Cellular Changes

It is difficult to assign a single cause for late virus-induced changes in infection. These may involve all kinds of secondary products or mechanisms which in one way or another may contribute to the final demise of the cell. One laboratory, however, has conducted a systematic and thorough study of a vaccinia virus cytotoxic product which can be detected in infected HeLa cells and which may contribute to cellular degeneration late in infection (Stephen *et al.*, 1974; Wolstenholme *et al.*, 1977; Burgoyne and Stephen, 1979). Burgoyne and Stephen (1979) present persuasive evidence that the toxic product in infected cells is the monomer of the surface tubule (ST) of the virion. The cytotoxic factor is a protein and its molecular weight (3–10 $\times$ 10^4 daltons) is similar to the monomer of surface tubules. It is synthesized late in infection similar to surface tubules, and when surface tubules are obtained from purified virions they are also cytotoxic. Furthermore, if hydroxyurea is present in infected cells, the detachment of cells, which is the index of cytotoxicity, is significantly delayed. These results suggest that surface tubules may play a role in late vaccinia virus cytopathology. Additional results of these workers showed that the cytotoxin did not cause early cell rounding, nor did it bring about inhibition of host cell protein and RNA synthesis (Burgoyne and Stephen, 1979). As we shall see later another report claims that surface tubules are involved in shut-off (Mbuy *et al.*, 1982).

2.2. Inhibition of Host Cell RNA Synthesis

Vaccinia virus infection inhibits the transport of newly synthesized ribosomal RNA (rRNA) to the cytoplasm (Becker and Joklik, 1964; Salzman *et al.*, 1964). This inhibition was studied in HeLa cells and it was detected 3–4 hr after infection. In another study where L cells were used, inhibition of rRNA synthesis occurred more rapidly (Jefferts and Holowczak, 1971). These authors showed that the cleavage of the 45 S ribosomal precursor RNA to 32 S and 18 S was inhibited by 2 hr after infection and the transport of rRNA to the cytoplasm was curtailed by 3 hr after infection. The methylation of the 45 S RNA was reduced in infected cells but was not completely abolished. Furthermore, inhibition of ribosomal protein synthesis occurred before a decline in the synthesis of rRNA. The product responsible for these changes has not been determined; however, it may be a structural protein since such peptides enter the host cell nucleus and become associated with the nucleoli (Jefferts and Holowczak, 1971). Vaccinia virus infection also causes inhibition of HeLa cell mRNA synthesis after 3 hr of infection (Becker and Joklik, 1964). In a more recent report using cloned DNAs of β-actin and α-tubulin mRNAs as probes, evidence is presented that these mRNAs in infected L cells are degraded (Rice and Roberts, 1983). The role of this degradation of host cell mRNA, in shut-off, if any, has not yet been determined.

2.3. Inhibition of Host Cell DNA Synthesis

Soon after infection with vaccinia virus, host cell DNA synthesis is markedly inhibited (Hanafusa, 1960a; Hanafusa, 1960b; Kit and Dubbs, 1962; Joklik and Becker, 1964). When heat or UV-inactivated virus is used to infect cells, host DNA synthesis is also inhibited to the same extent as with active virus (Hanafusa, 1960b; Jungwirth and Launer, 1968). Initial observations suggested that this inhibition may be caused by a component of the virion (Joklik and Becker, 1964). Degradation of host DNA was seen in infected cells (Walen, 1971; Oki *et al.*, 1971). It was later demonstrated that a virion-associated DNase enters the nucleus (Pogo and Dales, 1973) and acts on single-stranded DNA (Pogo and Dales, 1974). Thus, in the case of virus-induced inhibition of host DNA synthesis, it is clearly demonstrated that a virion-associated product (a DNase most active at neutral pH) is directly responsible for this inhibition.

2.4. Inhibition of Host Protein Synthesis

Vaccinia virus infection causes a rapid and drastic inhibition of host cell protein synthesis (shut-off) (Kit and Dubbs, 1962; Shatkin, 1963; Salzman and Sebring, 1967; Holowczak and Joklik, 1967; Moss and Salzman, 1968; Metz and Esteban, 1972). This rapid inhibition is followed by the degradation of host polyribosomes (Joklik and Merigan, 1966) and the formation of viral polyribosomes (Metz *et al.*, 1975). Shut-off can also occur under some restrictive conditions (for review, see Bablanian, 1975).

As mentioned earlier, several hypotheses have been proposed for shut-off. In order for these hypotheses to be viable, they must take into account the selective nature of shut-off in that host protein synthesis is inhibited while viral protein synthesis continues. The following are hypotheses which have been postulated for the inhibition of host protein synthesis by vaccinia virus:

1. Alterations in intracellular ionic environment favoring the initiation and translation of viral mRNAs.
2. Effects of virion component(s).
3. Effects of virus-induced or virus-directed RNAs.

2.4.1. Role of Alterations in Intracellular Ionic Environment

A general hypothesis proposed by Carrasco (1977) suggested that virus-induced shut-off is related to cell membrane leakiness developed during infection, which in turn allows an increase of intracellular sodium ions. It has been shown that many cytocidal viruses, either early or late in infection, alter the cell membrane permeability causing an imbalance in the sodium/potassium ion concentration gradient (for review see Carrasco and Lacal, 1984). It had been noted earlier that in several virus–cell systems viral mRNAs translated better than host mRNAs at elevated concentrations of monovalent ions (Saborio *et al.*, 1974; Nuss *et al.*, 1975; Carrasco *et al.*, 1979). In some virus cell systems there is strong evidence that ionic changes play a primary role in shut-off (Alonso and Carrasco, 1981). On the other hand, the major cause of shut-off in poliovirus infected cells does not seem to be the intracellular elevation of monovalent ions (Lacal and Carrasco, 1982) although poliovirus mRNAs are known to translate better at high salt concentrations (Saborio *et al.*, 1974).

Vaccinia virus infection causes an early and transient alteration in membrane permeability followed by a gradual resealing of the membrane leading to full restoration to control levels (Carrasco and Esteban, 1982). It is not clear whether this transient membrane leakiness contributes to shut-off. Furthermore, it has been recently demonstrated that vaccinia virus mRNAs have a much broader salt concentration optimum than host mRNAs when assayed in an *in vitro* reticulocyte cell-free system suggesting that alterations in the intracellular ionic environment are not necessary to favor viral over host mRNA translation (Benavente and Esteban, submitted for publication). Evidence is also presented that the ratio of intracellular sodium to potassium concentration is noticably altered only 13 hr postinfection, long after shut-off occurs in these cells (Norrie *et al.*, 1982).

Thus, the available data presented above indicate that in vaccinia virus infections the alteration of the intracellular ion concentrations does not play a primary role in the selective inhibition of host protein synthesis.

2.4.2. Role of Virion Component(s)

The viral products most commonly implicated in shut-off in many virus–cell systems have been structural components of the virus particles. Structural components of adenovirus (Pereira, 1960), frog virus-3 (Maes and Granoff, 1967), vesicular stomatitis virus (Baxt and Bablanian, 1976), herpesvirus type 1 (Fenwick and Walker, 1978), and vaccinia virus (Moss, 1968) have been implicated as the agents responsible for inhibition of host protein synthesis.

In examining shut-off in vaccinia virus-infected HeLa cells, Moss (1968) proposed that virion component is responsible for this effect. His evidence is based on the following data: (1) Infection of cells (10, 50, or 100 PFU/cell) in the presence of actinomycin D (5 μg/ml) results in a multiplicity-dependent inhibition of protein synthesis. (2) Cells infected with 25 PFU/cell in the presence of both actinomycin D (5 μg/ml) and cycloheximide (300 μg/ml) fail to resume protein synthesis when cycloheximide is washed out. (3) Ultraviolet light-inactivated, but not heat- or detergent-inactivated, virus also inhibits host cell protein synthesis. These data, along with the assumption that 5 μg/ml of actinomycin D is sufficient to prevent viral RNA synthesis, led Moss (1968) to conclude that a virion component is responsible for the inhibition. It has since been shown that 1–5 μg/ml of actinomycin D is insufficient to completely prevent the synthesis of virus-induced RNA synthesis (Metz and Esteban, 1972; Rosemond-Hornbeak and Moss, 1975; Schrom and Bablanian, 1979a) and, thus, the possibility exists that RNA transcripts synthesized under these conditions may play a role in shut-off. Further evidence that a virion component may be involved in shut-off comes from *in vitro* work where viral cores were added directly to cell-free protein synthesizing systems (Ben-Hamida and Beaud, 1978; Pelham *et al.*, 1978; Cooper and Moss, 1978). In the first case (Ben-Hamida and Beaud, 1978), a large amount of viral cores (4–8 A_{260} units/50 μl assay) was added to a reticulocyte cell-free protein-synthesizing system resulting in the inhibition of endogenous protein synthesis. Exogenous mRNA translation programmed by a plant virus mRNA in a wheat germ cell-free protein-synthesizing system was also inhibited. On this basis, these authors concluded that a structural component of the core was responsible for this inhibition. Cooper and Moss (1978) used a coupled transcription–translation cell-free system and showed that when low concentrations of cores were added to the system (1.2×10^9 cores/ml) transcription was followed by translation. However, when they increased the amount of cores ten-fold a greater amount of transcription occurred but protein synthesis was inhibited. These authors suggested that the inhibition seen with the higher amount of cores may be due to a component of the cores. Pelham et al. (1978), again using a coupled transcription–translation cell-free system rendered message-dependent, made an attempt to define the in vitro inhibition brought about by cores. These authors used boiled cores in the in vitro protein synthesizing system and demonstrated that these were just as inhibitory to protein synthesis as unboiled cores. From this result, they suggested that the inhibition could not be due to proteins. More recently, a component has been isolated from vaccinia virus cores which inhibits initiation of protein synthesis but with no discrimination (Ben-Hamida et al., 1983). This fraction contains basic phosphoproteins with a molecular weight between 9–15K and is micrococcal nuclease resistant. Unfortunately, the conditions of micrococcal treatment are not given in this paper. We shall see later that the in vitro viral transcripts which selectively inhibit protein synthesis in cell-free systems require treatment with higher concentrations of nuclease to lose activity (150 Ū/ml for 2 hr 23°C) than that usually used to abrogate reticulocyte lysate mRNA function (Coppola and Bablanian, 1983). It is, thus, possible that some core-associated RNA may have copurified with this fraction.

A structural polypeptide in the form of surface tubules has also been implicated in shut-off (Mbuy et al., 1982). Exposure of HEp-2 cells to high concentrations of surface tubules (equivalent to 10^4 particles/cell) resulted in a striking inhibition of protein synthesis. Under similar conditions of infection, host RNA and DNA synthesis were not inhibited. The translation of globin mRNA in a messenger-dependent reticulocyte lysate was also inhibited by these surface tubules. These workers performed a series of control experiments, including the use of specific antibodies for surface tubules, to show that shut-off by these tubules was specific. However, these tubules were obtained by extraction with a non-ionic detergent (Nonidet P40), and it is well known that this and similar detergents (Triton X) remain firmly bound to proteins and require extensive dialysis or other procedures to remove the tightly-bound detergents from hydrophobic proteins (Holloway, 1973; Helenius and Simons, 1975). No controls were provided by Mbuy *et al.* (1982) to exclude the possibility that Nonidet P40 did not copurify with the isolated surface tubules and may have caused this inhibition of protein synthesis both in the intact cells and in the cell-free protein-synthesizing system. Another group of investigators (Wolstenholme *et al.*, 1977; Burgoyne and Stephen, 1979) isolated surface tubule protein from both virions and infected cells and showed that this protein is cytotoxic when introduced into intact cells by hypertonic shock. In contrast to the results of Mbuy *et al.* (1982), these authors showed that the surface tubules did not cause shut-off in HeLa cells nor did they inhibit protein synthesis in the reticulocyte lysate cell-free system. These authors also used Nonidet P40 to isolate surface tubules from virions and make a point to mention that the surface tubules obtained by the non-ionic detergent were extensively dialyzed. Finally, Burgoyne and Stephen (1979) conclude that the surface tubules, which are proteins synthesized late in infection (Stern and Dales, 1976*a*), act from within the infected cells to cause degeneration of cells at late times after infection, and are not the cause of shut-off, which occurs soon after infection.

2.4.3. Role of Vaccinia Virus RNAs in Shut-Off

2.4.3a. Double-Stranded RNA in Shut-Off

Double-stranded RNA (dsRNA) has been implicated in picornavirus-induced shut-off (Ehrenfeld and Hunt, 1971); however, the inhibition of protein synthesis caused by poliovirus double-stranded RNA is nondiscriminatory with respect to host or viral mRNAs (Celma and Ehrenfeld, 1974). Double-stranded RNA is obtained during extraction of RNA from infected cells and can also be found in *in vitro* transcription. It has been proposed that this dsRNA is produced by vaccinia virus both *in vivo* (Duesberg and Colby, 1969) and *in vitro* (Colby *et al.*, 1971) as a result of symmetrical transcription. The role of vaccinia virus double-stranded RNA in shut-off has been investigated. To determine the relation of *in vivo*-synthesized vaccinia virus-induced double-stranded RNA to shut-off, we measured the percentage and amount of double-strand RNA formed in infected L929 cells treated with cycloheximide; conditions where severe inhibition occurs and a large amount of viral RNA is produced (Schrom and Bablanian, 1979a). Our results showed that double-stranded RNA was formed in infected cycloheximide-treated cells but much less than in infected cells. Thus, the involvement of double-stranded RNA in this inhibition cannot be definitely ruled out. It is possible that this small increase may act in a catalytic fashion to inhibit protein synthesis (Hunter et al., 1975). It has been shown that double-stranded RNA may inhibit protein synthesis through the phosphorylation of the small subunit (α) of the initiation factor eIF-2 (Kaempfer and Kaufman, 1973), and that ribosomal proteins are phosphorylated in vaccinia virus-infected cells (Kaerlein and Horak, 1976). We determined the extent of phosphorylation in infected and infected cycloheximide-treated cells and found that the severe inhibition seen in L929 infected-treated cells cannot be correlated with the extent of phosphorylation (Schrom and Bablanian, 1979a). It remains to be determined if phosphorylation of eIF-2 and of ribosomal proteins in vaccinia virus-infected cells plays a role in shut-off.

Baglioni et al. (1978) have presented evidence that in vitro-transcribed vaccinia virus mRNAs contain double-stranded forms which inhibit their own translation in the reticulocyte cell-free system (sensitive to inhibition by double-stranded RNA) but not in the wheat germ system (insensitive to inhibition by double-stranded RNA). These authors showed that if they used procedures that abrogate the inhibitory effect of the double-stranded RNA in the reticulocyte system such as heating and ice quenching and the use of high concentrations of double-stranded RNA (Hunter et al., 1975), they were able to show better translation of vaccinia virus mRNA. Using a coupled transcription–translation system, both Cooper and Moss (1978) and Pelham et al. (1978) observed an inhibition of protein synthesis after the initial translation of polypeptides. Both groups considered the possibility that double-stranded RNA was being formed causing this inhibition in the coupled system. Cooper and Moss (1978) concluded that if any double-stranded RNA was being formed, it was not the primary cause of inhibition seen in the coupled system. This conclusion was reached by the following evidence: (1) The ribonuclease-digested, self-annealed poly(A)-containing RNA from infected cells was a potent inhibitor of protein synthesis in the reticulocyte cell-free system (70% inhibition at 1 μg/ml). (2) This inhibition was reversed by 6 mM cyclic AMP and by high concentrations (20 μg/ml) of reovirus double-stranded RNA, conditions known to reverse the inhibitory effect of double-stranded RNA (Legon et al., 1974; Hunter et al., 1975). (3) The inhibition of protein synthesis seen in the coupled transcription–translation system was not alleviated by the addition of 6 mM cyclic AMP and 20 μg/ml of reovirus double-stranded RNA added to the system. Thus, these authors concluded that double-stranded RNA does not play a major role in this inhibition. Pelham *et al.* (1978), on the other hand, using the same coupled transcription–translation system, came to the conclusion that double-stranded RNA may be partly but not entirely responsible for this delayed inhibition of protein synthesis. Their evidence is based on the use of high concentrations of double-stranded RNA (20 μg/ml) which reduces but does not completely abrogate the delayed inhibition. These authors suggest that ''other factors'' than double-stranded RNA probably also contribute to delayed shut-off in the coupled system. We have observed a selective inhibition brought about by *in vitro* vaccinia virus transcripts in cell-free protein-synthesizing systems which will be discussed in greater detail below (Coppola and Bablanian, 1983). In order to determine if this inhibition is brought about by double-stranded RNA which may be present in the *in vitro* transcripts, we tested the effect of these transcripts on the translation of HeLa cell RNA and RNA obtained from infected cycloheximide-treated HeLa cells in the wheat germ protein-synthesizing system. It is well known that the wheat germ cell-free system is insensitive to double-stranded RNA (Grill *et al.*, 1976; Reijnders *et al.*, 1975; Baglioni *et al.*, 1978). Our results indicate that the discriminatory inhibition observed in the reticulocyte cell-free system was also evident in the wheat germ system indicating that double-stranded RNA is not the cause of this selective inhibition. From this discussion, it may be concluded that double-stranded viral RNA does not play a primary role in shut-off and is not selective. However, the possibility exists that double-stranded RNA may contribute to the overall process of inhibition of protein synthesis.

2.4.3b. Single-Stranded RNA and Shut-Off

The notion that vaccinia virus-induced RNA may be involved in shut-off was independently proposed by Rosemond-Hornbeak and Moss (1975) and Bablanian (1975). Rosemond-Hornbeak and Moss (1975) infected HeLa cell suspension cultures in the presence of 5–10 μg/ml of actinomycin D and showed that short poly(A)-rich RNAs were transcribed which sedimented a 5 S. Normal vaccinia virus transcripts sediment at 10–14 S (Becker and Joklik, 1964; Salzman *et al.*,

1964). In the presence of 5 μg/ml of actinomycin D, infected host polyribosomes were degraded and replaced by small polyribosomes inactive in protein synthesis. In view of these findings, these authors suggested that the small poly(A)-rich species of RNA transcribed in the presence of actinomycin D may compete with host mRNAs in translation and thus contribute to shut-off (Rosemond-Hornbeak and Moss, 1975). Work from our laboratory, on the other hand, established a correlation between the rate of synthesis of viral RNA in infected cycloheximide-treated cells and the degree of inhibition observed in various cell types (Bablanian, 1975; Bablanian *et al.*, 1978*b*; Schrom and Bablanian, 1979*a*). We measured the rate of RNA synthesis in the presence of cycloheximide in three cell types. In L cells and HeLa cell monolayers infected with 300 particles/cell, the rate of RNA synthesis increased with time, whereas, in LLC-MK$_2$ cells (a monkey kidney cell line), it remained constant for at least 4 hr. When higher multiplicities (900 and 2700 particles/cell) were used to infect LLC-MK$_2$ cells, the rate of RNA synthesis in the presence of cycloheximide did increase with time and was greater at the higher multiplicity. A direct relationship between the amount of RNA synthesized and the extent of shut-off was established. Inhibition of host cell protein synthesis in infected cells was strikingly prevented in HeLa cells in the presence of actinomycin D and cordycepin. This prevention of shut-off in HeLa cell monolayers was directly dependent on the concentration of actinomycin D. At an actinomycin D concentration of 20 μg/ml, virtually no shut-off was observed in infected cells (Bablanian *et al.*, 1978*b*). Furthermore, if actinomycin D (20 μg/ml) was added to cycloheximide-treated infected HeLa cells at various times after infection, the amount of inhibition of protein synthesis seen upon the removal of cycloheximide remained at the level seen before the addition of actinomycin D (Fig. 4) (Schrom and Bablanian, 1979*a*). This result indicated that the putative RNA inhibitor acts stoichiometrically rather than catalytically and, in addition, made it less likely that a preformed virion component was primarily responsible for shut-off. Previous results which have suggested the involvement of a virus component in shut-off have based their interpretations on data obtained by the use of heat and UV-inactivated virus and inhibitors of viral transcription (Shatkin, 1963; Moss, 1968; Esteban and Metz, 1973; Nevins and Joklik, 1975; Person and Beaud, 1978). But the conditions to restrict transcription in the above studies may not have been stringent enough to completely inhibit RNA synthesis.

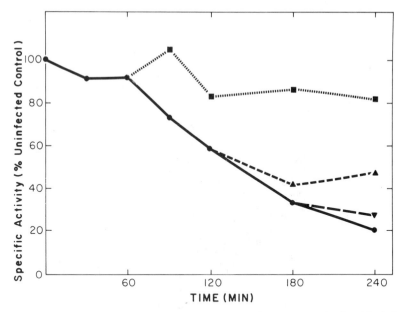

Fig. 4. Relationship between amount of cytoplasmic RNA synthesized in infected cells and inhibition of protein synthesis. HeLa cells were infected (900 particles/cell) in the presence of 300 μg/ml cycloheximide. Replicate cultures were treated with 20 μg/ml actinomycin D at 60, 120, or 180 min after infection. At the indicated times after infection, the cells were washed free of cycloheximide, labeled with [^{35}S]methionine for 30 min and specific activity (counts per min/mg of protein) was determined. Results are expressed as percentage of uninfected control. No actinomycin D, ●——●; actinomycin D added at 60 min, ■——■; 120 min, ▲——▲; 180 min, ▼——▼. From Schrom and Bablanian (1979a).

In order to analyze further the role of preformed viral components and the role of viral RNA in shut-off, we used vaccinia virus irradiated with increasing doses of ultraviolet light. This procedure reduces viral transcription stepwise (Moss, 1968; Bossart *et al.*, 1978; Pelham, 1977; Gershowitz and Moss, 1979). Such virus preparations were used to infect cells and the sequence of events leading to shut-off was studied (Bablanian *et al.*, 1981a). Several of the morphological and biochemical properties of these UV-irradiated virus preparations were examined. It was shown that vaccinia virus irradiated with increasing doses of UV synthesized progressively less RNA. Protein synthesis in cells infected with UV-irradiated virus was inhibited when virus irradiated at levels of 1920, 5760, and 9600 ergs/mm^2 were used, however, virus irradiated at 28,800 ergs/mm^2 no longer caused shut-off (Fig. 5). These results may indicate that shut-off either did

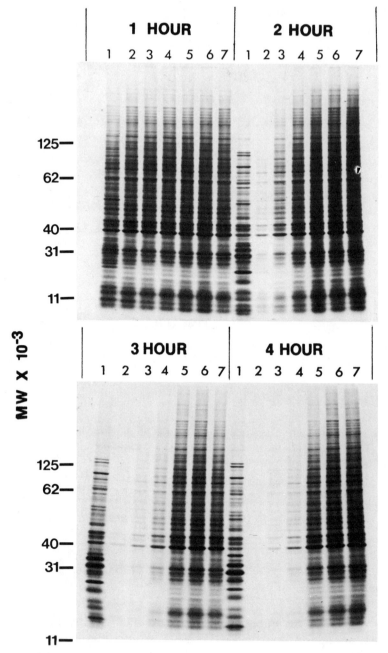

Fig. 5. Polyacrylamide–gel electrophoresis of proteins of HeLa cells infected with UV-irradiated vaccinia virus. HeLa cell monolayers were infected with 900 particles/ cell of UV-irradiated vaccinia virus. At hourly intervals, the samples were washed with methionine-free medium and pulse labeled for 30 min with 10 μCi/ml of [^{35}S]methionine. The radio labeled proteins were electrophoresed through a gradient (5–15%) polyacrylamide gel using the Laemmli buffer system. Lane assignments for the irradiation levels (ergs/mm^2) used are: lane 1, 0; lane 2, 1920; lane 3, 5760; lane 4, 9600; lane 5, 17,280; lane 6, 28,800; lane 7, cell control. From Bablanian et al. (1981a).

occur because transcription was completely abolished with the highest dose used or because some viral structural component, essential for inhibition, was damaged. Cells infected with UV-irradiated virus ranging from 1920–28,800 ergs/mm^2 had no apparent effect on the morphology of the virus as determined by electron microscopy, on the activity of several virus-associated enzymes, and on virus adsorption. Penetration and uncoating of UV-irradiated virus was not affected up to 9600 ergs/mm^2, but high doses of UV irradiation (28,800 ergs/mm^2) reduced the processing of virus to cores during the course of infection. This last result does not rule out that a virion component is not involved in shut-off because it can be argued that the exposure of viral component(s), unmasked when the virus is transformed to cores, is necessary for shut-off to occur (Bablanian et al., 1981a). Additional data, however, indicates that, in the presence of high concentrations of ACD where the virus is transformed to cores and viral RNA synthesis is not allowed to take place, the inhibition of host protein synthesis in infected cells is prevented (Fig. 6; Bablanian et al., 1981b).

The role of viral RNA in shut-off was further investigated by determining the size of virus-induced RNA in infected cells which can still cause inhibition. When cells were infected with virus which had been UV irradiated with a dose of 9600 ergs/mm^2, shut-off was delayed but still was marked. Under these conditions, RNAs 50–100 nucleotides long were found in the cytoplasm in these infected cells. These RNAs hybridized to EcoRI restriction fragments of vaccinia DNA indicating that they were of viral origin. Protein synthesis was not inhibited in cells infected with the same dose (9600 ergs/mm^2) of UV-irradiated virus and which were also treated with high doses of actinomycin D (20 and 30 µg/ml). When lower doses of actinomycin D were used (2, 5, and 10 µg/ml), however, there was only partial prevention of this inhibition. This strongly suggested that viral RNA, even as small as 50–100 nucleotides, was associated with shut-off (Bablanian et al., 1981b) As mentioned earlier, Rosemond-Hornbeak and Moss (1975) have shown that small RNA molecules synthesized in the presence of actinomycin D may contribute to shut-off. Similarly, Gershowitz and Moss (1979) using an in vitro system have suggested that such molecules may also be present in cells infected with UV-irradiated virus. This suggestion is in agreement with our in vivo findings (Bablanian et al., 1981b). Paoletti et al. (1980) have sought the significance of small RNA molecules synthesized in vitro by vaccinia virus when one of the four nucleotides is omitted from the sys-

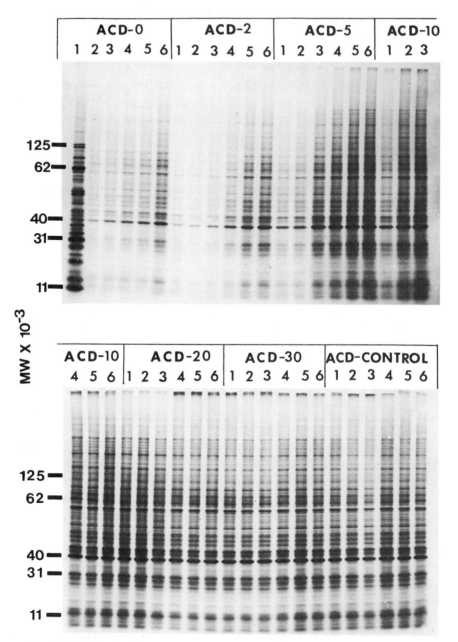

Fig. 6.   Polyacrylamide–gel electrophoresis of proteins of HeLa cells infected with UV-irradiated vaccinia virus in the presence of actinomycin D (ACD). HeLa cells monolayers were infected with 900 particles/cell of UV-irradiated vaccinia virus in the presence of various concentrations of ACD. At 4 hr postinfection, the monolayers were washed with methionine-free medium and labeled with 10 μCi/ml

tem. These small RNAs which are capped and methylated at the 5' terminus and polyadenylated at the 3' end (polyadenylated leaders) were termed PALS for short. These PALS inhibit the translation of vaccinia virus mRNA *in vitro* and they themselves do not translate (Paoletti *et al.*, 1980). These authors speculate that small RNA molecules may, in some unknown manner, be involved in shut-off.

2.4.4. Role of Vaccinia Virus *in Vitro* Transcripts in Selective Inhibition

On the basis of our previous results discussed above, early viral transcripts were primary candidates in the inhibition of host protein synthesis. Since *in vitro* transcription produces only the early species of viral transcripts (Kates and Beeson, 1970), we acquired these early species of viral RNA by means of *in vitro* transcription by viral cores and tested their effect on the translation of various exogenous mRNAs in *in vitro* cell-free systems rendered messenger dependent. The results from such experiments (Coppola and Bablanian, 1983) revealed that transcripts prepared *in vitro* of either 8–10 S or 4–7 S size classes inhibit globin, HeLa, and hamster cell mRNA translation in a reticulocyte cell-free protein-synthesizing system. Inhibition was observed not only under conditions where *in vitro* viral transcripts by themselves were capable of producing viral polypeptides, but also when low concentrations of *in vitro*-synthesized transcripts were used which were incapable of synthesizing polypeptides as determined by polyacrylamide–gel electrophoresis. In contrast, the transcripts synthesized *in vitro* by vaccinia virus cores had no inhibitory effect on the translation of cytoplasmic RNA obtained from vaccinia virus-infected cells at early times after infection. Therefore, this inhibitory effect, like that seen in vaccinia virus-infected cells, was selective. The vaccinia virus transcripts, synthesized *in vitro*, also inhibited encephalomyocarditis virus mRNA translation in the cell-free reticulocyte lysate system indicating that this inhibition does not require

[^{35}S]methionine. The radiolabeled proteins were electrophoresed through a gradient (5–15%) polyacrylamide–gel using the Laemmli buffer system. For each concentration of ACD indicated (μg/ml), the lane assignments for the irradiation level used on the infecting virus (ergs/mm^2) are as follows: lane 1, 0; lane 2, 1920; lane 3, 3840; lane 4, 5760; lane 5, 7680; lane 6, 9600. For the lanes marked ACD-control, no virus was used, but ACD was present in the following amounts (μg/ml): lane 1, 0; lane 2, 2; lane 3, 5; lane 4, 10; lane 5, 20; lane 6, 30. From Bablanian *et al.* (1981*b*).

a cap structure. Vaccinia virus *in vitro* transcripts also inhibited HeLa cell mRNA translation in the wheat germ cell-free system. The translation of cytoplasmic RNA obtained from vaccinia virus-infected cells, however, was not inhibited, again demonstrating the selective nature of this inhibition in another cell-free system. We next added cytoplasmic RNA obtained from vaccinia virus-infected cyclohexi-mide-treated HeLa cells (a condition which allows the synthesis of only early viral mRNAs) together with globin or HeLa cell mRNAs to the reticulocyte cell-free protein-synthesizing system. The result of this experiment showed an inhibition of both globin and HeLa cell protein synthesis without an effect on the translation of the vaccinia virus-infected cytoplasmic RNA. This total inhibition of HeLa cell protein synthesis without noticeable effect on the translation of cytoplasmic vaccinia virus RNA suggested that this inhibition was not due to straightforward competition. This result showed that early RNAs obtained from vaccinia virus-infected cells have an inhibitory effect, in the cell-free system, similar to that obtained by the *in vitro* transcripts synthesized by cores, suggesting but not proving a similar mode of action.

In order to demonstrate that the inhibitory property of vaccinia virus *in vitro* transcripts is RNA, we subjected these preparations to alkaline hydrolysis, and micrococcal nuclease digestion. The resulting hydrolysate and digest no longer had an inhibitory effect on HeLa cell mRNA function indicating that the integrity of the transcripts is necessary for this inhibition (Coppola and Bablanian, 1983).

Although the vaccinia virus *in vitro* transcripts exert their discriminatory inhibition in the reticulocyte cell-free system at concentrations far below that necessary to detect their translation by poly-acrylamide–gel electrophoresis and autoradiography, the possibility exists that undetectable amounts of a polypeptide may be synthesized in the cell-free system to cause this inhibition. To test this possibility, Coppola, Bablanian, and Banerjee (manuscript in preparation) preincubated vaccinia virus transcripts in reticulocyte lysates, and after 1 hr of incubation at 32°C, the lysates were treated with a high concentration of micrococcal nuclease (5000 Ū/ml). This high concentration was necessary to digest effectively the transcripts under standard conditions of cell-free protein synthesis. When these digested transcripts along with the preincubated reticulocyte lysate were added to a reticulocyte cell-free system programmed by HeLa cell RNA, protein synthesis was not inhibited. Control preincubated reticulocyte lysates also did not inhibit protein synthesis. However, when untreated *in vitro* viral transcripts were added along with either one of the preincubated lysates, inhibition was evident. All these experiments strongly suggest that this selective inhibition by vaccinia virus *in vitro* transcripts is brought about by undegraded viral RNA(s).

2.4.5. Role of Vaccinia Virus-Associated RNA in Selective Inhibition

It has been known for a while that virus-specific RNA is found in vaccinia virus particles with no particular known role (Planterose *et al.*, 1962; Roening and Holowczak, 1974). In view of previous reports that high concentrations of vaccinia cores inhibit protein synthesis in cell-free systems (Ben-Hamida and Beaud, 1978; Cooper and Moss, 1978; Person *et al.*, 1980; Ghosh-Dastidar *et al.*, 1981), it became of interest to see if core-associated RNA (a component of the virus particle) is the cause of this inhibition. By preincubating cores for 30 min in the absence of added triphosphates and an energy source, we were able to extract core-associated RNA (Coppola and Bablanian, 1983). The RNA extracted from the cores selectively inhibited HeLa and hamster cell polypeptide synthesis in the reticulocyte cell-free system (Fig. 7). Thus, the hypothesis that a component of the virion is responsible for shut-off (Sagot and Beaud, 1979; Person *et al.*, 1980), when very high concentrations of cores ($1A_{260}$ units/ml) are introduced into the cell-free systems, may be explained by our finding that RNA extracted from virus particles is, by itself, sufficient to cause a selective inhibition of cellular protein synthesis. Does this virion-associated RNA, then, have a role in virus replication? Under normal conditions of infection, probably very little virion RNA is introduced into cells to cause an effect. Our previous results (Bablanian *et al.*, 1978a) demonstrate that shut-off does not occur even when cells are infected with 900 particles per cell in the presence of actinomycin D (30 µg/ml), an inhibitor of transcription. More recent work has shown that another drug (2'O-methyladenosine), which specifically inhibits vaccinia virus production in infected cells, also prevents shut-off (Raczynski and Condit, 1983). Thus, we suggest that in infected cells, transcription is necessary to amplify the amount of inhibitor RNA before this inhibition becomes noticeable. This result may now reconcile the two divergent interpretations to explain shut-off, namely, *de novo* synthesis of virus-directed RNA as opposed to virion component, other than virion-associated RNA.

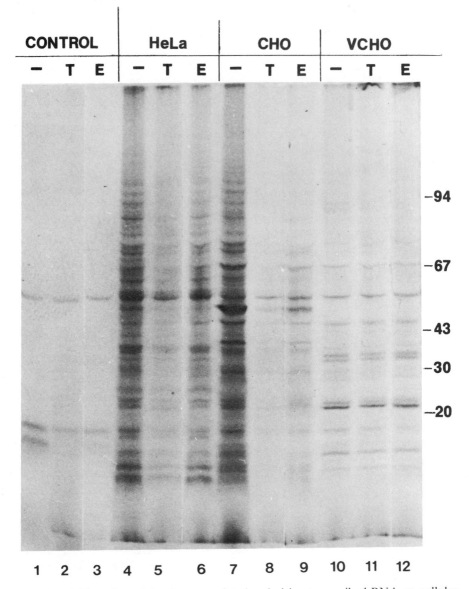

Fig. 7. Effect of vaccinia virus-associated and virion-transcribed RNA on cellular and early viral mRNA function in messenger-dependent reticulocyte lysates. Total cytoplasmic extracts of HeLa (20 μg; lane 4), Chinese hamster ovary (CHO; 10 μg; lane 7), or vaccinia virus-infected Chinese hamster ovary cells (VCHO; 4 μg; lane 10) were used to program 25 μl assays. As indicated, 1 μg of vaccinia virus RNA obtained from an *in vitro*-synthesizing reaction (T; lanes 5, 8, and 11) or extracted from purified virus (E; lanes 6, 9, and 12) was added in combination with the above amounts of RNA from cytoplasmic extracts. The control lanes (1, 2, and 3) contained no exogenous RNA, 1 μg of T, or 1 μg of E, respectively. The products of the nuclease-treated and exogenously-programmed reticulocyte lysates labeled with [35S]methionine (20 μl samples) were subjected to 7.5–15% polyacrylamide–gel electrophoresis and autoradiography. From Coppola and Bablanian (1983).

2.4.6. Mode of Action of Vaccinia Virus in Shut-Off

Attempts have been made to determine at what level of protein synthesis does vaccinia virus cause shut-off. The observation that purified vaccinia virus cores inhibit protein synthesis in the cell-free reticulocyte and wheat germ systems led Ben-Hamida and Beaud (1978) to investigate the mechanism of this inhibition. These authors demonstrated that the translation of poly(U) which does not require normal initiation, was not inhibited by vaccinia virus cores in a mouse ascites tumor cell-free system. On this basis, they postulated that the inhibition was at the level of polypeptide chain initiation. Furthermore, these authors demonstrated that vaccinia virus cores cause disagreggation of reticulocyte lysate polyribosomes, which indicates that polypeptide chain initiation is the major site of action of vaccinia virus cores in the reticulocyte lysate cell-free system. Additional evidence that this inhibition takes place at the level of initiation comes from the work of Schrom and Bablanian (1979*b*) and Person and Beaud (1980). Using separated supernatant, ribosome, and ribosomal salt wash fractions of infected and uninfected cells, Schrom and Bablanian (1979*b*) reconstituted protein-synthesizing systems programmed by encephalomyocarditis virus mRNA. The results of such experiments showed that infected cell ribosomes, supernatant, and salt wash fractions were all defective, but that the primary defect was in the puromycin KCl wash fraction of the infected ribosomes. This crude initiation factor fraction released from infected ribosomes may contain a defective or altered initiation factor(s) or an inhibitor of initiation. Person and Beaud (1980), using cytoplasmic lysates from vaccinia virus-infected cells treated with cordycepin, showed that binding of initiator tRNA to 40 S ribosomal subunits was diminished and the 40 S-Met-tRNA-Met complex formation was inhibited by purified vaccinia virus cores in the reticulocyte lysate cell-free system. Ben-Hamida *et al.* (1983) have recently isolated from vaccinia virus cores a partially purified fraction which inhibited protein synthesis in the reticulocyte system. This fraction also inhibited Met-tRNA-Met binding to 40 S ribosomes from reticulocyte lysates, was composed of basic phosphoproteins with a molecular weight between 9–15 K, and was resistant to micrococcal nuclease.

If this partially purified fraction does not selectively inhibit protein synthesis as was shown to be the case with the purified cores (Person and Beaud, 1980), then its function in vaccinia virus-induced shut-off is not clear. Its mode of action is certainly different from that of the vaccinia virus *in vitro* transcripts which selectively inhibit protein synthesis in the reticulocyte and wheat germ cell-free protein-synthesizing systems, mirroring the inhibition seen in vaccinia virus-infected cells (Coppola and Bablanian, 1983). Further, characterization of the inhibitory nature of both the *in vitro* vaccinia virus transcripts (Coppola and Bablanian, 1983) and the solubilized fraction isolated from virus cores (Ben-Hamida *et al.*, 1983) is needed to establish similarities and differences between these two inhibitory products.

An earlier work has suggested that discrimination in vaccinia virus-infected cells may be due to a complete cessation of protein synthesis which then may be reversed by an early viral protein (Moss and Filler, 1970). Such a discriminatory product has not yet been identified.

3. CONCLUDING REMARKS

It does not take long for the reader of this volume to realize the multitude and the complexity of the strategies employed by cytocidal viruses in causing inhibition of host macromolecular synthesis with the concomitant production of progeny virus. This phenomenon has been recognized for over 20 years, but it is during the past decade that some of the molecular mechanisms of these inhibitions have been clarified. Although many virus models are being used profitably to study the mechanisms of host shut-off as indicated in this volume, I shall conclude by comparing and contrasting the picornavirus and poxvirus models. In the case of the picornaviruses, represented by poliovirus, the putative viral protein (Steiner-Pryor and Cooper, 1973) responsible for shut-off has not been identified; however, its mechanism of action has been localized to the initiation factors of protein synthesis. Many of the steps leading to this inhibition have been elucidated (for reviews, see Shatkin, 1983; Ehrenfeld, this volume). In comparison, very little work has been done on the mode of action of poxviruses, represented by vaccinia virus. There are indications that this inhibition, like the one caused by polioviruses, also occurs at the level of initiation (Ben-Hamida and Beaud, 1978; Schrom and Bablanian, 1979*b*; Person and Beaud, 1980). Contrary to the situation found with polioviruses, however, where the product responsible for shut-off has not been determined in the case of vaccinia virus, two viral products have been identified as the initiators of shut-off: early vaccinia virus transcripts (Coppola and Bablanian, 1983) or partially purified basic phosphoproteins isolated from viral cores (Ben-Hamida *et al.*, 1983). Future experiments will determine if either one or both of these viral products is involved in the selective inhibition caused by vaccinia virus *in vivo*.

ACKNOWLEDGMENTS

I wish to thank Drs. Mariano Esteban and John Lewis for their valuable comments and critical reading of the manuscript. I also wish to thank Mr. Santo Scribani for his help in the preparation of the manuscript. Studies from our laboratory were supported in part by Grant PCM8022604 from the National Science Foundation.

4. REFERENCES

Alonso, M. A., and Carrasco, L., 1981, Reversion by hypotonic medium of the shut-off of protein synthesis induced by encephalomyocarditis virus, *J. Virol.* **37**:535.

Appleyard, G., Westwood, J. C. N., and Zwartouw, H. T., 1962, The toxic effect of rabbit poxvirus in tissue culture, *Virology* **18**:159.

Armstrong, J. A., Metz, D. H., and Young, M. R., 1973, The mode of entry of vaccinia virus into L cells, *J. Gen. Virol.* **21**:533.

Bablanian, R., 1968, The prevention of early vaccinia-virus-induced cytopathic effects by inhibition of protein synthesis, *J. Gen. Virol.* **3**:51.

Bablanian, R., 1970, Studies on the mechanism of vaccinia virus cytopathic effects: Effect of inhibitors of RNA and protein synthesis on early virus-induced cell damage, *J. Gen. Virol.* **6**:221.

Bablanian, R., 1975, Structural and functional alterations in cultured cells infected with cytocidal viruses, *Prog. Med. Virol.* **19**:40.

Bablanian, R., Esteban, M., Baxt, B., and Sonnabend, J. A., 1978a, Studies on the mechanisms of vaccinia virus cytopathic effects. I. Inhibition of protein synthesis in infected cells is associated with virus-induced RNA synthesis, *J. Gen. Virol.* **39**:391.

Bablanian, R., Baxt, B., Sonnabend, J. A., and Esteban, M., 1978b, Studies on the mechanisms of vaccinia virus cytopathic effects. II. Early cell rounding is associated with virus polypeptide synthesis, *J. Gen. Virol.* **39**:403.

Bablanian, R., Coppola, G., Scribani, S., and Esteban, M., 1981a, Inhibition of protein synthesis by vaccinia virus. III. The effect of UV-irradiated virus on inhibition of protein synthesis, *Virology* **112**:1.

Bablanian, R., Coppola, G., Scribani, S., and Esteban, M., 1981b, Inhibition of protein synthesis by vaccinia virus. IV. The role of low molecular weight virus RNA in the inhibition of protein synthesis, *Virology* **112**:13.

Baglioni, C., Lenz, J. R., Maroney, P. A., and Weber, L. A., 1978, Effect of double-stranded RNA associated with viral messenger RNA on *in vitro* protein synthesis, *Biochemistry* **17**:16.

Banerjee, A. K., 1980, 5'-Terminal cap structure in eucaryotic messenger ribonucleic acids, *Microbiol. Rev.* **44**:175.

Bauer, W. R., Ressner, E. C., Kates, J., and Patzke, J. V., 1977, A DNA nickling-closing enzyme encapsidated in vaccinia virus. Partial purification and properties, *Proc. Natl. Acad. Sci. USA* **74**:1841.

Baxt, B., and Bablanian, R., 1976, Mechanism of vesicular stomatitus virus-induced cytopathic effects. II. Inhibition of macromolecular synthesis by infectious and defective-interfering particles, *Virology* **72**:383.

Becker, Y., and Joklik, W. K., 1964, Messenger RNA in cells infected with vaccinia virus, *Proc. Natl. Acad. Sci. USA* **51**:577.

Benavente, J., and Esteban, M., Differential cell-free translation properties of early and late classes of RNA from untreated and interferon-treated vaccinia virus infected cells, submitted for publication.

Ben-Hamida, F., and Beaud, G., 1978, *In vitro* inhibition of protein synthesis by purified cores from vaccinia virus, *Proc. Natl. Acad. Sci. USA* **75**:175.

Ben-Hamida, F., Person, A., and Beaud, G., 1983, Solubilization of a protein synthesis inhibitor from vaccinia virions, *J. Virol.* **45**:452.

Berns, K. J., and Silverman, C., 1970, Natural occurrence of cross-linked vaccinia virus deoxyribonucleic acid, *J. Virol.* **5**:299.

Bolden, A., Noy, G. P., and Weissbach, A., 1979, Vaccinia virus infection of HeLa cells, II, Disparity between cytoplasmic and nuclear viral specific RNA, *Virology* **94**:138.

Bossart, W., Nuss, D. L., and Paoletti, E., 1978, Effect of UV-irradiation on the expression of vaccinia virus gene products synthesized in a cell-free system coupling transcription and translation, *J. Virol.* **26**:673.

Bratt, M. A., and Gallaher, W. R., 1969, Preliminary analysis of the requirements from within and fusion from without by Newcastle disease virus, *Proc. Natl. Acad. Sci. USA* **64**:536.

Brown, A., Mayyasi, S. A., and Officer, J. E., 1959, The toxic activity of vaccinia virus in tissue culture, *J. Infect. Dis.* **104**:193.

Burgoyne, R. D., and Stephen, J., 1979, Further studies on a vaccinia virus cytotoxin present in infected cell extracts. Identification as surface tubule monomer and possible mode of action, *Arch. Virol.* **59**:107.

Cabrera, C. V., and Esteban, M., 1978, A procedure for the purification of intact DNA from vaccinia virus, *J. Virol.* **25**:442.

Cairns, J., 1960, The initiation of vaccinia infection, *Virology* **11**:603.

Carrasco, L., 1977, The inhibition of cell functions after viral infection, a proposed general mechanism, *FEBS Lett.* **76**:11.

Carrasco, L., and Esteban, M., 1982, Modification of membrane permeability in vaccinia virus-infected cells, *Virology* **117**:62.

Carrasco, L., and Lacal, J. C., 1984, Permeabilization of cells during animal virus infection. *Pharmacol. Ther.*, in press.

Carrasco, L., Harvey, R., Blanchard, C., and Smith, A. E., 1979, Regulation of translation of eukaryotic virus mRNAs, in: *Modern Trends in Human Leukemia III*, (R. Neth, P. Hofsheneider, and K. Mannwiler, eds.), p. 189, Springer-Verlag, New York.

Celma, M. L., and Ehrenfeld, E., 1974, Effect of poliovirus double-stranded RNA on viral and host-cell protein synthesis, *Proc. Natl. Acad. Sci. USA* **71**:2440.

Centrella, M., and Lucas-Lenard, J., 1982, Regulation of protein synthesis in vesicular stomatitis virus-infected mouse L-929 cells by decreased protein synthesis initiation factor 2 activity, *J. Virol.* **41**:781.

Colby, C., and Duesberg, P. H., 1969, Double-stranded RNA in vaccinia virus infected cells, *Nature* **222**:940.

Colby, C., Jurale, C., and Kates, J. I., 1971, Mechanism of synthesis of vaccinia virus double-stranded RNA *in vivo* and *in vitro*, *J. Virol.* **7**:71.

Colby, D. S., Finnerty, V., and Lucas-Lenard, J., 1974, Fate of mRNA of L-cells infected with mengovirus, *J. Virol.* **13**:858.

Cooper, J. A., and Moss, B., 1978, Transcription of vaccinia virus mRNA coupled to translation *in vitro*, *Virology* **88**:149.

Coppola, G., and Bablanian, R., 1983, Discriminatory inhibition of protein synthesis in cell-free systems by vaccinia virus transcripts, *Proc. Natl. Acad. Sci. USA* **80**:75.

Dales, S., 1965, Effects of streptovitacin A on the initial events in the replication of vaccinia and reovirus, *Proc. Natl. Acad. Sci. USA* **54**:462.

Dales, S., Stern, W., Weintraub, S. B., and Huima, T., 1976, Genetically controlled surface modifications by poxviruses influencing cell–cell and cell–virus interactions, in: *Cell Membrane Receptors for Viruses, Antigens and Antibodies, Polypeptide Hormones, and Small Molecules*, Chap. 19, Raven Press, New York.

Duesberg, P. H., and Colby, C., 1969, On the biosynthesis and structure of double-stranded RNA in vaccinia virus-infected cells, *Proc. Natl. Acad. Sci. USA* **64**:396.

Ehrenfeld, E., and Hunt, T., 1971, Double-stranded poliovirus RNA inhibits initiation of protein synthesis by reticulocyte lysates, *Proc. Natl. Acad. Sci. USA* **68**:1075.

Ehrenfeld, E., and Lund, H., 1977, Untranslated vesicular stomatitis virus mRNA after poliovirus infection, *Virology* **80**:297.

Enders, J. F., 1954, Cytopathology of virus infections, *Annu. Rev. Microbiol.* **8**:473.

Ensinger, M. J., Martin, S. A., Paoletti, E., and Moss, B., 1975, Modification of the 5'-terminus of mRNA by soluble guanylyl and methyl transferases from vaccinia virus, *Proc. Natl. Acad. Sci. USA* **72**:3385.

Essani, K., and Dales, S., 1979, Biogenesis of vaccinia: Evidence for more than 100 polypeptides in the virion, *Virology* **95**:385.

Esteban, M., and Holowczak, J., 1977, Replication of vaccinia DNA in mouse L cells, I, *In vivo* DNA synthesis, *Virology* **78**:57.

Esteban, M., and Metz, K. H., 1973, Early virus protein synthesis in vaccinia virus-infected cells, *J. Gen. Virol.* **19**:201.

Esteban, M., Flores, L., and Holowczak, J., 1977, Model for vaccinia virus DNA replication, *Virology* **83**:467.

Fenwick, M. L., and Walker, M. J., 1978, Suppression of the synthesis of cellular macromolecules by herpes simplex virus, *J. Gen. Virol.* **41**:37.

Fernandez-Munoz, R., and Darnell, J. E., 1976, Structural difference between the 5' termini of viral and cellular mRNA in poliovirus-infected cells. Possible basis for the inhibition of host protein synthesis, *J. Virol.* **126**:719.

Garon, C. F., and Moss, B., 1971, Glycoprotein synthesis in cells infected with vaccinia virus, II, a glycoprotein component of the virion, *Virology* **46**:233.

Gershowitz, A., and Moss, B., 1979, Abortive transcription products of vaccinia virus are guanylylated, methylated, and polyadenylated, *J. Virol.* **31**:849.

Geshelin, P., and Berns, K. I., 1974, Characterization and localization of the naturally occurring cross-links in vaccinia virus DNA, *J. Mol. Biol.* **88**:785.

Ghosh-Dastidar, P., Goswami, B. B., Das, A., Das, P., and Gupta, N. K., 1981, Vaccinia viral core inhibits Met-tRNA$_f$. 40S initiation complex formation with physiological mRNAs, *Biochem. Biophys. Res. Commun.* **99**:946.

Gillies, S., and Stollar, V., 1982, Protein synthesis in lysates of Aedes albopictus cells infected with vesicular stomatitis virus, *Mol. Cell. Biol.* **2**:1174.

Grill, L. J., Sun, J. D., and Kandel, J., 1976, Effect of double-stranded RNA on protein synthesis in an *in vitro* wheat germ embryo system, *Biochem. Biophys. Res. Commun.* **73**:149.

Grinnell, B. W., and Wagner, R. R., 1983, Comparative inhibition of cellular transcription by vesicular stomatitis virus serotypes New Jersey and Indiana, role of each viral leader RNA, *J. Virol.* **48**:88.

Hanafusa, H., 1960*a*, Killing of L-cells by heat and UV-inactivated vaccinia virus, *Biken J.* **3**:191.

Hanafusa, T., 1960*b*, Alteration of nucleic acid metabolism by active and inactivated forms of vaccinia virus, *Biken J.* **3**:313.

Hansen, J. L., Etchison, D. O., Hershey, J. W. B., and Ehrenfeld, E., 1982, Localization of cap-binding protein in subcellular fractions of HeLa cells, *Mol. Cell. Biol.* **2**:1639.

Helenius, A., and Simons, K., 1975, Solubilization of membranes by detergents, *Biochim. Biophys. Acta* **415**:29.

Helentjaris, T. E., Ehrenfeld, E., Brown-Luedi, M. L., and Hershey, J. W. B., 1979, Alterations in initiation factor activity from poliovirus infected HeLa cells, *J. Biol. Chem.* **254**:10973.

Hewlett, M. J., Rose, J. K., and Baltimore, D., 1976, 5′-Terminal structure of poliovirus polyribosomal RNA is pUp, *Proc. Natl. Acad. Sci. USA* **73**:327.

Hill, T. M., Sinden, R. R., and Sadler, J. R., 1983, Herpes simplex virus types 1 and 2 induce shut-off of host protein synthesis by different mechanisms in Friend erythroleukemia cells, *J. Virol.* **45**:241.

Holloway, P. W., 1973, A simple procedure for removal of Triton X-100 from protein samples, *Anal. Biochem.* **53**:304.

Holowczak, J. A., 1982, Poxvirus DNA, in: *Current Topics in Microbiology and Immunology*, Springer-Verlag, Berlin, Heidelberg, New York, **97**:27.

Holowczak, J. A., and Joklik, W. K., 1967, Studies on the properties of vaccinia virus. I. Structural proteins of virions and cores, *Virology* **3**:717.

Hruby, D. E., Guarino, L. A., and Kates, J. R., 1979, Vaccinia virus replication. I. Requirement for the host-cell nucleus, *J. Virol.* **29**:705.

Hunter, T., Hunt, T., Jackson, R. J., and Robertson, H. D., 1975, The characteristics of inhibition of protein synthesis by double-stranded RNA in reticulocyte lysates, *J. Biol. Chem.* **250**:109.

Ichihashi, Y., and Dales, S., 1971, Biogenesis of poxviruses. Interrelationships between hemagglutinin production and polykaryocytosis, *Virology* **46**:533.

Inglis, S. C., 1982, Inhibition of host protein synthesis and degradation of cellular mRNAs during infection by influenza and Herpes simplex virus, *Mol. Cell. Biol.* **2**:1644.

Jaye, M. C., Godchaux, W., III, and Lucas-Lenard, J., 1982, Further studies on the inhibition of cellular protein synthesis by vesicular stomatitis virus, *Virology* **116**:148.

Jefferts, E. R., and Holowczak, J. A., 1971, RNA synthesis in vaccinia-infected L cells: Inhibition of ribosomes formation and maturation, *Virology* **46**:730.

Jen, G., and Thach, R. E., 1982, Inhibition of host translation in encephalomyocarditis virus-infected L cells, a novel mechanism, *J. Virol.* **43**:250.

Jen, G., Detjen, B. M., and Thach, R. E., 1980, Shut-off of HeLa cell protein synthesis by encephalomyocarditis virus and poliovirus, a comparative study, *J. Virol.* **35**:150.

Joklik, W. K., 1964, The intracellular uncoating of poxvirus DNA. II. The molecular basis of the uncoating process, *J. Mol. Biol.* **8**:277.

Joklik, W. K., and Becker, Y., 1964, The replication and uncoating of vaccinia DNA, *J. Mol. Biol.* **10**:452.

Joklik, W. K., and Merigan, T. C., 1966, Concerning the mechanism of action of interferon, *Proc. Natl. Acad. Sci. USA* **56**:558.

Jungwirth, C., and Launer, J., 1968, Effect of poxvirus infection on host cell DNA synthesis, *J. Virol.* **2**:401.

Kaempfer, R., and Kaufman, J., 1973, Inhibition of cellular protein synthesis by double-stranded RNA. Inactivation of an initiating factor, *Proc. Natl. Acad. Sci. USA* **70**:1222.

Kaerlein, M., and Horak, I., 1976, Phosphorylation of ribosomal proteins in HeLa cells infected with vaccinia virus, *Nature (London)* **259**:250.

Kates, J., and Beeson, J., 1970, RNA synthesis in vaccinia virus. I. The mechanism of synthesis and release of RNA in vaccinia cores, *J. Mol. Biol.* **50**:1.

Kates, J., and McAuslan, B. R., 1967a, Messenger RNA synthesis by a coated viral genome, *Proc. Natl. Acad. Sci. USA* **57**:314.

Kates, J., and McAuslan, B. R., 1967b, Poxvirus DNA-dependent RNA-polymerase, *Proc. Natl. Acad. Sci. USA* **58**:134.

Kaverin, N. V., Varich, N. L., Surgay, V. V., and Chernos, V. I., 1975, A quantitative estimation of poxvirus genome fraction transcribed as "early" and "late" mRNA, *Virology* **65**:112.

Kit, S., and Dubbs, D. R., 1962, Biochemistry of vaccinia-infected mouse fibroblasts (strain L-M). Effects on nucleic acid and protein synthesis, *Virology* **18**:274.

Lacal, J. C., and Carrasco, L., 1982, Relationship between membrane integrity and the inhibition of host translation in virus-infected mammalian cells. Comparative studies between encephalomyocarditis virus and poliovirus, *Eur. J. Biochem.* **127**:359.

La Colla, P., and Weisbach, H., 1975, Vaccinia virus infection of HeLa cells. I. Synthesis of vaccinia DNA in host cell nuclei, *J. Virol.* **15**:305.

Legon, S., Brayley, A., Hunt, T., and Jackson, R. J., 1974, The effect of cyclic AMP and related compounds on the control of protein synthesis in reticulocyte lysates, *Biochem. Biophys. Res. Commun.* **56**:745.

Lodish, H. F., and Porter, M., 1980, Translational control of protein synthesis after infection by vesicular stomatitis virus, *J. Virol.* **36**:719.

Lodish, H. F., and Porter, M., 1981, Vesicular stomatitis virus mRNA and inhibition of translation of cellular mRNA. Is there a P function in vesicular stomatitis virus? *J. Virol.* **38**:504.

Loh, P. C., and Payne, F. E., 1965, Effect of *p*-fluorophenylalanine on the synthesis of vaccinia virus, *Virology* **25**:560.

Maes, R., and Granoff, A., 1967, Viruses and renal carcinoma of Rana pipiens. IV. Nucleic acid synthesis in frog virus 3 infected BHK $21/13$ cells, *Virology* **33**:491.

Mbuy, G. N., Morris, R. E., and Bubel, H. C., 1982, Inhibition of cellular protein synthesis by vaccinia virus surface tubules, *Virology* **116**:137.

McCarron, R. J., Cabrera, C. V., Esteban, M., McAllister, W. T., and Holowczak, J. A., 1978, Structure of vaccinia DNA: Analysis of the viral genome by restriction endonucleases, *Virology* **86**:88.

McFadden, G., Pace, W. E., Purres, J., and Dales, S., 1979, Biogenesis of poxviruses. Transitory expression of Molluscum contagiosum early functions, *Virology* **94**:297.

Metz, D. H., and Esteban, M., 1972, Interferon inhibits viral protein synthesis in L cells infected with vaccinia virus, *Nature (London)* **238**:385.

Metz, H., Esteban, M., and Danielescu, G., 1975, The formation of polyribosomes in L cells infected with vaccinia virus, *J. Gen. Virol.* **27**:181.

Moss, B., 1968, Inhibition of HeLa cell protein synthesis by the vaccinia virion, *J. Virol.* **2**:1028.

Moss, B., 1974, Reproduction of poxviruses, in: *Comprehensive Virology*, Vol. 3 (H. Fraenkel-Conrat and R. R. Wagner, eds.), pp. 405–474, Plenum Press, New York.

Moss, B., 1978, Poxviruses, in: *Molecular Biology of Animal Viruses*, Vol. 2 (C. D. Nayak, ed.), pp. 849–890, Marcel Dekker, New York.

Moss, B., and Filler, R., 1970, Irreversible effects of cycloheximide during the early period of vaccinia virus replication, *J. Virol.* **5**:99.

Moss, B., and Salzman, N. P., 1968, Sequential protein synthesis following vaccinia virus infection, *J. Virol.* **2**:1016.

Moss, B., Gershowitz, A., Wei, C., and Boone, R., 1976, Formation of the guanylylated and methylated 5′-terminus of vaccinia virus mRNA, *Virology* **72**:341.

Moss, B., Rosenblum, E. N., and Garon, C. F., 1973, Glycoprotein synthesis in cells infected with vaccinia virus. III. Purification and biosynthesis of the virion glycoprotein, *Virology* **55**:143.

Munyon, W. H., and Kit, S., 1966, Induction of cytoplasmic RNA synthesis in vaccinia-infected LM cells during inhibition of protein synthesis, *Virology* **29**:303.

Munyon, W., Paoletti, E., and Grace, J. T., 1967, RNA polymerase activity in purified infectious vaccinia virus, *Proc. Natl. Acad. Sci. USA* **58**:2280.

Munyon, W., Paoletti, E., Ospina, J., and Grace, J. T., 1968, Nucleotide phosphohydrolase in purified vaccinia virus, *J. Virol.* **2**:167.

Nevins, J. R., and Joklik, W. K., 1975, Poly(A) sequences of vaccinia virus messenger RNA: Nature, mode of addition and function during translation *in vitro* and *in vivo*, *Virology* **63**:1.

Nishioka, Y., and Silverstein, S., 1978, Requirement of protein synthesis for the degradation of host mRNA in Friend erythroleukemia cells infected with Herpes simplex virus type 1, *J. Virol.* **27**:619.

Nomoto, A., Lee, Y. F., and Wimmer, E., 1976, The 5′ end of poliovirus mRNA is not capped with m^7G-(5′)ppp(5′)Np, *Proc. Natl. Acad. Sci. USA* **73**:375.

Norrie, D. H., Wolstenholme, J., Howcroft, H., and Stephen, J., 1982, Vaccinia virus-induced changes in (Na^+) and (K^+) in HeLa cells, *J. Gen. Virol.* **62**:127.

Nuss, D. L., Oppermann, H., and Koch, G., 1975, Selective blockage of initiation of host protein synthesis in RNA-virus-infected cells, *Proc. Natl. Acad. Sci. USA* **72**:1258.

Oda, K., and Joklik, W. K., 1967, Hybridization and sedimentation studies on "early" and "late" vaccinia messenger RNA, *J. Mol. Biol.* **27**:395.

Oki, T., Jujiwara, Y., and Heidelberger, C., 1971, Utilization of host-cell DNA by vaccinia virus replicating in HeLa cells irradiated nuclearly with tritium, *J. Gen. Virol.* **13**:401.

Paoletti, E., and Moss, B., 1972, Protein kinase and specific phosphate acceptor proteins associated with vaccinia virus cores, *J. Virol.* **10**:417.

Paoletti, E., Lipinskas, B. R., and Panicali, D., 1980, Capped and polyadenylated low-molecular weight RNA synthesized by vaccinia virus *in vitro*, *J. Virol.* **33**:208.

Payne, L. G., and Norrby, E., 1978, Adsorption and penetration of enveloped and naked vaccinia virus particles, *J. Virol.* **27**:19.

Pelham, H. R. B., 1977, Use of coupled transcription and translation to study mRNA production by vaccinia cores, *Nature* **269**:532.

Pelham, H. R. B., Sykes, J. M. M., and Hunt, T., 1978, Characteristics of a coupled cell-free transcription and translation system directed by vaccinia cores, *Eur. J. Biochem.* **82**:199.

Pennington, T. H., 1974, Vaccinia virus polypeptide synthesis: Sequential appearance and stability of pre- and post-replicative polypeptides, *J. Gen. Virol.* **25**:433.

Pereira, H. G., 1960, A virus inhibitor produced in HeLa cells infected with adenovirus, *Virology* **11**:590.

Person, A., and Beaud, G., 1978, Inhibition of host protein synthesis in vaccinia virus-infected cells in the presence of cordycepin (3'-deoxyadenosine), *J. Virol.* **25**:11.

Person, A., and Beaud, G., 1980, Shut-off of host protein synthesis in vaccinia virus-infected cells exposed to cordycepin, *Eur. J. Biochem.* **103**:85.

Person, A., Ben-Hamida, F., and Beaud, G., 1980, Inhibition of 40S-Met-tRNA met ribosomal initiation complex formation by vaccinia virus, *Nature (London)* **287**:355.

Planterose, D. N., Nishimura, C., and Salzman, N. P., 1962, The purification of vaccinia virus from cell cultures, *Virology* **18**:294.

Pogo, B. G. T., and Dales, S., 1969, Two deoxyribonuclease activities within purified vaccinia, *Proc. Natl. Acad. Sci. USA* **63**:820.

Pogo, B. G. T., and Dales, S., 1973, Biogenesis of poxvirus. Inactivation of host DNA polymerase by a component of the invading inoculum particle, *Proc. Natl. Acad. Sci. USA* **70**:1726.

Pogo, B. G. T., and Dales, S., 1974, Biogenesis of poxvirus. Further evidence for inhibition of host and virus DNA synthesis by a component of the invading inoculum particle, *Virology* **58**:377.

Raczynski, P., and Condit, R. C., 1983, Specific inhibition of vaccinia virus growth by 2'-*O*-methyladenosine. Isolation of a drug-resistant virus mutant, *Virology* **128**:458.

Reijnders, L., Aalbers, A. M. J., van Kammen, A., and Berns, A. J. M., 1975, The effect of double-stranded cowpea mosaic viral RNA on protein synthesis, *Biochim. Biophys. Acta.* **390**:69.

Rice, A. P., and Roberts, B. E., 1983, Vaccinia virus induces cellular mRNA degradation, *J. Virol.* **47**:529.

Roening, G., and Holowczak, J. A., 1974, Evidence for the presence of RNA in the purified virions of vaccinia virus, *J. Virol.* **14**:704.

Rose, J. K., Trachsel, H., Leong, K., and Baltimore, D., 1978, Inhibition of a specific initiation factor, *Proc. Natl. Acad. Sci. USA* **75**:2732.

Rosemond-Hornbeak, H., and Moss, B., 1975, Inhibition of host protein synthesis by vaccinia virus, fate of cell mRNA and synthesis of small poly(A)-rich polyribonucleotides in the presence of actinomycin-D, *J. Virol.* **16**:34.

Saborio, J. L., Pong, S. S., and Koch, G., 1974, Selective and reversible inhibition of initiation of protein synthesis in mammalian cells, *J. Mol. Biol.* **85**:191.

Sagot, J., and Beaud, G., 1979, Phosphorylation *in vivo* of vaccinia-virus structural protein found associated with the ribosome from infected cells, *Eur. J. Biochem.* **98**:131.

Salzman, N. P., and Sebring, E. D., 1967, Sequential formation of vaccinia virus proteins and viral DNA, *J. Virol.* **1**:16.

Salzman, N. P., Shatkin, A. J., and Sebring, E. D., 1964, The synthesis of a DNA-like RNA in the cytoplasm of HeLa cells infected with vaccinia virus, *J. Mol. Biol.* **8**:405.

Sarov, I., and Joklik, W., 1972, Characterization of intermediates in the uncoating of vaccinia virus DNA, *Virology* **50**:593.

Sarov, I., and Joklik, W., 1973, Isolation and characterization of intermediates in vaccinia virus morphogenesis, *Virology* **52**:223.

Schnitzlein, W. M., O'Banion, M. K., Poirot, M. K., and Reichman, M. B., 1983, Effect of intracellular vesicular stomatitis virus mRNA concentration on the inhibition of host cell protein synthesis, *J. Virol.* **45**:206.

Schrom, M., and Bablanian, R., 1979*a*, Inhibition of protein synthesis by vaccinia virus. II. Studies on the role of virus-induced RNA synthesis, *J. Gen. Virol.* **44**:625.

Schrom, M., and Bablanian, R., 1979*b*, Inhibition of protein synthesis by vaccinia virus. I. Characterization of an inhibited cell-free protein-synthesizing system from infected cells, *Virology* **99**:319.

Schrom, M., and Bablanian, R., 1981, Altered cellular morphology resulting from cytocidal virus infection, *Arch. Virol.* **70**:173.

Shatkin, A. J., 1963, Actinomycin D and vaccinia virus infection of Hela cells, *Nature (London)* **199**:357.

Shatkin, A. J., 1983, 1984, Molecular mechanisms of virus-mediated cytopathology, in press.

Sonenberg, N., Morgan, M. A., Merrick, W. C., and Shatkin, A. J., 1978, A polypeptide in eukaryotic initiation factors that crosslinks specifically to the 5'-terminal cap in mRNA, *Proc. Natl. Acad. Sci. USA* **75**:4843.

Sonenberg, N., Rupprecht, K. M., Hecht, S. M., and Shatkin, A. J., 1979, Eukaryotic mRNA cap binding protein purification by affinity chromatography on sepharose-coupled m^7 GDP, *Proc. Natl. Acad. Sci. USA* **76**:4345.

Stanners, C. P., Francoeur, A. M., and Lam, T., 1977, Analysis of VSV mutant with attenuated cytopathogenicity. Mutation in viral function, P, for inhibition of protein synthesis, *Cell* **11**:273.

Steiner-Pryor, A., and Cooper, P. D., 1973, Temperature-sensitive poliovirus mutants defective in repression of host protein synthesis are also defective in structural protein, *J. Gen. Virol.* **21**:215.

Stenberg, R. M., and Pizer, L. I., 1982, Herpes simplex virus-induced changes in cellular and adenovirus RNA metabolism in an adenovirus type 5-transformed human cell line, *J. Virol.* **42**:474.

Stephen, J., Birkbeck, T. H., Woodward, C. G., and Wolstenholme, J., 1974, Vaccinia virus cytotoxin(s), *Nature (London)* **250**:236.

Stern, W., and Dales, S., 1976a, Biogenesis of vaccinia. Isolation and characterization of a surface component that elicits antibody suppressing infectivity and cell–cell fusion, *Virology* **75**:232.

Stern, W., and Dales, S., 1976b, Biogenesis of vaccinia. Relationship of the envelope to virus assembly, *Virology* **75**:242.

Trachsel, H., Sonenberg, N., Shatkin, A. J., Rose, J. K., Leong, K., Bergmann, J. E., Gordon, J., and Baltimore, D., 1980, Purification of a factor that restores translation of vesicular stomatitis virus mRNA in extracts from poliovirus-infected Hela cells, *Proc. Natl. Acad. Sci. USA* **77**:770.

Walen, K. H., 1971, Nuclear involvement in poxvirus infections, *Proc. Natl. Acad. Sci. USA* **68**:165.

Weintraub, S., Stern, W., and Dales, S., 1977, Biogenesis of vaccinia. Effects of inhibitors of glycosylation on virus-mediated activities, *Virology* **78**:315.

Willems, M., and Penman, S., 1966, The mechanism of host cell protein synthesis inhibition by poliovirus, *Virology* **30**:355.

Wolstenholme, J., Woodward, C. G., Burgoyne, R. D., and Stephen, J., 1977, Vaccinia virus cytotoxin, *Arch. Virol.* **53**:25.

Woodson, B., 1967, Vaccinia mRNA synthesis under conditions which prevent uncoating, *Biochem. Biophys. Res. Commun.* **27**:169.

CHAPTER 9

Reovirus Cytopathology: Effects on Cellular Macromolecular Synthesis and the Cytoskeleton

Arlene H. Sharpe

Department of Microbiology and Molecular Genetics
Harvard Medical School
Boston, Massachusetts 02115

Department of Pathology
Brigham and Women's Hospital
Boston, Massachusetts 02115

Bernard N. Fields

Department of Microbiology and Molecular Genetics
Harvard Medical School
Boston, Massachusetts 02115

Department of Medicine
Brigham and Women's Hospital
Boston, Massachusetts 02115

1. INTRODUCTION

Despite extensive knowledge of the molecular biology of viral replication relatively little is known about how viruses alter the host cell during infection and produce cellular injury. Studies of the events following viral infection of animal cells have focused more on the synthesis of new virion components, than on viral effects on the host cell. Because viruses are obligate intracellular parasites, an understanding of viral infections requires not only a knowledge of the details of the viral replicative cycle, but also an understanding of how viruses modify and damage their host cells. An understanding of the mechanisms by which viruses modify and damage host cells is key to understanding viral diseases at the molecular level and provides a basis for the rational therapy of viral infections.

The study of reovirus–host cell interactions provides a unique opportunity to identify and study viral components that modify and damage host cells because reovirus genetics provides a means to identify the viral components involved in these processes. This chapter will summarize our current understanding of reovirus–host cell interactions. We will review the effects of reovirus on cellular metabolism and the cytoskeleton and will describe how reovirus genetics aids in the identification of specific viral components involved in the production of cytopathogenicity. Initially, we will summarize pertinent data concerning reovirus structure, genetics, and replication (for a more comprehensive discussion of reovirus structure, replication, and genetics, the reader is referred to *The Viruses: The Reoviridae*, Chapters 2, 4, 5, and 6; Joklik, 1983; Zarbl and Millward, 1983; Ramig and Fields, 1983; Sharpe and Fields, 1983).

1.1. Reovirus Structure

The mammalian reoviruses are divided into three serotypes (1, 2, and 3) on the basis of hemagglutination inhibition and antibody neutralization tests (Rosen, 1960). Reovirions contain a segmented, double-stranded RNA (dsRNA) genome surrounded by two concentric icosahedral protein shells (Gomatos *et al.*, 1962). There is an internal core containing the genome and a closely applied inner capsid surrounded by an outer capsid shell (Smith *et al.*, 1969; Fig. 1).

The reovirus genome consists of ten segments of double-stranded RNA (dsRNA) that fall into three size classes designated L (large), M (medium), and S (small; Bellamy *et al.*, 1967; Shatkin *et al.*, 1968; Millward and Graham, 1970; Furuichi *et al.*, 1975). There are three segments in the large size class (L1, L2, L3) with approximate molecular weights of $2.3–2.8 \times 10^6$ daltons, three middle-sized segments (M1, M2, and M3) with molecular weights ranging from $1.4–1.6 \times$

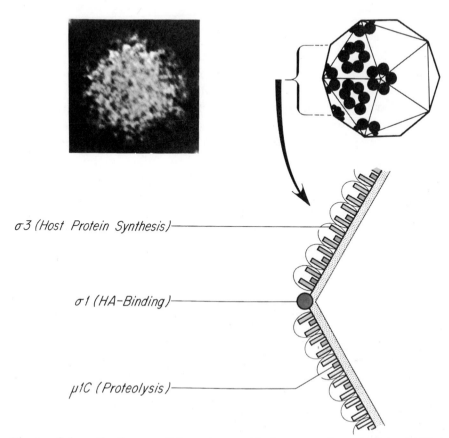

Fig. 1.   Schematic diagram of the outer capsid of a mammalian reovirus. (Left) A negative stain electron micrograph of an intact virion. (Top right) A schematic drawing of the icosahedral reovirion, showing the organization of capsomeres. (Bottom) A schematic drawing that indicates the organization of the three outer capsid proteins σ1, σ3, and μ1C. The σ1 protein is located at the vertices of the icosahedron while σ3 and μ1C are associated with each other (in ratio of one μ1C to two σ3) on the flat surface of the icosahedron. From Fields (1982), by permission of *Archives of Virology*.

10^6 daltons and four small-sized segments (S1, S2, S3, S4) with molecular weights of approximately $0.6–0.9 \times 10^6$ daltons. Each of the dsRNA segments encodes a unique mRNA that is translated into a primary polypeptide (Shatkin and Rada, 1967; Both *et al.*, 1975; Mustoe *et al.*, 1978*b*; McCrae and Joklik, 1978). Like the genome segments that encode them, the viral proteins fall into three molecular weight classes: large (λ), middle (μ), and small (σ). The L segments encode the λ-polypeptides (λ 1, 2, 3), the M segments encode the μ-polypeptides (μ 1, 2, NS), and the S segments encode the σ-poly-peptides (σ 1, 2, NS, 3; Mustoe *et al.*, 1978*b*; McCrae and Joklik, 1978). The outer capsid of the virus consists of three polypeptides, σ1 μ1C, and σ3 derived from the S1, M2, and S4 genome segments, respectively (Bellamy *et al.*, 1967; Mustoe *et al.* 1978*b*). The μ1C polypeptide is derived from the μ1 polypeptide by proteolytic cleav-age. Recent lactoperoxidase-labeling studies have shown that the λ2 polypeptide, part of the viral core, is exposed on the outer surface of the virus (Hayes *et al.*, 1981).

1.2. The Reovirus Replicative Cycle

The reovirus replicative cycle has been the subject of several recent, extensive reviews (Zarbl and Millward, 1983; Shatkin and Kozak, 1983; Silverstein *et al.*, 1976). For the purposes of this chap-ter, the reovirus multiplication cycle will be reviewed briefly. The reader is referred to the aforementioned reviews for further details.

Reoviruses are thought to bind to the surface of host cells through an interaction of the σ1 outer capsid protein with specific cell surface receptors (Weiner *et al.*, 1977, 1978, 1980; Lee *et al.*, 1981). The virions enter the cell via a phagocytic mechanism (Silverstein and Dales, 1968). Virions are transported to the interior of the cell within phagocytic vacuoles. Virus is not released from the vacuoles but re-mains within them until they fuse with primary or secondary lyso-somes. As a consequence of this fusion, virus particles are introduced into an environment rich in proteolytic enzymes. Lysosomal enzymes partially digest the outer capsid of the virus, leaving the genome intact within a subviral particle (SVP). These subviral particles have lost all of polypeptide σ1, σ3, and a fragment of μ1C (Silverstein *et al.*, 1972; Chang and Zweerink, 1971). The removal of viral polypeptides during the uncoating process results in the activation of the virion transcrip-tase and the capping enzymes.

Following activation, the viral transcriptase within SVPs begins to transcribe the parental dsRNA genome, forming ssRNAs, that serve as mRNAs for the translation of viral proteins and templates for the formation of progeny dsRNAs (Schonberg *et al.*, 1971; Sil-verstein *et al.*, 1970, 1976). The newly translated viral proteins sub-sequently assemble with mRNAs by an obscure mechanism, which leads to the incorporation of each of the ten mRNA species into nas-cent SVPs. These progeny SVPs have a replicase activity and produce complementary ($-$) ssRNA, yielding the dsRNA genome segments (Zweerink *et al.*, 1972). Completion of dsRNA synthesis within the SVP signals the initiation of ssRNA within the particle, thereby amplifying the system by which ssRNA is produced and leading to the synthesis of large quantities of viral mRNAs and polypeptides. The synthesis of ssRNAs in the particle terminates when the assembly of outer capsid proteins is completed (Astell *et al.*, 1972). Little is known as yet concerning the reovirus morphogenetic pathway that leads to the formation of mature virions.

1.3. Reovirus Genetics and the Genetic Approach to the Analysis of Reovirus Cytopathogenicity

The observation that cells infected with pairs of *ts* mutants of reovirus type 3 yields a high frequency of *ts*$^+$ progeny first suggested that reovirus RNA segments could recombine by a mechanism of independent reassortment of genome segments (Fields and Joklik, 1969; Cross and Fields, 1976). Analysis of progeny viruses generated by coinfection of cells with reoviruses of different serotypes revealed viruses containing genome segments derived from both serotypes present among the progeny and provided further proof of reassortment of genome segments as the mechanism of recombination (Sharpe *et al.*, 1978). Such "intertypic" reassortants, that can be analyzed readily by polyacrylamide–gel electrophoresis of dsRNA segments, have been used to map the location of reovirus *ts* mutations, to correlate genome segments among the three reovirus serotypes, and to identify the dsRNA segments encoding the viral polypeptides (Mustoe *et al.*, 1978a,b; Sharpe *et al.*, 1978; Ramig *et al.*, 1978; Ramig and Fields, 1983). In addition, these intertypic reassortants have enabled the identification of viral genome segments that are important in a number of stages of reovirus pathogenesis.

Although structurally quite similar, the three serotypes of the mammalian reoviruses interact differently with mammalian hosts and produce distinct patterns of disease (Kilham and Margolis, 1969; Margolis *et al.*, 1971; Raine and Fields, 1973; Stanley and Joske, 1975a,b). Our general approach to the study of reovirus pathogenesis is to compare the pattern of infection produced by reovirus laboratory strains of type 1 Lang, type 2 Jones, type 3 Dearing, or by field isolates of reovirus and to perform a genetic analysis of any differences observed between two strains. A biological property that differs between two serotypes can be mapped to a single gene or a group of genes by examining the behavior of reassortant viruses that have RNA segments derived from both parental strains for the biologic property in question. Because each dsRNA segment can be isolated against the background of another serotype, the effect of one particular viral gene on a biologic property can be studied unambiguously.

Studies with reoviruses type 2 and 3 have shown that both viruses alter host–cell macromolecular synthesis. Type 1 has not been reported to significantly alter host–cell metabolism.

2. REOVIRUS EFFECTS ON HOST CELL DNA SYNTHESIS

Reovirus type 3 Dearing inhibits cellular DNA synthesis in mouse L cells. DNA synthesis inhibition starts approximately at 8–10 hr postinfection, before the onset of viral cytopathic effects (Gomatos and Tamm, 1963). The inhibition reflects a true decrease in DNA synthesis and not an alteration of DNA precursor pools (Ensminger and Tamm, 1970; Shaw and Cox, 1973). It occurs without detectable degradation of cellular DNA or modification of the activity of DNA polymerase or enzymes involved in the conversion of thymidine to thymidine triphosphate (Ensminger and Tamm, 1969b). DNA autoradiographic techniques and sedimentation analyses in alkaline sucrose density gradients indicate that nascent DNA chains grow with normal kinetics under conditions where DNA synthesis is inhibited overall (Ensminger and Tamm, 1969a,b; Hand et al., 1971). DNA synthesis appears to be inhibited because the number of chromosomal regions involved in active DNA replication is reduced. Reovirus thus appears to block the multifocal initiation of new DNA chain synthesis on replication units without altering the rate of replication fork movement (Ensminger and Tamm, 1969b; Hand et al., 1971; Hand and Tamm, 1972, 1974).

Studies using synchronized L cells have shown that DNA synthesis in the late part of the S phase is inhibited strongly following infection with reovirus, whereas, DNA synthesis in the early part of the S phase is relatively unaffected (Ensminger and Tamm, 1970). Infected cells enter the S phase early during infection, but later the cells are blocked in G1 (Ensminger and Tamm, 1970; Cox and Shaw, 1974). Electron-microscopic examination of the ultrastructural changes in the nuclei has indicated that the initial decrease in DNA synthesis is accompanied by decompaction of condensed chromatin within nuclei. Later, nuclei show margination and clumping of heterochromatin (Chaly *et al.*, 1980).

Type 3 reovirus does not inhibit cellular DNA synthesis in all cell types. Shaw and his co-workers found that there is a differential sensitivity of normal and transformed human cells to the effect; inhibition occurs in SV40 transformed WI-38 cells but not in normal WI-38 cells (Duncan *et al.*, 1978). This differential inhibition correlates with a difference in cytopathology. Infection of transformed cells results in cell lysis by 96 hr postinfection, whereas, nontransformed cells are productively infected and continue to produce virus for days after infection without detectable cytopathology.

The initial inhibition of DNA synthesis is temporally related to the number of infectious particles infecting a cell, suggesting that the inhibitory process is specifically related to a virus-specific function. An examination of the effects of noninfective reovirus components on cellular DNA synthesis reveals that reovirions inactivated by ultraviolet light, so as to abolish their infectivity but not their transcriptase activity, inhibit DNA synthesis. However, reovirus cores, which have low infectivity but possess transcriptase activity, and reovirus empty capsids, do not inhibit cellular DNA synthesis. Reovirus oligoadenylates which have been adsorbed to cells in the presence of DEAE dextran, inhibit cellular DNA synthesis weakly (Hand and Tamm, 1973; Shaw and Cox, 1973; Lai *et al.*, 1973). Taken together, these results have been interpreted to suggest that none of the polypeptides of the parental virus cause the inhibition since neither empty capsid nor cores result in inhibition. Hand and Tamm have suggested a two-stage mechanism for inhibition of DNA synthesis, whereby initial inhibition could be mediated by UV-irradiated virus due to the combined effect of nonprotein components or a product of abortive transcription (possibly even a *de novo*-synthesized product). Sustained inhibition would require larger amounts of these inhibitory components. Also, it has been suggested that a nonstructural viral protein could mediate the inhibition.

Studies performed by several groups of investigators indicate that type 3 reovirus infection does not produce a concomitant inhibition of cellular protein synthesis, suggesting that the action of reovirus on DNA replication is direct and specific (Ensminger and Tamm, 1969*a,b*; Gomatos and Tamm, 1963). However, other investigators found similar kinetics of inhibition of cellular DNA and protein synthesis in suspension cultures of L cells and suggested that type 3 reovirus inhibits host cell DNA synthesis by preventing synthesis of essential host proteins (Joklik, 1974). Interestingly, the mechanism proposed for inhibition of cellular DNA synthesis, the reduction in number of multifocal initiation sites, is analogous to that observed in cells treated with inhibitors of protein synthesis. This similarity also has led to the suggestion that reovirus inhibition of DNA replication may be secondary to viral inhibition of host cell protein synthesis.

Other lines of investigation indicate that other mechanisms may be involved in the inhibition of DNA synthesis. Hand and co-workers have studied the affinity of reovirus proteins for DNA, postulating that such an affinity could provide a mechanism for inhibition (Shelton, *et al.*, 1981). They used DNA affinity chromatographic techniques to examine nuclear and cytoplasmic extracts of infected cells and observed that two classes of reovirus proteins, σ and μ, had affinity for native and denatured DNA. The σNS protein had the strongest affinity for DNA, and was considered as a candidate for producing an inhibition of DNA synthesis. However, the σ3 protein also had marked affinity. In addition, Hand had found that a small fraction of purified virus had affinity for DNA and, thus, could not rule out the possibility that other structural proteins might be involved in inhibiting DNA synthesis. The results of Hand and co-workers contrast with those of Huismans and Joklik (1976) that indicated that neither σ3 nor σNS could bind to native DNA. Possibly, the differences in the findings relate to the extracts used for the procedures. Hand and co-workers performed these studies with crude infected-cell extracts, whereas, Huismans and Joklik (1976) used partially purified viral proteins for their studies. The affinity of DNA-binding proteins can be altered by the presence of other proteins in solution.

Using a genetic approach, Sharpe and Fields (1981) also investigated reovirus inhibition of cellular DNA synthesis. They found that while type 3 reovirus inhibits cellular DNA synthesis in mouse L cells, type 1 reovirus exerts little or no effect on L cell DNA synthesis (Fig. 2). The different effects of type 1 and 3 reovirus on L cell DNA synthesis could not be explained by differences in the growth characteristics of these viruses in mouse L cells because both serotypes grow to the same extent and at the same rate in these cells. Furthermore, no cytopathic effects were observed in L cells at the time of inhibition of DNA synthesis.

A series of reassortant viruses containing genome segments of type 1 Lang and type 3 Dearing were examined for their ability to inhibit L cell DNA synthesis. The reassortant viruses behaved like the type 1 or type 3 parental viruses. No intermediate pattern of DNA

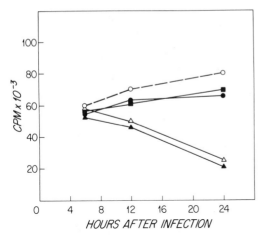

Fig. 2.   Effect of reovirus type 1 and type 3 and reassortant viruses on mouse L cell DNA synthesis. Subconfluent monolayers of mouse L cells were infected with 30 PFU/cell of reovirus type 1 (■), type 3 (△), reassortant clones 1HA3 (▲), or 3HA1 (●). Reassortant clone 1HA3 has all type 1 dsRNA segments except for the S1 dsRNA segment which is derived from type 3 reovirus. 3HA1 has all type 3 dsRNA segments except for the S1 dsRNA gene which is derived from type 1. The cells were infected and pulsed labeled with [³H]thymidine at varying times postinfection. Mock-infected L cells (○) were included as controls. The capacity of reovirus to inhibit host cell DNA synthesis maps to the S1 dsRNA segment. From Sharpe and Fields (1981), by permission of *Journal of Virology*.

synthesis inhibition was observed. Reassortants containing a type 3 S1 dsRNA segment inhibited L cell DNA synthesis, whereas, reassortants having a type 1 S1 dsRNA segment did not (Fig. 2). It was especially striking that a reassortant 1 HA3, that contains all type 1 dsRNA segments except for the S1 RNA segment that is derived from type 3, inhibits L cell DNA synthesis. The reciprocal reassortant, 3HA1, that contains all type 3 dsRNA segments except for the type 1 S1 dsRNA segment does not inhibit DNA synthesis. Thus, the type 3 S1 dsRNA segment enables reovirus type 3 to inhibit L cell DNA synthesis.

Identical results were obtained with live and UV-irradiated parental and reassortant viruses. Since inactivated as well as live reoviruses inhibit DNA synthesis, the S1 gene product and not dsRNA appears to be the viral component responsible for this inhibition.

The observation that irradiation of reassortant viruses did not impair their effect on host cell DNA synthesis extends the studies of Shaw and Cox (1973), who found that UV irradiation of reovirus did not abolish the inhibitory effect on host DNA synthesis, suggesting that either a component of the virus is capable of causing inhibition or that ribonucleic acid can function after inhibition to give rise to an inhibitory component. Possibly, an initial σ1 protein–cell interaction could exert the inhibitory effect. Alternatively, the transcription of limited amounts of S1 mRNA may lead to the production of newly synthesized σ1 protein that could mediate the inhibition. The possibility that newly synthesized σ1 molecules inhibit DNA synthesis is difficult to reconcile with the fact that reovirus cores, which lack σ1 but are infectious, do not cause inhibition of DNA replication (Cox and Shaw, 1974; Hand and Tamm, 1973). Thus, σ1 protein contained in the outer capsid may be more likely to mediate the inhibition. It should be noted that top component (empty capsids), even though it contains σ1, does not inhibit DNA synthesis (Hand *et al.*, 1971; Sharpe and Fields, 1981). The binding of top component to cells is aberrant with only 15% of the empty capsids binding during a 2-hr adsorption period, as opposed to 80% of the intact reovirus particles binding during the same time period. Hence, empty capsids may not have the proper configuration to interact with cell surface receptors.

In addition to the laboratory strains of type 1 and type 3, five field isolates collected by Rosen and co-workers of reovirus type 1 and 3 (Rosen and Abinanti, 1960; Rosen *et al.*, 1963) were examined for their ability to inhibit L cell DNA synthesis. Two type 1 human isolates did not inhibit L cell DNA synthesis, whereas, three type 3 isolates (two bovine and one mouse) did inhibit L cell DNA synthesis (Sharpe and Fields, 1981). Thus, the capacity of type 3 reovirus to inhibit DNA synthesis is a serotype-specific property and not just peculiar to the laboratory strain of reovirus type 3 Dearing. This finding is consistent with the identification of the S1 gene product as the determinant of serotype specificity (Weiner *et al.*, 1977, 1978).

Monoclonal antibodies directed against the reovirus type 3 hemagglutinin have been used to define functionally distinct antigenic domains on the protein (Burstin *et al.*, 1982). An analysis of the behavior of monoclonal antibodies in hemagglutination inhibition (HI) and neutralization tests identified three antigenically distinct regions of the σ1 polypeptide. One major site is involved in neutralization of viral infectivity (defined by monoclonal antibodies A2 and G5). A second site that blocks neutralization is defined by another group of monoclonals. Another group of monoclonals has no detectable neutralization or HI activity.

These monoclonal antibodies were used to further define the region of the σ1 protein involved in inhibition of DNA synthesis. The neutralization domain was found to be the portion of the protein-mediating inhibition of DNA synthesis (Sharpe and Fields, unpublished results).

The identification of the S1 gene as the viral component responsible for inhibition of cellular DNA synthesis may provide a clue to the mechanism of inhibition. The S1 gene product, the viral hemagglutinin, plays a key role in a number of virus–cell surface interactions. This protein is responsible for determining cell and tissue tropism and is the major antigen determining specificity in humoral and cellular immune responses (Weiner and Fields, 1977; Weiner *et al.*, 1977, 1980; Finberg *et al.* 1979; Greene and Weiner, 1980). Furthermore, it mediates viral binding to cell receptors and cellular subcomponents, such as microtubules (Babiss *et al.*, 1979). Inhibition of cellular DNA synthesis may be mediated similarly, through an interaction at the cell surface.

3. REOVIRUS EFFECTS ON HOST CELL RNA AND PROTEIN SYNTHESIS

Studies of reovirus effects on cellular protein synthesis have produced conflicting results that appear to relate in part to cell culture techniques and cellular growth state. Gomatos and Tamm (1963) observed no inhibition of L cell protein synthesis in reovirus type 3-infected L cells grown as monolayers for up to 16 hr postinfection. In contrast, inhibition of protein synthesis was seen in reovirus type 3-infected L cells grown as suspension cultures, beginning at 10 hr post-infection (Kudo and Graham, 1965). Ensminger and Tamm (1969a) investigated this disparity and attributed the inhibitory effect seen in suspension cultures to cell damage. L cells no longer excluded trypan blue at times when an inhibition of protein synthesis was observed.

All of the aforementioned studies examined total rather than host-specific protein synthesis. Joklik and his colleagues studied host cell-specific protein synthesis in mouse L cells and found that host cell protein synthesis continues relatively undiminished at early times after infection but gradually falls such that host-specific protein synthesis is inhibited 3–6 hr prior to the inhibition of total protein synthesis in suspension cultures (Zweerink and Joklik, 1970). This inhibition is dependent upon the multiplicity of infection and on the temperature of incubation. It correlates closely with the onset of viral messenger RNA (mRNA) synthesis, suggesting that host protein synthesis might be inhibited by direct competition between host and viral RNA for some component of the cell translational machinery. Other studies, however, have indicated that the relative rates of host and viral protein synthesis are variable, apparently due to the growth state of the cells and the capacity of the host translational machinery (Walden *et al.*, 1981). Furthermore, different cell types appear to differ in the degree of host-specific protein synthesis shut-off. Walden *et al.* (1981) found that reovirus type 3 inhibited protein synthesis by only 15% in mouse SC-1 cells at 12 hr postinfection. Reovirus replication is not defective in these cells, however. Although yields of approximately 1000–3000 pfu per infected cell were observed, the viral life cycle in SC-1 cells appears to be longer than in mouse L cells, with a maximum appearance of infectious virus between 24–48 hr postinfection. Findings similar to those with SC-1 cells have also been observed in monkey kidney CV-1 cells. Reovirus type 3 grows to good yields in these cells without an inhibition of cellular protein synthesis (Sharpe and Fields, unpublished observations).

In vitro studies suggest that reovirus mRNAs and cellular mRNAs can be distinguished by the host cell translational machinery. Interferon preferentially blocks viral protein synthesis, whereas, the histidine analogue L-histidinol preferentially inhibits host cell protein synthesis (Warrington and Wratten, 1977; Gupta *et al.*, 1974). Interestingly, not only can host and viral mRNAs be distinguished, but also individual species of reovirus mRNA can be distinguished from one another. Under optimal conditions, all ten genome segments are transcribed at the same rate (Skehel and Joklik, 1969). The number of molecules of any given transcript is inversely proportional to its molecular weight, implying that each segment is transcribed independently. The amount of each polypeptide translated, however, does not reflect the amount of each mRNA species present. Some mRNAs are translated more frequently than others, both *in vivo* (Zweerink and Joklik, 1970; Zweerink *et al.*, 1971; Fields *et al.*, 1972; Skup *et al.*, 1981) and *in vitro* (McDowell and Joklik, 1971). In addition, during the early stages of reovirus infection, there is selective transcription of the S1, M3, S3, and S4 genome segments (Nonoyama *et al.*, 1974). Cycloheximide prevents the expansion of transcription to all ten mRNA species (Watanabe *et al.*, 1968), suggesting that host protein(s) are involved in regulating viral transcription.

The relative efficiencies of translation of individual mRNA species appear to relate to the structure of the mRNA. The mRNA cap structures do not affect the frequency at which different mRNAs are transcribed in a cap-independent translation system (Samuel *et al.*, 1977; Skup and Millward, 1980*a,b*; Skup *et al.*, 1981). Thus, it appears that the unique sequence content or secondary structure of reovirus mRNAs provides the basis for transcriptional discrimination. Kozak (1981, 1982a,b) observed that translational frequency correlates with the sequences flanking the initiation codons. Thus, competition among reovirus mRNAs for components of the translational machinery seems to be important in determining mRNA translation frequency (Brendler et al., 1981a,b).

Thach and his colleagues suggest that competition among mRNAs may also play a role in regulating the translation of host and reovirus mRNAs in infected cells. They based this hypothesis on the finding that low doses of cycloheximide specifically stimulate the translation of reovirus mRNAs in type 3-infected cells. They found that a reduced rate of polypeptide elongation produced the differential effect of cycloheximide on the translation of individual mRNA species. In addition, they observed that viral-specific polysomes were smaller than those of comparable host polysomes, suggesting that viral mRNAs would compete poorly for limiting components of the translation system. Such an inability of viral mRNAs to compete effectively with host mRNAs could account for the high amount of viral mRNA present. Furthermore, it was suggested that competition of an excess of viral mRNA with host mRNAs for limiting initiation factors could account for the decline in host protein synthesis and the concomitant rise in viral protein synthesis. Thus, this model proposes that mRNAs must compete for a limiting message-discriminatory initiation factor in order to be translated, and that competitive inhibition of translation of one mRNA by other mRNAs might be important in regulating initiation rates.

The above data led Thach and co-workers to propose a model for translation control based upon competition of cellular and viral mRNAs for a "discriminatory factor." They suggested that a host cell-derived discriminatory factor that must be bound to mRNA prior to its recognition by the native 40 S ribosomal subunit exists in two states, bound to mRNA or free. Reducing the rate of elongation with cycloheximide would lead to an increase in the amount of time that the factor is in the free state. This, in turn, would increase the steady state of unbound discriminatory factor resulting in an increase in the probability that a low-affinity mRNA will bind to the factor and be translated.

The identity of the message discriminatory factor is as yet unknown. *In vitro* studies have suggested that message discrimination involves more than "cap" (5′-terminal m^7 Gppp N^mp) recognition

(Godefroy-Colburn, and Thach, 1981). Ray *et al.* (1983) developed an *in vitro* assay to detect mRNA discriminatory initiation factors. They established a competitive situation using reovirus and globin mRNAs and measured the ability of specific factors to relieve the competition. Eukaryotic initiation factor eif-4A and a "cap-binding protein" complex were identified as candidates for the mRNA discriminatory factor. Since the cap-binding protein complex contains a subunit similar or identical to initiation factor eiF-4A, probably only one form of eiF-4A is active *in vivo*. Both factors can relieve competition among uncapped reovirus mRNAs with the same specificity as for capped mRNAs, suggesting that unique features of mRNA, other than the cap, such as a particular nucleotide sequence or secondary structure, is recognized by the cap-binding protein complex and/or eiF-4A. These unique features of mRNA determine initiation efficiency of the mRNA.

In contrast to other investigators, Detjen *et al.* (1982) and Skup *et al.* (1981) have reported that the host-translational machinery undergoes a reovirus-induced transition from cap-dependent to cap-independent translation, resulting in the preferential translation of uncapped progeny transcripts. They developed a nuclease-free cell-free lysate from L cells that could efficiently translate exogenous mRNA with no negligible background translation (Skup and Millward, 1977). Cell-free extracts prepared from L cells infected with type 3 reovirus at a time of maximal virus-specific protein synthesis preferentially translate uncapped reovirus mRNAs and translate capped mRNAs at a reduced efficiency. In contrast, cell-free extracts prepared from uninfected L cells translate capped reovirus mRNAs with a higher efficiency and uncapped reovirus mRNAs with lower efficiency. The ability of cell-free extracts from reovirus-infected cells to translate uncapped mRNA preferentially increases as a function of time postinfection (Skup and Millward, 1980*a*), corresponding temporally to the period of maximal mRNA synthesis by progeny subviral particles. Interestingly, Skup and Millward (1980*b*) have shown that progeny subviral particles have masked capping enzymes and synthesize only uncapped RNAs *in vitro*.

Based on these results, Zarbl *et al.* (1980) proposed that a mechanism for the reovirus inhibition of cellular protein synthesis involves a virus-induced transition from cap-dependent to cap-independent translation. This transition would lead to a preferential translation of uncapped progeny mRNAs, while capped viral and host mRNAs would be excluded from polysomes. Therefore, this model suggests that at early times postinfection, parental subviral particles synthesize capped mRNAs that would be translated by the normal cap-dependent translational mechanism of host cells. As infection progresses, progeny subviral particles that contain an active viral polymerase and an inactive methyltransferase would become the primary producers of viral mRNAs. This mRNA would be uncapped and would be translated preferentially by the virally modified translational machinery. As a result, the relative ratio of viral to host proteins produced would rise as infection progressed, eventually resulting in only viral protein synthesis.

If such a mechanism were true, then the viral mRNAs synthesized late in infection should not be capped. Indeed, Skup *et al.* (1981) report that the bulk of viral mRNAs isolated from infected cells at late times postinfection are uncapped, since they are not translated in uninfected cell lysates. Host mRNAs were still present in the infected cells and could be translated in cell-free extracts from uninfected cells. Thus, these results are consistent with the hypothesis that late viral mRNAs are uncapped *in vivo*.

Further studies of Millward and his colleagues indicate that reovirus mRNAs associated with polysomes at late times postinfection are not capped and that the proportion of uncapped mRNA increases as the infection progresses (Skup *et al.*, 1981; Zarbl and Millward, 1983). The proportion of capped reovirus mRNAs decreases simultaneously. This transition from capped to uncapped mRNAs corresponds temporally with the synthesis of progeny transcripts and the capacity of cell-free extracts from infected cells to translate uncapped mRNAs.

The reovirus-induced transition of the host cell machinery from cap-dependent to cap-independent translation may be analogous to the modification of the translational machinery by poliovirus. Poliovirus causes a marked inhibition of host cell protein synthesis, soon after infection. Poliovirus mRNAs, which are naturally uncapped (Nomoto *et al.*, 1976; Hewlett *et al.*, 1976), are preferentially translated. Although capped host cell mRNA remains structurally intact and can be translated *in vitro*, it does not enter into initiation complexes (Leibowitz and Penman, 1971; Kaufman *et al.*, 1976; Fernandez-Munoz and Darnell, 1976). *In vitro* studies indicate that poliovirus inhibits the initiation of cellular protein synthesis by inactivating an initiation factor. In a cell-free translation system, the ribosomal salt wash from poliovirus-infected cells does not stimulate translation of capped host mRNAs but does stimulate the translation of poliovirus mRNA. Initiation factor preparations can restore the ability of cell-free extracts obtained from poliovirus-infected cells to translate capped mRNAs (Rose *et al.*, 1978). Trachsel *et al.* (1980) identified the cap-binding protein as the initiation factor inactivated by poliovirus infection.

The cap-binding protein (CBP) is a polypeptide of molecular weight 24,000, present among host cell initiation factors, that facilitates the binding of mRNAs to 40 S ribosomal subunits during the initiation of protein synthesis (Sonenberg and Shatkin, 1977; Sonenberg *et al.*, 1978, 1979*a,b*, 1980). The CBP binds specifically to 5'-terminal cap structures. Monoclonal antibodies directed against the CBP inhibit initiation complex formation with mRNAs having extensive secondary structure but only slightly inhibit initiation complex formation with mRNAs having less secondary structure (Sonenberg *et al.*, 1981*a*). Based on these results, Sonenberg *et al.*, (1981*a,b*) has hypothesized that the CBP stimulates the formation of initiation complexes through unwinding the secondary structure at the 5' end of mRNA, thereby permitting the attachment of the 40 S ribosomal subunit to the mRNA.

Trachsel *et al.* (1980) hypothesized that during the course of poliovirus infection, the CBP is inactivated. As a result, uncapped viral mRNAs are translated, but capped host mRNAs are not translated. Recent studies indicate that poliovirus infection leads to the inability of CBP to form active complexes with other initiating factors, thereby causing an inhibition of capped mRNA entry into initiation complexes (Brown *et al.*, 1982; Hansen *et al.*, 1982*a,b*).

Pursuing similar types of experiments with reovirus-infected L cells, Skup *et al.* studied the ability of CBP to restore translation of capped reovirus mRNAs. They found that CBP could partially restore activity to lysates from infected cells (Zarbl and Millward, 1983). Zarbl and Millward, however, point out that inactivation of CBP does not explain why uncapped reovirus mRNAs can be translated in infected cells because, unlike other uncapped mRNAs which can be translated in lysates from uninfected cells, uncapped reovirus mRNAs cannot be translated under these conditions (Skup and Millward, 1980*a,b*). Therefore, they propose that an additional factor, presumably viral, is necessary for the translation of reovirus uncapped mRNAs. In order to determine whether this is the case, Skup and Millward have examined the effect of antireovirus antibodies on the translation of capped and uncapped reovirus mRNAs *in vitro*. The antiserum specifically inhibits the translation of uncapped viral mRNA in cell-free extracts from infected cells but does not affect the translation of capped reovirus mRNAs in lysates prepared from uninfected cells. Thus, these findings suggest that a viral protein may be involved in the translation of uncapped viral mRNAs.

In recent studies, Lemieux and Millward (Zarbl and Millward, 1983) have examined fractionated translation system lysates for their ability to stimulate the translation of uncapped reovirus mRNAs in unfractionated lysates prepared from uninfected cells. Fractions from infected but not uninfected cells could stimulate this translation, indicating that a viral factor stimulates the translation. The bulk of the stimulation activity was present equally in the S-200 supernate and the ribosomal salt wash. The specific activity of the stimulatory factor was greatest in the ribosomal salt wash, that appears to be enriched for the σ3 outer capsid polypeptide. A monoclonal antibody to σ3 could specifically inhibit the translation of uncapped reovirus mRNAs in lysates prepared from infected cells.

In contrast to Millward and his colleagues, Detjen et al. (1982) find that protein synthesis in SC-1 cells remains cap dependent throughout infection and do not find a transition from cap-dependent to cap-independent translation in reovirus-infected L cells. These findings are consistent with their proposed mechanism of translation control by mRNA competition for a discriminatory factor. They suggest that a partial explanation for the difference of their results with those of Skup and Millward (1980a,b) may be the difference in experimental technique. They used m^7GTP inhibition to assess the extent of capping of mRNA species, which measures only mRNAs actively translated. Thus, their method would not detect the presence of uncapped mRNAs unless they were capable of translation. Thus, they cannot exclude the possibility that uncapped mRNAs are present but not translatable for some unclear reason. Millward and his coworkers also find that the translation of late reovirus mRNA is sensitive to m^7GTP but do not have an explanation for this observation (Zarbl and Millward, 1983). Clearly, further studies are necessary to explain the conflicts in the data reported by the laboratories of Millward and Thach.

Using a genetic approach, Sharpe and Fields (1982) also have examined the effects of the mammalian reoviruses on host cell protein synthesis. Reovirus type 2 was found to inhibit protein synthesis in L cell monolayers more rapidly and efficiently than reovirus type 1 or 3. Type 2 reovirus also inhibits protein synthesis in human amnion cells (Loh and Soergel, 1967). The capacity of type 2 reovirus to

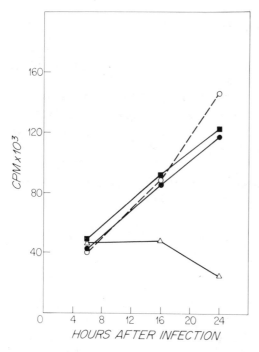

Fig. 3. Effect of reovirus types 1, 2, and 3 on L cell protein synthesis. Subconfluent monolayers of mouse L cells were infected with 80 PFU/cell of reovirus type 1 (●), type 2 (△), or type 3 (■). At various times postinfection, the cells were pulse-labeled with [^{35}S]methionine. Mock-infected L cells (○) were included as controls. The capacity of reovirus to inhibit host cell protein synthesis maps to the S4 dsRNA segment. From Sharpe and Fields (1982), by permission of *Virology*.

inhibit protein synthesis increases as the multiplicity of infection is increased. Inactivation of type 2 reovirus by ultraviolet irradiation destroys the ability of type 2 reovirus to inhibit cellular protein synthesis, suggesting that viral replication is required to mediate the inhibition.

Through an analysis of the capacities of viral reassortants to inhibit L cell protein synthesis, the type 2 S4 dsRNA segment, was shown to be responsible for the inhibition of cellular protein synthesis (Fig. 3). Reassortant viruses containing a type 2 S4 dsRNA segment greatly inhibit host cell protein synthesis, whereas, recombinants having a type 3 S4 dsRNA segment inhibit protein synthesis to a much lesser degree. Especially informative is a recombinant clone that contains all type 3 dsRNA segments except for the S4 dsRNA segment and inhibits protein synthesis. Similarly, the reciprocal reassortant, containing all type 2 dsRNA segments except for the type 3 S4 dsRNA segment, behaves like the type 3 parent.

The mechanism by which the S4 dsRNA segment inhibits cellular protein synthesis is not known. The σ3 polypeptide binds strongly to double-stranded regions of RNA (Huismans and Joklik, 1976). The binding of σ3 to double-stranded regions of rRNA, tRNA, or mRNA might be involved, perhaps by preventing RNA processing or maturation. In addition, recent studies suggest that σ3 may be the viral factor that stimulates the initiation of late uncapped reovirus mRNAs (Zarbl and Millward, 1983).

The capacity of the S4 gene product to inhibit cellular macromolecular synthesis may be related to its modulation in persistent nonlytic viral infection of L cells. Ahmed and Fields (1982) have observed that the S4 gene plays a crucial role in the establishment of persistent infections in mouse L cells. The presence of an S4 dsRNA segment that inhibits cellular macromolecular synthesis would be incompatible with a persistent noncytocidal interaction between reovirus and the host cell. Clearly, further studies are needed to provide more detailed insight into the biochemical mechanism, whereby the S4 gene product inhibits protein synthesis.

The identification of the S4 gene as the dsRNA segment responsible for the capacity of reovirus type 2 to inhibit L cell protein synthesis, however, does demonstrate that there is a difference in the way that reoviruses mediate the inhibition of cellular DNA and protein synthesis. The finding that two distinct reovirus genes mediate the inhibition of DNA and protein synthesis lends support to the idea that inhibition of cellular DNA synthesis is not secondary to the inhibition of host protein synthesis.

With respect to cellular RNA synthesis, virtually no inhibition of host transcription has been observed following the infection of L cells with reovirus type 3 (Gomatos and Tamm, 1963; Kudo and Graham, 1965; Sharpe and Fields, 1982). Host mRNAs are present in type 3 reovirus-infected L cells late in the infectious cycle, although they are not translated, indicating host mRNA stability in the infected cell (Skup *et al.*, 1981). If the mechanism of reovirus inhibition of cellular protein synthesis does indeed involve a shift of the host translational machinery from cap dependence to cap independence, then the inhibition of protein synthesis does not require that reovirus induce an inhibition of host mRNA synthesis since all host mRNAs are capped.

Few studies have examined the effects of reovirus type 1 and 2 on host cell macromolecular synthesis. Loh and Soergel (1965) observed that type 2 reovirus inhibits protein and DNA synthesis in human amnion cells but has little effect on RNA synthesis in these cells. Sharpe and Fields (1982) demonstrated that type 2 reovirus produces a decrease in the rate of total RNA synthesis in mouse L cell, whereas, type 3 causes little or no alteration in the rate of RNA synthesis in L cells. Experiments with reassortant viruses revealed that the ability of a virus to inhibit RNA synthesis correlated with the identity of the S4 dsRNA segment. The presence of a type 2 S4 dsRNA segment conferred the ability to inhibit RNA synthesis (Sharpe and Fields, 1982). The inhibition of RNA and protein synthesis occurred at approximately the same time during the course of infection in type 2-infected L cells. Further studies are needed to determine how host cell RNA synthesis is inhibited by type 2 reovirus.

4. REOVIRUS EFFECTS ON CYTOSKELETAL ORGANIZATION

Viruses may produce their cytopathologic effects not only through an interference with host cell metabolism, but also through a disruption of the host cell cytoskeleton. Microscopic studies of reovirus-infected cells indicate that cytological changes accompany reovirus multiplication and morphogenesis (Rhim *et al.*, 1962; Spendlove *et al.*, 1963).

Reoviruses enter cells within phagocytic vacuoles and remain within these vacuoles until they fuse with primary or secondary lysosomes (Dales *et al.*, 1965). Within lysosomes, reovirions are uncoated and the virion transcriptase is activated. The formation of nascent virions, the synthesis of ssRNAs and dsRNAs, and the maturation of complete virions occur within regions of the cytoplasm known as "viral factories." The intracytoplasmic route by which subviral particles move to these sites of replication is not clear. Furthermore, the mechanism of formation of viral factories is not understood.

Viral factories are first present as phase dense granular material scattered throughout the cytoplasm (Fig. 4). As infection progresses, these discrete granules coalesce and migrate toward the nucleus, eventually forming phase-dense perinuclear inclusions. The size and extent of the viral inclusions increase as the number of viral progeny increase. Eventually, at late times post-infection (48–72 hr), these inclusions form a network that spreads throughout the entire cytoplasm. The temporal formation of viral inclusions relates to the temperature of infection as well as the cell type studied. Inclusions form more slowly at 31°C than at higher temperatures (Fields *et al.*, 1971). They are observed earlier in mouse L cells than in monkey kidney cells infected at the same multiplicity and at the same temperature

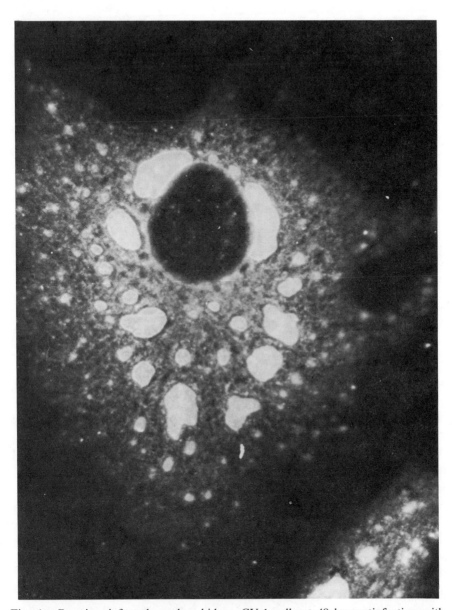

Fig. 4.   Reovirus-infected monkey kidney CV-1 cells at 48 hr postinfection with reovirus type 3. Cells were stained with rabbit antireovirus serum and fluorescein-conjugated goat and rabbit serum according to the method of Sharpe *et al.* (1982). Cytoplasmic inclusions are represented by the numerous globular white areas within the cytoplasm. Note the gradation of size of inclusions, from small to large, as the inclusions approach the nucleus (bar = 20 μm). From Sharpe *et al.* (1982), by permission of *Virology*.

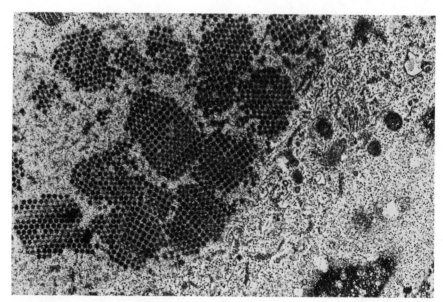

Fig. 5.   Electron micrograph of a type 3-infected CV-1 cell containing a viral inclusion. Note the kinky filaments present within the viral inclusion. These are exactly 100 Å in diameter (magnification x 7000). From Sharpe *et al.* (1982), by permission of *Virology*.

(Rhim *et al.*, 1962; Fields *et al.*, 1971). Recent studies of infected monkey kidney cells using time lapse photographic techniques reveal that inclusions are dynamic structures, constantly breaking apart and reforming. In the process of cell lysis, inclusions become quite active, rounding up before the cell lysis (Sharpe *et al.*, unpublished results).

Electron-microscopic techniques have partially characterized the viral inclusion (Fig. 5). Although it is a discrete cytoplasmic entity, the viral inclusion is not enclosed by a membrane. Ribosomes are not present within the viral factories, indicating that viral protein must be synthesized on polysomes outside the factory and then transported into the factory prior to assembly of mature virus. Therefore, viral mRNAs must be synthesized within the factory and transported to polysomes within the cytoplasm for translation.

Electron-microscopic studies have also identified the presence of several types of filaments within the viral factory (Dales, 1963; Dales *et al.*, 1965). In the electron microscope, viral factories are seen to contain viral particles aligned on parallel arrays of microtubules that are thought to be covered with viral proteins, possibly the viral hemagglutinin (Babiss *et al.*, 1979). Between these microtubules is a complex filamentous network consisting of masses of densely twisted "kinky" filaments having a diameter of 50–80 Å. These kinky filaments are observed to be in intimate contact with viral capsids and with viral protein-coating microtubules. Similar filaments occur regularly in wavy bundles in the cytoplasm of uninfected mouse L cells. Dales *et al.* (1965) have suggested that these filaments change from the wavy to the kinky form in areas where virus is being assembled and that such filaments may serve as the site of attachment of viral mRNAs or the site of viral morphogenesis.

The recent availability of antibodies to cytoskeletal components offers an additional means to study reovirus interactions with cytoplasmic filaments. Although electron-microscopic techniques can elucidate filament structure and distribution, the full extent of cytoplasmic filament organization was not known until indirect immunofluorescence techniques were developed (Weber *et al.*, 1975; Brinkley *et al.*, 1978; Lazarides, 1975). Immunofluorescence microscopic techniques indicate that cytoplasmic filaments form networks within the cell cytoplasm.

Sharpe *et al.* (1982) used immunofluorescent microscopic techniques to extend the analysis of reovirus–cytoskeletal interaction from a description of reovirus interactions with filaments in discrete regions of the cytoplasm to an analysis of the overall effects of reovirus on cytoskeletal organization. They examined how reovirus infection affected the three major filamentous systems in the cell cytoplasm: microtubules, microfilaments, and intermediate filaments. Reovirus infection produced by all three serotypes was found to cause a major disruption of vimentin filament (a type of intermediate filament) organization without causing an alteration of microtubule or microfilament organization in monkey kidney CV-1 cells (Fig. 6). At 12 hr after infection, before intracytoplasmic inclusions were discernable, the perinuclear organization sites of vimentin filaments disappeared. At later times postinfection, intermediate filaments were further disrupted such that wavy filaments with no apparent organization were present in the cytoplasm (Fig. 6). Reovirus infection not only disrupted vimentin filament organization, but also appeared to produce a reorganization of intermediate filaments. Viral inclusions contain filamentous structures that are detected by antivimentin antibody. Hence, these filamentous structures may be the complex filamentous network observed by Dales *et al.* (1965) in their electron-microscopic studies (Fig. 7).

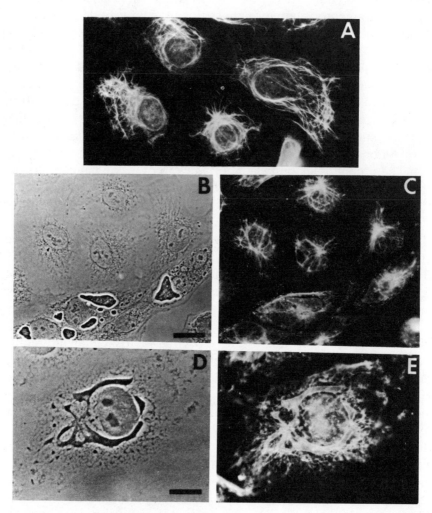

Fig. 6. (A) Organization of intermediate filaments in mock-infected and reovirus type 3-infected CV-1 cells at 48 hr postinfection. (B) and (D) Fluorescence micrograph demonstrating intermediate filament organization in mock-infected CV-1 cells. (C) and (E) Phase contrast micrographs of type 3-infected CV-1 cells showing phase-dense inclusions bodies. (A), (B), and (C) Fluorescence micrographs of the identical field of cells as in (B) and (D), showing intermediate filament organization in reovirus-infected CV-1 cells. Note that reovirus infection alters intermediate filament organization. Cells were subjected to indirect immunofluorescence microscopy using antibody against vimentin bar = 40 μm). From Sharpe *et al.* (1982), by permission of *Virology*.

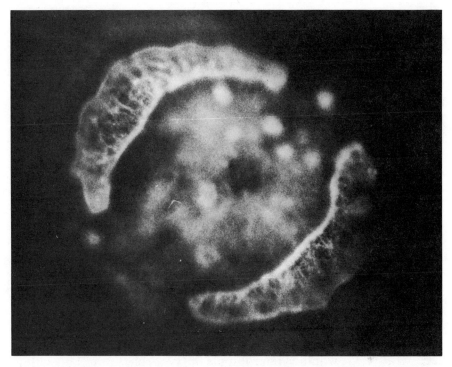

Fig. 7.   Fluorescence micrograph of a reovirus type 3-infected CV-1 cell demonstrating vimentin filament organization within viral inclusions. Cell was subjected to indirect immunofluorescence microscopy using antibody against vimentin (bar = 20 μm). From Sharpe *et al.* (1982), by permission of *Virology*.

The presence of vimentin filaments in viral factories suggests that these filaments may be participating in the formation of unique viral structures, distinct from vimentin filament organization in the cytoplasm of noninfected cells. Possibly, reovirus is using vimentin filaments to create a discrete cytoplasmic entity, the viral factory, that functions as the site of viral replication and assembly. Lazarides (1980) has proposed that intermediate filaments function to coordinate the organization of the contents of the cytoplasm. With such a role, intermediate filaments might be expected to play a crucial part in the regulation of cell shape and to be influenced by the cellular metabolic state. The finding of intermediate filaments within viral factories suggests that these filaments may play a crucial role in organizing viral replication and/or assembly. In fact, intermediate filaments may be one of the major cellular targets involved in producing reovirus cytopathic effects, e.g., rounding of cells and inclusion formation.

Whether intermediate filament disruption (and/or reorganization) plays a primary and central step in viral infection, leading ultimately to cell death, remains to be determined.

The disruption of vimentin filament organization by reovirus is especially interesting when considered together with the finding that agents which disrupt cellular protein synthesis, such as diphtheria toxin, Pseudomonas aeruginosa exotoxin A, and cycloheximide, disrupt vimentin filament organization without altering microtubule or microfilament organization (Sharpe *et al.*, 1980). Possibly, the disruption of intermediate filaments by reovirus is related to the capacity of reovirus to inhibit host cell protein synthesis.

In contrast to the marked disruption of intermediate filaments, reovirus infection did not disrupt microtubule organization. Antibodies to tubulin visualized microtubules coursing through regions of the cytoplasm containing viral factories, without interruption or distortion. These findings are consistent with the electron-microscopic studies that indicate that reovirions are aligned on parallel arrays of microtubules within viral factories (Dales *et al.*, 1965) and can bind to microtubules *in vitro* (Babiss *et al.*, 1979).

Since colchicine treatment of reovirus-infected cells does not reduce viral yield, whether microtubules play a role in viral growth is uncertain (Spendlove *et al.*, 1964). Colchicine treatment of virally infected cells, however, does alter the morphology of viral inclusions. Only small inclusions located at the cell periphery are seen in colchicine-treated infected cells. The large perinuclear inclusions usually observed in reovirus-infected cells are not seen. Thus, microtubules may be involved in the coalescence of viral inclusions and in inclusion movement toward the cell nucleus.

Reovirus infection not only produces a disruption of intermediate filaments, but also leads to a disorganization of mitochondrial distribution (Sharpe *et al.*, 1982). Vizualized with the fluorescent probe Rhodamine 123 (Johnson *et al.*, 1980; Walsh *et al.*, 1979), mitochondria have a characteristic discontinuous distribution in the CV-1 cell cytoplasm. Reovirus infections result in the aggregation of mitochondria around the nucleus with only occasional mitochondria present at the cell periphery. Mitochondria are not present within viral inclusions. Although reovirus infection affects mitrochondrial distribution, whether reovirus infection alters mitochondrial function is uncertain. Johnson *et al.* (1981) showed that the accumulation of Rhodamine 123 by mitochondria reflects the transmembrane potential. The accumulation of Rhodamine 123 is similar in infected and uninfected cells, suggesting that respiratory activity is unaffected by reovirus infection.

5. SUMMARY

During the course of viral infection, the mammalian reoviruses alter the host cells in several ways: reoviruses alter host cell metabolism and alter the organization of the cytoskeleton.

A number of studies with reovirus type 3 indicate that it inhibits cellular DNA synthesis, at the level of initiation by a reduction in the number of multifocal initiation sites without altering the rate of replication fork movement. Recent genetic studies indicate that the type 3 σ1 outer capsid protein is responsible for the inhibition of cellular DNA synthesis. Reovirus infection also produces an inhibition of cellular protein synthesis, the mechanism of which is currently not resolved. Studies of Millward and his co-workers suggest that this inhibition involves the transition of the host cell translational machinery from cap dependence to cap independence. The experiments of Thach and his co-workers suggest that this inhibition involves competitive inhibition of translation of mRNAs for limiting message-discriminatory factors, possibly the cap-binding protein or eiF-4A, as a means to regulate mRNA initation rates. Recent genetic studies indicate that type 2 reovirus causes a marked inhibition of cellular protein synthesis. The σ3 protein is involved in the inhibition of cellular protein synthesis. Recent biochemical studies also suggest a role for σ3 in stimulating the translation of late viral mRNAs.

Studies of the effects of reovirus infection on cytoskeletal organization have revealed that all three serotypes produce similar alterations of the cytoskeleton. Reovirus infection leads to a major disruption of intermediate filaments without producing a discernable effect upon the organization of microtubules or microfilaments. Reovirus inclusions, discrete sites of viral replication and assembly, appeared to contain vimentin filamentous structures, suggesting that reovirus infection not only disrupts intermediate filament organization but also leads to a reorganization and reutilization of vimentin. Mitochondrial distribution was also disrupted by reovirus infection, supporting the observation that vimentin filaments may play a role in determining mitochondrial distribution.

It is interesting to note that two distinct viral outer capsid proteins appear to be the viral components involved in mediating virus–host cell interactions. Capsid proteins of other cytolytic viruses also appear to be involved in inhibiting host cell macromolecular synthesis (Ginsberg *et al.*, 1967; Steiner-Pryor and Cooper, 1973).

ACKNOWLEDGMENTS

We gratefully acknowledge Rosemary Bacco for her outstanding assistance in the preparation of this review. This research was supported by grants from the National Institutes of Health (National Institute of Allergy and Infectious Diseases) and the Institute of Neurologic Diseases and Blindness.

6. REFERENCES

Ahmed, R., and Fields, B. N., 1982, Role of the S4 gene in establishment of persistent reovirus infection in L cells, *Cell* **28**:605.

Astell, C., Silverstein, S. C., Levin, D. H., and Acs, G., 1972, Regulation of the reovirus RNA transcriptase by a viral capsomere protein, *Virology* **48**:648.

Babiss, L. E., Luftig, R. B., Weatherbee, J. A., Weihing, R. R., Ray, U. R., and Fields, B. N., 1979, Reovirus serotypes 1 and 3 differ in their *in vitro* association with microtubules, *Virology* **30**:863.

Bellamy, A. R., Shapiro, L., August, J. T., and Joklik, W. K., 1967, Studies on reovirus RNA. I. Characterization of reovirus genome RNA, *J. Mol. Biol.* **29**:1.

Both, G. W., Banerjee, A. K., and Shatkin, A. J., 1975, Methylation-dependent translation of viral messenger RNAs *in vitro, Proc. Natl. Acad. Sci. USA* **72**:1189.

Brendler, T., Godefroy-Colburn, T., Carlill, R. D., and Thach, R. E., 1981*a*, The role of mRNA competition in regulating translation. II. Development of a quantitative *in vitro* assay, *J. Biol. Chem.* **256**:11,747.

Brendler, T., Godefroy-Colburn, T., Yu, S., and Thach, R. E., 1981*b*, The role of mRNA competition in regulating translation. III. Comparison of *in vitro* and *in vivo* results, *J. Biol. Chem.* **256**:11,755.

Brinkley, B. R., Fuller, G. M., and Highfield, D. P., 1978, Cytoplasmic microtubules in normal and transformed cells in culture: Analysis of tubulin antibody immunofluorescence, *Proc. Natl. Acad. Sci. USA* **72**:4981.

Brown, D., Jones, C. L., Brown, B. A., and Ehrenfeld, E., 1982, Translation of capped and uncapped VSV mRNAs in the presence of initiation factors from poliovirus-infected cells, *Virology* **123**:60.

Burstin, S. J., Spriggs, D. R., and Fields, B. N., 1982, Evidence for functional domains on the reovirus type 3 hemagglutinin, *Virology* **117**:146.

Chaly, N., Johnstone, M., and Hand, R., 1980, Alterations in nuclear structure and function in reovirus-infected cells, *Clin. Invest. Med.* **2**:141.

Chang, C.-T., and Zweerink, H. J., 1971, Fate of parental reovirus in infected cell, *Virology* **46**(3):544.

Cox, D. C., and Shaw, J. E., 1974, Inhibition of the initiation of cellular DNA synthesis after reovirus infection, *J. Virol.* **13**:760.

Cross, R. K., and Fields, B. N., 1976, Use of an aberrant polypeptide as a marker in three-factor crosses: Further evidence for independent reassortment as the mechanism of recombination between temperature-sensitive mutants of reovirus type 3, *Virology* **74:**345.

Dales, S., 1963, Association between the spindle apparatus and reovirus, *Proc. Natl. Acad. Sci. USA* **50:**268.

Dales, S., Gomatos, P. J., and Hsu, K. C., 1965, The uptake and development of reovirus in strain L cells followed with labelled viral ribonucleic acid and ferritin-antibody conjugates, *Virology* **25:**193.

Detjen, B. M., Walden, W. E., and Thach, R. E., 1982, Translational specificity in reovirus-infected mouse fibroblasts, *J. Biol. Chem.* **257:**9855.

Duncan, M. R., Stanish, S. M., and Cox, D. C., 1978, Differential sensitivity of normal and transformed human cells to reovirus infection, *J. Virol.* **28:**444.

Ensminger, W. D., and Tamm, I., 1969a, Cellular DNA and protein synthesis in reovirus-infected cells, *Virology* **39:**357.

Ensminger, W. D., and Tamm, I., 1969b, The step in cellular DNA synthesis blocked by reovirus infection, *Virology* **39:**935.

Ensminger, W. D., and Tamm, I., 1970, Inhibition of synchronized cellular deoxyribonucleic acid synthesis during Newcastle disease virus, mengovirus or reovirus infection, *J. Virol.* **5:**672.

Fernandez-Munoz, R., and Darnell, J. E., 1976, Structural differences between the 5' termini of viral and cellular mRNA in poliovirus-infected cells: Possible basis for the inhibition of host protein synthesis, *J. Virol.* **18:**726.

Fields, B. N., 1982, Molecular basis of reovirus virulence, *Arch. Virol.* **71:**95.

Fields, B. N., and Joklik, W. K., 1969, Isolation and preliminary genetic and biochemical characterization of temperature-sensitive mutants of reovirus, *Virology* **37:**335.

Fields, B. N., Raine, C. S., and Baum, S. G., 1971, Temperature-sensitive mutants of reovirus type 3: Defects in viral maturation as studied by immunofluorescence and electron microscopy, *Virology* **43:**569.

Fields, B. N., Laskov, R., and Scharff, M. D., 1972, Temperature-sensitive mutants of reovirus type 3: Studies on the synthesis of viral peptides, *Virology* **50:**209.

Finberg, R., Weiner, H. L., Fields, B. N., Benacerraf, B., and Burakoff, S. J., 1979, Generation of cytolytic T lymphocytes after reovirus infection: Role of S1 gene, *Proc. Natl. Acad. Sci. USA* **76:**442.

Furuichi, Y., Muthukrishnan, S., and Shatkin, A. J., 1975, 5'-Terminal m(7)G(5') ppp(5')G(m)p *in vivo*: Identification in reovirus genome RNA, *Proc. Natl. Acad. Sci. USA* **72:**742.

Ginsberg, H. S., Bella, L. J., and Levine, A. J., 1967, Control of biosynthesis of host macromolecules in cells infected with adenovirus, in *The Molecular Biology of Viruses* (J. S. Colter and W. Parenchych, eds.), pp. 547–572, Academic Press, New York.

Godefroy-Colburn, T., and Thach, R. E., 1981, The role of mRNA competition in regulating translation. IV. Kinetic model, *J. Biol. Chem.* **256:**11762.

Gomatos, P. J., and Tamm, I., 1963, Macromolecular synthesis in reovirus-infected cells, *Biochim. Biophys. Acta* **72:**651.

Gomatos, P. J., Tamm, I., Dales, S., and Franklin, R. M., 1962, Reovirus type 3: Physical characteristics and interaction with L cells, *Virology* **17:**441.

Greene, M. I., and Weiner, H. L., 1980, Delayed hypersensitivity in mice infected with reovirus. II. Induction of tolerance and suppressor T cells to viral specific gene products, *J. Immunol.* **125**:283.

Gupta, S. L., Graziadei, W. D., III, Weideli, H., Sopori, M. L., and Lengyel, P., 1974, Selective inhibition of viral protein accumulation in interferon-treated cells: Nondiscriminate inhibition of the translation of added viral and cellular messenger RNAs in their extracts, *Virology* **57**:49.

Hand, R., and Tamm, I., 1972, Rate of DNA chain growth in mammalian cells infected with cytocidal RNA viruses, *Virology* **47**:331.

Hand, R., and Tamm, I., 1973, Reovirus: Effect of noninfective viral components on cellular deoxyribonucleic acid synthesis, *J. Virol.* **11**:223.

Hand, R., and Tamm, I., 1974, Initiation of DNA synthesis in mammalian cells and its inhibition by reovirus infection, *J. Mol. Biol.* **82**:175.

Hand, R., Ensminger, W. D., and Tamm, I., 1971, Cellular DNA replication in infections with cytocidal RNA viruses, *Virology* **44**:527.

Hansen, J. I., Etchison, D., Hershey, J. W. B., and Ehrenfeld, E., 1982a, Association of cap-binding protein with eukaryotic initiation factor 3 in initiation factor preparations from uninfected and poliovirus-infected HeLa cells, *J. Virol.* **42**:200.

Hansen, J. I., Etchison, D. O., Hershey, J. W. B., and Ehrenfeld, E., 1982b, Localization of cap-binding protein in subcellular fractions of HeLa cells, *Mol. Cell. Biol.* **2**:1639.

Hayes, E. C., Lee, P. W. K., Miller, S. E., and Joklik, W. K., 1981, The interaction of a series of hybridoma IgGs with reovirus particles, *Virology* **108**:147.

Hewlett, M. J., Rose, J. K., and Baltimore, D., 1976, 5'-Terminal structure of poliovirus polyribosomal RNA is pUp, *Proc. Natl. Acad. Sci. USA* **73**:327.

Huismans, H., and Joklik, W. K., 1976, Reovirus coded polypeptides in infected cells: Isolation of two native monomeric polypeptides with affinity for single-stranded and double-stranded RNA respectively, *Virology* **70**(2):411.

Johnson, L. V., Walsh, M., and Chen, L., 1980, Localization of mitochondria in living cells with rhodamine 123, *Proc. Natl. Acad. Sci. USA* **77**:990.

Johnson, L. V., Walsh, M. L., Bockus, B. J., and Chen, L. B., 1981, Monitoring of relative mitochondrial membrane potential in living cells by fluorescence microscopy, *J. Cell. Biol.* **88**:526.

Joklik, W. K., 1974, Reproduction of reoviridae, in: *Comprehensive Virology,* Vol. 2 (H. Fraenkel-Conrat and R. Wagner, eds.), pp. 297–334, Plenum Press, New York.

Joklik, W. K., 1983, The reovirus particle, in: *The Viruses, The Reoviridae* (H. Fraenkel-Conrat and R. Wagner, eds.), pp. 9–78, Plenum Press, New York.

Kaufman, Y., Goldstein, E., and Penman, S., 1976, Poliovirus-induced inhibition of polypeptide initiation *in vitro* on native polyribosomes, *Proc. Natl. Acad. Sci. USA* **73**:1834.

Kilham, L., and Margolis, G., 1969, Hydrocephalus in hamsters, ferrets, rats and mice following inoculations with reovirus type 1, *Lab. Invest.* **21**:183.

Kozak, M., 1981, Possible role of flanking nucleotides in recognition of the AUG initiator codon by eukaryotic ribosomes, *Nucl. Acids Res.* **9**:5233.

Kozak, M., 1982a, Analysis of ribosome binding sites from the s1 message of reovirus. Initiation at the first and second AUG codons, *J. Mol. Biol.* **156**:807.

Kozak, M., 1982b, Sequences of ribosome binding sites from the large size class of reovirus mRNA, *J. Virol.* **42**:467.

Kudo, H., and Graham, A. F., 1965, Synthesis of reovirus ribonucleic acid in L cells, *J. Bacteriol.* **90**:936.

Lai, M. T., Werenne, J. J., and Joklik, W. K., 1973, The preparation of reovirus top component and its effect on host DNA and protein synthesis, *Virology* **54**:237.

Lazarides, E., 1975, Tropomyosin antibody: The specific localization of tropomyosin in nonmuscle cells, *J. Cell Biol.* **65**:549.

Lazarides, E., 1980, Intermediate filaments as mechanical integrators of cellular space, *Nature (London)* **283**:249.

Lee, P. W. K., Hayes, E. C., and Joklik, W. K., 1981, Protein σ1 is the reovirus cell attachment protein, *Virology* **108**:156.

Leibowitz, R., and Penman, S., 1971, Regulation of protein synthesis in HeLa cells. III. Inhibition during poliovirus infections, *J. Virol.* **8**:661.

Loh, P. C., and Soergel, M., 1965, Growth characteristics of reovirus type 2: Actinomycin D and the synthesis of viral RNA, *Proc. Natl. Acad. Sci. USA* **54**:857.

Loh, P. C., and Soergel, M., 1967, Macromolecular synthesis in cells infected with reovirus type 2 and the effect of ARA-C, *Nature (London)* **214**:622.

Margolis, G., Kilham, L., and Gonatos, N., 1971, Reovirus type III encephalitis; Observations of virus-cell interactions in neural tissues. I. Light microscopy studies, *Lab. Invest.* **24**:91.

McCrae, M. A., and Joklik, W. K., 1978, The nature of the polypeptide encoded by each of the 10 double-stranded RNA segments of reovirus type 3, *Virology* **89**:578.

McDowell, M. J., and Joklik, W. K., 1971, An *in vitro* protein synthesizing system from mouse L fibroblasts infected with reovirus, *Virology* **45**(3):724.

Millward, S., and Graham, A. F., 1970, Structural studies on reovirus: Discontinuities in the genome, *Proc. Natl. Acad. Sci. USA* **65**:422.

Mustoe, T. A., Ramig, R. F., Sharpe, A. H., and Fields, B. N., 1978a, A genetic map of reovirus. III. Assignment of dsRNA positive mutant groups A,B, and G to genome segments, *Virology* **85**:545.

Mustoe, T. A., Ramig, R. F., Sharpe, A. H., and Fields, B. N., 1978b, Genetics of reovirus: Identification of the dsRNA segments encoding the polypeptides of the μ and σ size classes, *Virology* **89**:594.

Nomoto, A., Lee, Y. F., and Wimmer, E., 1976, The 5' end of poliovirus mRNA is not capped with $m^7G(5')ppp(5')Np$, *Proc. Natl. Acad. Sci. USA* **73**:375.

Nonoyama, M., Millward, S., and Graham, A. F., 1974, Control of transcription of the reovirus genome, *Nucl. Acids Res.* **1**:373.

Raine, C. S., and Fields, B. N., 1973, Ultrastructural features of reovirus type 3 encephalitis, *J. Neuropathol. Exp. Neurol.* **32**:19.

Ramig, R. F., and Fields, B. N., 1983, Genetics of reoviruses, in: *The Viruses, The Reoviridae* (H. Fraenkel-Conrat and R. Wagner, eds.), pp. 197–228, Plenum Press, New York.

Ramig, R. F., Mustoe, T. A., Sharpe, A. H., and Fields, B. N., 1978, A genetic map of reovirus. II. Assignments of dsRNA negative mutant groups C, D, and E to genome segments, *Virology* **85**:531.

Ray, B. K., Brendler, T. G., Adya, S., Daniels-McQueen, S., Miller, J. K., Hershey, J. W., Grifo, J. A., Merrick, W. C., and Thach, R. E., 1983, Role of mRNA competition in regulating translation: Further characterization of mRNA discriminatory factors, *Proc. Natl. Acad. Sci. USA* **80**:663.

Rhim, J. S., Jordan, L. E., and Mayor, H. D., 1962, Cytochemical, fluorescent-antibody and electron microscopic studies on the growth of reovirus (ECHO 10) in tissue culture, *Virology* **17**:342.

Rose, J. K., Trachsel, H., Leong, K., and Baltimore, D., 1978, Inhibition of translation by poliovirus: Inactivation of a specific initiation factor, *Proc. Natl. Acad. Sci. USA* **75**:2732.

Rosen, L., 1960, Serologic grouping of reoviruses by hemagglutination-inhibition, *Am. J. Hyg.* **71**:243.

Rosen, L., and Abinanti, F. R., 1960, Natural and experimental infection of cattle with human types of reovirus, *Am. J. Hyg.* **71**:424.

Rosen, L., Abinanti, F. R., and Hovin, J. F., 1963, Further observations on the natural infections of cattle with reoviruses, *Am. J. Hyg.* **77**:38.

Samuel, C. E., Farris, D. A., and Levin, K. H., 1977, Biosynthesis of virus specified polypeptides: System-dependent effect of reovirus mRNA methylation on translation *in vitro* catalyzed by ascites tumor and wheat embryo cell-free extracts, *Virology* **81**:476.

Schonberg, M., Silverstein, S. C., Levin, D. H., and Acs, G., 1971, Asynchronous synthesis of the complementary strands of the reovirus genome, *Proc. Natl. Acad. Sci. USA* **68**:505.

Sharpe, A. H., and Fields, B. N., 1981, Reovirus inhibition of cellular DNA synthesis: Role of the S1 gene, *J. Virol.* **38**:389.

Sharpe, A. H., and Fields, B. N., 1982, Reovirus inhibition of cellular RNA and protein synthesis: Role of the S4 gene, *Virology* **122**:381.

Sharpe, A. H., and Fields, B. N., 1983, Pathogenesis of reoviruses, in: *The Viruses, The Reoviridae* (H. Fraenkel-Conrat and R. Wagner, eds.), pp. 229–286, Plenum Press, New York.

Sharpe, A. H., Ramig, R. F., Mustoe, T. A., and Fields, B. N., 1978, A genetic map of reovirus. I. Correlation of genome RNAs between serotypes 1, 2 and 3, *Virology* **84**:63.

Sharpe, A. H., Chen, L. B., Murphy, J. R., and Fields, B. N., 1980, Specific disruption of vimentin filament organization in monkey kidney CV-1 cells by diphtheria toxin, exotoxin A, and cycloheximide, *Proc. Natl. Acad. Sci. USA* **77**:7267.

Sharpe, A. H., Chen, L. B., and Fields, B. N., 1982, The interaction of mammalian reoviruses with the cytoskeleton of monkey kidney CV-1 cells, *Virology* **120**:399.

Shatkin, A. J., and Kozak, M., 1983, Biochemical aspects of reovirus transcription and translation, in: *The Viruses, The Reoviridae* (H. Fraenkel-Conrat and R. Wagner, eds.), pp. 79–106, Plenum Press, New York.

Shatkin, A. J., and Rada, B., 1967, Reovirus-directed ribonucleic acid synthesis in infected L cells, *J. Virol.* **1**:24.

Shatkin, A., Sipe, J. D., and Loh, P., 1968, Separation of ten reovirus segments by polyacrylamide gel electrophoresis, *J. Virol.* **2**:986.

Shaw, J. L., and Cox, D. C., 1973, Early inhibition of cellular DNA synthesis by high multiplicities of infectious and UV-irradiated reovirus, *J. Virol.* **12**:704.

Shelton, I. H., Kasupski, G. J., Oblin, C., and Hand, R., 1981, DNA binding of a nonstructural reovirus protein, *Can. J. Biochem.* **59**:122.

Silverstein, S. C., and Dales, S., 1968, The penetration of reovirus RNA and initiation of its genetic function in L-strain fibroblasts, *J. Cell. Biol.* **36**:197.

Silverstein, S. C., Schonberg, M., Levin, D. H., and Acs, G., 1970, The reovirus replicative cycle: Conservation of parental RNA and protein, *Proc. Natl. Acad. Sci. USA* **67**:275.

Silverstein, S. C., Astell, C., Levin, D. H., Schonberg, M., and Acs, G., 1972, The mechanisms of reovirus uncoating and gene activation *in vivo, Virology* **47**(3):797.

Silverstein, S. C., Christman, J. K., and Acs, G., 1976, The reovirus replication cycle, *Annu. Rev. Biochem.* **45**:375.

Skehel, J. J., and Joklik, W. K., 1969, Studies on the *in vitro* transcription of reovirus RNA catalyzed by reovirus cores, *Virology* **39**:822.

Skup, D., and Millward, S., 1977, Highly efficient translation of messenger RNA in cell-free extracts prepared from L-cells, *Nucl. Acids Res.* **4**:3581.

Skup, D., and Millward, S., 1980*a*, Reovirus induced modification of cap dependent translation in infected L cells, *Proc. Natl. Acad. Sci. USA* **77**(1):152.

Skup, D., and Millward, S., 1980*b*, mRNA capping enzymes are masked in reovirus progeny subviral particles, *J. Virol.* **34**:490.

Skup, D., Zarbl, H., and Millward, S., 1981, Regulation of translation in L-cells infected with reovirus, *J. Mol. Biol.* **151**:35.

Smith, R. E., Zweerink, H. J., and Joklik, W. K., 1969, Polypeptide components of virions, top component and cores of reovirus type 3, *Virology* **39**:791.

Sonenberg, N., and Shatkin, A. J., 1977, Reovirus mRNA can be covalently cross-linked via the 5′ cap to proteins in initiation complexes, *Proc. Natl. Acad. Sci. USA* **74**:4288.

Sonenberg, N., Morgan, M. A., Merrick, W. C., and Shatkin, A. J., 1978, A polypeptide in eukaryotic initiation factors that crosslinks specifically to the 5′-terminal cap in mRNA, *Proc. Natl. Acad. Sci. USA* **75**:4843.

Sonenberg, N., Morgan, M. A., Testa, D., Colonno, R. J., and Shatkin, A. J., 1979*a*, Interaction of a limited set of proteins with different mRNAs and protection of 5′-caps against pyrophosphatase digestion in initiation complexes, *Nucl. Acids Res.* **7**:15.

Sonenberg, N., Rupprecht, K. M., Hecht, S. M., and Shatkin, A. J., 1979*b*, Eukaryotic mRNA cap binding protein: Purification by affinity chromatography on Sepharose-coupled m⁷GDP, *Proc. Natl. Acad. Sci. USA* **76**:4345.

Sonenberg, N., Trachsel, H., Hecht, S., and Shatkin, A. J., 1980, Differential stimulation of capped mRNA translation *in vitro* by cap-binding protein, *Nature (London)* **285**:331.

Sonenberg, N., Guertin, D., Cleveland, D. R., and Trachsel, H., 1981*a*, Probing the function of the eukaryotic 5′ cap structure by using a monoclonal antibody directed against cap-binding proteins, *Cell* **27**:563.

Sonenberg, N., Skup, D., Trachsel, H., and Millward, S., 1981*b*, *In vitro* translation in reovirus- and poliovirus-infected cell extracts. Effects of anti-cap binding protein monoclonal antibody, *J. Biol. Chem.* **256**:4138.

Spendlove, R. S., Lennette, E. H., Knight, C. O., and Chin, J. H., 1963, Development of viral antigen and infectious virus in HeLa cells infected with reovirus, *J. Immunol.* **90**:548.

Spendlove, R. S., Lennette, E. H., Chin, J. N., and Knight, C. O., 1964, Effect of antimitotic agents on intracellular reovirus antigen, *Cancer Res.* **24**:1826.

Stanley, N. F., and Joske, R. A., 1975*a*, Animal model: Chronic murine hepatitis induced by reovirus type 3, *Am. J. Pathol.* **80**:181.

Stanley, N. F., and Joske, R. A., 1975b, Animal model: Chronic biliary obstruction caused by reovirus type 3, Am. J. Pathol. **80**:185.

Steiner-Pryor, A., and Cooper, P. P., 1973, Temperature-sensitive poliovirus mutants defective in repression of host protein synthesis are also defective in structural protein, J. Gen. Virol. **21**:215.

Trachsel, H., Sonenberg, N., Shatkin, A. J., Rose, J. K., Leong, K., Bergman, J. E., Gordon, J., and Baltimore, D., 1980, Purification of a factor that restores translation of VSV mRNA in extracts from poliovirus-infected HeLa cells, Proc. Natl. Acad. Sci. USA **77**:770.

Walden, W. E., Godefroy-Colburn, T., and Thach, R. E., 1981, The role of mRNA competition in regulating translation. I. Demonstration of competition in vivo, J. Biol. Chem. **256**:11,739.

Walsh, M. L., Jen, J., and Chen, L. B., 1979, Transport of serum components into structures similar to mitochondria, in: Cell Proliferation, Vol. 6, pp. 513–520, Cold Spring Harbor Press, New York.

Warrington, R. C., and Wratten, N., 1977, Differential action of L-histidinol in reovirus-infected and uninfected L-929 cells, Virology **81**:408.

Watanabe, Y., Millward, S., and Graham, A. F., 1968, Regulation of transcription of the reovirus genome, J. Mol. Biol. **36**:107.

Weber, K., Pollack, R., and Bibring, T., 1975, Antibody against tubulin: The specific visualization of cytoplasmic microtubules in tissue culture cells, Proc. Natl. Acad. Sci. USA **72**:459.

Weiner, H. L., and Fields, B. N., 1977, Neutralization of reovirus: The gene responsible for the neutralization antigen, J. Exp. Med. **146**:1305.

Weiner, H. L., Drayna, D., Averill, D. R., Jr., and Fields, B. N., 1977, Molecular basis of reovirus; Role of the S1 gene, Proc. Natl. Acad. Sci. USA **74**:5744.

Weiner, H. L., Ramig, R. F., Mustoe, T. A., and Fields, B. N., 1978, Identification of the gene coding for the hemagglutinin of reovirus, Virology **86**:581.

Weiner, H. L., Ault, K. A., and Fields, B. N., 1980, Interaction of reovirus with cell surface receptors. I. Murine and human lymphocytes have a receptor for the hemagglutinin of reovirus type 3, J. Immunol. **124**:2143.

Zarbl, H., and Millward, S., 1983, The reovirus multiplication cycle, in: The Viruses, The Reoviridae (H. Fraenkel-Conrat and R. Wagner, eds.), pp. 107–196, Plenum Press, New York.

Zarbl, H., Skup, S., and Millward, S., 1980, Reovirus progeny subviral particles synthesize uncapped mRNA, J. Virol. **34**:497.

Zweerink, H. J., and Joklik, W. K., 1970, Studies on the intracellular synthesis of reovirus-specified proteins, Virology **44**:501.

Zweerink, H. J., McDowell, M. J., and Joklik, W. K., 1971, Essential and nonessential noncapsid reovirus proteins, Virology **45**(3):716.

Zweerink, H. J., Ito, Y., and Matsuhisa, T., 1972, Synthesis of reovirus double-stranded RNA within virion-like particles, Virology **50**:349.

CHAPTER 10

Inhibition of Host Cell Macromolecular Synthesis following Togavirus Infection

Bunsiti Simizu

Department of Microbiology
School of Medicine
Chiba University
Chiba 280, Japan

1. INTRODUCTION

The family Togaviridae includes four genera alphavirus with 25 member viruses, flavivirus with approximately 50 members, rubivirus (rubella virus), and pestivirus (with several members, including mucosal diarrhea virus; Matthews, 1982). I will deal with only the first two genera, because viruses in these groups have common biological and epidemiological characteristics related to the arthropod-vertebrate-arthropod transmission cycle and include a number of important pathogens of humans and domestic animals (Schlesinger, 1980; Shope, 1980). Compared with alphaviruses and flaviviruses, which have been extensively studied, little is known about the structure, composition, and replication of rubiviruses and pestiviruses.

The replication of alphaviruses and flaviviruses is supported by both vertebrate and invertebrate cells in tissue culture. Infection by either virus group in permissive vertebrate cells produces infectious viruses and inhibits host cell macromolecular synthesis leading to eventual host cell death, whereas, in permissive arthropod cells, infectious progeny virus is produced without detectable inhibition of host cell growth. This difference between the cytolytic infection of vertebrate cells and noncytolytic infection of invertebrate cells is not absolute, since it is possible to establish noncytolytically infected vertebrate cell culture lines. Conversely, certain clones of mosquito cells exhibit a transient cytolytic response during acute infection (Stollar, 1980; Weiss and Schlesinger, 1981).

The replication of alphaviruses is relatively well understood at the molecular level and extensive review articles on these viruses have recently appeared (Strauss and Strauss, 1977, 1983; Kääriäinen and Söderlund, 1978; Schlesinger and Kääriäinen, 1980; Garoff *et al.*, 1982). In particular, an excellent review concerning the effects of alphavirus infection on host cell macromolecular synthesis has been written by Wengler (1980), and I will concentrate on recent progress in this field.

On the other hand, our knowledge of the molecular biology of flaviviruses is still primitive (Pfefferkorn and Shapiro, 1974; Westaway, 1980). Although we know little about the details of flavivirus transcription and translation, I will briefly discuss the currently available information concerning host cell responses to flavivirus infection.

2. ACUTE INFECTION BY ALPHAVIRUSES IN VERTEBRATE CELL CULTURES

Alphaviruses are particles approximately 70 nm in diameter consisting of an icosahedral nucleocapsid surrounded by a lipid envelope. The nucleocapsid contains the 42 S genomic RNA and a single species of nucleocapsid (C) protein with a molecular weight of about 30K. The envelope is composed of a host-derived lipid bilayer containing two virus-specific glycoproteins, called E1 and E2. The E1 and E2 proteins each have a molecular weight of about 50K and are present in the membrane as spikes. A third viral glycoprotein, E3, is associated with these spikes in the case of Semliki Forest virus, but is released into the culture fluid during infection by other alphaviruses (Garoff *et al.*, 1982; Strauss and Strauss, 1983).

Alphaviruses produce two mRNAs after infection in both vertebrate and invertebrate cells (Strauss and Strauss, 1977; Wengler *et al.*, 1978). One is identical to the genomic 42 S RNA and is translated into the nonstructural proteins. The other is a subgenomic 26 S RNA, identical to the 3'-terminal one-third of the genomic RNA, which is translated into the structural proteins of the virus. Both of these RNAs are capped and polyadenylated (Kennedy, 1980).

The nonstructural proteins of alphaviruses are translated from the 42 S genomic RNA as polyprotein precursors, which are processed by posttranslational cleavage. Studies of the polypeptides synthesized in Sindbis virus-infected cells suggest that four final nonstructural polypeptides are produced in the order 5'-ns60-ns89-ns76-ns72-3' (Strauss et al., 1983). Genetic analysis of temperature-sensitive (ts) mutants of the virus has shown that there are four complementation groups corresponding to these four nonstructural polypeptides which are involved in RNA synthesis (Strauss and Strauss, 1980). These polypeptides act as the replicase and/or transcriptase for RNA replication and subgenomic 26 S RNA transcription. Whether other functions are encoded in the nonstructural proteins as well is not known. Rice and Strauss (1981) have postulated that one of the nonstructural proteins may contain a protease activity for processing the polyproteins.

The structural proteins of the virus are translated from a subgenomic 26 S RNA. The 26 S RNA is produced in about threefold molar excess over the genomic RNA and much of the genomic 42 S RNA is quickly sequestered into nucleocapsids, where it cannot serve as messenger. The result is that 90% of the virus-specific mRNA is 26 S RNA, and 10% is genomic RNA and, thus, a large excess of structural over nonstructural polypeptides is produced (Strauss and Strauss, 1977, 1983).

Alphavirus replication requires three RNA synthesis activities: a minus strand replicase to produce full-length minus strands using the plus strand as a template, a plus-strand replicase to produce full length plus strands from the minus strands, and a transcriptase to produce the 26 S mRNA for the structural proteins. Each of these activities appears to be independently regulated and different recognition sites for the corresponding enzymes are utilized. These enzymatic activities apparently reside in four different nonstructural polypeptides (Keränen and Ruohonen, 1983; Strauss and Strauss, 1983; Strauss et al., 1983).

Most of the molecular biology of alphaviruses has been determined using a continuous baby hamster kidney (BHK) cell line or primary chicken embryo fibroblast (CEF) cells infected with either Sindbis (SIN) virus or Semliki Forest (SF) virus. Alphavirus–cell in-

teractions during acute and chronic infections in permissive vertebrate cells will be considered in Sections 2 and 3, respectively.

2.1. Effect on Cellular Protein Synthesis

It is well known that overall protein synthesis of permissive vertebrate cells is inhibited by alphavirus infections (Wengler, 1980).
When incorporation of radioactive amino acids into proteins is analyzed using SDS–polyacrylamide gel electrophoresis (PAGE), the decrease in host protein synthesis is concomitant with the increase in
the relative amounts of virus-specific polypeptides. At the end of the
exponential phase of virus growth, the majority of the newly synthesized proteins are virus specific (Pfefferkorn and Shapiro, 1974;
Strauss and Strauss, 1977). In general, the rate of the inhibition of
host protein synthesis depends upon the multiplicity of infection as
well as the time after infection.

Many fewer ribosomes are actively involved in synthesizing protein in infected cells compared to uninfected controls. However,
Wengler and Wengler (1976) analyzed the polysomes isolated from
SF virus-infected BHK cells on sucrose density gradients and determined that the inhibition of total protein synthesis could not be accounted for simply by the decrease in the number of ribosomes actively engaged in translation. Pulse-labeling experiments have also
shown that the half-lives of most newly synthesized cellular proteins
are not drastically altered. These data suggest that the inhibition is
not due to destruction of the protein-synthesizing machinery by infection.

It is reasonable to speculate that the inhibition of host protein
synthesis results from the replacement of cellular mRNAs by viral
mRNAs on the polysomes. Since viral structural polypeptides constitute the majority (see above) of the virus-specific polypeptides except at the earliest times after infection, 26 S mRNA should be the
predominant message associated with polysomes late in infection.
The relative amounts of 28 S rRNA, 18 S rRNA, and 26 S mRNA
present in polysomes have been determined at a time when most of
the protein being synthesized is virus specific, and it was found that
about 60% of the polysomes contain 26 S RNA as mRNA (Wengler
and Wengler, 1976). Tuomi *et al.* (1975) determined the number of
ribosomes present in HeLa and BHK cells and the amount of virus-
specific RNA accumulated in these cells after infection with SF virus

using a ^{32}P equilibrium-labeling method. They calculated that a BHK cell contains approximately 3×10^6 ribosomes and about 2×10^5 molecules of virus-specific 26 S RNA accumulated during virus replication. This number was comparable to the number of cellular mRNA molecules present in an uninfected cell. Since the 26 S RNA serves only as mRNA and is not sequestered into nucleocapsids, it does not exist to any significant extent in a nonpolysome-associated state (Mowshowitz, 1973; Strauss and Strauss, 1977). From these data we can conclude that most cellular mRNAs have been replaced by viral mRNAs in the polysomes, and that no apparent reduction in the number of polysomes occurs during infection.

It seems to be difficult to determine the molecular basis for the substitution of viral mRNA for cellular mRNA in the polysomes. Translation of the nonstructural and structural proteins of alphaviruses begins at an AUG codon approximately 60–80 nucleotides and 50 nucleotides from the 5′-terminal cap of 42 S RNA and subgenomic 26 S RNA, respectively (Ou *et al.*, 1983). The nucleotide sequence analysis of the 5′ ends of both 42 S and 26 S mRNA of alphaviruses has been determined (Garoff *et al.*, 1980; Rice and Strauss, 1981; Ou *et al.*, 1982*a*). Both mRNAs have a common sequence proposed for attachment to ribosomes (Ou *et al.*, 1982*a*). The site of ribosome attachment is an essential element of the scanning model recently proposed by Kozak (1981) to explain how ribosomes recognize a single initiation site in mRNA. This model appears to be applicable to most animal and viral mRNAs, therefore, it is unlikely that viral mRNAs have a significantly higher affinity for initiation of translation as compared to the majority of cellular mRNAs (Kozak, 1981). On the other hand, viral mRNAs differ from host mRNAs in their sensitivity to interferon (Metz, 1975; Yakobson *et al.*, 1977), high salt (England *et al.*, 1975; Nuss *et al.*, 1975; Carrasco and Smith, 1976; Cherney and Wilhelm, 1979), and certain antibiotics (Warrington and Wratten, 1977; Contreras and Carrasco, 1979; Ramabhadran and Thach, 1980). These differences may reflect those structural features of viral messages which distinguish them from cellular mRNAs and, thus, make it possible for viruses to take over the translational machinery of the host cell. For example, during poliovirus infection, the translation of capped, cellular mRNAs is inhibited by inactivation of a cap-binding protein, while translation of the uncapped poliovirus genomic RNA is unaffected (Ehrenfeld, 1982). Since both alphavirus 42 S and 26 S mRNAs have capped structures at their 5′ ends, such a mechanism would not explain selective translation of alphavirus messages after infection. However, van Steeg et al. (1981a,b) found in SF virus-infected neuroblastoma cells that host mRNAs and SF virus 42 S mRNA were sensitive to the initiation factor eIF-4B (cap-binding factor), relative to 26 S mRNA. These findings imply that translation of some capped mRNAs is inhibited in infected cells and that the recognition of late viral mRNA (26 S RNA) appears to be less stringent. This messenger may possess other features that promote its recognition by the protein-synthesizing machinery in infected cells.

Wengler and Wengler (1972) have shown that increased osmolarity of the growth medium inhibits the initiation of protein synthesis *in vivo*. Nuss et al. (1975) have also selectively blocked initiation of host protein synthesis in RNA virus-infected cells. Saborio et al. (1974) found that cellular protein synthesis is reversibly inhibited when cells are exposed to hypertonic medium. However, it has been reported that initiation of virus-specific protein synthesis is more resistant to a hypertonic initiation block (Carrasco and Smith, 1976). Carrasco (1977) also showed that treatment of encephalomyocarditis virus-infected HeLa cells with hypertonic medium stimulated the synthesis of virus-specific proteins and proposed that insertion of viral proteins into the cell membrane altered the ionic conditions within the cell to favor viral protein synthesis. The Carrasco hypothesis has been shown to apply to many cytolytic infections by animal viruses, although there are a few exceptions (Fenwick and Walker, 1978; Gray et al., 1983).

For alphaviruses, Garry et al. (1979a) and Contreras and Carrasco (1979) did similar experiments with SIN virus-infected CEF cells and SF virus-infected BHK cells, respectively. In SIN virus-infected CEF cells, Garry et al. (1979a) observed that the inhibition of host protein synthesis is temporally correlated with an increase in the intracellular Na^+ concentration and a decrease in the intracellular K^+ concentration after infection (Fig. 1). For uninfected CEF cells, either raising or lowering the NaCl concentration of the incubation medium inhibited protein synthesis, interfered with the initiation of translation of cellular mRNAs and changed the intracellular monovalent cation concentrations. Lowering the NaCl concentration of the extracellular medium decreases intracellular K^+, while raising the extracellular NaCl concentration increases both intracellular Na^+ and K^+.

The selective inhibition of host cell protein synthesis, both by infection and by media with altered NaCl, appears to result from

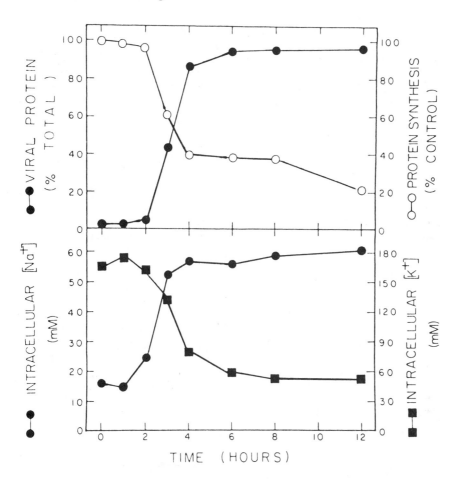

Fig. 1.   Levels of protein synthesis and intracellular Na$^+$ and K$^+$ concentrations in Sindbis virus-infected CEF cultures. (Upper panel) Infected and uninfected CEF cultures were labeled in triplicate with [^{3}H]leucine (1 μCi/ml) for 1 hr, ending at the times indicated on the graph. They were processed to determine acid-precipitable radioactivity. Total protein synthesis in infected cells (○) is expressed as a percentage of total incorporation into parallel uninfected cultures (range in controls: 2054–2179 cpm/culture). At each time interval, an additional infected culture was labeled with [^{35}S]methionine (25 μCi/ml) and the proteins were subjected to SDS–PAGE. Autoradiograms of the gels were analyzed in a densitometer. Bands corresponding to known alphavirus proteins were cut from the tracings and weighed. The percentage of the total incorporated label in viral proteins in infected cells was determined (●). (Lower panel) At various times after infection with Sindbis virus, duplicate CEF monolayers were washed with deionized distilled H$_2$O. The amount of intracellular Na$^+$ and K$^+$ was determined by atomic absorption spectroscopy. Protein, cell numbers, and cell volumes were determined in parallel cultures, and the intracellular cation concentrations were calculated. Intracellular Na$^+$ concentration (●); intracellular K$^+$ concentration (■). From Garry et al. (1979a), reprinted with permission.

changes in the ratio of intracellular Na^+ to K^+. To examine the role of Cl^- ions, Na-acetate was substituted for NaCl in isotonic medium (138 mM Na^+). Raising or reducing the Na-acetate concentration inhibited host protein synthesis but had little effect on virus replication (Garry et al., 1979a).

In normal cells, ion transport is regulated by the Na^+/K^+-dependent ATPase (sodium pump; Dahl and Hokin, 1974). Inhibition of the sodium pump should result in alteration of ionic environment inside infected cells. To test this hypothesis further, Garry et al. (1979b) studied the effect of ouabain, a specific inhibitor of the sodium pump. At a concentration of ouabain which inhibited protein synthesis in uninfected cells by 95% or more, viral protein synthesis was not affected. Ouabain treatment raises the intracellular Na^+ ion concentration and simultaneously decreases the intracellular K^+ ion concentration of uninfected cells. The effect of SIN virus infection is similar to that of ouabain treatment, suggesting that the sodium pump is indeed inhibited by virus infection.

Alterations in the permeability of the plasma membrane after infection cause not only changes in the ionic composition of the cytoplasm but also affect the distribution of other small metabolites (Condit, 1975; Ponta et al., 1976; Carrasco, 1977). Carrasco (1978) observed that the membrane of SF virus-infected cells became permeable to Gpp CH_2p, a nucleotide analogue that inhibits protein synthesis in vitro, at the onset of viral protein synthesis and that inhibition in vivo by Gpp CH_2p was specific for infected cells.

It is important to know whether viral-specified proteins are involved in the inhibition of host protein synthesis and ts mutants have been used to try to answer this question. RNA$^-$ ts mutants did not inhibit cellular protein synthesis at the restrictive temperature (Atkins, 1976). Under these conditions, RNA$^-$ ts mutants of alphaviruses make neither viral encoded proteins nor viral RNAs (Keränen and Kääriäinen, 1975; Hashimoto and Simizu, 1978). From these findings, we can conclude that initiation of viral RNA replication is necessary for inhibition of host protein synthesis and also that viral components introduced into cells by the infecting virions are not directly responsible for this inhibition.

On the other hand, RNA$^+$ ts mutants of alphaviruses are known to make defective proteins or uncleaved precursors at the restrictive temperature (reviewed in Strauss and Strauss, 1980). Mutants belonging to all three complementation groups of SIN RNA$^+$ ts mutants inhibit host protein synthesis at the restrictive temperature (Atkins,

1976). Thus, it appears that a normal complement of the final virion structural polypeptides E1, E2, and C is not required for the cellular protein inhibition, although intermediate cleavage products or an uncleaved precursor polypeptide may be involved. Which virus-specific polypeptide is responsible for damaging the sodium pump and altering the permeability of the host cell membrane can only be determined by further experimentation.

2.2. Effect on Cellular RNA Synthesis

Alphavirus infection also inhibits cellular RNA synthesis without a concomitant increase in the rate or extent of mRNA degradation. The rate of this inhibition depends on the multiplicity of infection.

Taylor (1965) showed that the amount of radioactivity incorporated into RNA is not significantly different in SF virus-infected CEF cells and in mock-infected cells early in infection. During the exponential phase of viral replication, however, about 70–90% of the newly synthesized RNA is virus specific, since it can be synthesized in the presence of actinomycin D. Mussgay *et al.* (1970) demonstrated almost the same inhibition pattern in CEF cells infected with SIN virus. Moreover, infection of BHK cells with WEE virus did not enhance cellular RNA degradation (M. Wagatsuma and B. Simizu, unpublished data).

These data lead to the conclusion that inhibition of host cell RNA synthesis occurs concomitantly with the increase in virus-specific RNA synthesis. Further evidence for this conclusion was obtained from the following experiments: (1) in interferon-treated CEF cells, SF virus-specific RNAs are not detected, and no inhibition of host cell RNA synthesis is observed (Taylor, 1965), and (2) ultraviolet-irradiated WEE virus loses the ability to inhibit host cellular RNA synthesis (M. Wagatsuma and B. Simizu, unpublished data). These data suggest that viral replication is necessary for the inhibition of host RNA synthesis, and that components of infecting virions are insufficient for this effect.

Inhibition of protein synthesis leads to a cessation of the appearance of newly synthesized ribosomes in the cytoplasm of eukaryotic cells (Hadjiolov and Nikolaev, 1976; Wengler, 1980). This phenomenon has not been characterized in detail, but is has been shown that synthesis and/or processing of the precursors to ribosomal RNAs is inhibited. Since alphavirus infection strongly inhibits host cell protein synthesis, as stated above, reduced levels of both ribosomal RNAs and cellular mRNAs may contribute to the observed inhibition of overall host cell RNA synthesis.

These data suggest that the alphavirus-induced inhibition of host cell RNA synthesis occurs at the transcriptional level. To test this possibility, the cell-free transcription system (Manley *et al.*, 1980) which has been developed for studying an inhibitory factor present in poliovirus-infected cells (Baron and Baltimore, 1982) would be useful. With such a system, it would be possible to determine whether inhibitory factor(s) are induced in alphavirus-infected cells or whether the inhibition results from a competition for the available pool of RNA precursors. The genetic approach is not possible at this time since no conditional lethal alphavirus mutant has been described which affects the regulation of host cell RNA synthesis.

2.3. Effect on Cellular DNA Synthesis

Infection of cells with cytocidal RNA viruses, including alphaviruses, often causes a rapid inhibition of host DNA synthesis. The mechanism or the significance of this inhibitory action remains obscure with all these viruses, though some suggestive findings have been provided. Hand and Tamm (1972) measured the rate of DNA chain growth in cells infected with Newcastle disease virus, mengovirus, or reovirus using DNA-fiber radioautography and concluded that the inhibition of DNA synthesis was due to a decrease in the frequency of chain initiation, rather than a reduction in the rate of chain growth. They also suggested that this inhibition might be a secondary effect of depression of necessary protein synthesis (Hand *et al.*, 1971). This explanation seems unlikely for alphaviruses based on the rapid and almost identical rates for inhibition of protein and DNA synthesis in WEE-infected cells (S. Koizumi and B. Simizu, unpublished data).

In alphavirus infection, the rate of this inhibition depends on the multiplicity of infection, and is closely related to viral replication. This inhibition does not result from degradation of cellular DNA, nor suppression of precursor uptake (Simizu *et al.*, 1976).

Recently, Koizumi *et al.* (1979*a,b*) have isolated a factor from BHK cells infected with SEE virus, which inhibits DNA synthesis *in vitro*. This factor could not be detected in uninfected cells. They showed that the factor is a nucleoside triphosphate phosphohydrolase (NTPase) which releases inorganic phosphate from all deoxyribonucleoside and ribonucleoside triphosphates *in vitro*. This enzyme is probably virus coded, since (1) the level of the NTPase did not decrease when infected cells were treated with actinomycin D, indicating that it did not originate from newly transcribed host cell mRNA, and (2) similar enzymes were produced in WEE, eastern equinine encephalitis, SIN virus-infected CEF, and mosquito cells but were not produced after vesicular stomatitis virus infection. The NTPase activity was associated with a fraction containing the 82K nonstructural protein of WEE virus when virus-specific proteins were analyzed by gradient centrifugation and SDS–PAGE (Ishida *et al.*, 1981). Therefore, the NTPase activity is probably a function of the nonstructural 82K polypeptide and may be involved in both viral RNA replication and inhibition of cellular DNA synthesis (Koizumi *et al.*, 1979*b*). A comparable activity has not yet been described for any of the four nonstructural proteins of SIN virus, the best characterized of the alphaviruses.

Another important question is whether the induced NTPase is related to the inhibition of DNA synthesis *in vivo*. In BHK cells infected with WEE, it has been shown that the higher the multiplicity of infection, the greater the inhibition of cellular DNA synthesis and the smaller the relative amount of dTTP in the intracellular nucleotide pool (Koizumi *et al.*, 1979*b*). These results imply that the induced NTPase degraded DNA precursors which, in turn, led to a reduced rate of DNA synthesis. However, at the same time, dTMP accumulated in the infected cells. This result could not be explained simply by the action of the NTPase, which hydrolyzes nucleoside triphosphates to diphosphates but does not hydrolyze diphosphates to monophosphates. However, dTMP could also accumulate if the synthesis of dTTP were blocked. Kielley (1970) reported that nucleoside diphosphates, particularly dADP and rADP, are strong inhibitors of dTMP kinase. Nucleoside diphosphates which accumulate in the cell due to the action of the NTPase could block the dTMP kinase, resulting in the accumulation of dTMP and the loss of dTTP, leading eventually to the inhibition of DNA synthesis. This hypothesis would also explain why DNA synthesis was inhibited without affecting viral RNA synthesis, although the NTPase dephosphorylated both ribonucleoside and deoxyribonucleoside triphosphates at nearly the same rates *in vitro*. It is well known that intracellular pools of rNTPs are much larger than those of dNTPs, and that the pool of rATP is particularly large (Brown, 1970; Skoog, 1970). If the NTPase lacks specificity for a particular triphosphate then the major product of the enzyme expected *in vivo* would be rADP, which would strongly block dTTP synthesis. In this way, partial hydrolysis of nucleoside triphosphates might be responsible for the inhibition of DNA synthesis without greatly affecting the pool of rNTPs.

On the other hand, Atkins (1976) has shown that $RNA^- ts$ mutants and those $RNA^+ ts$ mutants of SIN virus which make only the uncleaved precursor to the structural proteins at the restrictive temperature fail to inhibit DNA synthesis. These data suggest one or more of the structural proteins or intermediate cleavage products is responsible. Therefore, further studies are necessary to determine whether a virus coded nonstructual or structural protein is responsible for inhibition of host DNA synthesis.

2.4. Necessary Host Components for Viral Replication

In this section, I will discuss the evidence that host cell components are directly involved in the alphavirus replication cycle during acute cytolytic infection.

The entry of the virus particles into a host cell starts with binding of the virus through its spikes to a receptor present on the cell membrane. After binding, the virion is taking up into coated pits and transported inside coated vesicles to lysosomes where the acidic pH is thought to induce a change in the conformation of the spike glycoproteins causing fusion between the viral and the lysosomal membrane to occur (Marsh *et al.*, 1983). As a result, the virus nucleocapsid enters the cytoplasm and releases the RNA genome (Goldstein *et al.*, 1979; Bretscher *et al.*, 1980; Helenius *et al.*, 1980). These steps from entry to uncoating of alphaviruses appear to make use of a normal cellular transport system. These initial processes probably do not require *de novo* cellular protein synthesis, since they can occur equally well in cells treated with actinomycin D for long periods or in untreated cells (Baric *et al.*, 1983a,b).

Initial translation of the genomic RNA produces the nonstructural polypeptides for viral RNA replication at the rough endoplasmic reticulum (RER) membrane. RNA replication also occurs in association with the membranes of vacuoles which develop in the infected cells (Grimley *et al.*, 1968). These "cytopathic vacuoles" appear to be formed by fusion of various preexisting vacuoles. The viral RNA replication complex probably contains nonstructural proteins which have replicase and transcriptase activities including RNA chain elongation, as well as specific initiation factors (Gomatos *et al.*, 1980; Kennedy, 1980). The question of whether host components are associated with the replication complex is unresolved. However, there is evidence based on both biochemical (Clewley and Kennedy, 1976) and genetic (Kowal and Stollar, 1981) experiments that host-specific functions are required at some stage of alphavirus replication. One possible involvement is that host components form part of the viral replicase. Clewley and Kennedy (1976) found a host protein which copurified with the viral polymerase activity from SF virus-infected cells, although a requirement for this protein for proper polymerase function has not been demonstrated.

Another example of host cell involvement has been examined by Baric *et al.* (1983*a,b*) using SIN virus-infected hamster cells. Although alphavirus replication is insensitive to actinomycin D or α-amanitin added at infection, they found that treatment of cells with these inhibitors of host transcription before infection, using concentrations of the drugs which had little effect on the replication of vesicular stomatitis virus, reduced the yield of SIN virus to one-tenth or one-hundredth of normal. SIN virus replication was sensitive to α-amanitin in wild-type Chinese hamster ovary (CHO) cells but was resistant to the drug in α-amanitin-resistant CHO cells (which contain a resistant RNA polymerase II). They also isolated an α-amanitin-resistant SIN mutant after mutagenesis followed by selection in cells which had been treated with actinomycin D. This mutant grew normally not only in cells treated with actinomycin D but also in α-amanitin-treated cells. They speculate from these data that (1) the synthesis of cell mRNA (and presumably protein) is required for SIN virus replication at the stage after translation of the nonstructural proteins but before or during the synthesis of SIN virus negative strand RNA (Baric *et al.*, 1983*b*), (2) prior treatment with either drug affects the same aspect of SIN virus replication, and (3) mutations in the SIN genome can allow the virus to overcome the effect of inhibitors of host transcription.

Another example of host cell involvement in alphavirus replication comes from the results of Mento and Siminovitch (1981), who isolated mutant CHO cell lines in which SIN replicated without causing cytopathic effects. The mutants were selected in one step from mutagenized wild-type cells. In one of the mutant cell lines, the host cell block to virus replication appears to be at the level of virus mRNA translation. Although they could detect only very small amounts of viral antigens in the infected mutant cells, viral mRNAs made in these cells were fully functional and viral RNA was synthesized in amounts similar to those found in wild-type cells. These results indicate a direct or indirect host involvement in the alphavirus replication, either at the level of viral RNA synthesis or viral mRNA translation.

During the final stage of infection, viral membrane proteins are transported from their site of synthesis in the RER to the cell surface through the Golgi apparatus, and become modified during transport in several ways (Simons and Garoff, 1980; Hashimoto and Simizu, 1982). Carbohydrate units and fatty acids are added to both viral membrane proteins (Schlesinger and Kääriäinen, 1980), and the PE2 precursor polypeptide is cleaved to produce the E2 and E3 polypeptides (Simizu *et al.*, 1983). The chemical nature of these modifications has been well established (Garoff *et al.*, 1982). Transport and modification of viral membrane glycoproteins appear to utilize the same cellular organelles involved in synthesis and transport of secreted cellular glycoproteins (Green *et al.*, 1981). In an *in vitro* translation system, 26 S mRNA of alphaviruses is translated to produce normal capsid protein and a small amount of a membrane protein precursor. When the system is supplemented with RER membranes, the 26 S mRNA is translated into all the viral structural proteins (Bonatti and Blobel, 1979; Bonatti *et al.*, 1979). Thus maturation of the viral glycoproteins requires cytoplasmic membranes, but probably does not require *de novo* protein synthesis since cycloheximide does not interfer with these processes (Hashimoto and Simizu, 1976; Saraste *et al.*, 1980).

Garry *et al.* (1979*a*) have observed that under high Na^+ and low K^+ inside cells (the conditions after viral infection), two major classes of mRNAs which they have termed high-affinity mRNAs (HAMs) and low-affinity mRNAs (LAMs) are differentially translated. HAMs are efficiently translated and include viral mRNAs and a few host mRNAs such as those for IgG (Nuss and Koch, 1976), the β-chain of hemoglobin (Lodish, 1974), and chick actin (Garry *et al.*, 1979*a*). Since alphaviruses are known to be good interferon inducers in vertebrate cells (Lockart *et al.*, 1968), the mRNA for interferon also must belong to the HAM class. This would explain how interferon can be made under conditions where the synthesis of most host cell proteins is inhibited. Garry and Waite (1979) have also shown that HAM translation is inhibited in interferon-treated cells while LAM translation continues.

Thus, it is of note that some necessary host transcription and/or translation events continue even during lytic virus infection.

3. CHRONIC AND PERSISTENT INFECTION BY ALPHAVIRUSES IN VERTEBRATE CELL CULTURES

In the definition of persistent infection, a distinction can be made between persistence at the population level, in which only a small fraction of the cells are infected, and a persistent state in which most of the cells contain viral antigens (Walker, 1964). I will only discuss the latter case here.

It has been well known that many highly cytocidal RNA viruses are capable of long-term persistence and replication in vertebrate cell cultures (Friedman and Ramseur, 1979). The mechanisms for establishing and maintaining persistently infected cell lines are clear for rhabdoviruses (Holland *et al.*, 1982), but the factors mediating viral persistence with alphaviruses are less well understood. In general, alphaviruses are strongly cytopathic for vertebrate cells; however, several noncytocidal persistent infections have been established with alphaviruses, and events involved in the initiation and maintenance of chronically infected cultures have been characterized (Schwobel and Ahl, 1972; Eaton and Hapel, 1976; Weiss and Schlesinger, 1981). Three major factors have been implicated in the establishment and maintenance of alphavirus-persistent infections; (1) defective interfering (DI) particles, (2) mutants (particularly *ts* and plaque-type mutants of standard virus), and (3) production of interferon or other antiviral substances. The relative importance of each of these factors in persistence has not been determined.

Several investigators have suggested that DI particles mediate persistent infection by the alphaviruses in mammalian cells, although the molecular basis of the mechanism is not well understood (Schwobel and Ahl, 1972; Inglot *et al.*, 1973; Meinkoth and Kennedy, 1980). The ability of DI particles of SIN virus to interfere with the standard virus suggests that there is some similarity in the structure of the two RNAs which can be recognized by the replicase. This idea has been confirmed by the recent work of Ou *et al.* (1982*b*) and Monroe *et al.* (1982) showing that the first 20 nucleotides at the 3' end of SIN and its DI RNAs are highly homologous.

Weiss *et al.* (1980) established BHK cell lines persistently infected with SIN virus to examine the relative importance of the three factors named above in their establishment and maintenance. They initially established a persistent infection in BHK cells using a SIN virus inoculum greatly enriched in DI particles. A small fraction of the cells survived the initial infection and grew out to form a stable population of cells, most of which synthesized viral components. The presence of DI particles in the initial virus stock was required to establish this persistent state. These persistently infected cells released small-plaque, *ts* infectious virions as well as DI particles. When this *ts* virus viriant was isolated from the persistently infected cells and cloned to free it of detectable DI particles, it could initiate a persistent infection more quickly and with greater cell survival than the original stock of SIN virus-containing DI particles. They could show that this small-plaque, *ts* virus had a *ts* defect in RNA synthesis. They were unable to isolate either *ts*$^+$ or large-plaque revertants from the variant, indicating that it had acquired multiple mutations. BHK cells infected with the variant population could establish a persistent infection in the absence of DI particles. Since decreased cytopathogenicity of the infecting virus may be crucial for the initiation of persistent infection (Weiss and Schlesinger, 1981), it will be important to determine what kinds of mutations in the viral genome cause decreased cytopathogenicity. Eaton and Hapel (1976) have shown that confluent monolayers of mouse muscle cells can be persistently infected with Ross River virus in the absence of DI particles and that this noncytocidal infection produced low titers of small-plaque virus variants. Many selective pressures are probably operating in a persistent infection, all of which contribute to the types of virus mutants that survive in the culture. The most important pressures on the system are those which allow for the survival of both cells and virus (Holland *et al.*, 1979; Weiss and Schlesinger, 1981). The selection of new mutant viruses forced by the continued generation of DI particles provides a second role for DI particles different from the primary role they play in the initiation of a persistent virus infection (Weiss and Schlesinger, 1981). There are other examples in which the infectious virus recovered from persistently infected cultures have become resistant to the DI particles initially used in establishing the carrier state (Kawai and Matsumoto, 1977; Jacobson and Pfau, 1980; Brinton and Fernandez, 1983).

Endogenous interferon production also seems to play a part in initiating or maintaining some chronic infections. It may be, however, that in the persistently infected cultures interferon has been found to act in conjunction with at least one other factor stated above (Friedman and Ramseur, 1979).

4. INFECTIONS BY FLAVIVIRUSES IN VERTEBRATE CELL CULTURES

Flaviviruses are small, enveloped viruses, with cubic symmetry, approximately 45 nm in diameter, which replicate in vertebrate and invertebrate cells. The nucleocapsid contains the single-stranded plus RNA associated with a nucleocapsid protein, C, which has a molecular weight of 13K. The viral envelope consists of one large glycoprotein, E, of molecular weight approximately 55K, and a small 8K membrane-associated protein, M, which is not glycosylated (Westaway, 1980; Matthews, 1982). The structural components of flaviviruses have been reviewed by Russell *et al.* (1980) and the replication strategy of these viruses reviewed by Westaway (1980).

The flavivirus RNA with a molecular weight of 4–4.2 $\times$ 10^6 daltons, sediments at 42 S, is capped but lacks poly (A) (Wengler *et al.*, 1978; Cleaves and Dubin, 1979; Wengler and Wengler, 1981), and is infectious. The absence of a 3'-terminal poly(A) from the viral genome seems to be unique among the plus-stranded viruses (reviewed in Strauss and Strauss, 1983). Since the deproteinized RNA is infectious, the input genome must be translated to produce the viral replicase but the translation strategy of the flavivirus genome has not been definitively established. No evidence for a subgenomic RNA has been reported and since the only viral mRNA detected on polysomes is the 42 S RNA, it is generally accepted that the genomic RNA is the only messenger. Little is known about the viral replicase.

Westaway (1977) has proposed that three structural polypeptides, C, E, and M as well as five nonstructural polypeptides are separately initiated and terminated during translation, which would make the flavivirus mRNA unique since most other animal mRNAs studied to date have only a single translation initiation site, with a few exceptions (Kozak, 1981; Strauss and Strauss, 1983). On the other hand, Wengler *et al.* (1979) and Svitkin *et al.* (1981) reported that in an *in vitro* translation system only a single initiation site appeared to be used. The entire sequence of the flavivirus genome as well as amino acid sequence data for the polypeptides will be required to decide between these alternative translation strategies.

Most of the biochemical analysis and molecular biology of flaviviruses have been determined with dengue type 2 (DEN-2), Japanese encephalitis (JE), Kunjin (KJ), St. Louis encephalitis (SLE), and West Nile (WN) virus infections in BHK, CEF, and Vero cells. From a comparison of these data, it is clear that all flaviviruses share a common strategy of replication and elicit comparable responses in their host cells after infection (Westaway, 1980).

4.1. Acute Infection

The events leadng to adsorption, penetration, and uncoating of flaviviruses are poorly understood on a molecular level. After un-coating, replication of these viruses occurs entirely in the cytoplasm with a latent period of 12–16 hr before virions are released extracel-lularly. Early translation of the genome produces the virus-specific RNA polymerase, as yet unidentified. Transcription from 42 S RNA occurs on smooth membranes in the perinuclear region, probably in association with developing meshlike masses of coiled microtubules (Boulton and Westaway, 1972, 1976). During the latent period, RNase-resistant 20–26 S RNA appears in the cytoplasm (Trent *et al.*, 1969; Naeva and Trent, 1978). The 20 S RNA consists of the template and a nascent strand. Both plus and minus 42 S strands are synthe-sized in low and approximately equal amounts during the latent period in JE, KJ, and WN virus infections. Thereafter, the transcription of plus-strand RNA predominates.

Flavivirus-specific protein synthesis is difficult to detect up to 12 hr after infection, but subsequently increases, reaching a maximum rate at about 20–24 hr (Shapiro *et al.*, 1971; Trent and Qureshi, 1971). With KJ virus at 24 hr postinfection host protein synthesis is contin-uing, while viral protein synthesis constitutes a variable proportion of total cell protein synthesis, averaging about 30% (Westaway, 1973, 1975). In flavivirus-infected cells, the pattern of host proteins syn-thesized throughout the entire period of infection remains constant and is qualitatively equivalent to that seen in uninfected cells. Similar net electrophoretic profiles of virus-specific proteins, labeled and har-vested during successive periods postinfection, are obtained when ^{14}C-labeled mock-infected cell proteins are coelectrophoresed with ^{3}H-labeled infected cell proteins (Westaway, 1973, 1975).

At the late stage, host protein synthesis is inhibited rapidly con-comitant with the appearance of cellular vacuoles and other cyto-pathic changes. Virus is released when virion-containing vesicles empty into the extracellular medium by reverse phagocytosis or cell lysis. Budding figures are not seen, but virus may also be released by leakage from cisternae or lamellae contiguous with the plasma membrane. It is unclear whether inhibition of host protein synthesis at this late stage is due to the same mechanism as that observed for the alphavirus infection. Some experimental results, however, suggest that treatment of infected cells with hypertonic salt can be used to detect KJ virus proteins in Vero cells during the early period of infection (Nuss *et al.*, 1975). It has also been reported that control of translation of flavivirus RNA *in vitro* appears to be profoundly influenced by the concentration of K^+ ions in the incubation (Svitkin *et al.*, 1981; Monckton and Westaway, 1982).

Inhibition of host cell RNA and DNA synthesis during acute infection with flaviviruses has also been observed but no details have been reported. Some investigators have reported that actinomycin D does not interfere with flavivirus replication at any stage of replication (Pfefferkorn and Shapiro, 1974; Boulton and Westaway, 1976; Wengler *et al.*, 1978). On the other hand, some investigators have claimed that an early event in SLE virus replication requires host cell nuclear functions (Brawner *et al.*, 1979). Some of the early viral events appear to be intimately associated with the nucleus of the infected cell. Brawner *et al.* (1979) have speculated that a rapidly turned over protein is necessary for the synthesis of early viral RNA and that in the continual presence of actinomycin D, the synthesis of this protein is severely curtailed, resulting in reduced virus production. However, addition of the drug at 9 hr after infection had little effect on the final yield of virus (Brawner *et al.*, 1979).

The yield of JE virus from cells enucleated during the latent period was severely depressed, whereas, only a small effect was observed in similar experiments with SIN virus (Kos *et al.*, 1975). This inhibition, however, could result from loss of nuclear-associated cytoplasmic membranes during enucleation. Cleaves and Dubin (1979) suggest that DEN-2 RNA replication is wholly extranuclear, since viral RNA isolated during the logarithmic growth phase lacks internal m^6 adenine residues. It is obvious from these conflicting reports that the role of the cell nucleus in flavivirus replication needs to be investigated further.

4.2. Persistent Infection

Several mammalian cell lines persistently infected with flaviviruses have been established and possible mechanisms for establishing persistence have been proposed (Friedman and Ramseur, 1979; Schmaljohn and Blair, 1977, 1979; Mathews and Vorndam, 1982).

Brinton (1981) has established WN virus-persistent infections in mouse cell lines from mice genetically resistant or susceptible to flavivirus infection. Less incorporation of [^{3}H]-UR into genomic WN 42 S RNA was found in resistant cells than in susceptible cells, and fewer infectious virions were produced. While screening for *ts* WN mutants from one of these cultures, she isolated a non-*ts* mutant from a persistently infected resistant cell culture which replicated efficiently in resistant cells as measured by [^{3}H]-UR incorporation into 42 S RNA and production of infectious virus. This mutant also replicated somewhat more efficiently than the parental WN in susceptible cells. Brinton (1983) has also detected DI particles of WN virus in progeny obtained after acute infection of both congenic susceptible and genetically resistant mouse cell lines. The DI particles consistently represented a significantly larger proportion of the yield produced by the resistant cells than of that produced by the susceptible cells under the same conditions. These data strongly suggest that a host protein specified by the mouse flavivirus resistance gene may interact specifically at the level of the flavivirus RNA replication complex, affecting template–polymerase interactions (Brinton, 1983).

Schmaljohn and Blair (1979) cloned a number of cells from JE virus carrier cultures. These cells maintain viral persistence by permitting virus-specific RNA synthesis to continue even though viral antigens and infectious virions are not produced. The interpretation of this result is unclear, but they found no evidence for integration of some form of JE virus RNA into host DNA in this case as had been previously reported by another group (Zhdanov, 1975).

5. INFECTION BY TOGAVIRUSES IN INVERTEBRATE CELL CULTURES

Alphaviruses and flaviviruses are transmitted in nature by arthropod vectors, mosquitoes in the case of alphaviruses, and mosquitoes or ticks in the case of flaviviruses. At the present time, there are a number of cell lines derived from arthropods which are suitable for the growth of togaviruses (Stollar, 1980). The most widely used line of invertebrate cells for studying togaviruses is Singh's *Aedes albopictus* cells, which were derived originally from larvae (Singh, 1967) and grown in an undefined medium formulated by Mitsuhashi and Maramorosch (1964). Recently, *A. albopictus* cells have been adapted to Eagle's medium in some laboratories (see Stollar, 1980), making radioactive labeling and other biochemical studies feasible. This *A. albopictus* cell line can be grown easily either as a monolayer or in suspension culture at 28°C.

Infection of *A. albopictus* cells with alphaviruses leads to the synthesis of maximum titers of infectious virus by approximately 24 hr postinfection (acute phase). At the acute phase, up to 85% of the cells released infectious virus (Davey and Dalgarno, 1974), and the rate of synthesis of intracellular 42 S and 26 S viral RNA and structural proteins is maximal. By 48 hr postinfection, viral 26 S RNA and protein syntheses are inhibited, and the proportion of cells releasing virus is dramatically reduced (Davey and Dalgarno, 1974; Eaton, 1979). Inhibition of 42 S RNA synthesis occurs at 3 days after infection (start of chronic phase; see Fig. 2). The nature of the factor(s) responsible for inhibiting viral replication in infected mosquito cells at the chronic phase is not known. The fact that large amounts of 42 S RNA are made in infected mosquito cells 48 hr after infection, at a time when viral structural protein synthesis is inhibited, suggests that the "replication inhibition factor" may act initially at the level of viral protein synthesis (Eaton, 1979). Eaton demonstrated that viral structural protein synthesis is inhibited before a decrease in 26 S RNA synthesis is detected in SIN-infected *A. albopictus* cells. Thus, the biphasic nature of alphavirus infection is also observed in the case of flavivirus infection in mosquito cells (Paul *et al.*, 1969).

5.1. Acute Infection

Several investigators have reported that flaviviruses (DEN, JE, and WN) produce CPE on cultured Singh's *A. albopictus* cells (Paul *et al.*, 1969; Suitor and Paul, 1969; Djinawi and Olson, 1973). Infection of the same cultures with alphaviruses such as Chikungunya, SIN, and WEE is noncytopathic although the virus yields are comparable to those of flaviviruses (Stevens, 1970; Davey and Dalgarno, 1974; Esparza and Sanchez, 1975; Simizu and Maeda, 1981). These differences in the response to the two virus groups are not absolute and clones of *A. albopictus* cells have been isolated which show CPE after alphavirus infection (Igarashi, 1978; Logan, 1979). Sarver and Stollar (1977) have isolated several clones of variant *A. albopictus* cells which differ markedly in their response to infection with SIN virus. Certain clones showed no CPE (CPE resistant), while others manifested a severe CPE in less than 20 hr after infection (CPE susceptible).

The cytopathic effects manifested in mosquito cells after alphavirus infections differ from those seen in infected vertebrate cells. Aggregation and giant cell formation are prominent features of togavirus-infected mosquito cells (Suitor and Paul, 1969; Sarver and Stollar, 1977; Igarashi, 1978; Simizu and Maeda, 1981). The aggregation appears to be due to changes in the cell membrane (Impraim *et al.*, 1980). Even when extensive CPE occurs during togavirus infection, small foci of aggregated cells usually survive (Paul *et al.*, 1969) and eventually grow out into colonies. The cellular changes responsible for aggregation appear to enable some members of the infected population to overcome the acute infection and enter the persistent phase. The CPE depends on temperature. At 28°C, CPE-susceptible cells were frequently aggregated and in the form of extended fibroblasts, but at 34°C or 37°C the same infected cell population tended to show clear cut cell destruction with an accumulation of cellular debris (similar to the CPE seen in vertebrate cells) resulting in cell death. Infected CPE-resistant cells could not be distinguished by morphology from mock-infected cells at 28°C, or 37°C (Sarver and Stollar, 1977).

When *A. albopictus* cells are infected with an alphavirus, the major intracellular viral RNAs are single-stranded 26 S and 42 S RNAs, and a double-stranded species sedimenting at 22 S (Eaton and Randlett, 1978). Both the 26 S and 42 S RNA have been found associated with polysomes (Eaton and Regnery, 1975). Furthermore, Wengler *et al.* (1978) have confirmed that viral RNA species made in mosquito cells infected with flaviviruses did not differ from those made in BHK cells.

Sarver and Stollar (1977) compared host cell RNA synthesis in CPE-sensitive and CPE-resistant mosquito cells infected with SIN virus. Early in infection, prior to observed CPE, there was a marked depression of RNA synthesis in the CPE-susceptible cells, while CPE-resistant cultures showed only a slight transient decrease in host cell RNA synthesis. WEE and SF virus infection of CPE-susceptible cells also caused an extensive inhibition of host protein and DNA synthesis (Logan, 1979; Simizu and Maeda, 1981; Tooker and Kennedy, 1981).

From these data, one may infer that viral mRNAs replace the majority of cellular mRNAs on the polysomes in CPE-susceptible cells, a situation analogous to that described above for alphavirus-infected vertebrate cells. Two questions remain to be answered: (1) how do acutely infected cells recover quickly without any treatment, and (2) how can togaviruses replicate in CPE-resistant cells without depressing host macromolecular synthesis?

5.2. Persistent Infection

As previously noted, although infection with flaviviruses often causes an initial cytopathic response in *A. albopictus* cells, the cultures usually recover within a few days without any additional treatment, such as the addition of antivirus antiserum. In the case of alphavirus infection in CPE susceptible cells, the cultures recover easily as well. Thus, infection with either alpha- or flaviviruses easily leads to the establishment of persistently infected cultures (Peleg, 1969; Stollar and Shenk, 1973; Shenk *et al.*, 1974; Simizu and Maeda, 1981). Little is known about the properties of cultures persistently infected with flaviviruses, therefore, the following discussion will be devoted primarily to mosquito cells persistently infected with alphaviruses.

The establishment of mosquito cell cultures persistently infected with alphaviruses had been described in detail by Stollar (1980). In our laboratory, CPE-sensitive *A. albopictus* cells (strain C6/36) were infected with WEE virus at an input multiplicity of 5 PFU/cell. During the first few days, titers were 10^8-10^9 PFU/ml of culture fluid. Thereafter, the titer decreased quickly to a plateau value of 10^5 PFU/ml. After the acute phase (10 days after infection; Fig. 2), the cultures were split at weekly intervals and these persistently infected cultures were maintained and subcultured for more than 1 year (Maeda *et al.*, 1979; Simizu and Maeda, 1981; Simizu *et al.*, 1983). Similar results have been obtained with SIN and SFV infection in *A. albopictus* cells, suggesting that host factors act to regulate and control viral replication. Davey and Dalgarno (1974) found a marked shut-off of SF viral RNA synthesis within 48 hr after infection. Eaton (1979) showed that *A. albopictus* cells infected with SIN virus synthesized large amounts of viral structural proteins at approximately 24 hr after infection, but by 48 hr synthesis of virus-specific protein was barely detectable. At this time after infection, infected cells transiently synthesized large amounts of host nuclear proteins. Of five proteins detected (26K, 41K, 43K, 71K, and 80K), a major component (P43) was identified as heterogeneous nuclear ribonucleoprotein (hnRNP; Eaton, 1982). Eaton has speculated that if the inhibition of nuclear RNA synthesis accompanies virus replication then it might be expected that when viral replication and viral protein synthesis is de-

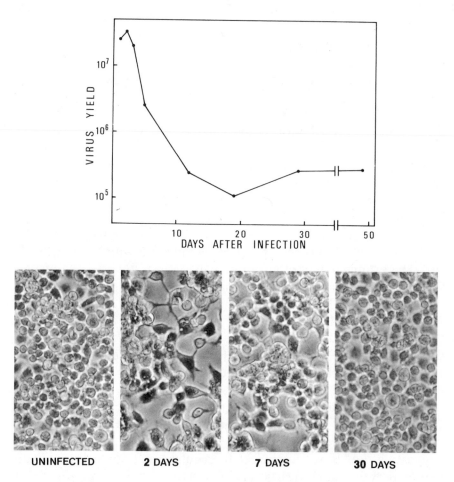

Fig. 2.   Virus growth curve of wild-type WEE virus in *A. albopictus* (strain C6/36) cells. Cultured monolayers were washed once with PBS and then infected with virus at 5 PFU/cell. After adsorption for 90 min at 28°C the inocula were removed, mono-layers were washed three times with PBS, and culture medium was added. Infected cultures were incubated at 28°C. The samples of culture fluid were taken at the indicated times for plaque assay on CEF monolayers at 34°C. After 10 days, the infected cultures were split weekly at 1:5 or 1:10. The lower panels show uninfected and WEE virus-infected C6/36 cells photographed with phase contrast light micros-copy (B. Simizu and S. Maeda, unpublished data).

creasing at 48 hr, cellular RNA synthesis may recommence. P43 synthesis at 2–4 days after infection may, therefore, accompany a burst of cellular hnRNA synthesis. Reigel and Koblet (1981) have also reported that two host proteins (54K and 64K) disappear from the culture fluid of SF virus-infected C6/36 cells during the period of maximal virus yield. When the cells entered the permanently infected state the two proteins reappeared. The amounts of these two secreted proteins are unchanged in the culture fluids from infected or uninfected CPE-resistant mosquito cells. The function of these proteins is not known but they may act as a CPE-depressing factor.

Another factor affecting the persistent infection of mosquito cells by togaviruses is the production of an interfering substance, although released virions or DI particles may also affect persistence early in infection (Igarashi *et al.*, 1977; Riedel and Brown, 1977). Interferon is not synthesized in invertebrate cells, but an antiviral factor has been found in the culture fluid of SIN virus-infected *A. albopictus* cells (Riedel and Brown, 1979*a,b*). The factor is rapidly inactivated on treatment with proteinase K or with heat, and is not inactivated by antibody prepared against extracts of SIN-infected BHK cells. Thus, the factor is both virus and cell specific, unlike interferon (Riedel and Brown, 1979*b*). The antiviral activity was produced at detectable levels by 3 days postinfection, and its concentration in the extracellular medium increased thereafter. It was a labile protein of low molecular weight which could specifically suppress the replication of the infecting virus but not heterologous virus during the acute phase of infection in mosquito cells. Newton and Dalgarno (1983) found a similar factor in *A. albopictus* cells persistently infected with SF virus, but they could not detect analogous antiviral factor in *A. albopictus* cells persistently infected with a flavivirus or bunyavirus.

Lee and Schloemer (1981*a,b*) have recently shown that mosquito cells infected with a flavivirus, Banzi virus, produce an antiviral factor. The factor is inactivated by anti-Banzi virus serum but not by anti-mosquito cell serum. The anti-Banzi viral factor also suppressed the replication of Banzi virus in BHK cells. They demonstrated that the antiviral factor is antigenically related to the viral membrane protein, M, and is of viral origin. This factor differs from the antiviral agent described above from SIN-infected cells, which was neither inactivated by anti-SIN antiserum nor affected the growth of SIN virus in BHK cells (Riedel and Brown, 1979*b*).

Finally, *ts* mutants are often produced during infection of mosquito cells by alphaviruses and may contribute to the persistent state

(Shenk *et al.*, 1974; Maeda *et al.*, 1979). Maeda *et al.* (1979) isolated and analyzed *ts* mutants from mosquito cells persistently infected with WEE virus. They found that single *ts* mutants (either RNA$^-$ or RNA$^+$) could be isolated during the acute phase, but that the predominant viruses produced after many passages in culture were multiple mutants of RNA$^-$ phenotype. Since these mutants arise long after the establishment of the chronic infection, it is likely that they have a significant selective advantage in these cells, even though they are not necessary for either the establishment or maintenance of persistent infection. Although wild-type or single *ts* mutants show typical CPE in C6/36 cells, the RNA$^-$ mutants with multiple *ts* lesions, isolated after many passages, did not show such detectable CPE.

Several observations indicate that the type and physiological state of host cell greatly influence virus replication. Tooker and Kennedy (1981) have isolated a number of *A. albopictus* cell clones which they classified as either high or low yielders of SF virus. The restriction of virus yield in the low-yielding clones seems to occur at the level of nonstructural protein synthesis or minus-strand RNA synthesis. A necessary host component for viral replication which is present in the high yielding clones may be absent or produced in insufficient amounts in the low-yielding clones. Kowal and Stollar (1981) isolated two mutants of SIN virus which were restricted in their ability to grow in mosquito cells at 34.5°C, but grew normally at this temperature in CEF cells. However, they were *ts* in CEF cells and failed to grow at 40°C. They belong to complementation group F, which is thought to encode the elongation function (see Strauss and Strauss, 1980).

In addition to a role in RNA synthesis, a host factor(s) may be involved at a later step in alphavirus maturation in mosquito cells, as some investigators have shown that SIN maturation is inhibited in *A. albopictus* cells treated with actinomycin D or low concentrations of trypsin (Scheefers-Borchel *et al.*, 1981; Adams and Brown, 1982).

Thus, the togaviruses are able to replicate in arthropod cells and in a wide range of vertebrate cells. This wide host range implies that any functions supplied by the host during replication must be common to a broad phylogenetic range of organisms. This is a unique characteristic of alpha- and flaviviruses.

6. CONCLUDING REMARKS

In this chapter, I have attempted to summarize the current state of our knowledge of the effects of togavirus infection on host cell macromolecular synthesis, considering both cytolytic and chronic infections in vertebrate and invertebrate cells. In the case of acute infection by alphaviruses, infection leads to an altered ionic environment within the cell cytoplasm which, in turn, strongly favors translation of viral messages, at the expense of host mRNAs; the host cell becomes a virus factory, producing viral components almost exclusively, leading to death of the host. In the case of chronic and inapparent infection by both alphaviruses and flaviviruses in vertebrate cells, and with both groups of viruses in most invertebrate cells, there is a balance between low levels of virus production and ongoing cellular synthesis and infected cultures can grow and divide for many generations. These two disparate modes of virus replication in culture probably reflect the events of viral transmission in nature. Acute infection of mammals and birds leads to sufficient viremia such that an arthropod vector receives an infecting dose of virus in a blood meal. Subsequently, an inapparent infection is established in the invertebrate vector, which then serves as a virus reservoir.

Togaviruses have necessarily evolved to replicate successfully under these two very different conditions and their genomic organization probably reflects the constraints imposed by the vertebrate-arthropod-vertebrate transmission cycle.

ACKNOWLEDGMENTS

I gratefully acknowledge the helpful comments, suggestions, and generous advice by Drs. Ellen G. Strauss and J. H. Strauss, California Institute of Technology during the preparation of this manuscript. I also thank Dr. Akiko Higa, Tsukuba University, Dr. K. Hashimoto, National Institute of Health in Tokyo, and Dr. Y. Tomita, Chiba University for their critical reading of this manuscript. The work of the author was done at the National Institute of Health in Tokyo while he stayed there and was supported by grants from the Ministry of Health and Welfare of Japan.

7. REFERENCES

Adams, R. H., and Brown, D. T., 1982, Inhibition of Sindbis virus maturation after treatment of infected cells with trypsin, *J. Virol.* **41**:692.
Atkins, G. J., 1976, The effect of infection with Sindbis virus and its temperature-sensitive mutants on cellular protein and DNA synthesis, *Virology* **71**:593.

Baric, R. S., Carlin, L. J., and Johnston, R. E., 1983a, Requirement for host transcription in the replication of Sindbis virus, *J. Virol.* **45**:200.

Baric, R. S., Linberger, D. W., and Johnston, R. E., 1983b, Reduced synthesis of Sindbis virus negative strand RNA in cultures treated with host transcription inhibitors, *J. Virol.* **47**:46.

Baron, M. H., and Baltimore, D., 1982, Purification and properties of a host cell protein required for poliovirus RNA replication *in vitro*, *J. Biol. Chem.* **257**:12351.

Bonatti, S., and Blobel, G., 1979, Absence of a cleavable signal sequence in Sindbis virus glycoprotein PE2, *J. Biol. Chem.* **254**:12261.

Bonatti, S., Cancedda, R., and Blobel, G., 1979, Membrane biogenesis: *In vitro* cleavage, core glycosylation and integration into microsomal membranes of Sindbis virus glycoproteins, *J. Cell. Biol.* **80**:219.

Boulton, R. W., and Westaway, E. G., 1972, Comparison of togaviruses: Sindbis virus (group A) and Kunjin virus (group B), *Virology* **49**:283.

Boulton, R. W., and Westaway, E. G., 1976, Replication of the flavivirus Kunjin: Proteins, glycoproteins, and maturation associated with cell membrane, *Virology* **69**:416.

Brawner, I. A., Trousdale, M. D., and Trent, D. W., 1979, Cellular localization of Saint Louis encephalitis virus replication, *Acta Virol.* **23**:284.

Bretscher, M. S., Thomson, J. N., and Pearse, B. M. F., 1980, Coated pits act as molecular filters, *Proc. Natl. Acad. Sci. USA* **77**:4156.

Brinton, M. A., 1981, Isolation of a replication-efficient mutant of West Nile virus from a persistently infected genetically resistant mouse cell culture, *J. Virol.* **39**:413.

Brinton, M. A., 1983, Analysis of extracellular West Nile virus particles produced by cell cultures from genetically resistant and susceptible mice indicates enhanced amplification of defective interfering particles by resistant cultures, *J. Virol.* **46**:860.

Brinton, M. A., and Fernandez, A. V., 1983, A replication-efficient mutant of West Nile virus is insensitive to DI particle interference, *Virology* **129**:107.

Brown, P. R., 1970, The rapid separation of nucleotides in cell extracts using high-pressure liquid chromatography, *J. Chromatogr.* **52**:257.

Carrasco, L., 1977, The inhibition of cell functions after viral infection. A proposed general mechanism, *FEBS Lett.* **76**:11.

Carrasco, L., 1978, Membrane leakiness after viral infection and a new approach to the development of antiviral agents, *Nature (London)* **264**:807.

Carrasco, L., and Smith, A. E., 1976, Sodium ion and the shut-off of host cell protein synthesis by picornaviruses, *Nature (London)* **264**:807.

Cherney, C. S., and Wilhelm, J. M., 1979, Differential translation in normal and adenovirus type 5 infected human cells and cell-free systems, *J. Virol.* **30**:533.

Cleaves, G. R., and Dubin, D. T., 1979, Methylation status of intracellular Dengue type 2 40S RNA, *Virology* **96**:159.

Clewley, J. P., and Kennedy, S. I. T., 1976, Purification and polypeptide composition of Semliki Forest virus RNA polymerase, *J. Gen. Virol.* **32**:395.

Condit, R. C., 1975, F factor-mediated inhibition of bacteriophage T7 growth: Increased membrane permeability and decreased ATP levels following T7 infection of male *E. coli*, *J. Mol. Biol.* **98**:45.

Contreras, A., and Carrasco, L., 1979, Selective inhibition of protein synthesis in virus-infected mammalian cells, *J. Virol.* **29**:1979.

Dahl, J. L., and Hokin, L. E., 1974, The sodium-potassium adenosine triphosphatase *Annu. Rev. Biochem.* **43**:327.

Davey, M. W., and Dalgarno, L., 1974, Semliki Forest virus replication in cultured *Aedes albopictus* cells: Studies on the establishment of persistence, *J. Gen. Virol.* **24**:453.

Djinawi, N. K., and Olson, L. C., 1973, Cell fusion induced by Germiston and Wesselsbron virus, *Arch. Ges. Virusforsch.* **43**:144.

Eaton, B. T., 1979, Heterologous interference in *Aedes albopictus* cells infected with alphaviruses, *J. Virol.* **30**:45.

Eaton, B. T., 1982, Transient enhanced synthesis of a cellular protein following infection of *Aedes albopictus* cells with Sindbis virus, *Virology* **117**:307.

Eaton, B. T., and Hapel, A. J., 1976, Persistent noncytolytic togavirus infection of primary mouse muscle cells, *Virology* **72**:266.

Eaton, B. T., and Randlett, D. J., 1978, Origin of the Actinomycin D insensitive RNA species in *Aedes albopictus* cells, *Nucl. Acids Res.* **5**:1301.

Eaton, B. T., and Regnery, R. L., 1975, Polysomal RNA in Semliki Forest virus infected *Aedes albopictus* cells, *J. Gen. Virol.* **29**:35.

Ehrenfeld, E., 1982, Poliovirus-induced inhibition of host-cell protein synthesis, *Cell* **28**:435.

England, J. M., Howett, M. K., and Tan, K. B., 1975, Effect of hypertonic conditions on protein synthesis in cells productively infected with simian virus 40, *J. Virol.* **16**:1101.

Esparza, J., and Sanchez, A., 1975, Multiplication of Venezuelan equine encephalitis (Mucambo) virus in cultured mosquito cells, *Arch. Virol.* **49**:273.

Fenwick, M. L., and Walker, M. J., 1978, Suppression of the synthesis of cellular macromolecules by herpes simplex virus, *J. Gen. Virol.* **41**:37.

Friedman, R. M., and Ramseur, J. M., 1979, Mechanism of persistent infections by cytopathic viruses in tissue culture, *Arch. Virol.* **60**:83.

Garoff, H., Frischauf, A.-M., Simons, K., Lehrach, H., and Delius, H., 1980, Nucleotide sequence of cDNA coding for Semliki Forest virus membrane glycoproteins, *Nature (London)* **288**:236.

Garoff, H., Kondor-Koch, C., and Riedel, H., 1982, Structure and assembly of alphaviruses, *Curr. Top. Microbiol. Immunol.* **99**:1.

Garry, R. F., and Waite, M. R. F., 1979, Na$^+$ and K$^+$ concentrations and the regulation of the interferon system in chick cells, *Virology* **96**:121.

Garry, R. F., Bishop, J. M., Parker, S., Westbrook, K., Lewis, G., and Waite, M. R. F., 1979*a*, Na$^+$ and K$^+$ concentrations and the regulation of protein synthesis in Sindbis virus-infected chick cells, *Virology* **96**:108.

Garry, R. F., Westbrook, K., and Waite, M. R. F., 1979*b*, Differential effects of ouabain on host- and Sindbis virus-specified protein synthesis, *Virology* **99**:179.

Goldstein, J. L., Anderson, R. G. W., and Brown, M. S., 1979, Coated pits, coated vesicules, and receptor-mediated endocytosis, *Nature (London)* **279**:679.

Gomatos, P., Kääriäinen, L., Keränen, S., Ranki, M., and Sawicki, D. L., 1980, Semliki Forest virus replication complex capable of synthesizing 42S and 26S nascent RNA chain, *J. Gen. Virol.* **49**:61.

Gray, M. A., Micklem, K. J., Brown, F., and Pasternak, C. A., 1983, Effect of vesicular stomatitis virus and Semliki Forest virus on uptake of nutrients and intracellular cation concentration, *J. Gen. Virol.* **64**:1449.

Green, J., Griffiths, G., Louvard, D., Quinn, P., and Warren, G., 1981, Passage of viral membrane proteins through the Golgi complex, *J. Mol. Biol.* **152**:663.

Grimley, P. M., Berezesky, I. K., and Friedman, R. M., 1968, Cytoplasmic structures associated with an arbovirus infection: Loci of viral ribonucleic acid synthesis, *J. Virol.* **2**:1326.

Hadjiolov, A. A., and Nikolaev, N., 1976, Maturation of ribosomal ribonucleic acids and the biogenesis of ribosomes, *Prog. Biophys. Mol. Biol.* **31**:95.

Hand, R., and Tamm, I., 1972, Rate of DNA chain growth in mammalian cells infected with cytocidal RNA viruses, *Virology* **47**:331.

Hand, R., Ensminger, W. D., and Tamm, I., 1971, Cellular DNA replication in infections with cytocidal RNA viruses, *Virology* **44**:527.

Hashimoto, K., and Simizu, B., 1976, Effect of cycloheximide on the replication of western equine encephalitis virus, Japan. *J. Med. Sci. Biol.* **29**:215.

Hashimoto, K., and Simizu, B., 1978, Isolation and preliminary characterization of temperature-sensitive mutants of western equine encephalitis virus, *Virology* **84**:540.

Hashimoto, K., and Simizu, B., 1982, A temperature-sensitive mutant of western equine encephalitis virus with an altered envelope protein E1 and a defect in the transport of envelope glycoproteins, *Virology* **119**:276.

Helenius, A., Kartenbeck, J., Simons, K., and Fries, E., 1980, On the entry of Semliki Forest virus into BHK-21 cells, *J. Cell. Biol.* **84**:404.

Holland, J., Grabau, E. A., Jones, C. L., and Semler, B. L., 1979, Evolution of multiple genome mutations during long term persistent infection by vesicular stomatitis virus, *Cell* **16**:495.

Holland, J., Spindler, K., Horodyski, F., Grabau, E., Nichol, S., and VandePol, S., 1982, Rapid evolution of RNA genomes, *Science* **215**:1577.

Igarashi, A., 1978, Isolation of a Singh's *Aedes albopictus* cell clone sensitive to dengue and chikungunya viruses, *J. Gen. Virol.* **40**:531.

Igarashi, A., Koo, R., and Stollar, V., 1977, Evolution and properties of *Aedes albopictus* cell cultures persistently infected with Sindbis virus, *Virology* **82**:69.

Impraim, C. C., Foster, K. A., Micklem, K. J., and Pasternak, C. A., 1980, Nature of virally mediated changes in membrane permeability to small molecules, *Biochem. J.* **186**:847.

Inglot, A., Albin, M., and Chudzio, T., 1973, Persistent infection of mouse cells with Sindbis virus role of virulence of strains, auto-interfering particles and interferon, *J. Gen. Virol.* **20**:105.

Ishida, I., Simizu, B., Koizumi, A., Oya, A., and Yamada, M., 1981, Nucleotide triphosphate phosphohydrolase produced in BHK cells infected with western equine encephalitis virus is probably associated with 82K dalton nonstructural polypeptide, *Virology* **108**:13.

Jacobson, S., and Pfau, C. J., 1980, Viral pathogenesis and resistance to defective interfering particles, *Nature (London)* **283**:311.

Kääriäinen, L., and Söderlund, H., 1978, Structure and replication of alpha-viruses, *Curr. Top. Microbiol. Immunol.* **82**:15.

Kawai, A., and Matsumoto, S., 1977, Interfering and noninterfering defective particles generated by a rabies small plaque variant virus, *Virology* **76**:60.

Kennedy, S. I. T., 1980, The genome of alphaviruses, in: *The Togaviruses*, (R. W. Schlesinger, ed.), pp. 343–349, Academic Press, New York.

Keränen, S., and Kääriäinen, L., 1975, Proteins synthesized by Semliki Forest virus and its 16 temperature-sensitive mutants, *J. Virol.* **16**:388.

Keränen, S., and Ruohonen, L., 1983, Nonstructural proteins of Semliki Forest virus: Synthesis, processing and stability in infected cells, *J. Virol.* **47**:505.

Kielley, R. K., 1970, Purification and properties of thymidine monophosphate kinase from mouse hepatoma, *J. Biol. Chem.* **245**:4204.

Koizumi, S., Simizu, B., Hashimoto, K., Oya, A., and Yamada, M., 1979a, Inhibition of DNA synthesis in BHK cells infected with western equine encephalitis virus, 1. Induction of an inhibitory factor of cellular DNA polymerase activity, *Virology* **94**:314.

Koizumi, S., Simizu, B., Ishida, I., Oya, A., and Yamada, M., 1979b, Inhibition of DNA synthesis in BHK cells infected with western equine encephalitis virus, 2. Properties of the inhibitory factor of DNA polymerase induced in infected cells, *Virology* **98**:439.

Kos, K. A., Osborne, B. A., and Goldsby, R. A., 1975, Inhibition of group B arbovirus antigen production and replication in cells enucleated with cytochalasin B, *J. Virol.* **15**:913.

Kowal, K. J., and Stollar, V., 1981, Temperature-sensitive host-dependent mutants of Sindbis virus, *Virology* **114**:140.

Kozak, M., 1981, Mechanism of mRNA recognition by eukaryotic ribosomes during initiation of protein synthesis, *Curr. Top. Microbiol. Immunol.* **93**:81.

Lee, C. H., and Schloemer, R. H., 1981a, Mosquito cells infected with Banzi virus secrete an antiviral activity which is viral origin, *Virology* **110**:402.

Lee, C. H., and Schloemer, R. H., 1981b, Identification of the antiviral factor in culture medium of mosquito cells persistently infected with Banzi virus, *Virology* **110**:445.

Lockart, R. Z., Jr., Bayliss, N. L., Toy, S. T., and Yin, F. H., 1968, Viral events necessary for the induction of interferon in chick embryo cells, *J. Virol.* **2**:962.

Lodish, H., 1974, Model for the regulation of mRNA translation applied to haemoglobulin synthesis, *Nature (London)* **251**:385.

Logan, K. B., 1979, Generation of defective interfering particles of Semliki Forest virus in a clone of *Aedes albopictus* (mosquito) cells, *J. Virol.* **30**:38.

Maeda, S., Hashimoto, K., and Simizu, B., 1979, Complementation between temperature-sensitive mutants isolated from *Aedes albopictus* cells persistently infected with western equine encephalitis virus, *Virology* **92**:532.

Manley, J. L., Fire, A., Cano, A., Sharp, P. A., and Gefter, M. L., 1980, DNA-dependent transcription of adenovirus genes in a soluble whole cell extract, *Proc. Natl. Acad. Sci. USA* **77**:3855.

Marsh, M., Bolzau, E., and Helenius, A., 1983, Penetration of Semliki Forest virus from acidic prelysosomal vacuoles, *Cell* **32**:931.

Mathews, J. H., and Vorndam, A. V., 1982, Interferon-mediated persistent infection of Saint Louis encephalitis virus in a reptilian cell line, *J. Gen. Virol.* **61**:177.

Matthews, R. E. F. (ed.), 1982, *Classification and Nomenclature of Viruses,* Fourth report of the international committee on taxonomy of viruses, Karger, Basel.

Meinkoth, J., and Kennedy, S. I. T., 1980, Semliki Forest virus persistence in mouse L929 cells, *Virology* **100**:141.

Mento, S. J., and Siminovitch, L., 1981, Isolation and preliminary characterization of Sindbis virus-resistant Chinese hamster ovary cells, *Virology* **111**:320.

Metz, D. H., 1975, The mechanism of action of interferon, *Cell* **6**:429.

Mitsuhashi, J., and Maramorosch, K., 1964, Leafhopper tissue: Embryonic, nymphal, and imaginal tissue from aseptic insects, *Contrib. Boyce Thompson Inst.* **22**:435.

Monckton, R. P., and Westaway, E. G., 1982, Restricted translation of the genome of the flavivirus Kunjin *in vitro*, *J. Gen. Virol.* **63**:227.

Monroe, S. S., Ou, J.-H., Rice, C. M., Schlesinger, S., Strauss, E. G., and Strauss, J. H., 1982, Sequence analysis of cDNA's derived from the RNA of Sindbis virions and of defective interfering particles, *J. Virol.* **41**:153.

Mowshowitz, D., 1973, Identification of polysomal RNA in BHK cells infected by Sindbis virus, *J. Virol.* **11**:535.

Mussgay, M., Enzmann, P.-J., and Horst, J., 1970, Influence of an arbovirus infection (Sindbis virus) on the protein and ribonucleic acid synthesis of cultivated chick embryo cells, *Arch. Ges. Virusforsch.* **31**:81.

Naeva, C. W., and Trent, D. W., 1978, Identification of Saint Louis encephalitis virus mRNA, *J. Virol.* **25**:535.

Newton, S., and Dalgarno, L., 1983, Antiviral activity released from *Aedes albopictus* cells persistently infected with Semliki Forest virus, *J. Virol.* **47**:652.

Nuss, D. L., and Koch, G., 1976, Variation in the relative synthesis of immunoglobulin G and nonimmunoglobulin G proteins in cultured MPC-11 cells with changes in the overall rate of polypeptide chain initiation and elongation, *J. Mol. Biol.* **102**:601.

Nuss, D. L., Oppermann, H., and Koch, G., 1975, Selective blockage of initiation of host protein synthesis in RNA-virus-infected cells, *Proc. Natl. Acad. Sci. USA* **72**:1258.

Ou, J.-H., Rice, C. M., Dalgarno, L., Strauss, E. G., and Strauss, J. H., 1982*a*, Sequence studies of several alphavirus genomic RNAs in the region containing the start of the subgenomic RNA, *Proc. Natl. Acad. Sci. USA* **79**:5235.

Ou, J.-H., Trent, D. W., and Strauss, J. H., 1982*b*, The 3'-non-coding regions of alphavirus RNAs contain repeating sequences, *J. Mol. Biol.* **156**:719.

Ou, J.-H., Strauss, E. G., and Strauss, J. H., 1983, The 5'-terminal sequence of the genomic RNAs of several alphaviruses, *J. Mol. Biol.* **168**:1.

Paul, S. D., Singh, K. R. P., and Bhat, U. K. M., 1969, A study on the cytopathic effect of arboviruses on cultures from *Aedes albopictus* cell line, *Ind. J. Med. Res.* **57**:339.

Peleg, J., 1969, Inapparent persistent virus infection in continuously grown *Aedes aegypti* mosquito cells, *J. Gen. Virol.* **5**:463.

Pfefferkorn, E. R., and Shapiro, D., 1974, Reproduction of togaviruses, in: *Comprehensive Virology*, Vol. 2. (H. Fraenkel-Conrat and R. R. Wagner, eds.), pp. 171-230, Plenum Press, New York and London.

Ponta, H., Altendarf, K. H., Schlesinger, M., Hirsch-Kaufmann, M., Pferning-Yeh, M. L., and Herrlich, P., 1976, *E. coli* membranes become permeable to ions following T7-virus infection, *Mol. Gen. Genet.* **149**:145.

Ramabhadran, T. V., and Thach, R. E., 1980, Specificity of protein synthesis inhibitors in the inhibition of encephalomyocarditis virus replication, *J. Virol.* **34**:293.

Reigel, F., and Koblet, H., 1981, Selective disappearance of two secreted host proteins in the course of Semliki Forest virus infection of *Aedes albopictus* cells, *J. Virol.* **39**:321.

Rice, C. M., and Strauss, J. H., 1981, Nucleotide sequence of the 26S mRNA of Sindbis virus and deduced sequence of the encoded virus structural proteins, *Proc. Natl. Acad. Sci. USA* **78:**2062.

Riedel, B., and Brown, D. T., 1977, Role of extracellular virus in the maintenance of the persistent infection induced in *Aedes albopictus* (Mosquito) cells by Sindbis virus, *J. Virol.* **23:**554.

Riedel, B., and Brown, D. T., 1979a, Novel antiviral activity found in the media of Sindbis virus persistent infection induced in *Aedes albopictus* (mosquito) cells by Sindbis virus, *J. Virol.* **23:**554.

Riedel, B., and Brown, D. T., 1979b, Novel antiviral activity found in the media of Sindbis virus-persistently infected mosquito (*Aedes albopictus*) cell culture, *J. Virol.* **29:**51.

Russell, P. K., Brandt, W. E., and Dalrymple, J. M., 1980, Chemical and antigenic structure of flaviviruses, in: *The Togaviruses* (R. W. Schlesinger, ed.), pp. 503–529, Academic Press, New York.

Saborio, J. L., Pong, S.-S., and Koch, G., 1974, Selective and reversible inhibition of initiation of protein synthesis in mammalian cells, *J. Mol. Biol.* **85:**195.

Saraste, J., Bonsdorff, C.-H., Hashimoto, K., Kääriäinen, L., and Keränen, S., 1980, Semliki Forest mutants with temperature-sensitive transport defect of envelope proteins, *Virology* **100:**229.

Sarver, N., and Stollar, V., 1977, Sindbis virus-induced cytopathic effect in clones of *Aedes albopictus* (Singh) cells, *Virology* **80:**390.

Scheefers-Borchel, U., Scheefers, H., Edwards, J., and Brown, D. T., 1981, Sindbis virus maturation in cultured mosquito cells is sensitive to actinomycin D, *Virology* **110:**292.

Schlesinger, M. J., and Kääriäinen, L., 1980, Translation and processing of alphavirus proteins, in: *The Togaviruses* (R. W. Schlesinger, ed.), pp. 371–392, Academic Press, New York.

Schlesinger, R. W., 1980, Virus-host interactions in natural and experimental infections with alphaviruses and flaviviruses, in: *The Togaviruses* (R. W. Schlesinger, ed.), pp. 83–106, Academic Press, New York.

Schmaljohn, C. S., and Blair, C. D., 1977, Persistent infection of cultured mammalian cells by Japanese encephalitis virus, *J. Virol.* **24:**580.

Schmaljohn, C. S., and Blair, C. D., 1979, Clonal analysis of mammalian cell cultures persistently infected with Japanese encephalitis virus, *J. Virol.* **31:**816.

Schwobel, W., and Ahl, R., 1972, Persistence of Sindbis virus in BHK-21 cell cultures, *Arch. Ges. Virusforsch.* **38:**1.

Shapiro, D., Brandt, W. E., Cardiff, R. D., and Russell, P. K., 1971, The proteins of Japanese encephalitis virus, *Virology* **44:**108.

Shenk, T. E., Koshelnyk, K. A., and Stollar, V., 1974, Temperature-sensitive virus from *Aedes albopictus* cells chronically infected with Sindbis virus, *J. Virol.* **13:**439.

Shope, R. E., 1980, Medical significance of togaviruses: An overview of diseases caused by togaviruses in man and in domestic and wild vertebrate animals, in: *The Togaviruses* (R. W. Schlesinger, ed.), pp. 47–82, Academic Press, New York.

Simizu, B., and Maeda, S., 1981, Growth patterns of temperature-sensitive mutant of western equine encephalitis virus in cultured *Aedes albopictus* (mosquito) cells, *J. Gen. Virol.* **56:**349.

Simizu, B., Wagatsuma, M., Oya, A., Hanaoka, F., and Yamada, M., 1976, Inhibition of cellular DNA synthesis in hamster kidney cells infected with western equine encephalitis virus, *Arch. Virol.* **15:**251.

Simizu, B., Hashimoto, K., and Ishida, I., 1983, A variant of western equine encephalitis virus with nonglycosylated E3 protein, *Virology* **125:**99.

Simons, K., and Garoff, H., 1980, The budding mechanisms of enveloped animal viruses, *J. Gen. Virol.* **50:**1.

Singh, K. R. P., 1967, Cell culture derived from larvae of *Aedes albopictus* (Skuse) and *Aedes aegypti* (L), *Curr. Sci.* **36:**506.

Skoog, L., 1970, An enzymatic method for the determination of dCTP and dGTP in picomole amounts, *Eur. J. Biochem.* **17:**202.

van Steeg, H., van Grinsven, M., van Mansfeld, F., Voorma, H. O., and Benne, R., 1981a, Initiation of protein synthesis in neuroblastoma cells infected by Semliki Forest virus: A decreased requirement of late viral mRNA for e1F-4B and cap binding protein, *FEBS Lett.* **129:**62.

van Steeg, H., Thomas, A., Verbeek, S., Kasperaitis, M., Voorma, H. O., and Benne, R., 1981b, Shutoff of neuroblastoma cell protein synthesis by Semliki Forest virus: Loss of ability of crude initiation factors to recognize early Semliki Forest virus and host mRNA's, *J. Virol.* **38:**728.

Stevens, T. M., 1970, Arbovirus replication in mosquito cell lines (Singh) grown in monolayer or suspension culture, *Proc. Soc. Exp. Biol. Med.* **134:**356.

Stollar, V., 1980, Togaviruses in cultured arthropod cells, in: *The Togaviruses* (R. W. Schlesinger, ed.), pp. 583–621, Academic Press, New York.

Stollar, V., and Shenk, T. E., 1973, Homologous viral interference in *Aedes albopictus* culture chronically infected with Sindbis virus, *J. Virol.* **11:**592.

Strauss, J. H., and Strauss, E. G., 1977, Togaviruses, in: *The Molecular Biology of Animal Viruses* (D. P. Nayak, ed.), pp. 111–166, Marcel Dekker, New York.

Strauss, E. G., and Strauss, J. H., 1980, Mutants of alphaviruses: Genetics and physiology, in: *The Togaviruses* (R. W. Schlesinger, ed.), pp. 393–426, Academic Press, New York.

Strauss, E. G., and Strauss, J. H., 1983, Replication strategies of the single stranded RNA viruses of eukaryotes, *Curr. Top. Microbiol. Immunol.* **105:**1.

Strauss, E. G., Rice, C. M., and Strauss, J. H., 1983, The sequence coding for the alphavirus nonstructural proteins is interrupted by an opal termination codon, *Proc. Natl. Acad. Sci. USA* **80:**5271.

Suitor, E. C., Jr., and Paul, F. J., 1969, Syncytia formation of mosquito cell cultured by type 2 dengue virus, *Virology* **38:**482.

Svitkin, Y. V., Ugarova, T. Y., Chernovskaya, T. V., Lyapustin, V. N., Lashkevich, V. A., and Agol, V. I., 1981, Translation of tick-borne encephalitis virus (flavivirus) genome *in vitro*: Synthesis of two structural polypeptides, *Virology* **110:**26.

Taylor, J., 1965, Studies on the mechanism of action of interferon, 1. Interferon action and RNA synthesis in chick embryo fibroblasts infected with Semliki Forest virus, *Virology* **25:**340.

Tooker, P., and Kennedy, S. I. T., 1981, Semliki Forest virus multiplication in clones of *Aedes albopictus* cells, *J. Virol.* **37:**589.

Trent, D. W., and Qureshi, A. A., 1971, Structural and nonstructural proteins of Saint Louis encephalitis virus, *J. Virol.* **7:**379.

Trent, D. W., Swenson, C. C., and Qureshi, A. A., 1969, Synthesis of St. Louis encephalitis virus ribonucleic acid in BHK 21/13 cells, *J. Virol.* **3:**385.

Tuomi, K., Kääriäinen, L., and Söderlund, H., 1975, Quantitation of Semliki Forest virus RNAs in infected cells using ^{32}P equilibrium labeling, *Nucl. Acids Res.* **2**:555.

Walker, D. L., 1964, The viral carrier state in animal cell cultures, *Prog. Med. Virol.* **6**:111.

Warrington, R. C., and Wratten, N., 1977, Differential action of L-histidinol in reovirus-infected and uninfected L-929 cells, *Virology* **81**:408.

Weiss, B., and Schlesinger, S., 1981, Defective interfering particles do not interfere with the homologous virus obtained from persistently infected BHK cells but do interfere with Semliki Forest virus, *J. Virol.* **37**:840.

Weiss, B., Rosenthal, R., and Schlesinger, S., 1980, Establishment and maintenance of persistent infection by Sindbis virus in BHK cells, *J. Virol.* **33**:463.

Wengler, G., 1980, Effects of alphaviruses on host cell macromolecular synthesis, in: *The Togaviruses,* (R. W. Schlesinger, ed.), pp. 459–472, Academic Press, New York.

Wengler, G., and Wengler, G., 1972, Medium hypertonicity and polyribosome structure in HeLa cells. The influence of hypertonicity of the growth medium on polyribosomes in HeLa cells, *Eur. J. Biochem.* **27**:162.

Wengler, G., and Wengler, G., 1976, Localization of the 26S RNA sequence in the viral genome type 42S RNA isolated from SFV-infected cell, *Virology* **73**:190.

Wengler, G., and Wengler, G., 1981, Terminal sequences of the genome and replicative-form RNA of the flavivirus West Nile virus: Absence of poly (A) and possible role in RNA replication, *Virology* **113**:544.

Wengler, G., Wengler, G., and Gross, H. J., 1978, Studies on virus-specific nucleic acids synthesized in vertebrate and mosquito cells infected with flaviviruses, *Virology* **89**:423.

Wengler, G., Beato, M., and Wengler, G., 1979, *In vitro* translation of 42 S virus-specific RNA from cells infected with the flavivirus West Nile virus, *Virology* **96**:516.

Westaway, E. G., 1973, Proteins specified by group B togaviruses in mammalian cells during productive infections, *Virology* **51**:454.

Westaway, E. G., 1975, The proteins of Murray Valley encephalitis virus, *J. Gen. Virol.* **27**:283.

Westaway, E. G., 1977, Strategy of the flavivirus genome: Evidence for multiple initiation of translation of proteins specified by Kunjin virus in mammalian cells, *Virology* **80**:320.

Westaway, E. G., 1980, Replication of flaviviruses, in: *The Togaviruses* (R. W. Schlesinger, ed.), pp. 531–581, Academic Press, New York.

Yakobson, E., Prives, C., Hartman, J. R., Winocour, E., and Revel, M., 1977, Inhibition of viral protein synthesis in monkey cells treated with interferon late in simian virus 40 lytic cycle, *Cell* **12**:73.

Zhdanov, V. M., 1975, Integration of viral genomes, *Nature (London)* **256**:471.

Cumulative Contents

VOLUME 1

Descriptive Catalogue of Viruses

VOLUME 2

Reproduction: Small and Intermediate RNA Viruses

VOLUME 3

Reproduction: DNA Animal Viruses

VOLUME 4

Reproduction: Large RNA Viruses

VOLUME 11

Regulation and Genetics: Plant Viruses

VOLUME 12

Newly Characterized Protist and Invertebrate Viruses

VOLUME 14

Newly Characterized Vertebrate Viruses

VOLUME 17

Methods Used in the Study of Viruses

VOLUME 19

Viral Cytopathology: Cellular Macromolecular Synthesis and Cytocidal Viruses

Cytopathic Effects of Viruses: A General Survey

Robert R. Wagner

Transcription by RNA Polymerase II

Ulla Hansen and Phillip A. Sharp

Regulation of Eukaryotic Translation

Raymond Kaempfer

Picornavirus Inhibition of Host Cell Protein Synthesis

Ellie Ehrenfeld

Rhabdovirus Cytopathology: Effects on Cellular Macromolecular Synthesis

Robert R. Wagner, James R. Thomas, and John J. McGowan

Adenovirus Cytopathology

S. J. Flint

The Effects of Herpesviruses on Cellular Macromolecular Synthesis

Michael L. Fenwick

Poxvirus Cytopathogenicity: Effects on Cellular Macromolecular Synthesis

Rostom Bablanian

Reovirus Cytopathology: Effects on Cellular Macromolecular Synthesis and the Cytoskeleton

Arlene H. Sharpe and Bernard N. Fields

Inhibition of Host Cell Macromolecular Synthesis following Togavirus Infection

Bunsiti Simizu

Cumulative Contents
Index
Cumulative Author Index
Cumulative Subject Index

Index

Actin filament
 Newcastle disease virus and increase of, 48
 VSV and reduction of, 48
Actinomycin D
 picornavirus cytopathology and, 19–20
 vaccinia virus cytopathology and, 23, 24
Adenosine triphosphate, role in protein synthesis, 87, 120, 201
Adenovirus
 antigens, 30
 cellular gene expression in cells infected by, 330–334
 cytomorphological characteristics of, 29
 cytopathology, 29–33, 297–346
 DNA synthesis inhibition by, mechanism of, 309, 311, 313–314
 effect of HSV-1 on RNA synthesis in cells infected with, 363
 in HeLa cells, 29
 messenger RNA, 102
  inhibition of, 325–330
 microtubules and, 47
 mutagenicity, 321
 poliovirus interference with growth of, 182
 production cycle, 298–304
 transcription, 84, 302
 translation of viral messenger RNA in cells infected by, 337–343
Adenovirus E1A, 300

Adenovirus E1B, 300
Alfalfa mosaic virus RNA, translation in *Escherichia coli*, 105
Alphavirus
 defective-interfering particles, 479
 DNA synthesis inhibition by, 474–476
 messenger RNA produced by, 466
 protein synthesis inhibition by, 468
 replication of, 465, 466, 467
  host components for, 476–478
 RNA synthesis inhibition by, 473–474
 selective translation of after infection, 469
 structure, 465
Amino acid deprivation, effect on translation, 143
Amnion cell, reovirus inhibition of protein synthesis in humans, 447
Antigen(s), adenovirus, 30
Azauridine, inhibition of RNA synthesis by, 32

Cap structure, *see* 5'-Terminal cap structure
Cap-binding protein, *see* Protein, cap-binding
Cell
 cycle
  DNA synthesis inhibition and, 275
  effect on viral cytopathology, 21–22
 cytoskeletal structures of, effect of viruses on, 46–49, 450–457

515

Cumulative Author Index

Indexed below are authors of chapters in the series. Volume numbers are in *italic* type; chapter numbers are in roman type.

Cumulative Subject Index

Indexed below are subjects covered in the series. Volume numbers are in *italic* type; chapter numbers are in roman type.